Phase Separation
in Cuprate Superconductors

E. Sigmund K. A. Müller (Eds.)

# Phase Separation in Cuprate Superconductors

Proceedings of the second international workshop on "Phase Separation in Cuprate Superconductors" September 4 – 10, 1993, Cottbus, Germany

With 196 Figures

Springer-Verlag
Berlin Heidelberg GmbH

Professor Dr. Ernst Sigmund

Technische Universität Cottbus, Postfach 101344
D-03013 Cottbus, Germany

Professor Dr. h.c. mult. K. Alex Müller

Zürich Research Laboratory, IBM Research Division, Säumerstrasse 4
CH-8803 Rüschlikon, Switzerland

ISBN 978-3-642-78807-9 ISBN 978-3-642-78805-5 (eBook)
DOI 10.1007/978-3-642-78805-5

Library of Congress Cataloging-in-Publication Data. Phase separation in cuprate superconductors: proceedings of the second international workshop on "phase separation in cuprate superconductors," September 4-10, 1993, Cottbus, Germany / E. Sigmund, K. A. Müller, eds. p. cm. ISBN 3-540-57681-9 (Berlin: alk. paper). – ISBN 0-387-57681-9 (New York: alk. paper) 1. High temperature superconductors–Congresses. 2. Copper oxide superconductors–Congresses. 3. Liquation–Congresses. I. Sigmund, E. (Ernst), 1946- . II. Müller, K. A. (Karl A.), 1927- . QC611.98.C64P48 1994 537.6'236–dc20 94-25887


Originally published by Springer-Verlag Berlin Heidelberg New York in 1994
Softcover reprint of the hardcover 1st edition 1994



Typesetting: Camera ready copy from the authors/editors using a Springer LaTeX macro package
SPIN: 10125111 55/3140 - 5 4 3 2 1 0 - Printed on acid-free paper

# Preface

Phase separation has become a fascinating subject in the discussion of cuprate superconductors. All these materials have layered structures containing $CuO_2$ planes as the most important building blocks. They are coupled only weakly so that the electronic properties show a nearly two-dimensional behaviour. Due to correlations the undoped compounds are insulators of the Mott-Hubbard type exhibiting long-range antiferromagnetic order. Upon doping a rich scenario of physical phenomena appears: Even at low hole concentrations the antiferromagnetic ordering temperature is reduced drastically and spin-glass behaviour as well as a hopping type conductivity can be observed. Further doping leads to metallic-like conductivity and below $T_c$ to superconductivity. In this doping regime antiferromagnetic fluctuations are still observed. At very high charge carrier densities superconductivity is lost and the systems show pure metallic conduction without magnetic correlations.

One of the most interesting phenomena in high-$T_c$ research is the interplay between magnetism and conductivity or superconductivity. Especially the behaviour of charge carriers in the antiferromagnetic background raises a number of open questions. Two scenarios become possible: the carriers tend to delocalize over the whole crystal forming a homogeneous state with band-like structure or they separate into hole-rich (conducting, superconducting) and hole-poor (insulating, antiferromagnetic) phases leading to an inhomogeneous structure.

In recent years it has been established that the picture of electronic phase separation is a characteristic feature of models of hole-doped antiferromagnets. The idea of electronic phase separation in connection with high-$T_c$ cuprates has been brought up and discussed by independent theoretical groups (Hizhnyakov and Sigmund, 1988, Emery and Kivelson, 1990, Di Castro and coworkers, 1991) and preliminary experimental data supporting this idea have been obtained (Hammel et al., 1990, Kremer et al., 1992).

This development stimulated one of us (K. A. Müller) to bring together scientists with a common interest in phase separation in a small workshop. Consequently the first workshop in this field took place in May 1992 in Erice (First International Workshop on "Phase Separation in Cuprate Superconductors", May 6 – 12, 1992, Erice, Sicily, organized by K. A. Müller and G. Benedek). This first and very successful meeting initiated a variety of different research activities especially on the experimental side and we felt that these new and promising results should be discussed in a second meeting. Therefore we organized from September 6 – 12, 1993, together with S. Ewert the "Second International Workshop on Phase Separation in Cuprate Superconductors", this time in Cottbus, Germany.

In the majority of the contributions the possible appearance of electronic phase separation was directly addressed. Mössbauer spectroscopy, neutron scattering and electron paramagnetic resonance turned out to be most relevant in probing the local arrangement around single charge carriers. From the titles of the various presentations it becomes apparent that discussions of the phase separation of structural or chemical origin were also emphasized. The influence of electron-phonon interactions as well as anharmonicity effects of the apex oxygen was discussed and, in addition, the role of phase separation for the pairing mechanism was addressed in several contributions.

We hope that these proceedings will be a representative publication on the subject and on the state of art in the field.

Cottbus, Stuttgart, Zürich
June, 1994

*E. Sigmund, K. A. Müller*

# Acknowledgments

"Phase Separation in Cuprate Superconductors" was the 119th seminar of the Dr. Wilhelm Heinrich and Else Heraeus – Foundation (WE-Heraeus-Stiftung) and was held in Cottbus, Germany.

The editors gratefully acknowledge the substantial efforts of the coworkers of the Institute of Physics of the Technical University of Cottbus and of the Institute of Theoretical Physics of the University of Stuttgart in preparing the workshop and in assembling and editing the proceedings.

The workshop was supported by the WE-Heraeus-Stiftung, the DAAD ("Deutscher Akademischer Austauschdienst") by means of funds from the "Bundesministerium für Forschung und Technologie" and the "Ministerium für Wissenschaft, Forschung und Kultur des Landes Brandenburg".

# Table of Contents

## Part II Electronic Results I

## Part VI Electron-Phonon Interaction I

## Part VII Electron-Phonon Interaction II

## Part VIII Further Theoretical Concepts

# Recipe for High-$T_c$ Transition-Metal Chalcogenide Superconductors

K. Alex Müller[1,2]

[1] IBM Research Division, Zürich Research Laboratory, 8803 Rüschlikon,
[2] University of Zürich, Physics Department, 8801 Zurich, Switzerland

**Summary.** The conditions for the occurrence of superconductivity in cuprates are enumerated. There is a minimum of seven, which I will discuss in some detail. They may be taken as a recipe in the search for superconductors at elevated temperatures with other transition-metal chalcogenide compounds. Furthermore, it is apparent that in any theory one is involved in minimizing the many-body energy in a multiparameter space outlined by the conditions enumerated in this paper.

## 1. Introduction

Superconductivity is a cooperative phenomenon as are ferro- or antiferromagnetisms as well as the gas-liquid transition [1]. Whereas the latter can be dealt with easily in the van der Waals model, some magnetic cases are considerably more complex. In addition to the exchange constants that appear in the Hamiltonian, the structure and site occupancy of the magnetic atoms have to be known [2]. In a metal also the density of carriers and their screening at the energy of possible localized magnetic states play a role. With the above, I want to express that in a magnet or an antiferromagnet, there are a number of essential quantities that need to be just right in order that a specific magnetic phase occurs.

Clearly, using the most simple Hamiltonian, say, an Ising one, the essential aspects of a gas-liquid or a paramagnetic to ferro- or antiferromagnetic phase transition, including critical phenomena, can be deduced. In superconductivity, the Ginzburg-Laundau functioncal [3] and the seminal microscopic BCS theory [4] (from which the former can be deduced) have played this role. With the BCS theory, it was possible to link the various experimental observations of the normal to superconducting phase transition in simple metals such as aluminum, tin etc. The BCS theory has been expanded to more difficult metallic superconductors. The debate still continues whether the BCS formalism is sufficient to describe the phenomenon in the new, so-called "exotic" superconductors such as Chevrel phases, organic materials, cuprates [5] and fullerenes [6], to name but those more frequently investigated. It might well be that another formalism is needed for one or more of the new materials. Independent of this, I want to point out, by analogy with the magnetism example sketched above, which conditions appear to be important so that superconductivity occurs in the cuprates. The latter are still the compounds that exhibit the highest transition temperatures [7]. They

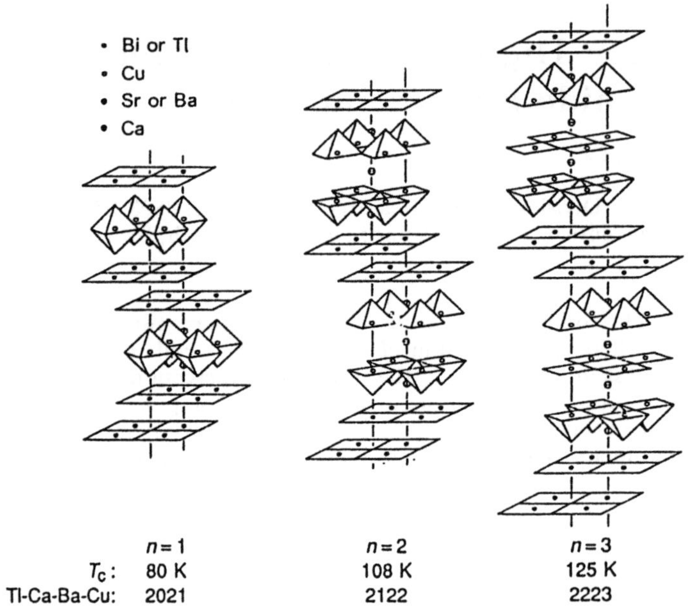


**Fig. 1.1.** Schematic lattice structure of the $n = 1, 2, 3$ bismuthates and thallates showing octahedral, pyramidal and four-fold oxygen coordination of $Cu^{2+}$. After S.S.P. Parkin. Copyright 1989 by International Business Machines Corporation, reprinted with permission.

have been investigated to such great detail that the conditions for high-$T_c$ superconductivity can now be listed, and in this presentation such an attempt is undertaken. It appears that there are at least seven of them, maybe eight. If one is not fulfilled, the phenomenon might not occur with, say, a $T_c > 100$ K. One can therefore regard these conditions as the ingredients to be used in a cooking recipe for other transition-metal chalcogenide compounds exhibiting superconductivity at elevated temperatures, possibly produced by epitaxy on suitable substrates or with high pressure.

## 2. The Properties

1. From the first [8] and second compound [9] to the one with the highest $T_c$ [10], all **structures are layered** [11] and *contain* $CuO_2$ planes typical of the *perovskite lattice*, i.e., in the plane a metal (here copper) is surrounded by a square of oxygens. Each such quincunx of atoms is corner sharing with the next one, i.e., every other oxygen square does not contain a metal ion. Recent careful structural analyses seem to indicate that in most — but not all — of the superconducting compounds a slight orthorhombic or incommensurate distortion of the tetragonal structure is present [5]. As this may be, the above-mentioned planar oxygen square

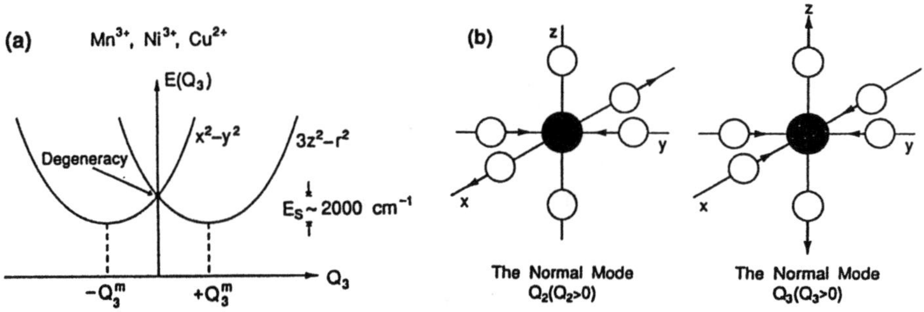


**Fig. 2.1.** (a) The Jahn-Teller energy $E(Q)$ for two electronic wave functions $x^2-y^2$ and $3z^2-r^2$ as a function of the normal coordinate $Q_3$ shown in (b) together with the coordinate $Q_2$. The total J-T energy near $\boldsymbol{Q}=0$ is given by $\mathcal{H}_{\mathrm{JT}} = A^{\Gamma_3}\begin{pmatrix}-Q_3 & Q_2 \\ Q_2 & Q_3\end{pmatrix}$. In all known octahedral distortions the unit expands, i.e. $Q_3 > 0$, probably due to second-order effects as indicated by a lower potential in (a). From [12].

corner-sharing surroundings of the copper ions are found to be octahedral [8], pyramidal [9] or have fourfold oxygen coordination [10]. See, for example, Fig. 1.1.

2. Five- and fourfold anion coordination is found for those transition-metals ions that have an orbital ground-state degeneracy in cubic symmetry and therefore undergo the *Jahn-Teller effect* [12]. The latter removes the orbital degeneracy by lowering the cubic symmetry either by distortion of the octahedron [8] or by losing one (pyramidal) [9] or two opposite chalcogenides [10], which yields a square planar coordination, see Fig. 2.1. The lowering of coordination occurs if the metal-chalcogenide binding induces a **strong Jahn-Teller effect**, for example, if occupied $\sigma$-bonds are present. A typical example for this effect is $Cu^{2+}$, which exhibits six-, five- or fourfold coordination. The latter two only occur in transition metals with a strong Jahn-Teller (J-T) effect in the oxides, as Raveau *et al.* discuss [13]. For example, $Mn^{3+}$ is such a ion and is found in several compounds in pyramidal configuration [13]. As we know that the $T_c$'s are higher for corner-sharing pyramidal compounds, and highest for those containing in addition square planar $CuO_2$ units, it is obvious that the presence of a transition-metal ion with an orbital degeneracy in the starting coordination with substantial $\sigma$-bonds is rather crucial. Whether the dynamic local J-T effect is relevant for the pairing of the carriers is another question related to the possible anharmonic potentials of the ligands [8]. This is in part discussed in the other papers of this workshop.
3. It is by now well known that the insulating layered cuprates *all are antiferromagnets* (AFM) with exchange constants of $J \simeq 500$ - $1000$ K [5]. They are all AFM Mott insulators with quasi-two dimensionality, reaching their highest $T_c$ with dopant concentrations of roughly 0.2 per copper ion. Overdoping them decreases their $T_c$. From this observation alone, the near *2d doped AFM appears to be very crucial*, see Fig. 2.2. There

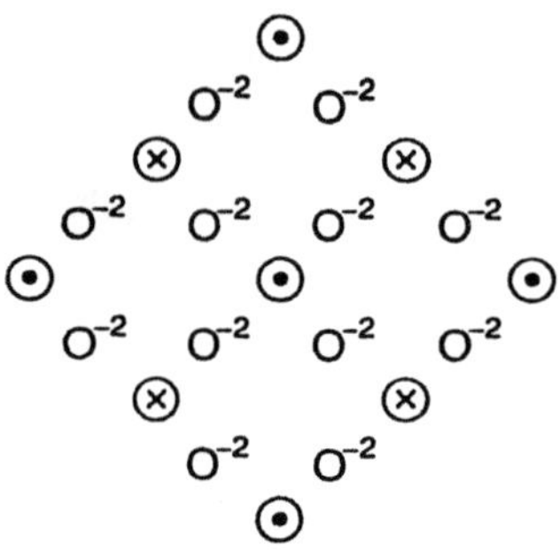


**Fig. 2.2.** Antiferromagnetic lattice in a $CuO_2$ plane. The Cu spins marked by ⊗ have opposite spin to those marked by ⊙.

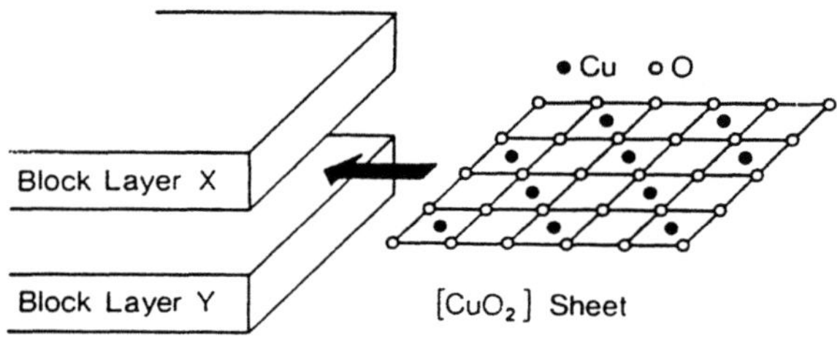


**Fig. 2.3.** Classification of cuprates after Tokura [18].

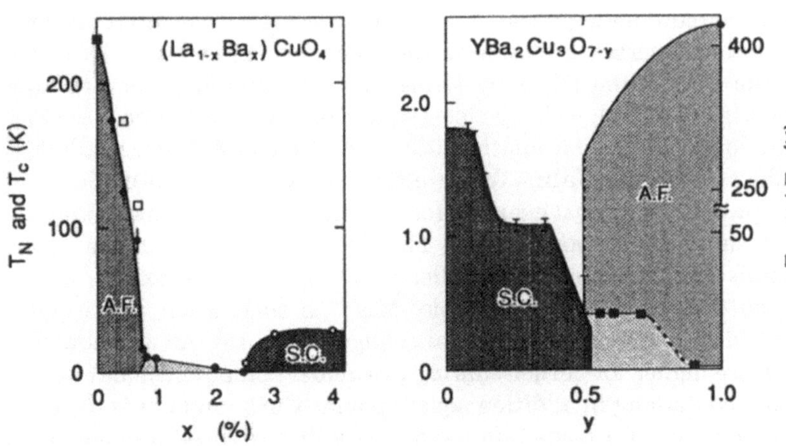


**Fig. 2.4.** Hole-temperature phase diagram of doped $La_2CuO_4$ and $YBa_2Cu_3O_7$ communicated privately by J. Rossat-Mignon.

is ample experimental evidence for this condition, and I am not going to enumerate all of it here. Coming back to Condition 2, the necessity of AFM eliminates the Jahn-Teller ion $Mn^{3+}$ as a candidate, since its coupling is via ferromagnetic oxygen.

4. The afore-mentioned $CuO_2$ planes are doped to such a percentage that they show a substantial conductivity. The doping is either by holes or electrons. An essential property is that **the doping** is due to acceptors

or donors present in structural layers or chains **adjacent to the $CuO_2$ planes**, see Fig. 2.3. In the superconducting cuprates the $CuO_2$ planes are structurally nearly perfect and the scattering or localization potentials of the donors or acceptors are considerably reduced owing to the high dielectric constant [14]. Thus, in these cuprates nature gives us for free what in high-mobility semiconductor research and applications is achieved by careful epitaxial growth, first realized for AlGaAs on GaAs such that the latter dopes carriers in the quasi-two-dimensional layer of the former [15]. Similarly, in the cuprates the acceptors or donors are sited in the next or the next-nearest oxide layers: in the LaO one in $La_{2-x}Sr_xCuO_4$, with $Sr^{2+}$ replacing $La^{3+}$, or as interstitial $O^{2-}$ in $La_2SrCuO_{4+\delta}$ [16] and in the chains in $YBa_2Cu_2O_{7-x}$ or $Ca^{2+}$ substitutional for $Y^{3+}$ [17]. In the bismuthates and thallates, the doping results from the $Bi_2O_2$, $Tl_2O_2$ and TlO layers, to list a few examples [18]. The important difference with respect to the high-mobility semiconductors is that the charge comes from atomically defined sites. With this knowledge, new oxide superconductors may be "engineered" as shown by Tokura, who used structural considerations [18]. In addition, with layer-by-layer epitaxy, also novel "artificial" dopants may be found. The well-known AFM to superconducting phase diagrams for $La_{2-x}Sr_xCuO_4$ and $YBa_2Cu_3O_{7-y}$ are shown in Fig. 2.4.

Upon doping one distinguishes an under- as well as an overdoped regime in the cuprates. The former occurs for concentrations $n < n_c$ and the latter for $n > n_c$. For the overdoped regime, both magnetism and superconductivity decrease, as has been beautifully shown for a number of such compounds [19]. The behavior approaches Fermi-liquid behavior with a resistivity $\rho \propto T^2$, whereas in the optimum $n_c$ doped case, $\rho \propto T$ [20].

5. The highest transition temperatures observed so far, 133 K, have been found in cuprates that are doped with **holes**: The $T_c$'s are a factor of four to five higher than in those doped with electrons [21]. Very soon after the discovery of the $La_2CuO_4$ and $YBa_2Cu_3O_7$ compounds, Bianconi *et al.* [22] in Rome deduced from their X-ray absorption spectroscopy data that the **main hole density is on oxygen p-orbitals**, forming $O^-$-type ions, rather than on the copper, i.e., $Cu^{3+}$ ($3d^8$). These very early results were beautifully evidenced with electron loss spectroscopy by Fink and collaborators in Karlsruhe [23]. *Conduction-carrier presence mainly on chalcogenides rather than on the metal is very infrequent* in condensed matter science, but is found for instance in CuS or $Ni_{1-x}Li_xO$ [24]. The reason for its occurrence in the cuprates was pointed out very early, and independently, by Emery [25]: It is primarily the large, of the order of 5 to 10 eV, Coulomb repulsion energy on the transition-metal ion. Its magnitude is close to that given for the bare atoms. This Hubbard energy $U$ increases from the transition-metal ions with few d-electrons on the left-hand side of the periodic table to the ions on the right-hand side. Thus it is not an accident that superconductivity is found in the cuprates:

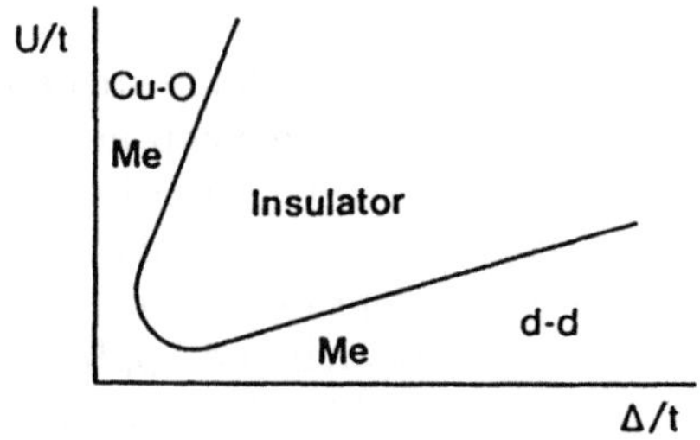


**Fig. 2.5.** The electron correlation energy $U$ versus the charge transfer energy $\Delta$, both normalized by the transfer matrix element $t$ between the Cu d and oxygen p-level. Adapted from [27].

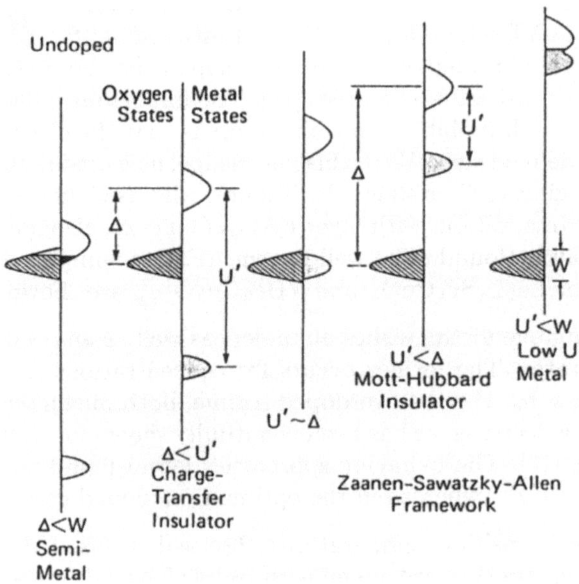


**Fig. 2.6.** Schematic diagram of the relative energies of the oxygen and metal states in the ZSA framework. Five examples are shown with different relative values of $\Delta$, $U'$ and $W$ (which are also defined in the figure). From [26].

Cu terminates the 3d series. However, for the larger charge states of that row, $Me^{3+}$ or $Me^{4+}$, Coulomb repulsion energies as large as those found for $Cu^{2+}$ occur also for elements located more towards the middle of the 3d row [26]. Hence, further search with other transition-metal elements may not be fruitless.

6. There is a second condition for the occurrence of holes on the chalcogenides: the near matching of the chalcogenide atomic $p$ and the transition-metal $d$ levels, the so-called charge transfer energy $\Delta$. **For holes to present on the chalcogenides, $\boldsymbol{\Delta < U}$.** This condition was recognized already in 1985 by Zaanen, Sawatzky and Allen [27], see Fig. 2.5. For normal metallic hole conductivity to occur, the width $w$ of the hybridized

electronic bandwidth with mainly oxygen $p$ and metal $d$ character was earlier thought to be larger than $\Delta$, $w > \Delta$. However, as we shall discuss in **8)**, the recent insight gained upon doping is that there is a novel kind of conduction altogether. Thus we do not consider the condition $w > \Delta$ as quite so relevant, but note that Torrance *et al.* have made a systematic study of $U$ vs. $\Delta$ (at constant $w$) for various transition-metal oxides [26], see Fig. 2.6. In the cuprates, the electron energy $\Delta$ is of the order of 2.5 eV and results from the difference in $O^{2-}$ ionization energy, $E_I(O^{2-})$, the electron affinity of the $Cu^{2+}$, $E_A(Cu^{2+})$, and the Madelung potential, $V_M$, at the two sites as well as correction terms due to screening.

$$\Delta = E_I(O^{2-}) - E_A(Cu^{2+}) + V_M(Cu) + V_M(O) + E_{correction}(\Delta).$$

Especially for the layer structure, the resulting Madelung energies are relevant, as calculations by Maekawa and collaborators have shown [28]. For example, the condition $\Delta < U$ is not fulfilled for the strong J-T ion $Ni^{3+}$, as has been shown by a group in Osaka who investigated $La_{2-x}Sr_xNiO_4$, with $x$ near 1 [29].

7. In the above equation, **the polarizability $\alpha$ of the anion** plays an important role in the correction terms. For oxygen $\alpha$ is of the order of 2 - 3 $Å^{-3}$. What has so far not been taken into account sufficiently is the two-dimensional **perovskite structure**. It enhances the collective polarizability of a compound substantially. Because of this combination of $O^{2-}$ polarizability and perovskite structure, the ferroelectrics with the highest transition temperatures are titanate and niobate perovskites or derivates thereof [30]. Recently a very high dielectric constant has been measured in $La_2CuO_4$ [14]. This has also a direct effect on the Hubbard energy $U$, especially in three-dimensional cases: Consider for example vanadium $V^{2+}$ with $U = 9.9$ eV [26]. In $SrTiO_3$ — a quantum paraelectric with a valence-conduction band gap of 3.2 eV — four stable valency states of vanadium, $V^{5+}$,$V^{4+}$, $V^{3+}$ [31] and $V^{2+}$ [32], are found in *that gap* by electron paramagnetic resonance (EPR) and optical methods. That is, on average between two of these states $U$ is reduced by polarization effects to the order of 1 eV, i.e., by one order of magnitude! Similarly, $Fe^{2+}$, $Fe^{3+}$, $Fe^{4+}$ and $Fe^{5+}$ have also been observed in $SrTiO_3$ [31]. However, $Fe^{2+}$ has a rather small atomic $U(Fe^{2+})$ of 6.1 eV. Incidentally, the large oxygen polarizability certainly also plays a substantial role in the screening of the dopants, and, according to Hirsch [33], possibly also for polaron pairing because of its *anharmonic* potential. One might now look for an anion with a still higher polarizability or, equivalently, a covalency like $N^{3-}$. However, here the chemical binding is so strong that it locks the atom and its shell to a cation [34]. For instance, no nitrogen ferroelectrics are known. Perhaps $S^{2-}$ might be a better candidate [11, 24]. But here again oxygen so far seems the optimum chalcogenide anion.
8. A question that has been posed for the past three to four years is what happens microscopically upon doping the 2D AFM. The majority of re-

searchers assumes that upon adding one hole, a $^1A_1$ Zhang-Rice singlet is formed with the $Cu^{2+}$ $S = 1/2$ spin antiparallel to the hole spin with $s = 1/2$ owing to the large exchange energy [24]. But what happens if a substantial concentration of holes is present? One school assumes the existence of a highly correlated many-body phase, the view held by the majority of the community. The other school considers the existence of a phase separation into metallic droplets and AFM domains. The latter possibility has been discussed extensively in a recent workshop [35], and is the topic of the present one.

An alternative to singlet formation is a kind of magnetic polaron described by Hizhnyakov and Sigmund [36] in their first paper. In it the gain in kinetic energy of one hole, due to a considerable transfer matrix element $t$ between oxygen sites, is such that the carrier becomes itinerant if the loss in exchange energy $J$ is comparatively smaller. In this case, several $Cu^{2+}$ become aligned ferromagnetically. Only then can the hole with antiparrallel spin 1/2 be itinerant. Upon doping these polarons percolate and become superconducting at lower temperatures. This has recently been shown by quenching $Sr^{2+}$ or $O^{2-}$ doped $La_2CuO_4$ and measuring the susceptibility upon warming. The polarons can be aligned by an external magnetic field of 3 T, whereby the superconducting volume is enhanced [37].

The existence of magnetic polarons is also indicated by electron paramagnetic resonance: Wübbeler and Schirmer [38] have observed a strong EPR signal upon quenching oxygen-doped $La_2CuO_4$ from high temperatures. The $g$-value observed, namely 2.14, is between those of $Cu^{2+}$ with 2.30 and $O^-$ with 2.04. Upon warming, the signal disappears towards 200 K, owing to percolation of the magnetic polarons, and the superconducting volume recorded upon slow cooling recovers again in *exact proportion* to the EPR intensity disappearing! This can be interpreted as Fermi glass [39] behavior of the unpercolated polarons observed by EPR, whereas the percolated ones have too short a relaxation time to be observable, but yield superconductivity.

If in the cuprates, as mentioned, a $^1A_1$ singlet results upon doping an electron or hole, because either of them as well as $Cu^{2+}$ have a spin 1/2 coupled antiferromagnetically — *then an eighth condition is necessary.* To discover a new superconductor, the transition-metal candidate also needs an effective spin-1/2 ground state. On the other hand, if a magnetic polaron is formed and the doped carrier density is present near several aligned magnetic ions, then the condition $S = 1/2$ can be relaxed, and transition-metal ions with $S > 1/2$ coupled antiferromagnetically in the undoped regime also become potential candidates. Essential for high $T_c$ are then a high enough AFM coupling constant $J$ (or the Néel temperature) and also that the transfer matrix element $t$ be sufficient to yield a high transition temperature.

## 3. Conclusions

In this paper a minimum of seven properties that all cuprate superconductors possess have been enumerated: 1) *They are layered*, containing slightly distorted perovskite $CuO_2$ sheets. 2) The $Cu^{2+}$ ions show a *strong Jahn-Teller effect*, therefore pyramidal and plane-square coordinations of the transition-metal ion are present. 3) *The antiferromagnetism of the undoped $CuO_2$ sheets is quite crucial,* with an optimum doping near 0.2 per $Cu^{2+}$. 4) *The doping* of the insulating $CuO_2$ layers with carriers *occurs from ions not sited in the $CuO_2$ layers.* 5) The highest $T_c$'s are observed *for hole doping*, with the holes mainly (80%) *on oxygen p-orbitals*, which can be traced to a large, 5 - 10 eV, atomic Coulomb repulsion energy $U$ on the transition-metal $Cu^{2+}$. 6) The size of this *Hubbard U* has to be *larger than the transfer energy $\Delta$* from the oxygen $p$ to the transition-metal $d$-level, $U > \Delta$. 7) The anion ought *to be highly polarizable.* The oxygen $O^{2-}$ with its anharmoninicity appears to be the best species. The polarizability in the perovskite or perovskite-derived structure enhances this property. 8) Future progress will show whether the $S = 1/2$ spin of the $Cu^{2+}$ is essential or whether transition-metal ions with $S > 1/2$ are allowed.

I have enumerated the above properties, mainly employing experimental observations and using a minimum of theory. The reader may take these conditions as a recipe to find or generate other high-$T_c$ superconductors (do any exist?) or to relax the one or the other of the above properties to better understand why the rather stable superconducting phenomenon occurs in the cuprates. The conditions also point to the multiparameter space in many-body theory that one is dealing with and why it is so difficult to find a proper solution.

*Acknowledgement.* I thank E. Sigmund for a critical reading of this manuscript.

## References

1. H.E. Stanley, *Introduction to Phase Transitions and Critical Phenomena*, Clarendon, Oxford, England, 1971.
2. C. Kittel, *Introduction to Solid State Physics*, John Wiley and Sons, New York, 1961.
3. V.L. Ginzburg and L.D. Landau, *Zh. Eksp. Teor. Fiz.* **20**, 1064 (1950); L.P. Gorkov, *Zh. Eksp. Teor. Fiz.* **36**, 1918 (1959) [*JETP* **9** 1364 (1959)].
4. J. Bardeen, L.N. Cooper and J.R. Schrieffer, *Phys. Rev.* **108**, 162 and 1175 (1957).
5. *Earlier and Recent Aspects of Superconductivity*, Int'l. School Mater. Sci. and Techn., Erice, July 1-15, 1989, J.G. Bednorz and K.A. Müller (eds.), *Springer Series in Solid State Sciences* **90**, Springer, Heidelberg, Berlin, New York, 1990.

6. K. Tanigaki, T.W. Eddesen, S. Saito, J. Mizuki, J.S. Tsai, Y. Kubo and S. Kuroshima, *Nature* **352**, 222 (1991); S.P. Kelty, Chia-Chun Chen and C.M. Lieber, *ibid.*, 223.
7. In fact, the cuprate $Tl_2B_2Ca_2Cu_3O_{10}$ exhibits a transition temperature of 124 K (see [10]) — nearly four times as high as that of $Ba_{1-x}K_xBiO_3$ [R.J. Cava *et al.*, *Nature* **333**, 814 (1988)] and the fullerene $K_2CsC_{60}$ (see [6]), both with a $T_c$ of approx. 32 K.
8. J.G. Bednorz and K.A. Müller, *Z. Phys. B* **64**, 189 (1986); J.G. Bednorz, M. Takashige and K.A. Müller, *Europhys. Lett.* **3**, 379 (1987).
9. M.K. Wu, J.R. Ashburn, C.J. Torng, P.H. Hor, R.L. Meng, L. Gao, Z.J. Huang, Y.Q. Wang and C.W. Chu, *Phys. Rev. Lett.* **58**, 908 (1987); P.M. Grant, R.R. Beyers, E.M. Engeler, G. Lin, S.S.P. Parkin, M. Ramirez, V.Y. Lee, A.I. Nazzal, J.E. Vasquez and R.J. Savoy, *Phys. Rev. B* **35**, 7242 (1987).
10. Z.Z. Sheng and A.M. Herman, *Nature* **332**, 138 (1988); R.B. Beyers, S.S.P. Parkin, V.Y. Lee, A.I. Nazzal, R.J. Savoy, G.L. Gorman, T.C. Huang, S.J. La Placa, *IBM J. Res. Develop.* **33**, 228 (1989); A. Schilling, M. Cantoni, J.D. Guo and H.R. Ott, *Nature* **363**, 56 (1993).
11. We note that Ginzburg [*High Temperature Superconductivity*, Nauka, Moskow, 1977. English transl.: Plenum, New York, 1982, Chapters I and VI, V.L. Ginzburg and D.A. Kirzhnits (Eds.)] had advocated the use of layered structures to yield high-$T_c$ superconductors via excitonic pairing.
12. H.A. Jahn and E. Teller, *Proc. R. Soc. London, Ser. A* **161**, 220 (1937); R. Englman, *The Jahn-Teller Effects in Molecules and Crystals*, Wiley, New York, 1972.
13. B. Raveau, C. Michel, M. Hervieu and D. Groult, *Crystal Chemistry of High-$T_c$ Superconducting Copper Oxides*, Springer-Verlag, Berlin Heidelberg New York, 1991.
14. D. Reagor, E. Ahrens, S-W. Cheong, A. Migliori and Z. Fisk, *Phys. Rev. Lett.* **62**, 2048 (1989).
15. J.M. Woodall, H. Rupprecht and W. Reuter, *J. Electrochem. Soc.* **116**, 899 (1969).
16. J.-C. Grenier, A. Wattiaux and M. Pouchard, in: *Proc. 1st Workshop on "Phase Separation in Cuprate Superconductors,"* Erice, Italy, May 6-12, 1992, K.A. Müller and G. Benedek (eds.). World Scientific, Singapore, 1993,
17. J.B. Torrance, Y. Tokura, A.. Nazzal and S.S.P. Parkin, *Phys. Rev. Lett.* **60**, 542 (1988).
18. Y. Tokura and T. Arima, *Jpn. J. Appl. Phys.* **29**, 2388 (1990).
19. M.W. Shafer and T. Penney, *Europ. J. Solid State Chem.* **27**, 191 (1990).
20. B. Batlogg, in: *High Temperature Superconductivitiy: The Los Alamos Symposium - 1989*, K. Bedell, D. Coffey, D. Meltzer, D. Pines and J.R. Schrieffer (eds.). Addison Wesley Publishing Co., p. 37.
21. Y. Tokura, H. Takagi and S. Uchida, *Nature* **337**, 345 (1989).
22. A. Bianconi, A. Congiu Castellano, M. De Santis, P. Rudolf, P. Lagarde and A.M. Flank, *Solid State Commun.* **63**, 1009 (1987).
23. N. Nücker, J. Fink, J.C. Fuggle, P.J. Durham and W.M. Temmerman, *Phys. Rev. B* **37**, 5158 (1988).
24. G.A. Sawatzky, in [5], p. 345.
25. V.J. Emery, *Phys. Rev. Lett.* **58**, 2794 (1987).
26. J.B. Torrance, P. Lacorre, C. Asararoengchai and R.M. Metzger *Physica C* **182**, 351 (1991).
27. J. Zaanen, G.A. Sawatzky and J.W. Allen, *Phys. Rev. Lett.* **55**, 418 (1985).
28. Y. Ohta, T. Tohyama and S. Maekawa, *Phys. Rev. B* **43**, 2968 (1991).
29. M. Kato, Y. Maeno and T. Fujita, *Physica C* **176**, 533 (1991).

30. P. Villars and J.C. Phillips, *Phys. Rev. B* **37**, 2345 (1988).
31. K.A. Müller, *J. Phys. (Paris)* **42**, 551 (1981).
32. K.A. Müller, M. Aguilar, W. Berlinger and K.W. Blazey, *J. Phys. Cond. Matter* **2**, 2735 (1990).
33. J.E. Hirsch, University of California at San Diego, La Jolla, preprint, September 1992.
34. A. Simon, Max Planck Institute Stuttgart, Germany, private communication, 1992.
35. *Proc. 1st Workshop on "Phase Separation in Cuprate Superconductors,"* Erice, Italy, May 6-12, 1992, K.A. Müller and G. Benedek (eds.). World Scientific, Singapore, 1993.
36. V. Hizhnyakov and E. Sigmund, *Physica C* **156**, 665 (1988).
37. R.K. Kremer, E. Sigmund, V. Hizhnyakov, F. Hentsch, A. Simon, K.A. Müller and M. Mehring, *Z. Phys. B* **86**, 319 (1992); R.K. Kremer, V. Hizhnyakov, E. Sigmund, A. Simon and K.A. Müller, preprint (1992).
38. G. Wübbeler and O.F. Schirmer, *Physica Stat. Solidi b* **174**, K21 (1992).
39. K.A. Müller, T. Penney, M.W. Shafer and W.J. Fitzpatrick, *Phys. Rev. Lett.* **47**, 138 (1981).

# Phase Separation as a Possible Scenario for High $T_c$ Superconductors: A Particular Overview

C. Di Castro and M. Grilli

Dipartimento di Fisica, Universitá La Sapienza, Piazzale Aldo Moro, 00185 Roma, Italy

**Summary.** After a few considerations on the present understanding of the normal and superconducting phases in high temperature superconductors, we discuss phase separation as a possible scenario for developping pairing and an anomalous normal state. A specific model is discussed, which allows for phase separation arising from two distinct mechanisms associated to spin and charge degrees of freedom. The two physical mechanisms may drive density fluctuations along two distinct channels, so that the variation of relative weight of these two types of fluctuations may produce a gradual change with doping of the characteristics of the metal in the normal phase. On the other hand, superconductivity can take place over an extended range of doping as a consequence of the attractive forces produced by the tendency to phase separate driven by the combination of the two mechanisms.

## 1. Introduction

The specific order associated with a superfluid transition, i.e. the macroscopic occupation of a "single-particle state", has found its complete formulation only in two limiting cases: i) The Bose-Einstein condensation with onset at the degeneracy temperature of the preexisting bosons ($k_B T_c^{BE} \simeq \hbar^2 n^{2/d}/m^*$, $d > 2$); ii) The BCS theory leading to the coherent superposition of pairs of fermions in $k$ space with large spatial overlap.

The condensation criterion associated to the Off-Diagonal-Long-Range-Order (ODLRO) of the one-particle reduced density matrix for bosons is extended to the two-particle reduced density matrix for fermions and defines naturally the order parameter for superconductivity, to which all the superconductivity effects are associated via the Ginzburg-Landau theory. ODLRO gives also the anomalous factorization property of the two-particle reduced density matrix leading to the generalized Hartree-Fock formulation of the BCS theory.

A macroscopic occupation of a single quantum state follows from ODLRO provided the fermions form bound pairs each extending over a finite coherence distance $\xi_0$. The critical temperature is then given by

$$k_B T_c \simeq \Delta_0 \simeq \hbar v_F \xi_0^{-1} \simeq \frac{E_F}{k_F \xi_0} \tag{1.1}$$

where $\Delta_0$ is the (zero-temperature) superconductivity gap and $v_F$ is the Fermi velocity. In ordinary superconductors the critical temperature is reduced from the degeneracy Fermi temperature ($T_F = E_F/k_B \simeq \hbar^2 n^{2/d}/(m^* k_B) \simeq 10^4$ K) to a much smaller value of the order of 10 K by the large ratio between the pair size $\xi_0$ (of the order of $10^4$ Å) and the average distance between Fermi particles in the normal phase ($\simeq 2\pi/k_F$).

The ordinary superconductors provide an example of a natural compromise between the tendency to form local tightly bound pairs (thereby increasing the critical temperature towards the one of preformed boson condensation) and the necessity to extend the size of the pairs to avoid strong fluctuations, which, together with a large effective mass $m^*$, would drastically decrease the critical temperature.

In the high temperature superconductors [1] $T_c$ from Eq. 1.1 is of the correct order of magnitude because although the Fermi energy is reduced by an order of magnitude [2], the coherence distance $\xi_0$ is strongly reduced ($\xi_0 \simeq$ 10–20 Å, $k_F \xi_0 \simeq 10$). However, there is general consensus that in the high temperature superconductors the $CuO_2$ planes are the prominent structural constituents. This quasi two-dimensionality favors the effects of fluctuations, which tend to destroy the ODLRO. In addition the coherence distance is so short that, because of fluctuations, the range of applicability of the simple BCS mean field theory should be taken with care, whatever the pairing mechanism may be.

On the other hand, the high value of the critical temperature, being incompatible with a large effective mass $m^*$ (experimentally $m^* \simeq$ 2–10$m_0$), makes any explanation based on tightly-bound real space pairs of bipolaron type very unplausible even though it appears more and more that phonons can not be fully excluded from the game.

In this intermediate situation ($k_F \xi_0 \simeq 10$) between the two previous categories of superfluidity the question arises whether we need a completely new theory or is it enough to reconsider the BCS theory or its Eliashberg extension, possibly with new pairing mediators and an appropriate symmetry of the order parameter (s- or d-wave pairing). In this latter case a full inclusion of thermodynamic fluctuations only near the transition would be required for a proper description of the behavior of a system belonging to a two component order parameter universality class. However, due to the smallness of $\xi_0$ the fluctuations should also be relevant far away from the critical temperature [3]. If they are not relevant, as the validity of Eq. (1.1) seems to imply, an important open issue is to understand the mechanism restricting the fluctuations in these quasi two-dimensional systems.

The difficulty in handling the superconductive phase has a similar counterpart in the normal phase. Usually in the normal phase the high energy processes are eliminated to form the Landau picture of the normal Fermi liquid as a gas of fermionic quasiparticles. A small residual interaction for states around the Fermi level leads to a superconductive instability, which can be handled with a mean-field approximation. This simple representation does not work in the high temperature superconductors due to the presence of various comparable energy scales of different nature. In fact electron spectroscopy in the copper oxides has shown that the systems with one hole per unit $CuO_2$ cell are charge-transfer insulators with the holes mostly residing in the $3d_{x^2-y^2}$ orbitals of the copper atoms. Additional holes go instead into states with large amplitude on the oxygen planar 2p orbitals. The system can then be conveniently described by a three band Hubbard model, which contains the large $U_{dd}$ repulsion on the same Cu site, the strong p-d covalency between copper and the planar oxygen and the Charge-Transfer (CT) energy difference between the effective Cu and O atomic levels. At this point we can ask ourselves if this strongly correlated quasi two-dimensional fermion system is compatible with a normal Fermi liquid picture, even though photoemission explicitly indicates the existence of a large Fermi surface satisfying the Luttinger theorem. Many physical quantities, like the in-plane resistivity, indeed deviate from the behaviour predicted by the Landau theory: The antiferromagnetic insulating behavior at and close to half-filling turns upon doping into a strange metal, which at optimal doping displays a linear $T$-dependence of the in-plane resistivity in the normal phase. Magnetic correlations with a possible spin gap are present in the metallic state in the underdoped regime. At still higher doping the strange metal gradually turns into a normal metal with strongly reduced magnetic correlations. No sharp change in the metallic behavior is experimentally observed by varying the charge carriers or by lowering the temperature. The absence of a clear experimental indication of a transition to a new state for the normal phase, (which could otherwise suggest the form of the symmetry of this anomalous state) and, on the theoretical side, the stability of the normal liquid in standard perturbation theory in 2D as in 3D, make the search for a new metallic phase very hard.

In this context much interest has been devoted to anomalous phases appearing in one-dimensional systems. Due to the close AF phase at low doping and the persistence of AF correlations well inside the metallic phase, Anderson proposed [4],[5] to extend to 2D systems the Resonating Valence Bond (RVB) form of the wavefunction, which gives the solution of the Heisenberg model in 1D in terms of a combination of spin singlets. Peculiar to this solution is the charge and spin decoupling in the excitations. The hypothesis behind this approach [5] is that, away from half-filling, the solution is never a Fermi liquid. In 2D, doping the interacting fermions should lead to a new metallic phase similar to the Luttinger metal occurring in 1D. In this latter metallic phase there is no discontinuity of the occupation number at the Fermi level, but an algebraic singularity, with a non universal exponent. This singularity still defines a Fermi surface satisfying the Luttinger theorem. The marginal Fermi liquid [6], proposed as a phenomenological theory of the normal phase of the cuprates, corresponds to a particular case for which the discontinuity at the Fermi surface goes to zero as the inverse of the logarithm of $|k - k_F|$ rather than algebraically. In the Luttinger liquid the single quasiparticle excitations do not exists at the Fermi surface and the spectrum is made of collective excitations with separate charge and spin modes. The thermodynamic quantities, instead, like specific heat, compressibility, and spin susceptibility, associated with specific limits of the response functions, are those proper of a *stable* metal. The singularities leading to the anomalous behavior of the single particle properties and to the charge and spin separation must therefore undergo very delicate cancelations in these thermodynamical quantities in order to give rise to a stable metallic phase. Conservation laws, peculiar to the 1D, where the Fermi surface is reduced to two points ($\pm k_F$) and the related Ward identities [7], accomplish this task.

An analogous conservation law is valid approximately in $1 < d < 2$, with an increasing accuracy as the Fermi surface is approached. This allows one to sum to all orders in the coupling the relevant small momentum transfer processes leading to the Luttinger liquid in 1D. In this way it is shown that the Luttinger liquid behavior crosses over into a Fermi liquid one as soon as $d > 1$ [8], even though the inverse lifetime of the quasiparticles scales as $|k - k_F|^d$ instead of the square law valid for $d \geq 2$. A breakdown of the Fermi liquid is therefore not obtained in a one band model with short range interactions. As soon as $d > 1$ the Fermi liquid is a very stable phase: additional degrees of freedom or singular non-perturbative phenomena leading to singular interactions are necessary but not sufficient requirements to obtain a different metallic phase.

It seems to us that these results indicate some difficulties inherent in all models for superconductivity, where pairing arises from a normal phase displaying charge and spin separation in the $CuO_2$ planes [9].

Many other proposals which could possibly lead to non-Fermi liquid behavior have been left out of this short analysis, e.g. gauge field theories in 2D and the multichannel Kondo (lattice) model. In any case we feel that to settle the question of non-Fermi liquid behavior in 2D has intrinsic difficulties. It is therefore worthwhile to pursue another route, which has a generic validity and makes no drastic assumptions on the normal phase to start with.

The richness of the phase diagram of the high temperature superconducting compounds is a clear indication of the presence of various competing energy terms of the same order of magnitude. As we have already stressed, these terms, on the one hand, make it difficult to handle the different problems separately (like antiferromagnetism, transport in the normal phase and superconducting pairing). On the other hand, in each region of doping and temperature these competing energy terms will favor a tendency towards Phase Separation (PS) [10], a phenomenon which we will discuss extensively in the rest of the paper.

Section 2. is devoted to the presentation of the PS problem. In Section 3. PS will be discussed in the framework of an extended three band t-J model. A discussion of the physical implications of the results will be given in Section 3. together with our concluding remarks.

## 2. Phase separation: a relevant issue

While on the one hand the solution of the Fermi-liquid vs non-Fermi-liquid dilemma is an intriguing issue, which still deserves intensive investigation, on the other hand the analysis summarized in the previous section clearly indicates that a non-Fermi liquid state is not easily realized in more than one dimension. Therefore it is surely appropriate to investigate different possibilities in order to account for the peculiar normal and superconducting properties of the copper oxides. The general tendency has been the investigation of the $CuO_2$ planes. In this latter context, besides the mechanisms involving the presence of (strong) van Hove singularities in the band structure of these oxides [11], an enormous variety of proposals has been put forward, ranging from an enhancement of the $q$-dependent scattering due to the nesting of the Fermi surface [12], to the scattering with magnetic excitations, both traditional like paramagnons [13] or "exotic", like chiral gauge field fluctuations [14]. Phonons [15] and excitons [16] have also been proposed, witnessing how open (and unfortunately confused) the situation still is. As already stressed, in this situation the need of general unifying paradigms, which are capable of accounting for most properties of the real materials without making too specific assumptions is particularly compelling. In the rest of this paper we will specifically focus on PS [10], which both experimentally and theoretically seems to satisfy this general prerequisite, so that, in the last two years it has become a relevant issue in the context of high $T_c$ superconductivity.

First of all there has been uncontroversial experimental evidence that PS occurs in some high $T_c$ systems. In particular this phenomenon takes place in superoxygenated $La_2CuO_{4+x}$ [17] whereas its occurrence has been suggested but not confirmed in $La_{2-x}Sr_xCuO_4$ [17] and in $ErBa_2Cu_3O_{6+x}$ [18]. A different but maybe related phenomenon is the superstructure formation in Bi 2212 and 2232 compounds [19]. Therefore some kinds of inhomogeneities surely occur in most high temperature superconducting materials. The electronic origin of the phenomena is uncertain, though photoinduced conductivity experiments [20] and NMR studies of local structure in metallic $La_2CuO_{4+x}$ [21] strongly suggest that the holes due to photodoping or to chemical doping tend by themselves to separate.

PS is also appealing from the theoretical point of view. In fact, whereas the possibility of PS in the Hubbard model was hypothesized some time ago [22], its occurrence has recently been found in the t-J model using a large-N expansion [23] and exact diagonalizations of small clusters [24]. This stimulated the theoretical investigation of the instabilities of purely electronic models. Since then it was found that PS seems to be the rule rather than the exception in the context of models with strongly interacting electrons in the absence of long range Coulomb forces. In fact a slave boson analysis of various models has revealed the occurrence of PS in some parameter region both in models with magnetic interactions like the single band t-J model [25], the three band t-J model [26],[27] and the Kondo-lattice model [28], as well as in models with nearest neighbor Coulombic repulsion [29],[30]. A common feature of these models is the presence of strong correlations between electrons, as reviewed in Ref. [31]. These correlations lead to a substantial reduction of the quasi-particle kinetic energy. As a consequence the stabilizing effect of the kinetic energy, which would lead to the formation of extended uniformly distributed electronic states, is depressed, opening the way to the localizing effects of the short range interactions. This fact is made apparent in those multiband models [26],[29],[30], where it is possible to observe the beginning of PS in the phase diagram region close to the metal-insulator transition. It is in this low doping region where the strong correlations most effectively depress the stabilizing effect of the particle motion and favor the particle segregation.

The analysis of these models, both with magnetic and with nearest neighbor (nn) Coulombic repulsion has revealed suggestive connections between PS and superconductivity. In particular it was found in the context of the infinite-$U$ three-band extended Hubbard model (in this case the short range interaction is due to nn copper-oxygen hole repulsion) [29] that a Cooper instability takes place in the parameter region where the system is a stable Fermi liquid, but close to the line where the compressibility $\kappa$ diverges. Superconductivity then appears as a kind of precursor effect of PS (at least to leading order in the large-$N$ expansion).

A second PS-superconductivity connection was found [28] in the Kondo-lattice model (here the short range interaction is of magnetic nature), where a mean field analysis indicated that pair formation gives rise to a state with lower energy than the separated phase, showing that the occurrence of superconductivity can stabilize the system preventing PS. This indicates that superconductivity is not necessarily spoiled by PS, as it was implicitly assumed when PS was first presented as a drawback of the one-band t-J model [24]. On the contrary, tendency to PS may favor the pairing formation [28],[29]. Of course the experimental observation of such a phenomenon is strictly related to the diffusional dynamics of charged ions in the systems. This subject has been the object of intensive and interesting experimental investigation [32].

All these phenomena naturally occur in models without a long range Coulomb repulsion, which would otherwise prevent the segregation of thermodynamically large clusters of charged particles. Nevertheless dynamical slow density fluctuations can surely survive the presence of Coulomb long range repulsion [33]. This latter interaction is, in fact, most effective at small $q$'s where it prevents the divergence of the static compressibility $\kappa$, but not necessarily the dynamical charge susceptibility, which may be large at finite (short) distances (finite large $q$'s).

The presence of long range interactions could explain why copper oxides only rarely display a thermodynamic PS. In particular PS would only be allowed in those systems where a high mobility of oxygen negative ions is present. In this case the negative ions could follow the positively charged holes compensating for the charge unbalance, thereby making Coulomb repulsion ineffective. In this respect superoxigenated La cuprates and 123 compounds are obvious candidates for the observation of a static thermodynamical PS of electronic origin. On the other hand the presence of common structural elements like the $CuO_2$ planes suggests that the tendency of an electronic PS should also be present in those high $T_c$ materials where it is not observed, due to the presence of long range Coulomb forces, which are not suppressed by the mobility of the negative ions.

From the theoretical point of view the presence of long range Coulomb forces opens the way to many interesting possibilities. On the one hand it has been suggested [33] that low frequency charge density fluctuations could be modeled by means of pseudospin variables thereby introducing a "mesoscopic" description of the system, which is claimed to lead to a mapping into a two-channel Kondo model. This latter model seems to have a similar behavior to the normal properties of the superconducting cuprates. This suggests the possibility [33] that the unusual normal state properties of these systems arise from the scattering of the carriers from soft density modes.

On the other hand the introduction of long range Coulomb forces further enriches the possibilities of fully microscopic models. We mentioned above the existence of the Cooper instability in the proximity of the infinite-compressibility ($\kappa = \infty$) line in the extended infinite-$U$ three band Hubbard model. However, the Maxwell construction reduces the accessible (stable) region around the $\kappa = \infty$ line. The introduction of long range interactions would instead eliminate the small-$q$ instabilities thereby making accessible to the system a much larger region. In particular it could happen that only a small region of a static finite-$q$ instability would remain unaccessible, where incommensurate charge density waves would occur [30]. Close to this region strong attractive forces would be present similar to those occurring at small $q$ near the $\kappa = \infty$ line in the neutral hole case. Of course the long-range repulsion tends to oppose attraction. In this case only a full dynamical treatment would answer the question on the subtle $q$-dependent screening and cancelation between repelling long range forces and attractive forces arising close to the instabilities. Unfortunately such a detailed microscopic treatment involving both the role of Coulomb forces, total density and CT fluctuations is overwhelmingly difficult and does not seem to be viable in the short term.

From the above discussion it should be remembered that i) PS is a common *tendency* of models with strong local (Hubbard-like) repulsion and additional short range interactions of various nature; ii) this tendency is expected to be present regardless of the actual existence of a static thermodynamic PS, which depends on the existence of mechanisms (e.g. oxygen diffusion) ruling out the action of the long-range Coulomb forces; iii) two main microscopic mechanisms have so far been found as possible sources of this tendency, namely the magnetic spin-spin interaction and the copper-oxygen nn Coulomb repulsion. Of primary importance is to recognize that, even though this second mechanism also occurs close to the metal-charge-transfer-insulator transition, it acts most effectively in the finite, large doping regime, forming (dynamic or static) regions of high and moderate doping. On the contrary, the magnetic mechanism is more effective at low doping giving rise to half-filled antiferromagnetic insulating regions and moderately doped metallic regions. The two mechanisms seem therefore dominant in different filling regimes. The simultaneous consideration of these two mechanisms provides a natural and unified description of the effects of the various competing energy terms present in these strongly correlated systems, leading to different phases by varying the charge carriers and the temperature.

Then the question naturally arises whether a real system exists, where both mechanisms play a role. In this case a crucial prerequisite is the possibility of doping the system on a wide range, so as to explore the *tendency* to phase separate by means of a magnetic mechanism at low doping and the *tendency* to phase separate at large doping due to the nn Coulombic repulsion. Obvious candidates are the compounds of the 123 class, where the oxygen content can run from $O_6$ to $O_7$. Last but not least these systems display a rather large oxygen mobility which could at least partially counterbalance the effect of the long range Coulomb forces. Far from being necessary to allow the *tendency* to phase separate, this high mobility of the negative ions could facilitate the experimental detection of the "desire" of PS present in the system.

Interestingly enough the existence of a double (possibly dynamical in nature) PS [18] has already been proposed from inelastic neutron scattering experiments performed on $ErBa_2Cu_3O_{6+x}$ [34], [35].

Whereas the electronic origin of this double (dynamical) PS has to be proven, this system is undoubtably appealing as a possible candidate to see the two, magnetic and nn Coulombic, mechanisms at work, in support to our general presentation. This point of view will be further elucidated by the analysis of a specific model [36] in the next section.

## 3. A specific example

We consider a three band extended t-J model in order to explicitly show the possibility of the two mechanisms discussed in the previous section coexisting in the same model. In particular we stress that *we are not interested in distinguishing between the occurrence of a truly thermodynamical static PS or of a PS, which is dynamical in nature.* In particular we will not include long range Coulomb interactions in our model, so that the PSs inherently present should not necessarily be intended as static.

The starting point [36] is represented by the following Hamiltonian for the $CuO_2$ planes,

$$\begin{aligned}\mathcal{H} &= (\varepsilon_{\mathrm{p}}-\mu)\sum_{i,l,\sigma} p^{\dagger}_{i+l,\sigma}p_{i+l,\sigma} + (\varepsilon_{\mathrm{d}}-\mu)\sum_{i,\sigma} d^{\dagger}_{i,\sigma}d_{i,\sigma} + \\ &+ \sum_{i,l,\sigma}\left[t^{l}_{\mathrm{pd}}d^{\dagger}_{i,\sigma}p_{i+l,\sigma}+h.c.\right] + \sum_{i,l,l',\sigma}\left(t^{ll'}_{\mathrm{pp}}p^{\dagger}_{i+l,\sigma}p_{i+l',\sigma}+h.c.\right) + \\ &+ \frac{U_{\mathrm{pp}}}{4}\sum_{il,\sigma} n^{\mathrm{p}}_{i+l,\sigma}n^{\mathrm{p}}_{i+l,-\sigma} + V\sum_{i,l,\sigma,\sigma'} n^{\mathrm{d}}_{i,\sigma}n^{\mathrm{p}}_{i+l,\sigma'} + \\ &+ J\sum_{\langle ij\rangle,\sigma\sigma'} d^{\dagger}_{i\sigma}d_{i\sigma'}d^{\dagger}_{j\sigma'}d_{j\sigma} + U_{\mathrm{dd}}\sum_{i} n^{\mathrm{d}}_{i,\uparrow}n^{\mathrm{d}}_{i,\downarrow} \end{aligned} \tag{3.1}$$

where $d^{\dagger}_{i\sigma}$ creates a hole with spin $\sigma$ in the $\mathrm{d}_{x^2-y^2}$ orbital on the copper site $i$, $p^{\dagger}_{i+l,\sigma}$ ($l=\pm x/2, \pm y/2$) creates instead a hole in the $\mathrm{p}_x$ or $\mathrm{p}_y$ orbitals on the oxygen site $i+l$ (we take Cu-Cu distance to be unit). The density operators are defined by $n^{\mathrm{d}}_{i\sigma}=d^{\dagger}_{i\sigma}d_{i\sigma}$ and $n^{\mathrm{p}}_{i+l,\sigma}=p^{\dagger}_{i+l,\sigma}p_{i+l,\sigma}$.

The Hamiltonian (3.1) includes the atomic level energies $\varepsilon_{\mathrm{p}}$ and $\varepsilon_{\mathrm{d}}$ on oxygen and copper sites respectively, and the hopping integrals $t^{l}_{\mathrm{pd}}=\pm t_{\mathrm{pd}}$ between the d and p orbitals and $t^{ll'}_{\mathrm{pp}}=\pm t_{\mathrm{pp}}$ between the p orbitals [37]. This tight-binding part reproduces therefore the p-d covalent nature of the $CuO_2$ planes. The interaction terms are, instead, represented by the Coulombic interaction $V$ between nn Cu and O holes, by the Cu-Cu superexchange term $J$ and by the Hubbard repulsion $U_{\mathrm{dd}}$ between holes on the same copper site. This last interaction is by far the largest energy scale ($U_{\mathrm{dd}}$ ~10eV) in copper oxides so that the $U_{\mathrm{dd}}\to\infty$ limit was considered by means of a standard slave-boson technique [38].

The above model is particularly suitable to describe the many different behaviors occurring upon doping in the copper oxides. In these systems the antiferromagnetic insulator which is present in the stoichiometric and lightly doped compound turns into the *strange metal* with magnetic correlations, as it was discussed in Section 1.. At still higher doping the strange metal gradually turns into a normal metal with strongly reduced magnetic correlations. The spin degrees of freedom therefore play a major role at low doping, when charge degrees of freedom are essentially frozen out due to the presence of strong correlations. Charge degrees of freedom become gradually more relevant in the intermediate and high doping regimes. This interplay of strong correlations in different degrees of freedom is well schematized by the above model.

It should be noted that the model in Eq. (3.1) gives rise to a Cu-Cu superexchange interaction even in the absence of the term proportional to $J$. However, due to the technical complexity of the model a mean-field approximation was used. Since this approximation neglects the virtual processes leading to the Cu-Cu superexchange interaction, the term proportional to $J$ was explicitly added to keep into account this important piece of physics and the value of this coupling was fixed according to the usually accepted experimental value ($J=0.1\,t_{\mathrm{pd}}$ with $t_{\mathrm{pd}}=1.3eV$).

It is worth stressing that the mean field approximation within the slave boson scheme is very different from the conventional Hartree-Fock procedure and, contrary to this latter, is even able to reproduce the Mott-Hubbard charge-transfer insulating character of the $CuO_2$ planes at half-filling. On the other hand the much weaker interactions $V$, $U_{pp}$ and $J$ can be safely decoupled via Hartree-Fock factorizations. To investigate the normal and the superconducting phases on equal footing both normal and superconducting decouplings were considered, to which the subscript "1" and "2" were appended respectively.

For the sake of simplicity, the magnetic interaction was treated neglecting that decoupling giving rise to spin long range order. This choice was dictated by the fact that in the real high temperature superconductors antiferromagnetism is rapidly destroyed by doping, so that one can directly focus on a nn singlet decoupling $\Delta_1^{ij} = (J/2)\sum_\sigma \langle d^\dagger_{i,\sigma} d_{j,\sigma}\rangle$. This latter has been widely used to suitably describe at mean-field level the magnetic correlations present in the metallic phases [39],[40]. Within this scheme the antiferromagnetic insulator is mimicked by the so called dimer phase having the same unit cell doubling, where the parameter $\Delta_1^{ij}$ assumes different values on the bonds pointing in different directions ($\Delta_1^x \neq \Delta_1^{-x} \neq \Delta_1^{\pm y}$). The magnetically correlated metal is, instead, represented by the uniform phase where all the bonds have the same value ($\Delta_1^x = \Delta_1^{-x} = \Delta_1^{\pm y} = \Delta_1$). The metallic phase without magnetic correlations is characterized by a vanishingly small value of $\Delta_1$.

On the other hand, the anomalous order parameter $\Delta_2$ arises as a consequence of the attractive superexchange interaction $J$ and represents the occurrence of a superconducting phase in this mean field approximation [41]. The possible occurrence of s-wave as well as d-wave superconductivity was investigated. This latter was always found to be more stable. Thus, superconducting phases of d-wave type only will be considered in this context.

The numerical solution of the self-consistency equations together with the condition for the chemical potential fixing to $(1+\delta)$ the number of particles per $CuO_2$ cell yields the phase diagram of the model for various values of the physical parameters [36].

The $T = 0$ phase diagram is shown in Fig. 3.1. A metal-insulator transition is found at half filling by varying the distance between the atomic levels $\varepsilon_p - \varepsilon_d$. This transition takes place between a magnetically uniform metal ($\Delta_1 \sim J, \Delta_2 = 0$) and an insulating phase with dimer magnetic symmetry. At low doping the system is unstable due to the presence of the dimer phase (or of the antiferromagnet), which tends to eject doped holes leading the system towards PS. Moreover, when $V$ is sufficiently large, the system is driven towards the charge-transfer (CT) instability at intermediate doping. To our knowledge the CT mechanism is the only one which provides a PS in the intermediate- and high-doping regimes. It must be emphasized that *these two mechanisms for phase separation do not exclude each other, but*

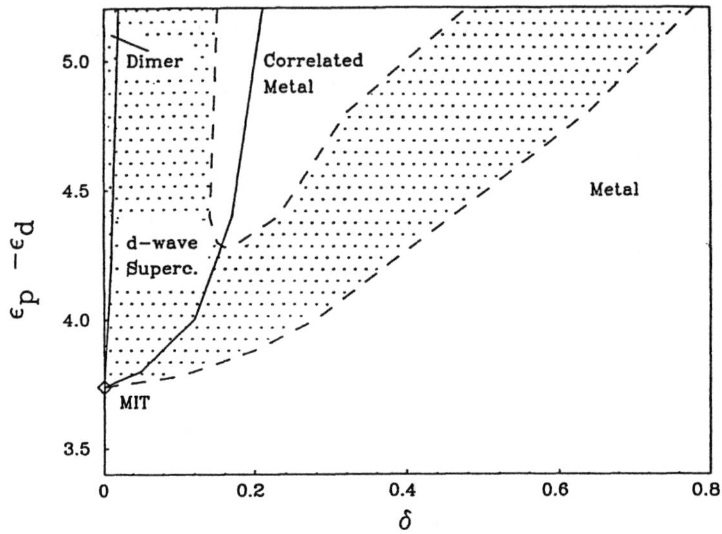


**Fig. 3.1.** Phase diagram $(\varepsilon_p - \varepsilon_d)$ vs. the doping $\delta$ with $T = 0.002$ eV, $t_{pd} = 1.4$ eV, $t_{pp} = 0.4$ eV, $J = 0.2$ eV, $V = 3.0$ eV and $U_{pp} = 0.5$ eV. Solid lines indicate the phase transition lines. The shaded area delimited by the dashed lines indicates the phase separation region determined via the Maxwell construction. The diamond indicates the metal-insulator transition.

*play their role in different regions of doping.* The two PS regions merge in the vicinity of the metal-insulator transition.

A d-wave pairing phase $(\Delta_1, \Delta_2 \sim J)$ is stable in the intermediate-doping regime and at large CT gaps, with the correlated metal $(\Delta_1 \sim J, \Delta_2 = 0)$ nearby. This latter smoothly crosses over to the uncorrelated metal $(\Delta_1 \simeq 0, \Delta_2 = 0)$, which is the stable phase at high doping.

The large overlap between the phase separation region and the magnetically induced superconductivity region strongly suggests that the two phenomena are related. As it was previously recalled, a similar occurrence was found to occur in the Kondo-lattice model [28], where a magnetism mediated attractive force in the particle-hole channel leads to phase separation and, at the same time, makes possible a stable homogeneous superconducting phase. Contrary to the superexchange term $J$, the nn interaction $V$, being repulsive, cannot induce pairing in mean field approximation. However, the analyzes in Refs. [29] and [30] have shown that attraction can appear close to the CT phase separation instability when CT fluctuations, enhanced by $V$, are considered. This mechanism therefore extends to higher doping and lower bare CT gaps the superconducting phase shown in the phase diagram of Fig. 3.1. Moreover, also in this case, superconductivity could stabilize the system with respect to phase separation, thus reducing the phase separation region.

According to the discussion of Section 2., phase separation provides a possible scenario for the developing of an exotic way to induce pairing and, at the same time, an anomalous normal state. The example discussed here indicates that various mechanisms for PS can coexist. In particular, the simultaneous presence of magnetic and CT terms in the Hamiltonian allows for two distinct tendencies towards PS. In turn the two physical mechanisms drive density fluctuations occurring along two distinct channels. A first channel involves density fluctuations between the AF insulator and the intermediate-doping metal mainly driven by magnetism, whereas the charge-transfer driving force would be responsible for the slow fluctuations between the intermediate-doping metal and the high-doping metal.

The smooth change of the relative weight of these fluctuations of two different types may produce a gradual change with doping of the characteristics of the metal in the normal phase. On the other hand, the attractive forces produced by the tendency to phase separate driven by the combination of the two mechanisms may lead to superconductivity over an extended range of doping.

*Acknowledgement.* This work was supported by the European Economic Community under contract no. SC1* 0222-C(EDB).

## References

1. J.G. Bednorz, and K.A. Müller, Z. Phys. B 64, 189 (1986).
2. For a review on experiments see e.g.*Proceedings of the International $M^2S$ HTSC Conference*, Kanazawa, Japan, July 22-26 1991, Physica C 185-189, (1991).
3. A. Kapitulnik, M.R. Beasley, C. Castellani, and C. Di Castro, Phys. Rev. B 37, 537 (1988).
4. P.W. Anderson, Science 235, 1196 (1987).
5. P.W. Anderson, Phys. Rev. Lett. 64, 1839 (1990); Phys. Rev. Lett. 65, 2306 (1990).
6. C.M. Varma, P.B. Littlewood, S. Schmitt-Rink, E. Abrahams, and A.E. Ruckenstein, Phys. Rev. Lett. 63, 1996 (1989).
7. C. Di Castro, and W. Metzner, Phys. Rev. Lett. 67, 3853 (1991); W. Metzner, and C. Di Castro, Phys. Rev. B 47, 16107 (1993).
8. C. Castellani, C. Di Castro, and W. Metzner, Phys. Rev. Lett. 72, 316 (1994).
9. S. Chakravarty, A. Sudbø, P.W. Anderson, S. Strong, Science 261, 337 (1993).
10. For a review on PS in high $T_c$ cuprates see e.g.*Proceedings of the Workshop on Phase Separation in Cuprate Superconductors*, Erice, Italy, 6-10 May 1992, edited by K.A. Müller and G. Benedek, World Scientific.
11. C.C. Tsuei, C.C. Chi, D.M. Newns, P.C. Pattnaik, and M. Däumling, Phys. Rev. Lett. 69, 2134 (1992); D.M. Newns, Comments Cond. Mat. Phys. 15, 273 (1992).
12. J. Ruvalds, and A. Virosztek, Phys. Rev. B 43, 5498 (1991).
13. T. Moriya, Y. Takahashi, and K. Ueda, Physica C 185-189, 114 (1991).

14. N. Nagaosa, and P. Lee, Phys. Rev. Lett. 64, 2450 (1990); L. Ioffe, and G. Kotliar, Phys. Rev. B 42, 10348 (1990).
15. Ju.H. Kim, K. Levin, R. Wentzcovitch, A. Auerbach, Phys. Rev. B 40, 11378 (1989); Ju.H. Kim, K. Levin, R. Wentzcovitch, A. Auerbach, Phys. Rev. B 44, 5148 (1991).
16. C .M. Varma, Int.J. of Mod. Phys. B 3, 2083 (1989).
17. J.D. Jorgensen B. Dabrowski, Shiyou Pei, D. G: Hinks, L. Soderholm, B. Morosin, J.E. Schirber, E.L. Venturini and D.S. Ginley, Phys. Rev. B 38, 11337 (1988).
18. J. Mesot, P. Allenspach, U. Staub, A. Furrer, and H. Mutka, Phys. Rev. Lett. 70, 865 (1993); A. Furrer, J. Mesot, P. Allenspach, U. Staub, F. Fauth, M. Guillaume, these Proceedings.
19. see, e.g., A. Bianconi in Ref.[10] and this Proceedings and references therein.
20. G. Yu, C.H. Lee and A.J. Heeger, in Ref. [10] and references therein.
21. P.C. Hammel, A.P. Reyes, S-W. Cheong, Z. Fisk, and J.E. Schirber, Phys. Rev. Lett. 71, 440 (1993)
22. P.B. Visscher, Phys. Rev. B 10, 943 (1974).
23. M. Marder, N. Papanicolau and G.C. Psaltakis, Phys. Rev. B 41, 6920 (1990).
24. V.J. Emery, S.A. Kivelson, and H.Q. Lin, Phys. Rev. Lett. 64, 475 (1990).
25. M. Grilli, C. Castellani, and G. Kotliar, Phys. Rev. B 45, 10805 (1992).
26. C. Castellani, M. Grilli, and G. Kotliar, Phys. Rev. B 43, 8000 (1991).
27. S. Caprara, and M. Grilli, Phys. Rev. B 49 (1994).
28. N. Cancrini, S. Caprara, C. Castellani, C. Di Castro, M. Grilli and R. Raimondi, Europhys. Lett. 14, 597 (1991).
29. M. Grilli, R. Raimondi, C. Castellani, C. Di Castro, and G. Kotliar, Phys. Rev. Lett. 67, 259 (1991); International Journal of Modern Physics B Vol. 5, 309 (1991).
30. R. Raimondi, C. Castellani, M. Grilli, Y. Bang, and G. Kotliar, Phys. Rev. B 47, 3331 (1993).
31. C. Di Castro and M. Grilli, in Ref. [10]
32. R.K. Kremer, V. Hizhnyakov, E. Sigmund, and K.A.Müller, Z.Phys. B-Condensed Matter, 91, 191 (1993).
33. V.J. Emery, and S.A. Kivelson, in Ref.[10] and V.J. Emery, and S.A. Kivelson, unpublished.
34. J. Mesot, P. Allenspach, U. Staub, A. Furrer, H. Mutka, R. Osborn and A. Taylor, Phys. Rev. B 47, 6027 (1993).
35. A two PS phase diagram has also been proposed on the basis of transport and structure experiments in $La_2CuO_{4+x}$ with $0 \leq x \leq 0.09$ in the paper by J.-C. Grenier, A. Wattiaux and M. Pouchard in Ref. [10]. In particular, whereas a first spinodal decomposition occurs for $x < 0.06$ between an AF insulating phase and an intermediate doping metallic phase, for $x > 0.06$ two different metallic species appear to be present. While the intermediate-doping metallic phase has a superconducting critical temperature of 32 K, the second high-doping metallic phase becomes superconducting at $T_c = 38$ K.
36. S. Caprara, C. Di Castro, and M. Grilli, unpublished.
37. For the choice of the phase convention determining the sign of the hoppings, see, e.g. Ref.[40]
38. S.E. Barnes, J. Phys. F 6, 1375 (1976); P. Coleman, Phys. Rev. B 29, 3035 (1984); N. Read, and D.M. Newns, J. Phys. C 16, 3273 (1983); N. Read, J. Phys. C 18, 2651 (1985); A.J. Millis, and P.A. Lee, Phys. Rev. B 35, 3394 (1987); G. Kotliar and A.R. Ruckenstein, Phys.Rev. Lett. 57, (1986) 1362.
39. I. Affleck, and B. Marston, Phys. Rev. B 37, 3774 (1988); and G. Kotliar, Phys. Rev. B 37, 3664 (1988).

40. For more complete description of a slave boson treatment of an extended three band t-J model, within a large-N expansion see, e.g. Ref.[27] and references therein.
41. G. Kotliar and J. Liu, Phys. Rev. B 38, 5142 (1988).

# Point Contact Spectroscopy in the High $T_c$ Oxides

G. Deutscher

School of Physics and Astronomy, Sackler Faculty of Exact Sciences, Tel Aviv University, Ramat Aviv, Tel Aviv, Israel

**Summary.** We review recent Point Contact Spectroscopy results obtained on the high $T_c$ oxides YBCO and LSCO, in the superconducting state. For Andreev type contacts, the shape of the $V(I)$ characteristics at low voltages is consistent with an s wave symmetry of the superconducting state and not with a d wave symmetry. The enhanced conductance at voltages smaller than the gap indicates a Fermi velocity in the superconducting electrode that is within a factor of 2 of that of the normal Au point. This result is interpreted as reflecting strong correlation effects. The structure of the second derivative of the characteristics at voltages in the 100 meV range fits well the known phonon spectrum for LSCO, which we interpret as a sign of strong electron-phonon interaction. The possibility of phase separation in systems where strong eléctron phonon and strong correlation interactions coexist is discussed.

## 1. Introduction

Point contact spectroscopy has been known for many years as a useful tool for the study of metals both in the normal and in the superconducting state. In normal metals [1], it has been used extensively for the study of the electron-phonon interaction, and in superconductors principally for the determination of the energy gap using the Andreev reflection phenomenon at a superconductor-normal metal interface [2]. However, it is fair to say that this method has not yet reached the accepted status of the classical Giaever-Rowell-McMillan tunneling method [3] for an in depth study of the superconducting state including the mechanism for superconductivity. The reasons for this are partly historical — the point contact method was mostly developed in the Soviet Union and tunneling in Western countries at a time when there was not too much contact between physicists on both sides — but they are also substantial from the technical and physics point of view. The method involves pressing a fine metal point against a larger size counter-electrode, lending a contact whose size and exact nature is difficult to control and to reproduce; at voltages of the order of the superconducting energy gap, the current density flowing through the contact is of the order of the critical current density, which may result in heating effects; the shape of the obtained $V(I)$ characteristics is quite variable from one contact to another, rendering their interpretation difficult. These problems are in contrast with the properties of the classical Aluminium oxide tunneling barrier first used by Giaever, for which there exists a well defined protocol that lends reproducible results (even though the exact nature of the barrier is not necessarily simple).

Yet, we feel that there are some excellent reasons for using and developing the point contact method for the study of the high $T_c$ oxides and other "exotic" superconductors. First of all, it is based on the realization of a metallic contact which is much easier to achieve than a thin dielectric barrier (Aluminium oxide cannot be used in practice in the case of the high $T_c$ oxides, due to metallurgical problems; more universal artificial barriers remain to be developed). Second, it can easily be used along defined directions of single crystal or textured samples. Third, a theoretical way to understand and analyze the different shapes of the $V(I)$ characteristics that can be obtained is available since the work of Blonder, Tinkham and Klapwijk (BTK) [2]. And finally — and most importantly for our purpose here — the point contact offers over tunneling the advantage that the shape of the characteristics is sensitive to the mismatch between the electronic wave functions in the two electrodes. This involves the ratio between their respective Fermi velocities. This information is of little relevance for the case of conventional superconductors for which the normal state electronic properties are well understood. But it is obviously of great interest for the high $T_c$ oxides. This advantage of the point contact method had apparently been missed before our work.

One of our main conclusions is that in the two oxides that we have studied so far ($YBa_2Cu_3O_7$ [4] and $La_{1.85}Sr_{0.15}CuO_4$[5]) the in plane electron velocity is at least equal to half that in Gold. This result is remarkable in several respects. First, it implies a large Fermi surface and a broad band in agreement with recent angle resolved photoemission results [6], but in contrast with earlier ideas involving narrow bands and large effective masses (not to speak of more exotic theories of the electronic structure of the oxides). But, as we shall see, it also implies the existence of large renormalization effects in the properties of the quasi-particles, pointing out particularly to the existence of strong correlations effects. This is discussed in the first section, drawing on some recent remarks of Eliashberg [7] on the possible existence of a BCS state in the presence of strong correlations. We also show that the shape of the characteristics is not consistent with a d wave symmetry of the superconducting state.

In the second section we discuss the $V(I)$ characteristics in the phonon energy range obtained on $La_{1.85}Sr_{0.15}CuO_4$. They give direct strong evidence for a substantial electron phonon interaction up to 50 to 60meV. This, together with the total strength of the renormalization effects obtained from the low voltage part of the characteristics, leads us to an estimate of the respective strengths of the electron phonon and correlation strength parameters, which we find to be comparable.

In the last section we discuss some possible general consequences of such a situation. We conjecture that the decrease in $T_c$ in the overdoped regime may reflect a simultaneous decrease of both interactions, and from this perspective we compare the behavior of the high $T_c$ oxides with that of granular Aluminium. We also discuss phase separation in systems characterized by

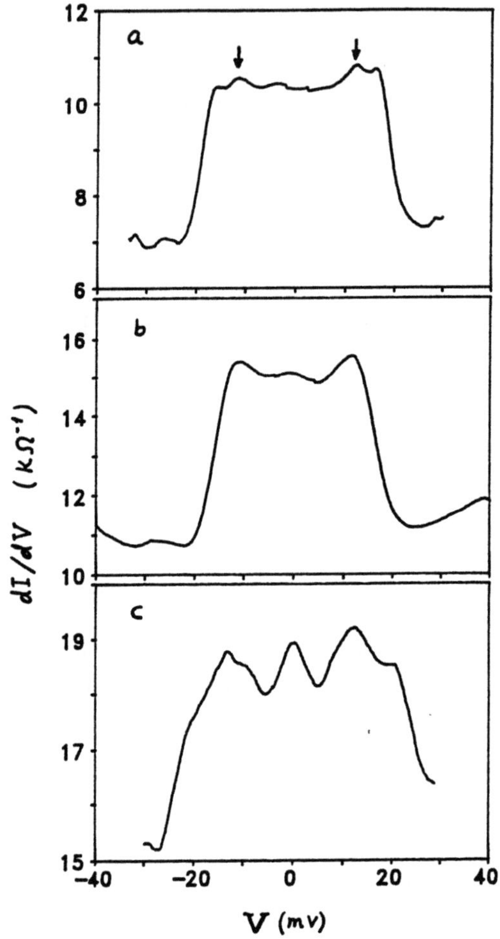


**Fig. 1.1.** Some Andreev characteristics of a Au-YBCO contact. The larger gap is close to 20 meV but there are indications of a smaller gap at 12 meV. Note the small peaks in conductance at the gap edges.

simultaneous strong electron phonon and correlation interactions. We show that in this case the gain in free energy obtained by condensation into the superconducting state can be of the same order as that gained by magnetic ordering. A phase diagram containing regions of coexistence of superconducting and magnetic phases, or superconducting and normal phases is a natural consequence of such a situation.

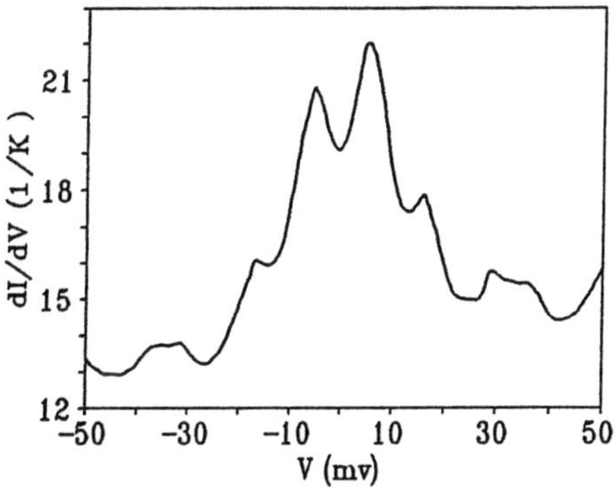


**Fig. 1.2.** An Andreev Au-LSCO characteristics. Note the more pronounced peaks at the gap edges.

## 2. Energy gap and Fermi velocity determination. The Eliashberg interpretation

We have reported earlier on the observation of Andreev reflections when a fine Au needle is pressed against $YBa_2Cu_3O_7$ (Fig. 1.1) and $La_{1.85}Sr_{0.15}CuO_4$ (Fig. 1.2) textured samples, with its axis oriented parallel to the CuO planes. For contact resistances of the order of a few 10 W, we observe a flat or somewhat increasing conductance up to an energy that we identify as the gap (about 20 meV in YBCO and 6 meV in LSCO) followed by a sharp decrease to a lower conductance. This decrease is by a factor of about 1.5 in YBCO and 1.3 in LSCO.

The shape of the characteristics has been analyzed in the framework of the BTK theory [2]. In the case of a perfect match of the electronic wave functions at the interface (no dielectric barrier, equal Fermi velocities in the S and N electrodes), the only reflections taking place at the interface are of the Andreev type, where an incoming particle on the N side (say an electron) is reflected (as a hole) while a Cooper pair is transmitted in the S side. In this case the change in the dynamical conductance above the gap value is equal to 2. The case where there exists a mismatch at the interface is treated by BTK, with a mismatch parameter $Z$ equal to 0 in the pure Andreev limit described above and going to infinity in the opposite Giaever limit where the mismatch is very strong and there are no Andreev reflections at all, only normal reflections, and the current is a tunneling current. The zero bias conductance then goes to zero. For $Z = 1$, the conductance at zero bias is equal to the conductance above the gap, and exhibits a peak at an applied voltage equal to the gap. From a fit of the characteristics to the BTK theory one can extract a value for $Z$, which is found to be close to 0.3 for YBCO and LSCO.

In the BTK theory the value of Z is given by:

$$Z = \left[ Z_{\rm b}^2 + \frac{(1-r)^2}{4r} \right]^{1/2} \tag{2.1}$$

where $Z_{\rm b}$ represents the contribution to the mismatch of a dielectric barrier, and the second term inside the bracket that of the mismatch of the Fermi velocities given by their ratio $r$.

From $Z = 0.3$ one deduces $v_{\rm FN} > v_{\rm FS} > 0.5\, v_{\rm FN}$ (or $v_{\rm FN} < v_{\rm FS} < 2v_{\rm FN}$, but we rule this out on physical grounds). Using the known Fermi velocity for Au, one concludes that the Fermi velocity in YBCO and LSCO is at least equal to $0.7 \cdot 10^8$ cm/s.

Before we discuss the implication of this result, we note that the shape of the experimental characteristics rules out non s wave superconductivity. Kirtley et al. have calculated the shape of these characteristics for different cases of d wave superconductivity [9]. The conductance then shows a peak like structure of width $2\Delta$ centered at zero bias, rather than a plateau shape of such width, as we observe. Moreover, we observe conductance peaks at the edges of the plateau, another characteristic feature of the s wave symmetry that appears as soon as $Z$ is different from zero. Zero bias peak like structures do occur in YBCO when the Au tip is oriented along the c axis, but along the CuO planes they only appear (if at all) as a small zero bias anomaly superimposed on the major structure. We conclude that in YBCO and LSCO the large majority of the carriers is certainly condensed in an s BCS state. In some heavy Fermion superconductors peak shaped conductance curves have been observed by point contact, and interpreted in terms of d wave superconductivity [9].

Returning now to the Fermi velocity value obtained above, we remark that it is much larger than that which we calculate from the BCS expression for the coherence length:

$$\xi = \frac{h\, v_{\rm F}}{{\rm k_B} T_{\rm c}} \tag{2.2}$$

using for $\Delta$ the gap values obtained from the point contact measurements and for $\xi$ values obtained from upper critical field measurements (15 Å for YBCO and 40 Å for LSCO). One then obtains Fermi velocities about five times smaller than that derived from the point contact measurements. This discrepancy is much larger than the experimental uncertainties combined.

An explanation has been proposed by Force and Bok [10, 11] in terms of the large anisotropy in the Fermi velocity that arises when the Fermi level lies close to a van Hove singularity: its maximum value is then significantly larger than the averaged one that enters in the expression for $\xi$. However, averaging of the logarithmically singular velocity over a width $2\Delta$ around the Fermi surface reduces the difference between the maximum and the average to a factor of 2 at most [12]. This is insufficient to explain the data.

Another interpretation is based on the recent Eliashberg description of a BCS ground state in the presence of strong correlations [7]. Eliashberg points out that one must distinguish between the particle velocity $v_{\rm PC}$, measured

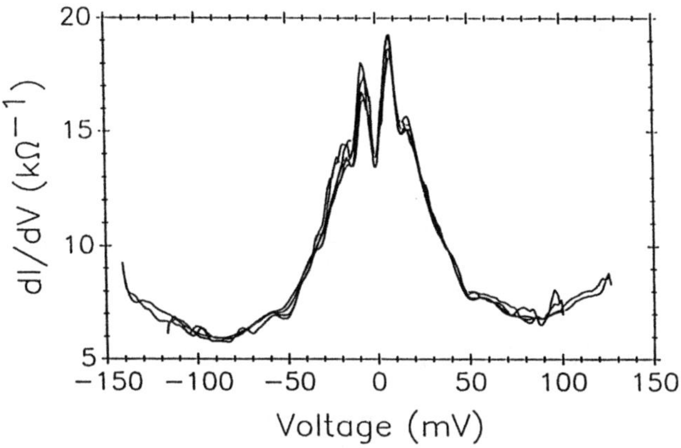


**Fig. 2.1.** Conductance of a Au-LSCO Andreev contact on the phonon energy scale

by experiments where particles are injected in or extracted from the sample, and the quasi particle velocity $v_{PQ}$ which is the relevant one when one measures properties of the excited states. The two velocities are related by the expression:

$$v_{PQ} = \frac{v_{PC}}{1 + \lambda_{ep} + \lambda_c} \tag{2.3}$$

where $\lambda_{ep}$ and $\lambda_c$ are respectively the electron-phonon and correlation parameters. The velocity that enters into the expression for $\xi$ is $v_{PQ}$, since this expression is derived from the uncertainty principle considering excited states that lie within a width $2\Delta$ around the Fermi level. We shall assume that the velocities that enter into Eq. 2.1 are the particle velocities, since a particle is transferred in the experiment from one side to another. We then obtain from Eq. 2.2 and Eq. 2.3 the condition:

$$\lambda_{ep} + \lambda_c > 4 \tag{2.4}$$

We shall see in the next section that for LSCO a reasonable estimate for *lep* is a value similar to that of Pb (about 1.5). Thus, for this oxide at least, the two interaction parameters are not very different from each other. Theories based on neglecting either the electron phonon interaction or correlation effects are not justified.

## 3. The electron phonon interaction in $La_{1.85}Sr_{0.15}CuO_4$

The study of the characteristics of Au-LSCO point contacts [13] shows significant structure at voltages up to 10 times the gap value. Moreover the general trend of the conductance is down for Andreev type contacts ($Z < 1$, Fig. 2.1) and up for Giaever type contacts ($Z > 1$, Fig. 2.2). This behavior is consistent with the occurrence of strong inelastic scattering, as discussed by

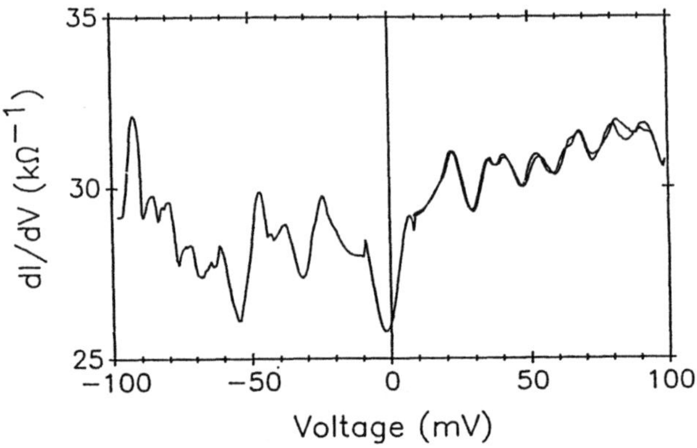


**Fig. 2.2.** Conductance of a Au-LSCO Giaever contact on the phonon energy scale

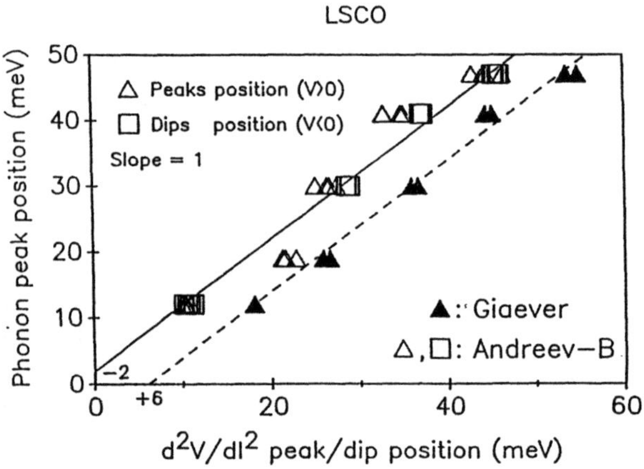


**Fig. 3.1.** Second derivative $d^2V/dI^2$ peak positions of an Andreev and a Giaever like contact compared to the phonon DOS peak positions. Note the shifts in energy in both cases.

Kirtley [14], who has predicted this sign reversal of the general trend of the conductance when $Z$ crosses the value unity. A detailed comparison of the positions in energy of the structures seen in both cases shows that they are consistent with a shift downwards for $Z < 1$ and a shift upwards for $Z > 1$, the shifts being of the order of the gap, as predicted by Kirtley [14].

This general consistency between the behavior of the conductance at low and at high voltages with the BTK and Kirtley's point contact theories has encouraged us to push the analysis further and to compare our high voltage data with the major structures of the phonon spectrum, as obtained from neutron scattering data [15]. As shown Fig. 3.1, the correspondence between peaks in the phonon DOS and the peaks of the second derivative $d^2V/dI^2$ is really very good, taking into account the upward and downward shifts in energy as discussed above. We thus conclude that the bosons responsible for the inelastic scattering are most likely to be phonons.

A detailed analysis to extract $\lambda_{ep}$ from the data is in progress, but we can already give some preliminary estimates. Most of the high voltage conductance decrease of an Andreev type contact (Fig. 2.1) occurs between 10 and 50 meV. This is also the energy range where most of the phonon DOS lies [15]. It is roughly 4 to 5 times higher than that of Pb. This ratio is almost the same as that of the critical temperatures. Hence, we ascribe to LSCO a $\lambda_{ep}$ not much larger than that of Pb, which is equal to 1.5, if we assume a similar value of $\mu^\star$ in the two superconductors. Now, this is probably not quite correct. First $\lambda_c$ is, we have argued, significantly higher in the cuprates than in ordinary metals. Second the screening factor $\ln(E_F/\mathrm{h}\,\omega_D)$ is not as effective in the cuprate: $E_F$ is smaller and $\mathrm{h}\,\omega_D$ is higher. In the limit of strong Coulomb effects:

$$\mu^\star = \ln^{-1}(\frac{E_F}{\mathrm{h}\,\omega_D}) \tag{3.1}$$

Assuming $E_F = 1$ eV and $\mathrm{h}\,\omega_D = 30$ meV, we obtain $\mu^\star$ between 0.2 and 0.3, rather than 0.1 in low $T_c$ superconductors. Thus $\lambda_{ep}$ must be somewhat larger than in Pb. Following Eq. 2.4 we thus estimate for LSCO, $\lambda_c \sim \lambda_{ep} \sim 2$. Again, this is just a preliminary estimate and more exact values must await a detailed analysis of the $V(I)$ characteristics.

Another indication for strong coupling is given by the ratio $(2\Delta/k_B T_c) = 4.5$, but with a rather large error of 0.5 (due mostly to the fact that the studied textured sample had a broad transition). A more accurate determination is necessary to make use of the detailed calculations of Combescot [16] on solutions of the Eliashberg equations in the strong coupling limit, to obtain allowed windows for $\lambda_{ep}$ and $\mu^\star$. But we feel that such more quantitative studies should not modify our main conclusion, which is that superconductivity in the studied cuprates is s type and due to the electron phonon interaction, although in a very non trivial way because the strong electron phonon coupling seems to go hand in hand with strong correlation effects.

The role of the apical oxygen in reaching high critical temperatures has been much discussed [17]. According to the calculations of Pickett [18] the energy of the related phonon is at 46 meV. At this energy we indeed see a very strong and reproducible structure of $(\mathrm{d}^2V/\mathrm{d}\,I^2)$ (Fig. 3.2). This is in fact the highest energy structure that is clearly identified. It was also observed by Ekino [19] on polycrystalline samples. Evidently there is also strong coupling to lower energy modes as seen in our data, but strong coupling up to this high energy is remarkable. Again, a more quantitative analysis of the characteristics is necessary before one can establish the exact weight of the coupling to this mode in reaching the high $T_c$ — but we would assume it to be substantial.

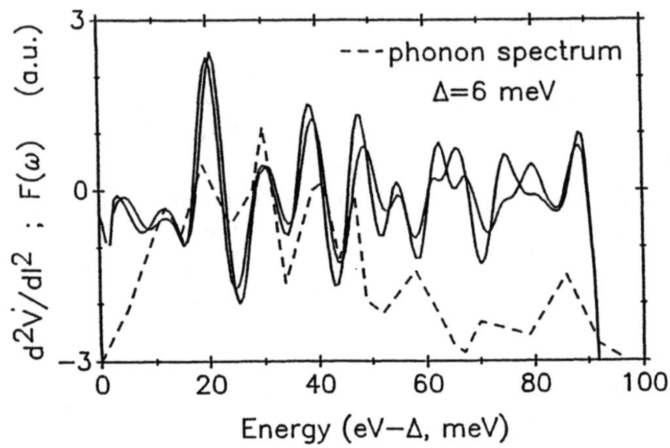


**Fig. 3.2.** Second derivative $\mathrm{d}^2V/\mathrm{d}I^2$ of a Giaever like contact. Note the large structure below 50 meV.

## 4. Conjectures on the strong electron phonon and correlations interactions scenario - and some consequences regarding phase separation

As we have already discussed, the enhanced conductance at low voltage and the conductance peaks at $+\Delta$ are within the BTK theory consistent with s wave superconductivity. These peaks are of particular importance to establish that this enhancement conductance is not due to some zero bias anomaly of unknown origin. Also within the BTK theory, the characteristics are not consistent with d wave superconductivity. Such a symmetry would give a broad conductance peak centered at $V = 0$, as already mentioned.

We have interpreted the high Fermi velocity deduced from the data as due to strong correlation effects. We have also argued that the structure of the characteristics at high voltage reflects a strong electron phonon interaction. This strong electron phonon strong correlation interactions scenario is consistent with Eliashberg recent ideas [7] according to which a BCS ground state can survive strong correlations.

This scenario may provide a framework for the understanding of the decrease of $T_c$ in the overdoped regime. As we move away from the metal insulator transition, $lc$ decreases and so does the particle velocity $v$, which in turn would lead to a decrease of $\lambda_{ep} = (v/v_0)$ where $v_0$ does not depend on the characteristics of the electron spectrum except for $kF$ [7].

It is tempting to apply these same general ideas to the behavior of another "exotic" superconductor, granular Aluminium [20]. Pure Aluminium has a $T_c$ of 1.2 K, but when driven near a metal insulator transition by the addition of other elements such as O or Ge, its $T_c$ increases several times. $\alpha^2\mathrm{F}(\omega)$ has been obtained from tunneling measurements both for granular and pure

Aluminium by Dayan [6] using the Rowell-McMillan inversion method [3], with the conclusion that the increase in $T_c$ is due entirely to an increase of $\lambda_{ep}$within the experimental accuracy. A decrease in the tunneling density of states near the Fermi level has been reported in granular Aluminium [21] and attributed to correlation effects. Thus the simultaneous increase in both $\lambda_{ep}$and $\lambda_c$near the metal insulator transition is well documented for this case.

A peculiar property of a high $T_c$ material in which $\lambda_{ep}$and $\lambda_c$are both large and comparable is that the gain in free energy by condensation into the superconducting state can be comparable to that obtained (at the same temperature) by magnetic ordering. In the case of antiferromagnetic ordering, the gain in free energy is of the order of $k_B T_N$ per spin. In a superconductor the condensation energy per carrier is reduced by the ratio $(\Delta/E_F)$, which is very small ($10^{-3}$ or less) for the low $T_c$ case. But in the strongly coupled, high $T_c$ case this ratio may not be so small. A direct way to see this is that critical fluctuations, in the heat capacity for instance, have been clearly observed for YBCO. The width of the critical region is determined by the ratio of the condensation energy per coherence volume to $k_B T_c$ (Ginzburg criterion); the experimentally observed width suggests that this ratio is of order unity [21]. With an in plane coherence length of 15 Å there are only a few Cooper pairs per coherence volume. Thus the condensation energy per pair is a fraction of $k_B T_c$. If $T_c$ and $T_N$ are comparable, and the respective condensation energies also, we must expect a phase diagram with coexistence regions between different phases, just as happens in a metallurgical phase diagram.

*Acknowledgement.* This review summarizes previous work done with N. Hass, D. Ilsycer, Y. Yagil at Tel Aviv University, A. Revcolevski and G. Dhalenne at the Orsay Center of the Universite Paris-Sud, and G. Desgardins and I. Monot at the University of Caen, whom I all wish to thank warmly. The theoretical discussion on correlation effects owes much to an enlightening exchange with G.M. Eliashberg. I have also greatly benefited from discussions with K. Kitazawa and K.A. Müller. This work was supported in part by NEDO and by the Oren Family Chair for Experimental Solid State Physics.

## References

1. I.K. Yanson and A.V.Khotkevich, in "Atlas of microcontact pectra of the electron-phonon interaction in metals", Kiev, Naukova Dumka, 1986
2. G.E. Blonder, M. Tinkham and T.M. Klapwijk, Phys. Rev. B25, 4515 (1982).
3. W.L. McMillan and J.M. Rowell, in "Superconductivity", ed. R.D. Parks (Marcel Dekker, New York 1969).
4. N. Hass, D. Ilsycer, G. Deutscher, G. Desgardins, I. Monot and M. Weger, J. of Superconductivity 5, 191 (1992).
5. N. Hass, PhD thesis Tel Aviv 1993 and N. Hass, Y. Yagil and G. Deutscher to be published.

6. D.S. Dessau et al., to be published.
7. G.M. Eliashberg, to appear in J. of Superconductivity (Proceedings of the MOS Conference, Oregon (USA) July 1993.
8. M. Dayan, J. of Low Temp. Phys. 32, 643 (1978).
9. K. Hasselbach, J.R. Kirtley and P. Lejay, Physica B186-188, 201 (1993).
10. J. Bok and L. Force, Physica C185-189, 1449 (1991).
11. J. Bok, Physica C209, 107 (1993).
12. L. Force, PhD Thesis, University of Paris 6, 1993.
13. G. Deutscher, N. Hass, Y. Yagil, A. Revcolevski and G. Dhalenne, to appear in J. of Superconductivity (Proceedings of the MOS Conference, Oregon (USA) July 1993).
14. J.R. Kirtley, Phys. Rev. B47, 11379 (1993).
15. B. Renker et al., Z. Phys. B67, 15 (1987).
16. R. Combescot, private communication.
17. K.A. Müller, Z. Phys. Condensed Matter 193 (1990); for recent results see also: "Apex anharmonicity observed by Raman scattering in $O_{18}$ substituted $YBa_2Cu_3O_{6+x}$" by G. Ruani, C. Taliani, M. Muccini, K. Conder, E. Kaldis, H. Keller, D. Zech and K.A. Müller, to be published.
18. W. Pickett, Phys. Rev. B47, 1002 (1993).
19. T. Ekino, Physica C198, 338 (1993).
20. G. Deutscher, H. Fenichel, M. Gershenson, E. Grunbaum and Z. Ovadyahu, J. Low Temp. Phys. 10, 231 (1973).
21. For a discussion of the width of the critical region in YBCO, see for instance G. Deutscher in "Novel Superconductivity", ed. S. Wolf and V. Kresin, p.293 (Plenum Press, 1987).

# Cuprate Superconductors: Universal Properties and Trends; Evidence for Bose-Einstein Condensation

T. Schneider and M.H. Pedersen

IBM Research Division, Zurich Research Laboratory, 8803 Rüschlikon, Switzerland

**Summary.** We explore the compatibility of empirical trends in various thermodynamic properties of cuprate superconductors with the Bose-Einstein condensation scenario. These trends include the relations between transition temperature, hole concentration and superfluid density, the rise and the upper limit of the transition temperature, the dependence of pressure and isotope coefficients on the transition temperature, as well as the observed critical behavior, reminiscent of three-dimensional systems with a scalar complex order parameter and short-range interactions. For this purpose we consider an interacting charged Bose gas. Owing to the high polarizability of the cuprates, the Coulomb interaction is strongly screened. Hence, the problem of calculating thermodynamic properties becomes essentially equivalent to that of the uncharged gas with short-range interactions. This problem, however, has not been solved either. Nevertheless, in the dilute limit the problem reduces to the ideal Bose gas treated by Schafroth, while in the dense regime condensation and superfluidity are suppressed, because bosons of finite extension fill the available volume. This limiting behavior provides an interpolation scheme for the dependence of both transition temperature and zero temperature superfluid density on boson density. On this basis and by relating the hole concentration in the cuprates which corresponds to the boson density and the superfluid density to the square of the inverse London penetration depth, the compatibility of the empirical trends in the cuprates with the Bose gas behavior can be verified. Our analysis reveals remarkable agreement between these trends and the corresponding Bose gas behavior. There is even strong evidence of the most striking implication of this scenario, the dependence of the transition temperature on the zero temperature superfluid density, resembling the outline of a fly's wing. This evidence emerges from recent muon spin resonance ($\mu$SR) data for $Tl_2Ba_2CuO_{6+\delta}$ and kinetic inductance measurements for $La_{2-x}Sr_xCuO_4$ films, revealing that the penetration depths of under and overdoped samples with $T_c \simeq 15$ K do not differ much. In view of this we have provided considerable evidence for the nature of the superconducting transition in the cuprates, without invoking any specific pairing mechanism.

## 1. Universal Properties and Trends

For various cuprate superconductors the measured thermodynamic properties clearly reveal remarkable consistency with three-dimensional $xy$-critical point behavior reminiscent of the normal fluid-superfluid transition in $^4$He. These properties include the specific heat singularity

$$C^{\pm} = \frac{A^{\pm}}{\alpha}|t|^{-\alpha} \simeq \frac{A^{\pm}}{\alpha} + B^{\pm} - A^{\pm}\ln|t| \qquad t = 1 - \frac{T}{T_c} \tag{1.1}$$

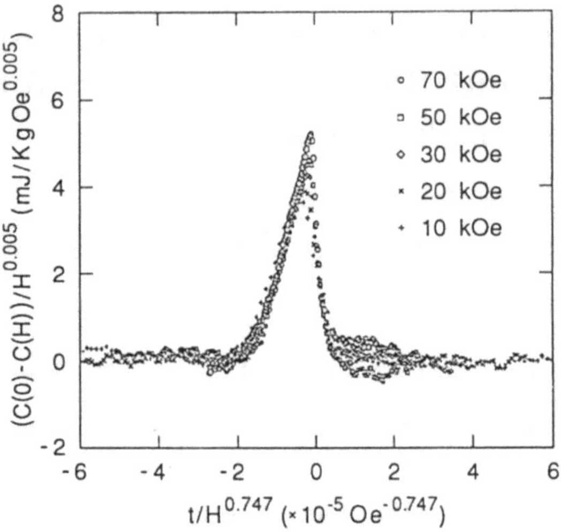


**Fig. 1.1.** Scaled heat capacity data for $YBa_2Cu_3O_{7-x}$ from Salamon *et al.* [1].

and the universal scaling properties of the temperature and magnetic field dependence of specific heat

$$\Delta C(t,H)H^{\alpha/2\nu} = C(x)\,, \qquad x = \frac{t}{H_{\perp}^{1/2\nu}} \tag{1.2}$$

and magnetization

$$\frac{m}{H^{1/2}} = M\left(\frac{t}{H^{1/2\nu}}\right)\,. \tag{1.3}$$

To illustrate the evidence for 3d-$xy$-critical behavior we reproduced in Fig. 1.1 experimental specific heat data of Salamon *et al.* [1] for $YBa_2Cu_3O_{7-x}$, scaled according to Eq. (1.2) with the critical exponents $\alpha$ and $\nu$ appropriate for the 3d-$xy$-critical point. The collapse of the data on a single curve is excellent and confirms the scaling properties. For a recent review of 3d-$xy$-critical behavior in cuprate superconductors and the resulting implications we refer to Refs. [2] and [3]. The essential points might be summarized as follows.

1. The estimates for the critical exponents, the scaling behavior of the magnetization and specific heat, the consistency of the critical amplitudes with the universal relations, the experimental evidence for the resulting trends and the decrease of $T_c$ with decreasing slab thickness, corresponding to a 3d-2d-$xy$ crossover, provide considerable evidence of 3d-$xy$-critical point behavior.
2. The volume of the phase correlation length amplitudes $(\xi^{\varphi}_{\parallel 0})^2\xi^{\varphi}_{\perp 0}$ is comparable to that of superfluid helium and several orders of magnitude smaller than in conventional superconductors (Table 1.1).

**Table 1.1.** Measured values of $T_c$ and experimental estimates for $A^+ \simeq A^-$. The volume of the amplitudes of the phase correlation length was derived from the universal amplitude relation $R_\varphi^3 = A^-(\xi_{\|0}^\varphi)^2\xi_{\perp 0}^\varphi \simeq (0.8)^3$. Here, $\xi_{\|,\perp}^\varphi = \xi_{\|,\perp 0}^\varphi t^{-\nu}$ is the phase correlation length, where $\|$ denotes a direction parallel and $\perp$ one perpendicular to the layers.

| | $T_c$ [K] | $1/A^- \simeq 1/A^+$ [Å$^3$] | $R_\varphi^3/A^- = (\xi_{\|0}^\varphi)^2\xi_{\perp 0}^\varphi$ [Å$^3$] |
|---|---|---|---|
| $DyBa_2Cu_3O_7$ | 91.4 | 1130 | 579 |
| $YBa_2Cu_3O_7$ | 91.7 | 1790 | 916 |
| $^4He$ | 2.18 | 600 | 307 |

Combined with the large amplitude $\lambda_{\|,\perp 0}$ of the penetration depth, diverging as $\lambda_{\|,\perp} = \lambda_{\|,\perp 0}t^{-\nu/2}$, the cuprates turn out to be extreme type II superconductors as long as $\kappa_{\|,\perp} = \lambda_{\|,\perp}/\xi_{\|,\perp} \gg 1$. In these superconductors, the resulting effective dimensionless charge $\tilde{e}_{\|,\perp} = 1/\kappa_{\|,\perp}$ is very small. For this reason the thermodynamic properties associated with the fluctuations of the order parameter are essentially those of an uncharged superfluid ($^4$He) or $xy$ model. (iii) As the universal relation $R_\varphi^3 = A^-(\xi_{\|0}^\varphi)^2\xi_{\perp 0}^\varphi \simeq (0.8)^3$ indicates, the critical amplitudes are not independent of one another. Indeed, there is also the relation

$$k_B T_c = \frac{\hbar^2}{4M_\|} n_{s0}\xi_{\|0}^\varphi = \frac{\hbar^2}{4M_\perp} n_{s0}\xi_{\perp 0}^\varphi = \frac{\hbar^2 c^2}{16\pi e^2}\frac{\xi_{\|,\perp 0}^\varphi}{\lambda_{\|,\perp 0}^2}, \tag{1.4}$$

connecting $T_c$ with the superfluid density, $n_s = n_{s0}|t|^{-\nu}$, and the amplitude of the phase correlation length. To obtain the last term in Eq. (1.4) we used London's expression for the penetration depth. Thus, the transition temperature, the amplitudes of the London penetration depth and phase correlation length are not independent of one another.

In this sense it is not surprising that $\mu$SR measurements revealed a remarkable correlation between the transition temperature $T_c$ and the zero temperature value of the $\mu$SR relaxation rate $\sigma(0)$ [2]–[4] which is proportional to the square of the inverse zero temperature penetration depth [5]

$$\sigma = \left(\frac{2754B}{\lambda}\right)^2 . \tag{1.5}$$

The value of $\lambda$ is measured in Å and $\sigma$ in $\mu s^{-1}$. In sintered materials we have $B = 0.813$ and in single crystals $B = 1$ [5]. In extreme type II materials the London expression

$$\frac{1}{\lambda^2} = \frac{4\pi n_{FS}e^2}{Mc^2} \tag{1.6}$$

relates the penetration depth $\lambda$ to the number density $n_{FS}$ of paired and superfluid fermions with mass $M$. Moreover, extended studies of the pressure and isotope mass[2]–[4] dependence of $T_c$ in various cuprates, measured in terms of the coefficients

$$\alpha = \frac{1}{T_c}\frac{dT_c}{dP} \quad \text{and} \quad \beta = -\frac{m}{T_c}\frac{dT_c}{dm}, \tag{1.7}$$

revealed remarkable generic trends. Indeed, for various compounds, $\alpha$ and $\beta$ fall into common $T_c - \alpha$ and $T_c - \beta$ regions, respectively, forming two branches, one for systems with positive – and the other for those with negative – pressure or isotope coefficients. The magnitudes of the two coefficients $\alpha$ and $\beta$ decrease with increasing $T_c$ and nearly vanish at the maximum $T_c$. These trends appear to be a common feature of cuprate superconductors, which share the rather unique dependence of the transition temperature on the square of the inverse zero temperature penetration depth ($\sigma(0) \propto 1/\lambda^2(0)$). Indeed, the assumption of a parabolic maximum in $T_c = T_c(\sigma(0))$ yields the simple scaling form[2]–[4]

$$\overline{T}_c = 2\overline{\sigma}\left(1 - \frac{\overline{\sigma}}{2}\right), \qquad \overline{T}_c = \frac{T_c}{T_c^m}, \qquad \overline{\sigma} = \frac{\sigma(0)}{\sigma^m(0)}, \tag{1.8}$$

which accounts well for the trends observed in the pressure and isotope coefficients[2]–[4]. Thus, plotting $T_c$ versus $\sigma(0)$, a particular family forms a unique parabolic branch characterized by its maximum transition temperature $T_c^m$ and the associated $\sigma^m(0)$. As it should be, $T_c$ vanishes at $\sigma(0) = 0$, implying a vanishing superfluid number density. A plot of $T_c$ versus $\sigma(0)$ is shown in Fig. 1.2 for various cuprate superconductors. With the exception of $Tl_2Ba_2CuO_{6+\delta}$ [6], the experimental data follows the parabolic behavior for $T < T_c^m$ remarkably well. In Fig. 1.3 we have collected the resulting estimates for $T_c^m$ and $\sigma^m(0)$.

There is also an empirical relation between $T_c$ and the concentration $p$ of holes per $CuO_2$ and per $CuO_2$ layer [7],[8]. The data for $YBa_2Cu_3O_x$ and $La_{2-x}Sr_xCuO_4$ shown in Fig. 1.3 are well described by [7]

$$\overline{T}_c = \frac{T_c}{T_c^m} = 1 - \gamma(p - p_m)^2 = \overline{x}(2 - \overline{x}), \qquad \overline{x} = \frac{p - p_0}{p_m - p_0}, \tag{1.9}$$

where $p_m \simeq 0.16$, $\gamma \simeq 82.6$ and $p_0 = p_m \pm 1/\sqrt{\gamma}$. The hole concentration $p_m$ corresponds to optimum doping, while $p > p_m$ and $p < p_m$ refer to the overdoped and underdoped regimes, respectively. Combining Eqs. (8) and (9), the doping dependence of $\sigma$ is given by $\overline{\sigma} = \overline{x}$ and materials with $\overline{x} = \sigma(0)/\sigma^m(0) > 1$ and $\overline{x} = \sigma(0)/\sigma^m(0) < 1$ are supposed to be over and underdoped, respectively. This behavior is fully consistent with the $\mu$SR data for underdoped and mildly overdoped compounds (Fig. 1.2). It contradicts, however, the $Tl_2Ba_2CuO_{6+\delta}$ data [6]. In this compound the overdoped regime becomes accessible as the oxygen content $\delta$ increases [8],[9] ($\overline{x} > 1$), while $\sigma(0)$

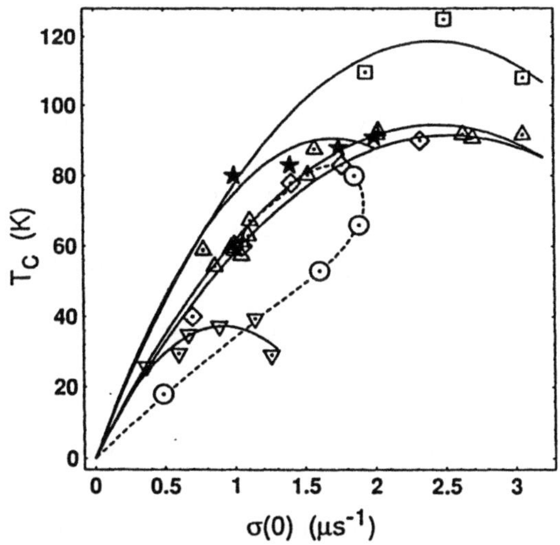


**Fig. 1.2.** $T_c$ versus $\mu$SR relaxation rate $\sigma(0)$ taken from Refs. [2] and [6]. The solid lines are fits to Eq. (1.8) and the dashed line is a guide to the eye.
⊡: $Tl_2Ba_2Ca_2Cu_3O_{10}$, $Tl_{0.5}Pb_{0.5}Sr_2Ca_2Cu_3O_9$, $Bi_{2-x}Pb_xSr_2Ca_2Cu_3O_{16}$
◇: $Y_{1-x}Pr_xBa_2Cu_3O_{6.97}$, △: $YBa_2Cu_3O_x$, ▽: $La_{2-x}Sr_xCuO_4$
⋆ : $Bi_2Sr_2Ca_{1-x}Y_xCu_2O_{8+\delta}$, ⊙: $Tl_2Ba_2CuO_{6+\delta}$.

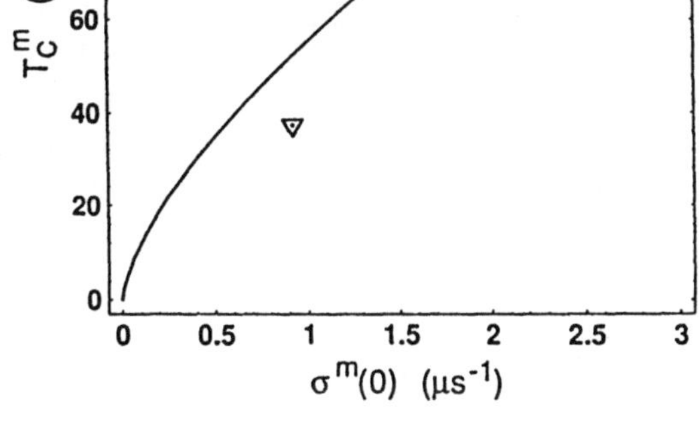


**Fig. 1.3.** $T_c^m$ versus $\sigma^m(0)$ for the data shown in Fig. 1.2. The curve is a fit to $T_c^m \propto (\sigma^m(0))^{2/3}$.

decreases with decreasing transition temperature (Fig. 1.2). Thus, overdoped $Tl_2Ba_2CuO_{6+\delta}$ contradicts $\overline{\sigma} = \overline{x}$ for $\overline{x} > 1$ and the applicability of ansatz (8) appears to be limited to underdoped and slightly overdoped systems. Indeed, with the exception of $Tl_2Ba_2CuO_{6+\delta}$, the $\mu$SR data shown in Fig. 1.2 refers to underdoped and mildly overdoped compounds. Combining ansatz (8), applicable to underdoped and slightly overdoped systems, with the $\sigma(0)$ data for overdoped $Tl_2Ba_2CuO_{6+\delta}$, the generic shape of the $T_c$ versus $\sigma(0)$ curve is expected to resemble the outline of a fly's wing, as indicated in Fig. 1.2. This shape is further supported by kinetic inductance measurements for $La_{2-x}Sr_xCuO_4$ films, revealing that the penetration depths of under and overdoped samples with $T_c \simeq 15$ K do not differ much [10].

## 2. Evidence for Bose-Einstein Condensation

In this work we explore the compatibility of these empirical trends with the Bose-Einstein scenario. In the ideal gas, treated by Schafroth [11], transition temperature and superfluid density per unit mass $\rho_{BS}(T = 0)$ are given by [2],[3],[11]:

$$T^0_{CB} = a\frac{n_B^{2/3}}{M_B}, \qquad a \simeq \frac{3.313\hbar^2}{k_B}, \qquad \rho_{BS}(T=0) = \frac{n_B}{M_B}, \tag{2.1}$$

where $n_B$ denotes the density and $M_B$ the mass of the bosons. The experimental evidence for three-dimensional $xy$-critical behavior [2],[3] requires, however, the inclusion of the interaction. Due to the high polarizability of the cuprates, the Coulomb repulsion is strongly screened and in turn is of short range. For this reason, the problem of calculating thermodynamic properties becomes tantamount to calculating the uncharged gas with short-range repulsive interaction. Thus, the ideal gas expressions (10) are valid within the dilute limit ($n_B \rightarrow 0$) while in the dense regime significant deviations from the ideal gas behavior will occur, because bosons with finite extension fill the available volume and become immobile at some density $n_B = n_{BC}$, where $T_{CB}$ and $\rho_{BS}(T = 0)$ vanish [12],[13]. In the intermediate density regime, however, the exact dependence of $T_{CB}$ and $\rho_{BS}(T = 0)$ on boson number density $n_B$ is not available. For this reason we use the self-consistent Hartree-Fock expressions [14],[15]

$$T_{CB} = T^0_{BC}\frac{M_B}{M_B^\star} = a\frac{n_B^{2/3}}{M_B^\star}, \qquad \frac{1}{M_B^\star} = \frac{1}{M_B}(1 - bn_B), \qquad b = \frac{2M_BV(0)}{\hbar^2 k_s^2} \tag{2.2}$$

and

$$\rho_{BS}(T=0) = \frac{n_B}{M_B^\star} = \frac{n_B}{M_B}(1 - bn_B) \tag{2.3}$$

as interpolation formulae. The term $V(k)$ is the Fourier transform of the interaction and $k_s$ the inverse screening length. The superfluid number density per unit mass, or helicity modulus, is obtained from the change in ground-state energy due to twisting the phase of the order parameter along one boundary [16]. Then using the ground-state energy resulting from the self-consistent energy spectrum, $\varepsilon_k = \hbar^2 k^2/2M_{\mathrm{B}}^{\star}$ [14],[15], we obtain expression (2.3) for the superfluid density per unit mass. In the dilute limit the interpolation formulae reduce to the corresponding ideal gas expressions (2.1), while $T_{\mathrm{CB}}$ and $\rho_{\mathrm{BS}}(T=0)$ vanish at $n_{\mathrm{BC}} = 1/b$ in the dense regime and both $T_{\mathrm{CB}} = T_{\mathrm{CB}}(n_{\mathrm{B}})$ and $\rho_{\mathrm{BS}}(T=0) = \rho_{\mathrm{BS}}(T=0, n_{\mathrm{B}})$ have a maximum in-between. This behavior is consistent with the hard-core boson model [12],[13], the attractive Hubbard model in the strong coupling regime [17] and the experimental HeI-HeII equilibrium curve in the $T_{\mathrm{CB}} - n_{\mathrm{B}}$ plane, where $\mathrm{d}T_{\mathrm{CB}}/\mathrm{d}n_{\mathrm{B}}$ becomes negative [18]. Thus, leaving out quantitative details, the interpolation formulae (2.2) and (2.3) appear to be well suited to clarifying the compatibility of the empirical trends in cuprate superconductors with the Bose-Einstein condensation. For this comparison it is convenient to rewrite Eq. (2.2) in the form

$$\overline{T}_{\mathrm{CB}} = \frac{T_{\mathrm{CB}}}{T_{\mathrm{CB}}^{\mathrm{m}}} = \frac{5}{3}\overline{n}_{\mathrm{B}}^{2/3}\left(1 - \frac{2}{5}\overline{n}_{\mathrm{B}}\right), \tag{2.4}$$

where

$$\overline{n}_{\mathrm{B}} = \frac{n_{\mathrm{B}}}{n_{\mathrm{B}}^{\mathrm{m}}}, \qquad T_{\mathrm{CB}}^{\mathrm{m}} = \frac{3a}{5M_{\mathrm{B}}k_{\mathrm{B}}}(n_{\mathrm{B}}^{\mathrm{m}})^{2/3}, \qquad n_{\mathrm{B}}^{\mathrm{m}} = \frac{2}{5b}, \tag{2.5}$$

where $T_{\mathrm{CB}}^{\mathrm{m}}$ is the maximum transition temperature at $n_{\mathrm{B}} = n_{\mathrm{B}}^{\mathrm{m}}$. As hole concentration $\overline{x}$ (Eq. (1.9)) and boson density $\overline{n}_{\mathrm{B}}$ are expected to match closely, we included in Fig. 1.3 the plot $\overline{T}_{\mathrm{B}}$ versus $\overline{n}_{\mathrm{B}}$. There is remarkable agreement up to $\overline{T}_{\mathrm{CB}} \simeq 1$, while for higher densities Eq. (2.4) underestimates the reduction of the transition temperature. In Fig. 1.3 we also included the interpolation formula

$$\overline{T}_{\mathrm{CB}} = \left(\overline{n}_{\mathrm{B}}(2 - \overline{n}_{\mathrm{B}})\right)^{2/3}, \tag{2.6}$$

where

$$n_{\mathrm{B}}^{\mathrm{m}} = \frac{1}{2b}, \qquad T_{\mathrm{c}}^{\mathrm{m}} = \frac{a}{M_{\mathrm{B}}}\left(\frac{n_{\mathrm{B}}^{\mathrm{m}}}{2}\right)^{3/2}. \tag{2.7}$$

Since $\overline{T}_{\mathrm{CB}}(\overline{n}_{\mathrm{B}}/2) = \overline{T}_{\mathrm{CB}}(1 - \overline{n}_{\mathrm{B}}/2)$, this exhibits particle-hole symmetry, a characteristic property of the attractive Hubbard model with nearest-neighbor hopping [17]. Recall that $\overline{x} > 1$ and $\overline{x} < 1$ refer to the overdoped and underdoped regime, respectively. The qualitative agreement between Eqs. (1.9),(2.4) and (2.6) (Fig. 2.1) implies that $\overline{T}_{\mathrm{c}} \simeq \overline{T}_{\mathrm{CB}}$ and $\overline{x} \simeq \overline{n}_{\mathrm{B}}$. A family is then characterized as a class of compounds sharing nearly the same maximum transition temperature $T_{\mathrm{CB}}^{\mathrm{m}}$ and associated boson number density $n_{\mathrm{B}}^{\mathrm{m}}$ (Eqs. (2.5) and (2.7)), and the variation of the transition temperature is controlled by doping in terms of the boson density.

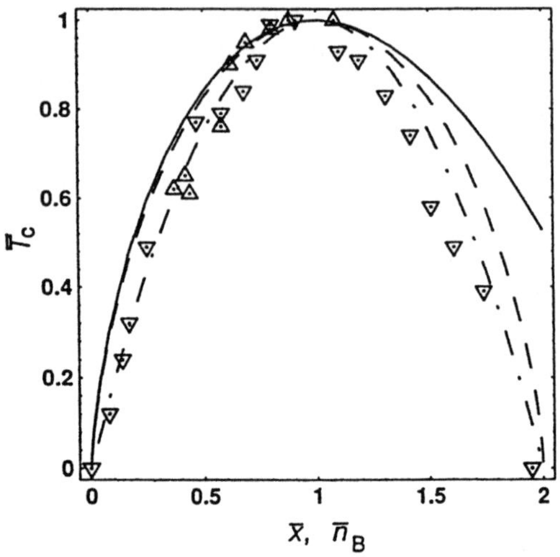


**Fig. 2.1.** $T_c$ versus hole concentration $p$ in terms of $\overline{T}_c$ versus $\overline{x}$ according to Eq. (1.9) (dash-dotted curve); $\overline{T}_{CB}$ versus $\overline{n}_B$ for the Bose gas according to Eq. (2.3) (solid curve) and Eq. (2.6) (dashed curve), and experimental data for $YBa_2Cu_3O_x$ ($\triangle$), $La_{2-x}Sr_xCuO_4$ ($\triangledown$) according to Ref. [7].

To explore the compatibility of the empirical trends in the properties of the cuprates with the Bose-Einstein condensation further, we turn to the relationship between $T_c$ and the zero temperature superfluid density per unit mass. The empirical behavior of the cuprates is shown in Fig. 1.2, in terms of $T_c$ versus $\sigma(0) \propto \rho_{FS}(0) = n_{FS}(0)/M$. The term $\rho_{FS}$ denotes the superfluid density per unit mass of paired fermions, and $\rho_{BS}(0) = 1/4\rho_{FS}(0) = n_{BS}/M_B$ is the corresponding superfluid boson density. Using Eqs. (1.5),(1.6) and (2.3) we find the relation between the $\mu$SR-relaxation rate (superfluid density per unit mass) and the boson density to be

$$\sigma(0) = d\rho_{BS}(0) = d\frac{n_B}{M_B}(1 - bn_B)\,, \qquad d = \frac{(2754B)^2 4\pi e^2}{c^2}\,. \tag{2.8}$$

Invoking Eq. (2.3) the transition temperature can be expressed in terms of the zero temperature $\mu$SR-relaxation rate $\sigma(0) = d\rho_{BS}(0)$ as

$$\overline{T}_{CB} = \left(\frac{4}{5}\right)^{1/3} \frac{\overline{\sigma}}{\left(1 \pm \sqrt{1 - (24/25)\overline{\sigma}}\right)^{1/3}} \tag{2.9}$$

with

$$\overline{\sigma} = \frac{\sigma(0)}{\sigma^m(0)} = \frac{5}{3}\overline{n}_B\left(1 - \frac{2}{5}\overline{n}_B\right)\,, \qquad \sigma^m(0) = \frac{3dn_B^m}{5M_B}\,, \tag{2.10}$$

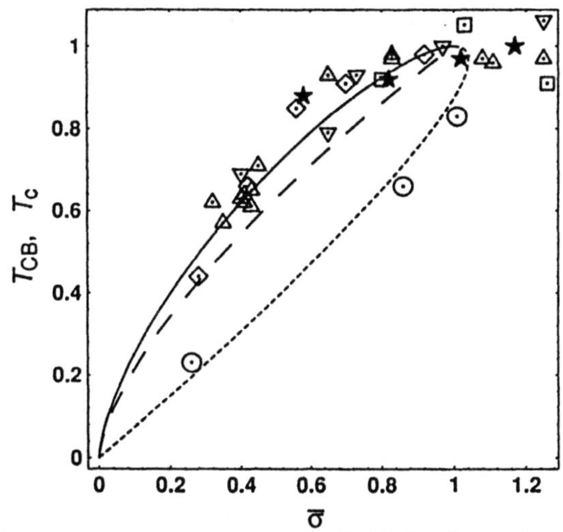


**Fig. 2.2.** Transition temperature versus superfluid density in terms of $\overline{T}_{\rm CB}$ versus $\overline{\sigma}$ for the Bose gas: Eq. (2.9): underdoped (solid curve) and overdoped (short-dashed curve), Eq. (2.12): collapsed underdoped and overdoped branches (long-dashed curve), and $\overline{T}_{\rm c}$ versus $\overline{\sigma}$ for the $\mu$SR data shown in Fig. 1.2.

$$T_{\rm CB}^{\rm m} = \frac{3a}{5}\left(\frac{5}{3d}\right)^{2/3} (\sigma^{\rm m}(0))^{2/3} . \tag{2.11}$$

For the interpolation formula with particle-hole symmetry (Eq. (2.6)) we find

$$\overline{T}_{\rm CB} = \overline{\sigma}^{2/3} \tag{2.12}$$

with

$$\overline{\sigma} = \overline{n}_{\rm B}(2-\overline{n}_{\rm B}) , \qquad \sigma^{\rm m}(0) = \frac{d}{M_{\rm B}}\frac{\overline{n}_{\rm B}}{2} , \qquad T_{\rm CB}^{\rm m} = \frac{a}{M_{\rm B}}\left(\frac{M_{\rm B}}{d}\right)^{2/3} (\sigma^{\rm m}(0))^{2/3} . \tag{2.13}$$

The plot $\overline{T}_{\rm CB}$ versus $\overline{\sigma}$ is shown in Fig. 2.2. If particle-hole symmetry is broken, Eq. (2.9) is valid, and the resulting curve resembles the outline of a fly's wing. In fact, it is close to the dashed curve in Fig. 1.2, which is assumed to describe the generic behavior of the cuprates, covering both the under and overdoped regimes. If particle-hole symmetry is maintained, however, the underdoped and overdoped branches of the wing collapse (Eq. (2.12)). The $\mu$SR data included in Fig. 2.2 confirms these theoretical curves remarkably well. Indeed, the experimental points of overdoped $Tl_2Ba_2Cu_{6+\delta}$ are close to the outline of the lower part of the wing, while the remaining data, referring to under and mildly overdoped materials, is close to the upper branch. This behavior is further confirmed by kinetic inductance measurements for $La_{2-x}Sr_xCuO_4$ films, revealing that the penetration depths of under and overdoped samples with $T_{\rm c} \simeq 15$ K do not differ much [10]. Indeed, the shape of $\overline{T}_{\mathrm{CB}} = \overline{T}_{\mathrm{CB}}(\overline{n}_{\mathrm{B}})$ for $La_{2-x}Sr_xCuO_4$ (Fig. 2.1) points to the presence of particle-hole symmetry and implies the collapse of the underdoped and overdoped branches of the wing. Thus, the overall picture emerging from the $\mu$SR data and the kinetic inductance measurements is remarkably consistent with the Bose-Einstein condensation scenario, leading to two branches, one referring to underdoped and the other to overdoped systems.

Another implication of this scenario is the behavior of the maximum transition temperature, $T_{\mathrm{CB}}^{\mathrm{m}} \propto (\sigma^{\mathrm{m}}(0))^{2/3}$ (Eqs. (2.11) and (2.13)), which is consistent with the $\mu$SR estimates collected in Fig. 1.3. However, an upper limit for the transition temperature arises from the bound, $n_{\mathrm{B}}^{\mathrm{m}} \leq p_{\mathrm{m}} m/(2abc)$, of the boson density. Here $p_{\mathrm{m}} \simeq 0.16$ is the optimum hole concentration, $m$ the number of $CuO_2$ layers per unit cell, and $a, b$ and $c$ the lattice constants.

Finally we consider the pressure and isotope coefficients. Neglecting the influence of isotope substitution on $T_{\mathrm{CB}}^{\mathrm{m}}$ and $n_{\mathrm{B}}^{\mathrm{m}}$, we obtain from Eqs. (1.7)

$$\alpha = \frac{1}{T_{\mathrm{c}}}\frac{\mathrm{d}T_{\mathrm{c}}}{\mathrm{d}P} \simeq \frac{1}{\overline{T}_{\mathrm{c}}}\frac{\mathrm{d}\overline{T}_{\mathrm{c}}}{\mathrm{d}\overline{n}_{\mathrm{B}}}\alpha_{\mathrm{P}}\,, \qquad \beta = -\frac{m}{T_{\mathrm{c}}}\frac{\mathrm{d}T_{\mathrm{c}}}{\mathrm{d}m} \simeq \frac{1}{\overline{T}_{\mathrm{CB}}}\frac{\mathrm{d}\overline{T}_{\mathrm{CB}}}{\mathrm{d}\overline{n}_{\mathrm{B}}}\beta_{\mathrm{m}}\,, \tag{2.14}$$

$$\alpha_{\mathrm{P}} = \frac{\mathrm{d}\overline{n}_{\mathrm{B}}}{\mathrm{d}P}\,, \qquad \beta_{\mathrm{m}} = -m\frac{\mathrm{d}\overline{n}_{\mathrm{B}}}{\mathrm{d}m} \tag{2.15}$$

with $\overline{T}_{\mathrm{CB}}$ given by Eqs. (2.4) or (2.6). Therefore, the plots of pressure and isotope effect coefficients versus $\overline{T}_{\mathrm{c}} = T_{\mathrm{c}}/T_{\mathrm{c}}^{\mathrm{m}}$ probe essentially $(\mathrm{d}\overline{T}_{\mathrm{c}}/\mathrm{d}\overline{n}_{\mathrm{B}})$ and accordingly will exhibit identical behavior. In keeping with the behavior of $\overline{T}_{\mathrm{BC}} = \overline{T}_{\mathrm{BC}}(\overline{n}_{\mathrm{B}})$ shown in Fig. 2.2, the coefficients vanish in the limit $\overline{T}_{\mathrm{c}}^{\mathrm{m}} \to 1$, their magnitude increases with reduced $\overline{T}_{\mathrm{c}}^{\mathrm{m}}$ and there are two branches, one for underdoped and the other for overdoped systems. In the cuprates $\alpha_{\mathrm{P}}$ and $\beta_{\mathrm{m}}$ might adopt family-specific values. Plots of $\beta$ versus $\overline{T}_{\mathrm{c}}$ shown in Fig. 2.3 fit the experimental data for the oxygen isotope coefficient in $La_{1.85}Sr_{0.15}Cu_{1-x}Ni_xO_4$ [19], $YBa_{2-x}La_xCu_3O_z$ [20] and $Y_{1-x}Pr_xCa_yBa_2Cu_3O_7$ [21] very well and the shift accounts for the isotope coefficient at $T_{\mathrm{c}}^{\mathrm{m}}$ which has been omitted in Eq. (2.15). In agreement with Eq. (2.15), analogous behavior has been established in the pressure coefficient of various cuprate superconductors [2]–[4],[22] and both negative $\beta$ [23] and $\alpha$ [24] values have been observed in several compounds.

## 3. Conclusion

To summarize we have shown that the Bose gas with short-range repulsive interaction accounts remarkably well for the empirical trends in various properties of cuprate superconductors. These trends include the relations between transition temperature, hole concentration and superfluid density per unit mass, the rise and the upper limit of $T_{\mathrm{c}}$, the dependence of the pressure and isotope coefficients on $T_{\mathrm{c}}$ and the observed critical behavior, reminiscent of

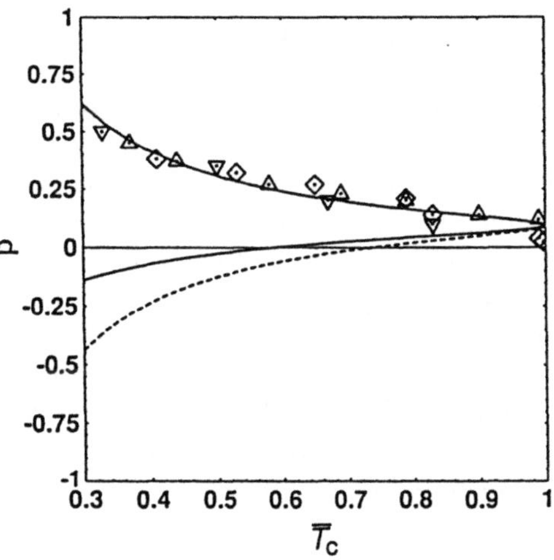


**Fig. 2.3.** Isotope effect coefficient $\beta$ vs. $\overline{T}_c = T_c/T_c^m$ for $La_{1.85}Sr_{0.15}Cu_{1-x}Ni_xO_4$ (△)[19], $YBa_{2-x}La_xCu_3O_z$ (◇)[20] and $Y_{1-x}Pr_xCa_yBa_2Cu_3O_7$ (▽)[21]. The curves correspond to Eq. (2.15) with $\overline{T}_{CB}$ given by Eq. (2.4) (solid line) and (2.6) (dashed line) with $\beta_m \simeq 0.0064$, $\beta(T_c^m) \simeq 0.095$ and $\beta_m \simeq 0.0085$, $\beta(T_c^m) \simeq 0.091$, respectively.

three-dimensional systems with a scalar complex order parameter and short-range interaction [2],[3]. In view of this we provided considerable evidence of the nature of the superconducting transition in the cuprates, without invoking any specific pairing mechanism. However, several experimental observations remain to be explained, among them the fermionic properties, including the static spin susceptibility, NMR relaxation rate and density of states. In this context it should be recognized that rather little is known about how the normal (nonsuperconducting) state of a system of fermions with attractive interaction evolves from a Fermi-liquid-like regime to a Bose-Einstein regime as a function of increasing interaction. In the Hubbard model with near-neighbor hopping $t$ and negative on-site interaction $U$ this problem is closely related to the crossover from BCS superconductivity to a condensate of pre-formed pairs [25],[26]. In the strong coupling regime ($t/|U| \ll 1$), the thermodynamic properties with energy scale $k_B T_c \propto t^2/|U|$ are essentially those of the hard core boson model which belongs to the 3d-$xy$-universality class, while in the weak coupling regime the standard BCS and Fermi-liquid treatments apply. In the strong coupling regime one has tightly bound pairs which contribute to the normal state properties only when they are ionized. The energy gap associated with the preformed pairs defines a pairing scale $k_B T_P \simeq U + \mu$, where the ionization and formation of the pre-formed pairs affect the normal state properties. For the values of $U$ considered, this pairing scale is much larger than $k_B T_c \propto t^2/|U|$ [17],[26]. For this reason anomalous temperature dependencies in the normal state properties are expected to be dominated by the formation and ionization of pairs without coherence rather than by thermal fluctuations. This expectation is confirmed by recent Quantum Monte Carlo simulations revealing that even for intermediate coupling strengths $U$ the normal state properties deviate from canonical Fermi-liquid behavior. Indeed, for temperatures $T > T_c$ the NMR relaxation rate $1/(T_1T)$ was found to track the static spin susceptibility $\chi_s(T)$, so that $T_1T\chi_s(T) \simeq$ const, which is very different from that expected in a Fermi liquid, where $1/(T_1T) \propto \chi_s^2(T)$, where $\chi_s(T)$ is temperature independent [27]. Moreover, there is a low temperature drop in $\chi_s(T)$ due to pair formation in the normal state. These results are strikingly similar to the anomalous behavior of the NMR Knight shifts and to the relaxation rates observed above $T_c$ in many cuprate superconductors [28]. Thus, Bose-Einstein condensation of pre-formed pairs also offers an explanation of the anomalous temperature dependencies in the normal-state properties.

*Acknowledgement.* The authors are grateful to J.G. Bednorz, D. Baeriswyl, H. Beck, J.I. Budnick, H. Keller, J.-P. Locquet, K.A. Müller, Ch. Niedermeier and J.J. Rodriguez for valuable discussions.

## References

1. M.B. Salamon, J. Shi, N. Overend and M.A. Howson, Phys. Rev. B **47**, 5520 (1993).
2. T. Schneider and H. Keller, Physica C **207**, 366 (1993).
3. T. Schneider and H. Keller, Int'l. J. Mod. Phys. B (in press).
4. T. Schneider and H. Keller, Phys. Rev. Lett. **69**, 3374 (1993).
5. B. Pümpin *et al.*, Phys. Rev. B **42**, 8019 (1990).
6. Ch. Niedermeier, C. Bernhard, U. Binninger, H. Glückler, J.L. Tallon, E.J. Ansaldo and J.I. Budnick, PSI Nuclear and Particle Newsletter, Annual Report 1992, p. 73.
7. J.L. Tallon and N.E. Flower, Physica C **204**, 237 (1993).
8. H. Zhang and H. Sato, Phys. Rev. Lett. **70**, 1697 (1993).
9. Y. Shimakawa, Y. Kubo, T. Manako and H. Igarashi, Phys. Rev. B **43**, 7875 (1991).
10. Y. Jaccard *et al.*, private communication.
11. M.R. Schafroth, Phys. Rev. **100**, 463 (1955).
12. E. Lieb, Ann. Phys. **16**, 407 (1961).
13. T. Onogi and Y. Murayama, preprint.
14. M. Luban, Phys. Rev. **128**, 965 (1962).
15. A. L. Fetter and J. D. Walecka, *Quantum Theory of Many-Particle Systems* (McGraw-Hill Book Company, New York, 1971) p. 259.
16. M. E. Fisher, M. N. Barber and D. Jasnov, Phys. Rev. A **8**, 1111 (1973).
17. T. Schneider, H. Beck, D. Bormann, T. Meintrup, S. Schafroth and A. Schmidt, preprint.
18. W.H. Keesom and A.P. Keesom, Physica **1**, 128 (1934).
19. N. Babushkina *et al.*, Physica C **185-189**, 901 (1991).
20. H.J. Bornemann and D.E. Morris, Phys. Rev. B **44**, 5322 (1991).

21. J.P. Franck *et al.*, Physica C **162-164**, 733 (1989), C **185-189**, 1379 (1991), Phys. Rev. B **44**, 5318 (1991).
22. N. Môri, in: *The Physics and Chemistry of Oxide Superconductors*, edited by Y. Iye and H. Yasuoka, *Springer Proceedings in Physics*, Vol. 60 (Springer Verlag, Berlin, 1992), p. 191.
23. H.J. Bornemann *et al.*, preprint.
24. J.E. Schriber, D.L. Overmyer, E.L. Venturini, D.S. Ginley and B. Morosin, Physica C **193**, 126 (1992).
25. P. Nozières and S. Schmitt-Rink, J. Low. Temp. Phys. **59**, 195 (1985).
26. R. Micnas, J. Ranninger and S. Robazkiewicz, Rev. Mod. Phys. **62**, 113 (1990).
27. M. Randeira, N. Trivedi, A. Moreo and R. Scalettar, Phys. Rev. Lett. **69**, 2001 (1992).
28. M. Mehring, Appl. Mag. Reson. **3**, 383 (1992).

# Polaron Formation and Percolative Phase Separation in HTSC

V. Hizhnyakov[1,2,4], E. Sigmund[3,4], and G. Seibold[3,4]

[1] Institute of Physics, Estonian Academy of Sciences, Riia 142, Tartu, Estonia
[2] Physics Department, Tartu University, Tähe 4, Tartu, Estonia
[3] Institut für Theoretische Physik, Universität Stuttgart, Pfaffenwaldring 57, 70550 Stuttgart, Germany
[4] Institut f. Physik, Technische Universität Cottbus, PBox 101344, 03013 Cottbus, Germany

**Summary.** When doping the antiferromagnetically (AF) ordered high-$T_c$ cuprates with holes magnetic polarons (spin-polarized clusters) are formed. We use the slave-boson formulation in the saddle-point approximation to study the properties of these clusters. At higher hole concentrations and due to cluster diffusion a phase separation is formed by establishing large fractal or percolative clusters. This percolation picture allows to understand experimentally obtained results for phase separation as well as magnetic and conductive phase diagrams. The second part intensively discusses the difference between one- and three-band Hubbard models. It turns out that in the latter local hole-induced ferromagnetism is strongly favoured also in one dimension. Additionally, the formation of correlated two-particle states in the CT-gap is suppressed in the low doping regime in a three-band Hubbard model.

## 1. Introduction

Doping or oxidation of the antiferromagnetically (AF) ordered perovskite compounds like $La_2CuO_4$ and $YBa_2Cu_3O_6$, which are the parent materials for high-$T_c$ superconductors, leads to the creation of holes in the $CuO_2$ planes. The dynamics of the charge carriers is governed by strong correlation effects which causes a redistribution of spin- and charge-densities in the vicinity of the excess holes. Due to these effects the antiferromagnetic order in the $CuO_2$ planes disappears quickly with doping and at higher doping concentrations superconductivity is established below the critical temperature.

A large number of experiments show that upon doping the high-$T_c$ cuprates a separation into small charge carrier enriched domains and carrier depleted regions takes place [1, 2, 3, 4]. The results of these investigations give support to an electronic mechanism of the phase separation in contrast to a chemical phase separation where the excess oxygen ions should cluster in domains. The simultaneous observation of a diamagnetic signal below $T_c$ and a Néel temperature of $T \approx 250K$ in $La_2CuO_{4+\delta}$ samples [1] strongly indicates a separation mechanism of percolative type as proposed in the model of percolative phase separation [5]. This model is based on the idea that doping of the $CuO_2$-planes with holes leads to the creation of small ferromagnetically ordered clusters (magnetic polarons, ferrons). These clusters have only low mobility whereas the holes inside the clusters can move freely. As a result, when increasing the hole concentration and due to cluster diffusion the clusters start to overlap and a (fractal) percolation network is built up. This leads to the destruction of the AF order and to the appearance of metallic-like conductivity or below $T_c$ to superconductivity within the percolation network [5]. By this mechanism a strongly inhomogeneous electronic state is established (at least in the low doping regime) with a microscopic characteristic length. This results e.g. in a strong enhancement of the interaction of the holes with long wave optical phonons leading to a remarkable increase of $T_c$ [6]. This model allows to explain the phase diagrams of $La_{2-x}Sr_xCuO_4$ and $YBa_2Cu_3O_{6+\delta}$ in quantitative agreement with experiment [7] giving support to spin cluster (magnetic polaron) formation in these materials.

The spin cluster model also allows to explain recent experimental data on percolative phase separation in weakly doped $La_2CuO_{4+\delta}$ and $La_{2-x}Sr_xCuO_4$ samples [1, 8]. Also the magnetic susceptibility measurements [1] show the existence of small ferromagnetic (superparamagnetic) particles associated with holes. Recently, the existence of spin-polarized clusters in $La_2CuO_{4+\delta}$ has been demonstrated by EPR experiments [9]. In addition, magnetic resonance signals measured in $RBa_2Cu_3O_x$ (R=Y,Gd) have been attributed to clusters with magnetic ordering [10]. A review on experiments about phase separation is given in [11].

In the present article we study the formation of a spin-polarized cluster in the $CuO_2$-planes. Our considerations are based on the three-band Hubbard model which is treated in two different ways. First, by means of a special unitary transformation we project out the hopping terms from the hamiltonian leading to double occupancy of Cu-sites. Then we introduce hybridized Cu-O states and obtain an effective one-band hamiltonian describing the spin-conditional hopping of a hole. The hole dynamics described by this hamiltonian differs essentially from the one described by the one-band Hubbard model. For example, in our model (in the strong Hubbard repulsion limit) the ferromagnetic ordering of spins around a hole is energetically favorable in all dimensions including the one-dimensional case (d=1). In the one-band Hubbard model such effect exists for d=2,3, but not for d=1. Second, we treat the three-band Hubbard model in a mean-field slave-boson approximation.

## 2. Hole dynamics in $CuO_2$ planes

Our considerations are based on the three band Hubbard model (see e.g. Emery [12]) which takes the $d_{x^2-y^2}$ band of the Cu ions and the $p_\sigma$ bands of the oxygen ions into account. The model hamiltonian reads

$$H = \sum_\sigma H_{0\sigma} + H_{\text{int}} \tag{2.1}$$

**Table 2.1.** The binding energy and the Cu-spin polarization for the undoped $\langle n_\uparrow \rangle, \langle n_\downarrow \rangle$ and doped $\langle n_\uparrow^l \rangle, \langle n_\downarrow^l \rangle$ cases with one and to turned spins in dependence of U and $\epsilon$ (Hartree-Fock approach).

| | | | | 1 turned spin, 1 hole | | | 2 turned spins,1 hole | | | 2 turned spins,2 holes | | |
|---|---|---|---|---|---|---|---|---|---|---|---|---|
| $\frac{U}{T}$ | $\frac{\epsilon}{T}$ | $\langle n_\uparrow^0 \rangle$ | $\langle n_\downarrow^0 \rangle$ | $\langle n_\uparrow^l \rangle$ | $\langle n_\downarrow^l \rangle$ | $-\Delta E$ | $\langle n_\uparrow^l \rangle$ | $\langle n_\downarrow^l \rangle$ | $-\Delta E$ | $\langle n_\uparrow^l \rangle$ | $\langle n_\downarrow^l \rangle$ | $-\Delta E$ |
| 4 | 2 | 0.66 | 3E-2 | 0.31 | 0.57 | 0.12 | 0.12 | 0.65 | 0.04 | 0.32 | 0.55 | 0.38 |
| 6 | 2 | 0.84 | 5E-3 | 0.29 | 0.76 | 0.30 | 0.11 | 0.82 | 0.24 | 0.29 | 0.75 | 0.63 |
| 8 | 2 | 0.91 | 2E-3 | 0.25 | 0.86 | 0.37 | 0.11 | 0.89 | 0.37 | 0.26 | 0.86 | 0.74 |
| 6 | 3 | 0.77 | 8E-3 | 0.25 | 0.68 | 0.22 | 0.10 | 0.75 | 0.11 | 0.25 | 0.67 | 0.51 |
| 8 | 3 | 0.88 | 2E-3 | 0.21 | 0.83 | 0.31 | 0.09 | 0.86 | 0.30 | 0.21 | 0.83 | 0.64 |
| 10 | 3 | 0.93 | 8E-4 | 0.18 | 0.90 | 0.34 | 0.08 | 0.92 | 0.37 | 0.18 | 0.90 | 0.68 |

where

$$H_{0\sigma} = \epsilon_{\rm d} \sum_n n_{n\sigma}^{\rm d} + \epsilon_{\rm p} \sum_m n_{m\sigma}^{\rm p} + T^{\pm} \sum_{nm} \left( d_{n\sigma}^{\dagger} p_{m\sigma} + h.c. \right) \tag{2.2}$$

$$H_{\rm int} = U \sum_n n_{n\sigma}^{\rm d} n_{n-\sigma}^{\rm d} \tag{2.3}$$

$n_{n\sigma}^{\rm d}$ and $n_{m\sigma}^{\rm p}$ are the electronic occupation numbers of the $\rm d_{x^2-y^2}$ and $\rm p_\sigma$ orbitals, respectively, $d^\dagger(d)$ and $p^\dagger(p)$ are electronic creation (annihilation) operators in the corresponding Cu and O orbitals. Band structure calculations indicate the following parameter values: U $\approx$ 8eV, $\epsilon = \epsilon_{\rm p} - \epsilon_{\rm d} \approx$ 3eV, T $\approx$1eV. In reference [11, 15] we have used a selfconsistent Hartree-Fock mean-field approach to determine the polarization and binding energy of the spin-polarized clusters (see table (2.1)). However, the renormalization of the band widths due to the strong correlation effects in the considered systems is not taken into account by this approach. This effect can be incorporated by using the slave-boson method in the saddle-point approximation, introduced by Kotliar and Ruckenstein [13], which is similar to the Gutzwiler variational approach.

We enlarge the original Hilbert space by introducing four boson fields for every copper site, respectively. $e_i^\dagger$ creates an empty state, $s_{i,\uparrow}$ and $s_{i,\downarrow}$ singly occupied states with spin up and down and finally $d_i^\dagger$ stands for the creation of a doubly occupied state. Since there are only four possible states per Cu site, the unphysical states are eliminated by the following conditions

$$e_i^\dagger e_i + \sum_\sigma s_{i,\sigma}^\dagger s_{i,\sigma} + d_i^\dagger d_i = 1 \tag{2.4}$$

$$c_{i,\sigma}^{d\dagger} c_{i,\sigma}^{d} - s_{i,\sigma}^\dagger s_{i,\sigma} - d_i^\dagger d_i = 0 \tag{2.5}$$

where $c_{i,\sigma}^{d\dagger}$ stands for the creation of a hole with spin $\sigma$ on the Cu site i.

In the saddle-point approximation the constraints (2.4,2.5) are only satisfied on the average by the Lagrange parameters $\lambda_\sigma$ and $\kappa$, respectively.

Long range antiferromagnetic order is incorporated by doubling the unit cell and introducing two operators $c^{1d}_{n,\sigma}$ and $c^{2d}_{n,\sigma}$ which create holes with spin $\sigma$ on the neighbouring Cu sites of the corresponding unit cell n. The symmetry of the AF ordered ground state provides the following relations

$$\lambda^1_\uparrow = \lambda^2_\downarrow \tag{2.6}$$
$$\langle n_{1,\sigma}\rangle = \langle n_{2,-\sigma}\rangle \tag{2.7}$$

This ends up with the Gutzwiler variational expression for the total energy

$$NE = \sum_\sigma \langle H^{\rm eff}_\sigma\rangle + UNd^2 - N[\lambda_\uparrow\langle n_\uparrow\rangle + \lambda_\downarrow\langle n_\downarrow\rangle] \tag{2.8}$$

$$\begin{aligned} H^{\rm eff}_\sigma &= \sum_n \epsilon_1 c^{\dagger 1d}_{n,\sigma} c^{1d}_{n,\sigma} + \sum_n \epsilon_2 c^{\dagger 2d}_{n,\sigma} c^{2d}_{n,\sigma} + \sum_m \epsilon_{\rm p} c^{\dagger \rm p}_{n,\sigma} c^{\rm p}_{n,\sigma} \\ &\quad + T_\sigma \sum_{n,m} [c^{\dagger 1d}_{n,\sigma} c^{\rm p}_{m,\sigma} + h.c.] + T_{-\sigma} \sum_{n,m} [c^{\dagger 2d}_{n,\sigma} c^{\rm p}_{m,\sigma} + h.c.] \end{aligned} \tag{2.9}$$

with

$$\epsilon_1 = \epsilon^{\rm d} + \lambda_\uparrow \tag{2.10}$$
$$\epsilon_2 = \epsilon^{\rm d} + \lambda_\downarrow \tag{2.11}$$
$$T_\sigma = T\frac{\sqrt{(1-\langle n^{\rm d}\rangle + d^2)(\langle n^d_\sigma\rangle - d^2)} + d\sqrt{\langle n^{\rm d}\rangle - \langle n^d_\sigma\rangle - d^2}}{\sqrt{\langle n^d_\sigma\rangle(1-\langle n^d_\sigma\rangle)}} \tag{2.12}$$

The summation in the transfer terms is restricted to nearest neighbours. The diagonalization of (2.9) can now be performed in the same way as in ref. [11].

## 2.1 Doped system

We will add now an additional hole to the system and show that a spin-flip process at site e.g. 0 will lead to a lower ground state energy. Such a spin reversion affects the Lagrange parameters $\lambda_{0,\sigma}$, the double occupancy parameter $\mathrm{d}_0^2$ and additionally the transfer to the next nearest oxygen atoms. Introducing symmetric combinations of these oxygen states the problem reduces to a $2\times 2$ perturbation matrix for each spin direction

$$V_\sigma = \begin{pmatrix} V^\lambda_\sigma & V^T_\sigma \\ V^{\rm T}_\sigma & 0 \end{pmatrix} \tag{2.13}$$

in the basis of $|c_{0,\sigma}\rangle$ and $|q\rangle = \frac{1}{2}\sum_{i=1}^4 |p_i\rangle$ where the summation is restricted to the four next nearest oxygen ions of site 0. The perturbations are denoted by

$$V^\lambda_\sigma = \lambda^\wr_\sigma - \lambda_\sigma \tag{2.14}$$
$$V^{\rm T}_\sigma = 2(T^\wr_\sigma - T_\sigma) \tag{2.15}$$

**Table 2.2.** The binding energy and the Cu-spin polarization for the undoped $\langle n_\uparrow \rangle, \langle n_\downarrow \rangle$ and doped $\langle n^l_\uparrow \rangle, \langle n^l_\downarrow \rangle$ case with one turned spin in dependence of U and $\epsilon$ (Slave-boson approach).

| | | | | 1 turned spin, 1 hole | | |
|---|---|---|---|---|---|---|
| $\frac{U}{T}$ | $\frac{\epsilon}{T}$ | $\langle n^0_\uparrow \rangle$ | $\langle n^0_\downarrow \rangle$ | $\langle n^l_\uparrow \rangle$ | $\langle n^l_\downarrow \rangle$ | $-\Delta E$ |
| 6 | 2 | 0.83 | 8E-3 | 0.34 | 0.66 | 0.28 |
| 7 | 2 | 0.88 | 4E-3 | 0.35 | 0.71 | 0.36 |
| 8 | 2 | 0.91 | 2E-3 | 0.35 | 0.73 | 0.41 |
| 6 | 3 | 0.76 | 2E-2 | 0.35 | 0.55 | 0.23 |
| 7 | 3 | 0.84 | 6E-3 | 0.34 | 0.64 | 0.30 |
| 8 | 3 | 0.88 | 2E-3 | 0.34 | 0.68 | 0.37 |

Following the Green's functions formalism we can calculate the disturbed Green's functions at the perturbed Cu site from the Dyson equation

$$G^\sigma_{00} = \frac{G^{0,\sigma}_{00}}{\mathrm{DET}_\sigma(E)} \tag{2.16}$$

$$\begin{aligned} \mathrm{DET}_\sigma(E) &= 1 + \frac{V^{\mathrm{T}}_\sigma}{T_\sigma}\left(1 + \frac{V^{\mathrm{T}}_\sigma}{4T_\sigma}\right) \\ &- G^{0,\sigma}_{00}[V^\lambda_\sigma + \frac{V^{\mathrm{T}}_\sigma}{T_\sigma}(E - \epsilon_\sigma)\left(1 + \frac{V^{\mathrm{T}}_\sigma}{4T_\sigma}\right)] \end{aligned} \tag{2.17}$$

Then we have to minimize the change of energy due to the three parameters $\lambda_{0,\uparrow}, \lambda_{0,\downarrow}$ and $d_0^2$. For every parameter set we have additionally to calculate the new Cu polarization values using the perturbed Green's function (2.15). We obtain two localized levels in the CT gap corresponding to the zeros of $\mathrm{DET}_\sigma(E)$, one arising from the lower Hubbard band whereas the other comes from the first oxygen type band.

Table (2.2) shows the results obtained for various values of U and $\epsilon = E_p - E_d$. As one can readily see, the incorporation of fluctuations within the scope of a slave-boson approximation reduces the polarization of the spin-cluster state to 10 – 20% in comparison to the Hartree-Fock approach. On the other hand we now obtain a larger value of the binding energy. As already remarked by Olés and Zaanen [14] the MF and Gutzwiler approach yield the same results in the strong coupling limit when charge fluctuations can be neglected. In the intermediate U-regime, which is interesting for real systems, the slave-boson approach in the MF approximation leads to a much better description of the ground state.

So far we have confined ourselves to a symmetry breaking solution of the translational invariant three-band Hubbard hamiltonian. However, as in the theory of strong coupling polarons (Pekar-type solutions) one should in principle perform a Bloch-superposition of these wave-functions to guarantee the translational invariance of the solution. This ends up with a dispersion relation for the quasi-particles i.e. in our case the spin-polarized clusters. To keep the calculation tractable we have confined ourselves to hopping processes of a magnetic polaron, formed by turning one Cu spin, to nearest and next nearest neighbours. The dispersion relation of the polaron is strongly anisotropic in k-space and yields the maximum of the effective mass in $\langle 11 \rangle$-direction. For U=8T and $\epsilon = 3$T we obtain a value of 10 bare masses.

## 3. Two-particle states

In the preceding section we have seen that within two mean-field approximations of the three band Hubbard model doping of holes leads to symmetry breaking solutions i.e. magnetic polarons with a considerable gain of energy with respect to the corresponding homogeneous band state. However, another model has been proposed for the doped hole states namely the formation of Zhang-Rice singlets which are built up by the Cu spins and the total symmetrical combination of the surrounding doped oxygen spins. The question is how these states can be included in our effective one-particle band picture. The straightforward generalization of our calculation would be the incorporation of higher-order Green's functions beyond the Hartree-Fock factorization. However, attempts from Hubbard [16] or L. Roth [17] for example suffer from the fact that they either don't yield antiferromagnetic spin-wave solutions for the ground state at half filling or are restricted to a two-pole approximation of the density of states. To avoid these difficulties we will consider here the two-particle Green's function $\ll c_{i,\uparrow}c_{j,\downarrow}; c_{m,\downarrow}c_{l,\uparrow} \gg$ which will be seen to include doped correlated two-particle states in a natural way. The calculation is first illustrated within a one-dimensional Hubbard-chain. As was pointed out by Sawatzky et al. [18] doping of the half-filled Hubbard model with electrons (holes) leads to the formation of correlated two-particle states in the Mott-Hubbard gap which split off from the upper (UHB) and lower (LHB) Hubbard band respectively. We will see that this feature is reproduced in our calculation of the two-particle GF but that the behaviour is considerably different in a multi-band Hubbard model.

Starting point is the one-band Hubbard model with doubled unit cell:

$$\begin{aligned} H &= T\sum_{n,\sigma}(c^{+}_{1,n,\sigma}c_{2,n,\sigma} + c^{+}_{1,n,\sigma}c_{2,n-1,\sigma} + h.c.) \\ &+ U\sum_{n} n_{1,\sigma}n_{1,-\sigma} + U\sum_{n} n_{2,\sigma}n_{2,-\sigma} \end{aligned} \tag{3.1}$$

The AF-order is imposed by the condition

$$< n_{1,\uparrow} > = < n_{\uparrow} > = < n_{2,\downarrow} > \tag{3.2}$$

$$< n_{1,\downarrow} > = < n_{\downarrow} > = < n_{2,\uparrow} > \tag{3.3}$$

To calculate the two-particle excitation spectrum we investigate the poles of the Green's functions (GF) $\ll c^{s}_{i,\uparrow}c^{t}_{j,\downarrow}; c^{+u}_{m,\downarrow}c^{+v}_{l,\uparrow} \gg$ where $\{s,t,u,v\} = 1, 2$. The GF obeys the equation of motion

$$\omega \ll c^{s}_{i,\uparrow}c^{t}_{j,\downarrow}; c^{+u}_{m,\downarrow}c^{+v}_{l,\uparrow} \gg \; = \; < [c^{s}_{i,\uparrow}c^{t}_{j,\downarrow}, c^{+u}_{m,\downarrow}c^{+v}_{l,\uparrow}]_{-} > + \ll [c^{s}_{i,\uparrow}c^{t}_{j,\downarrow}, H]_{-}; c^{+u}_{m,\downarrow}c^{+v}_{l,\uparrow} \gg \quad (3.4)$$

Consider the special GF $\ll c^{1}_{i,\uparrow}c^{1}_{j,\downarrow}; c^{+1}_{m,\downarrow}c^{+1}_{l,\uparrow} \gg$. Its equation of motion reads as

$$\begin{aligned} \omega \ll c^{1}_{i,\uparrow}c^{1}_{j,\downarrow}; c^{+1}_{m,\downarrow}c^{+1}_{l,\uparrow} \gg &= \delta_{il}\delta_{jm} - < c^{+1}_{l,\uparrow}c^{+1}_{i,\uparrow} > - < c^{+1}_{m,\downarrow}c^{+1}_{j,\downarrow} > \\ &+T\left(\ll c^{1}_{i,\uparrow}c^{2}_{j,\downarrow}; c^{+1}_{m,\downarrow}c^{+1}_{l,\uparrow} \gg + \ll c^{1}_{i,\uparrow}c^{2}_{j-1,\downarrow}; c^{+1}_{m,\downarrow}c^{+1}_{l,\uparrow} \gg\right) \\ &+T\left(\ll c^{2}_{i,\uparrow}c^{1}_{j,\downarrow}; c^{+1}_{m,\downarrow}c^{+1}_{l,\uparrow} \gg + \ll c^{2}_{i-1,\uparrow}c^{1}_{j,\downarrow}; c^{+1}_{m,\downarrow}c^{+1}_{l,\uparrow} \gg\right) \\ +U\delta_{ij} \ll c^{1}_{i,\uparrow}c^{1}_{j,\downarrow}; c^{+1}_{m,\downarrow}c^{+1}_{l,\uparrow} \gg &+U \ll (n^{1}_{i,\downarrow} + n^{1}_{j,\uparrow})c^{1}_{i,\uparrow}c^{1}_{j,\downarrow}; c^{+1}_{m,\downarrow}c^{+1}_{l,\uparrow} \gg \end{aligned} \quad (3.5)$$

In order to get a closed set of equations we apply the following factorization to the higher order GF on the rhs of eq.(3.5)

$$\begin{aligned} &\ll (n^{1}_{i,\downarrow} + n^{1}_{j,\uparrow})c^{1}_{i,\uparrow}c^{1}_{j,\downarrow}; c^{+1}_{m,\downarrow}c^{+1}_{l,\uparrow} \gg \approx \\ &(1-\delta_{ij})(< n^{1}_{i,\downarrow} > + < n^{1}_{j,\uparrow} >) \ll c^{1}_{i,\uparrow}c^{1}_{j,\downarrow}; c^{+1}_{m,\downarrow}c^{+1}_{l,\uparrow} \gg \end{aligned} \quad (3.6)$$

It has been shown in [19] that this approximation behaves correctly in the limit U → 0 and T → 0, respectively. Also in the limit of low concentration the decoupling (3.6) is exact and yields the result of Kanamori [20]. Eq.(3.5) contains another type of GF which obeys the equation of motion

$$\begin{aligned} \omega \ll c^{1}_{i,\uparrow}c^{2}_{j,\downarrow}; c^{+1}_{m,\downarrow}c^{+1}_{l,\uparrow} \gg &= - < c^{+1}_{m,\downarrow}c^{+2}_{j,\downarrow} > \delta_{il} \\ &+T\left(\ll c^{1}_{i,\uparrow}c^{1}_{j,\downarrow}; c^{+1}_{m,\downarrow}c^{+1}_{l,\uparrow} \gg + \ll c^{2}_{i,\uparrow}c^{2}_{j,\downarrow}; c^{+1}_{m,\downarrow}c^{+1}_{l,\uparrow} \gg\right) \\ &+T\left(\ll c^{1}_{i,\uparrow}c^{1}_{j+1,\downarrow}; c^{+1}_{m,\downarrow}c^{+1}_{l,\uparrow} \gg + \ll c^{2}_{i-1,\uparrow}c^{2}_{j,\downarrow}; c^{+1}_{m,\downarrow}c^{+1}_{l,\uparrow} \gg\right) \\ &+U \ll (n^{1}_{i,\downarrow} + n^{2}_{j,\uparrow})c^{1}_{i,\uparrow}c^{2}_{j,\downarrow}; c^{+1}_{m,\downarrow}c^{+1}_{l,\uparrow} \gg \end{aligned} \quad (3.7)$$

Similar to eq.(3.5) we factorize the higher order GF in (3.7) in the following way

$$\begin{aligned} &\ll (n^{1}_{i,\downarrow} + n^{2}_{j,\uparrow})c^{1}_{i,\uparrow}c^{2}_{j,\downarrow}; c^{+1}_{m,\downarrow}c^{+1}_{l,\uparrow} \gg \approx \\ &(< n^{1}_{i,\downarrow} > + < n^{2}_{j,\uparrow} >) \ll c^{1}_{i,\uparrow}c^{2}_{j,\downarrow}; c^{+1}_{m,\downarrow}c^{+1}_{l,\uparrow} \gg \end{aligned} \quad (3.8)$$

The obtained closed set of equations can be solved by standard GF techniques. Additionally we take the low density limit of the GF but with the imposed long range spin-density order obtained by Hartree-Fock. The resulting two-particle excitation spectrum at half filling consists of three bands of which the lower (upper) one describes the excitation of two particles in the LHB (UHB) whereas the band between describes the separate excitation of these particles in the LHB and UHB respectively. Fig. (3.1) shows the imaginary part of the two-particle GF for hole and electron doping respectively in comparison with the same GF corresponding to the Hartree-Fock approximation (solid line). In the half-filled case the perturbation

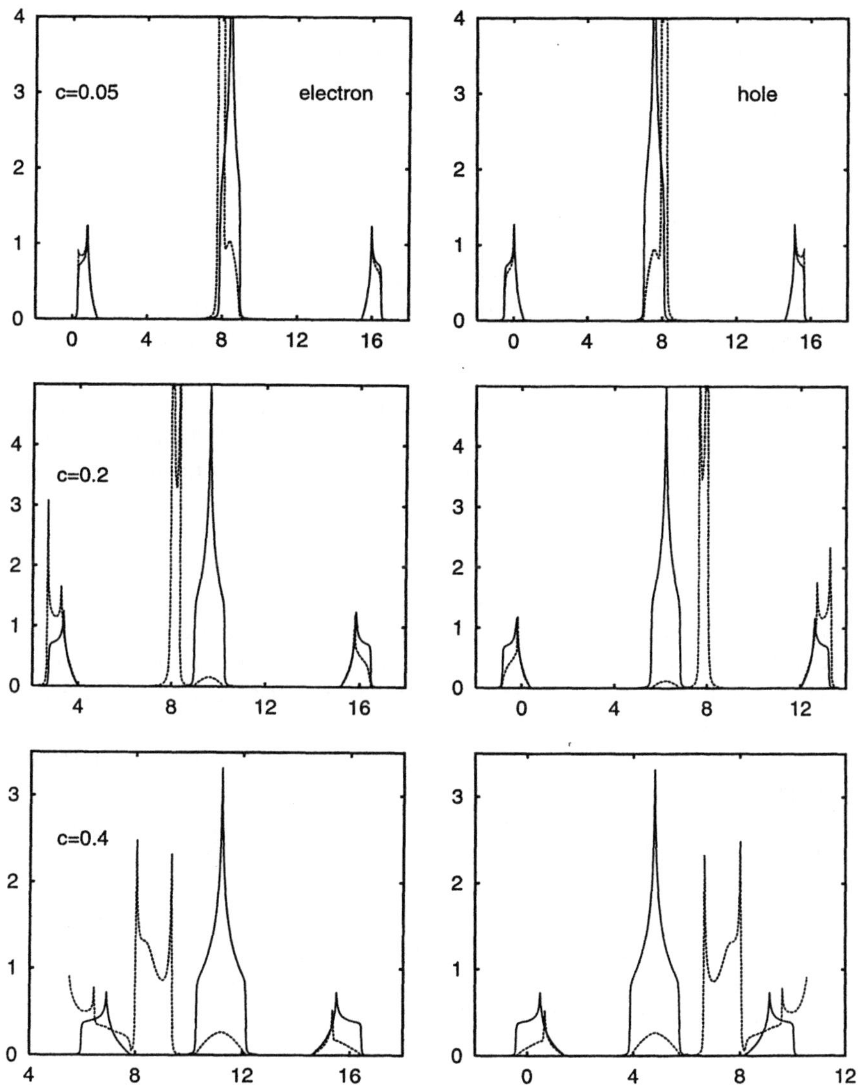


**Fig. 3.1.** Two-particle excitation spectrum in the case of electron (left) and hole (right) doping. The solid line corresponds to the Hartree-Fock spectrum. Parameters: $U = 8T$, $c = 0.05$, 0.2, and 0.4.

$V = U(1- < n_\uparrow > - < n_\downarrow >)$ vanishes so that both spectrums are equal. For small concentrations of electrons the second band is broadened caused by a shift of density towards lower energies compared with the Hartree-Fock result. At a critical concentration a split-off band emerges at the low energy side of the second band corresponding to a correlated band below the UHB, in accordance with our considerations at the beginning of this section. In the case of hole doping the band splits off from the upper band edge of the second band which is related to the occurrence of correlated states above the LHB. The symmetry between electron and hole doping is due to the fact that for the perturbation the relation holds $V_{hole} = -V_{electron}$. Increasing further the doping concentration leads to a broadening of the correlation band in the Mott-Hubbard gap. At a critical concentration the gap is filled up whereas in the Hartree-Fock case it is still persistent.

We thus have seen that there indeed exist two-particle excitations in the one-band Hubbard model which are not described within a Hartree-Fock approximation and which have lower energy compared with the excitation due to the Mott-Hubbard gap. It is also obvious that these correlated states only are important for the ground state properties at higher doping concentrations.

A different situation arises for the formation of correlated states in the high-$T_c$ cuprates. As was argued from Sawatzky et al. [18] the hybridization between Cu and oxygen states should lead to an admixture of Cu states of the UHB to the second hole band. Upon doping the system with holes correlated states should emerge at the bottom of the first oxygen type hole band called Zhang-Rice singlets in this case. It has been shown in [18] that in a 4 unit cell system the situation is indeed similar to the model in the first section. However, from our previous considerations it becomes evident that in these systems a critical doping concentration is necessary to induce these states because due to the hybridization between oxygen and copper the hole density on the Cu sites at half filling is always lower than one. As the perturbation $V = U(1- < n_\uparrow >^{\mathrm{Cu}} - < n_\downarrow >^{\mathrm{Cu}})$ has to be negative to cause correlated singlet states below the first oxygen band we need at least a critical concentration of c= $0.2 - 0.3$ for relevant physical parameters. So the scenario described in ref. [18] suffers from the fact that in a four unit cell system the minimal concentration one can reach is 0.25.

Thus we have seen that there is an important difference in the electronic structure of one- and multi-band Hubbard models upon doping them away from half-filling. This difference is more extensively worked out in the next section by transformational methods.

## 4. Conditional hopping in one- and three band Hubbard models

We use the fact that the relations $T/U$ and $T/(U-\varepsilon)$ can be considered as small parameters. Therefore, in the considered case of half-(or less) filled Cu $\mathrm{d}_{\mathrm{x}^2-\mathrm{y}^2}$ band the effect of the upper Cu-band caused by virtual double occupancy of Cu-sites by electrons is small. This means that the Cu-O interaction $\mathrm{H_T}$ can be divided into two terms: the large term,

$$H_{\mathrm{T}1} = \sum_{(mm')} \sum_{\sigma} (1 - n^{\mathrm{d}}_{m-\sigma}) d^{+}_{m\sigma} p_{m'\sigma} + h.c. \ , \qquad (4.1)$$

which excludes double occupancy of Cu-sites, and the small term,

$$H_{T2} = \sum_{(mm')} \sum_{\sigma} n^{\mathrm{d}}_{m-\sigma} d^{+}_{m\sigma} p_{m'\sigma} + h.c. \ , \tag{4.2}$$

which accounts such occupancies. The term $H_{T2}$ can be accounted approximately. One convenient method for such consideration is given by the unitary transformation technique [7]. Performing the following two-step unitary transformation

$$\overline{H} = e^{S_2} \tilde{H} e^{-S_2} = e^{S_2} e^{S_1} H e^{-S_1} e^{-S_2} \tag{4.3}$$

with

$$S_1 = \frac{T}{U_1} \sum_{(mm')} \sum_{\sigma} n^{\mathrm{d}}_{m-s} (d^{+}_{m\sigma} p_{m'\sigma} - h.c.) \tag{4.4}$$

$$S_2 = \frac{T}{U_1 U_2} \sum_{(mm_1)} \sum_{\sigma} n^{\mathrm{d}}_{m-d} (d^{+}_{m\sigma} d_{m_1\sigma} - h.c.) \tag{4.5}$$

($U_1 = U - \varepsilon$, $U_2 = U + 8T^2/U_1$) and keeping all terms of the order $T^2$ and only these terms of the orders $\sim T^3$ and $\sim T^4$ which have a non-zero mean value for the half-filled d-band, one obtains [7]

$$H = H_{\mathrm{e}} + H_{\mathrm{T1}} + H^{\mathrm{SS}}_{\mathrm{k}} + H^{\mathrm{SS}}_{\mathrm{dd}} \tag{4.6}$$

where

$$H^{\mathrm{SS}}_{\mathrm{k}} = 8 \frac{T^2}{U_1} \sum_{m} s^{\mathrm{d}}_{m} s^{\mathrm{P}}_{m} \tag{4.7}$$

$$H^{\mathrm{SS}}_{\mathrm{dd}} = J \sum_{mm_1} s^{\mathrm{d}}_{m} s^{\mathrm{d}}_{m_1} \tag{4.8}$$

Here

$$J = \frac{2T^4}{U_1^3} (1 + \frac{2U_1}{U_2}) \tag{4.9}$$

$s^{\mathrm{d}}_{m}$ and $s^{\mathrm{P}}_{m}$ are Cu- and O-spin operators introduced as usual according to

$$d^{+}_{m\uparrow} d_{m\downarrow} = s^{\mathrm{d}}_{xm} + i s^{\mathrm{d}}_{ym} \ , \quad d^{+}_{m\uparrow} d_{m\uparrow} = \frac{1}{2} n^{\mathrm{d}}_{m} + s^{\mathrm{d}}_{zm} \tag{4.10}$$

$$P^{+}_{m\uparrow} P_{m\downarrow} = s^{\mathrm{P}}_{xm} + i s^{\mathrm{P}}_{ym} \ , \quad P^{+}_{m\uparrow} P_{m\uparrow} = \frac{1}{2} N^{\mathrm{d}}_{m} + s^{\mathrm{P}}_{zm} \tag{4.11}$$

(4.12)

$$P^{+}_{m\sigma} = \frac{1}{2} \sum_{(m')} P^{+}_{m'\sigma} \tag{4.13}$$

is a totally symmetric oxygen electron operator of a $CuO_4$ plaquette, $(m')$ denotes the sum over four oxygen sites which are nearest to the m-th Cu site.

| | | | |
|---|---|---|---|
| Cu | ↑ | ↑ | ↓ |
| O | 0↓ | 0↓ | 0↓ |
| Cu | ↑ | ↓ | ↓ |
| | a) | b) | c) |

**Fig. 4.1.** Hopping possibilities of an oxygen hole for various Cu-spin configurations

The term $\mathrm{H}_{\mathrm{k}}^{\mathrm{SS}}$ describes a Kondo-type Cu-O spin-spin coupling; it differs from zero if there is a hole on an oxygen ion. The term $\mathrm{H}_{\mathrm{dd}}^{\mathrm{SS}}$ describes a Cu-Cu spin-spin interaction which causes AF order in the undoped system. The term $\sim n_{m\uparrow}^{\mathrm{d}} n_{m\downarrow}^{\mathrm{d}}$ is omitted, as it does not contribute to the energy of the system in the considered case of single (or less) occupancy of Cu-sites.

The main interaction term $\mathrm{H}_{\mathrm{T1}}$ describes the spin-conditioned hopping of the electron (hole) between nn Cu and O ions. This process here differs essentially from an analogous process in the one-band Hubbard model (and in the t-J model, being the large U limit of the one-band Hubbard model). Indeed, in our case the single hole motion described by $\mathrm{H}_{\mathrm{T1}}$ depends on spin orientation on Cu sites already in the first order, while such dependence in the t-J model appears first in the fourth order [21]. Indeed, in considering here a three-band model, a hole can hop from an oxygen ion only to such nearest Cu ion whose spin is opposite to that of the oxygen with the hole. E.g. in the case a) (see fig. 4.1) a hole can hop both up and down, in the case b) only up and in the case c) it cannot hop at all. Taking into account that the hole motion leads to its delocalization and to the lowering of its energy we come to the conclusion that in the $U \to \infty$ limit a hole will have the lowest energy if the Cu spins are ordered ferromagnetically.

For large but finite U, in addition to $\mathrm{H}_{\mathrm{T1}}$, the higher-order terms $\mathrm{H}_{\mathrm{k}}^{\mathrm{SS}}$ and $\mathrm{H}_{\mathrm{dd}}^{\mathrm{SS}}$ should be taken into account. The term $\mathrm{H}_{\mathrm{k}}^{\mathrm{SS}}$ also favors parallel (ferromagnetic) alignment of the Cu-spins nearest to the O-hole. Indeed, in the case a) (see Fig. 3.1) this interaction gives a negative contribution to the energy, in the case b) it gives no contribution, while in the case c) it gives a positive contribution.

The interaction term $\mathrm{H}_{\mathrm{dd}}^{\mathrm{SS}}$ is weaker than $\mathrm{H}_{\mathrm{T1}}$ and $\mathrm{H}_{\mathrm{k}}^{\mathrm{SS}}$ but unlike the last two terms it differs from zero in the undoped case and it causes the AF order in undoped crystals. In the presence of a single hole this interaction keeps the AF order apart from the hole but it cannot prevent the destruction of this order around the hole. As a result the hole forms around itself a ferromagnetically ordered cluster of a finite size and is self-trapped in this cluster. This type hole state is called spin polaron or ferron [22]. The size of the cluster depends on actual values of the large parameters $U/T$ and $U_1/T$.

## 4.1 One-band model for spin-conditioned hopping

In this section we consider a simplified model which allows for the motion of a hole in the highest occupied band and which takes into account the effect of spin-conditioned hopping between Cu and O sites. Using the Wannier operators

$$P'_{m\sigma} = \frac{1}{N}\sum_{\mathbf{k}} e^{-i\mathbf{k}m} P'_{\mathbf{k}\sigma} \tag{4.14}$$

$$Q_{m\sigma} = \frac{1}{N}\sum_{\mathbf{k}} e^{-i\mathbf{k}m} Q_{m\sigma} \tag{4.15}$$

with

$$P'_{\mathbf{k}\sigma} = \frac{1}{2}\beta_{\mathbf{k}}[P_{x,\mathbf{k}\sigma}(1+e^{ik_y}) + P_{y,\mathbf{k}\sigma}(1+e^{-ik_x})] \tag{4.16}$$

$$Q_{\mathbf{k}\sigma} = \frac{1}{2}\beta_{\mathbf{k}}[P_{x,\mathbf{k}\sigma}(1+e^{ik_y}) - P_{y,\mathbf{k}\sigma}(1+e^{-ik_x})] \tag{4.17}$$

$$P_{\alpha,\mathbf{k}\sigma} = \sum_m e^{i\mathbf{k}m} P_{\alpha,m\sigma} \tag{4.18}$$

$$\beta_{\mathbf{k}} = [1+\frac{1}{2}(\cos k_x + \cos k_y)]^{1/2} \tag{4.19}$$

(indexes $\alpha$ =x,y describe two oxygen states in the m-th cell) the zeroth-order Hamiltonian $H_0 = H_\mathrm{e} + H_{\mathrm{T}1}$ can be presented in the form [7]

$$H_0 \approx h_0 + h_{\mathrm{T}1} \tag{4.20}$$

where

$$h_0 = \sum_{m\sigma}(\varepsilon_\mathrm{d} d^+_{m\sigma} d_{m\sigma} + \varepsilon_\mathrm{p} P'^+_{m\sigma} P'_{m\sigma} + 2\lambda_0 T(1-n^\mathrm{d}_{m\,-\sigma})(d^+_{m\sigma}P'_{m\sigma} + h.c.)) \tag{4.21}$$

describes the on-site energy and

$$h_{\mathrm{T}1} = \lambda_1 T \sum_{(mm')}\sum_{\sigma}(d^+_{m\sigma}P'_{m'\sigma} + h.c.)(1-n^\mathrm{d}_{m\,-\sigma}) \tag{4.22}$$

describes the semiconditional hopping of electrons (holes).

$$\lambda_0 \simeq 0.96\,, \quad \lambda_1 \approx 0.14 \tag{4.23}$$

In the considered case of single (or less) occupancy of Cu sites the operator $h_0$ can be diagonalized by means of the operators

$$c_{m\sigma} = (\sin\alpha)d_{m\sigma} + (\cos\alpha)P'_{m\sigma} \tag{4.24}$$

$$c_{1m\sigma} = (\cos\alpha)d_{m\sigma} - (\sin\alpha)P'_{m\sigma} \tag{4.25}$$

describing Cu-O hybridized excitations. Here

$$\cos\alpha = 2\lambda_0 T/[\Omega(2\Omega-\varepsilon)]^{1/2} \tag{4.26}$$

$$\Omega = (4\lambda_0^2 T^2 + \varepsilon^2/4)^{1/2} \tag{4.27}$$

Taking into account that in the considered case the operator $\sim n^{\rm d}_{m\sigma}n^{\rm d}_{m-\sigma}$ does not give any contribution to the energy of the system, the on-site Hamiltonian may be presented in the form

$$h_0 = \sum_{m\sigma}\{(1-n^{\rm d}_{m-\sigma})[(\Omega-\varepsilon/2)c^+_{m\sigma}c_{m\sigma} - (\Omega+\varepsilon/2)c^+_{1m\sigma}c_{1m\sigma}] + \\ +\varepsilon_{\rm p}n^{\rm d}_{m-\sigma}P'^+_{m\sigma}P'_{m\sigma} + \varepsilon_{\rm p}Q^+_{m\sigma}Q_{m\sigma}\} \quad (4.28)$$

In the same representation the hopping hamiltonian equals

$$h_{\rm T1} = \sum_{(mm')}\sum_{\sigma}(1-n^{\rm d}_{m-\sigma})[t(c^+_{m\sigma}c_{m'\sigma} + c^+_{1m\sigma}c_{1m'\sigma}) + t_1 c^+_{m\sigma}c_{1m'\sigma}] + h.c. \quad (4.29)$$

where $t = \lambda_1 T \sin 2\alpha$, $t_1 = 2\lambda_1 T\cos 2\alpha$. We can see that the term $\sim t_1 c^+_{m\sigma}c_{1m'\sigma} + h.c.$ describing the mixing of bands is much smaller than the energy difference of the bands $\Omega$ and it can be neglected. In this approximation the hamiltonian describes 5 separate bands being totally filled in the undoped case. An extra hole can be considered to occupy the top of the upper band which is described by the following effective hamiltonian

$$h_{\rm eff} = t\sum_{(mm')}\sum_{\sigma}(1-n^{\rm d}_{m-\sigma})c^+_{m\sigma}c_{m'\sigma} + h.c. \quad (4.30)$$

(here the energy is counted from the middle of the upper band). This hamiltonian can also be presented in the form

$$h_{\rm eff} = t\sum_{(mm')}\sum_{\sigma}c^+_{m\sigma}c_{m'\sigma}(1-\frac{1}{2}n^{\rm d}_{m-\sigma} - \frac{1}{2}n^{\rm d}_{m'-\sigma}) + h.c. \quad (4.31)$$

This one-band hamiltonian leads to essentially different hole dynamics than the usual conditional hopping hamiltonian

$$h_{\rm ch} = t\sum_{(mm')}\sum_{\sigma}c^+_{m\sigma}c_{m'\sigma}(1-n_{m-\sigma})(1-n_{m'-\sigma}) \quad (4.32)$$

describing the $U/t \to \infty$ limit of a one-band Hubbard model, and the t-J model. Indeed, in our case every site-situated hole has a label — the spin, which had the Cu ion in this site before the hole creation; the holes described by the hamiltonian $h_{\rm ch}$ have no such labels. As a result, in our model a hole "remembers" the value of the spin which a Cu ion had before a hole creation at the given site, while no such spin memory exists in the case described by the hamiltonian $h_{\rm ch}$. We remind that this label comes from the spin of the oxygen with one hole situated nearest to the corresponding Cu ion. This difference results already in the single hopping probability. Indeed, the hole hopping probability between the sites with parallel Cu spins is equal to t (here $n^{\rm d}_{m-\sigma} + n^{\rm d}_{m'-\sigma} = 0$) while in the case of antiparallel Cu spins it equals $t/2$ (here $n^{\rm d}_{m-\sigma} + n^{\rm d}_{m'-\sigma} = 1$) and in the case of random spin orientation on

**Table 4.1.** F - ferromagnetic, R - random, AF - antiferromagnetic spin orders

| $m_l/t^l$, condit. hopping | | | |
|---|---|---|---|
| $l$ | F | R | AF |
| 2 | 0.25 | 0.25 | 0.25 |
| 4 | 0.1406 | 0.1172 | 0.1094 |

| $m_l/t^l$, spin-condit. hopping | | | |
|---|---|---|---|
| $l$ | F | R | AF |
| 2 | 0.25 | 0.156 | 0.0625 |
| 4 | 0.1406 | 0.0549 | 0.0068 |

the average it equals $3t/4$. In the usual conditional hopping model the hopping probability for all these cases equals t; the effect of the spin order for the hole motion in this model appears starting from the fourth-order terms $\sim t^4$ [21]. This difference in the models results in different energetical efficiency of the spins alignment around a hole: in the considered spin-conditional hopping model the ferromagnetic ordering of the spins around the hole is energetically much more favorable than in the usual conditional hopping model (for the same t value). Quantitatively this difference is illustrated in table 4.1 where the first two non-zero moments of the energetical band of a single hole in a two-dimensional square lattice are presented (the rules for the moments calculation see, e.g. in [23]). From table 4.1 one can see that the bandwidth in a spin-conditioned hopping model strongly depends on the spin configuration, while in a conditional hopping model this dependence is much weaker. Indeed, in the conditional hopping model the second moment is independent of the spin configuration, while in a spin-conditioned hopping model in the F(ferromagnetic) case it is 4 times larger than in the AF(antiferromagnetic) case; and 1.6 times larger than in the R (random; metal-like) case. Correspondingly in a spin-conditioned hopping model the bandwidths ratios in the F,R and the AF case are approximately equal 1:0.8:0.5. In the conditional hopping model, however, these ratios are 1:0.85:0.77 (in this case the bandwidths' difference comes mainly from the dependence of the 4th moment on the spin order). Consequently in both models the energy of the hole ground state is lowest for the ferromagnetic spin order but the energy gain due to this ordering is more than two times larger in a spin-conditional hopping model than in a conditional hopping model. We also mark that in the spin-conditioned hopping case the bandwidth difference between the ferromagnetic and the random case is remarkably smaller than between ferro- and antiferromagnetic and random and antiferromagnetic ones. For large but finite U the Cu-Cu spin-spin interaction term plays a very essential role; it keeps AF order apart from the hole (as it was already mentioned above). Therefore the spin-conditional hopping term and the Kondo-type interaction term $H_k^{SS}$ as well (which is weaker in the large U limit) establish ferromagnetic order only locally in the vicinity of the hole. As a result a spin-polarized hole cluster of finite size is created. This conclusion agrees with our numerical calculations presented above (see also [5, 15]).

For finite hole concentrations when hole clusters overlap in space the ferromagnetically ordered state becomes unstable with respect to spin-depolarization. As a result metal-like multi-hole clusters are formed. The mechanism for this process is provided by the Pauli exclusion principle. Indeed, according to this principle in the F case every state is singly occupied by a hole, while in the R state a double occupancy is possible. As a result the Fermi energy of the F state at sufficiently high hole concentrations will be higher than that of the R (metal-like) state. This mechanism of metalization is further supported by the small energy differences between ferromagnetic and random case when doping a single hole (see table 4.1).

## 5. Concluding remarks

We have shown that in the three-band Hubbard model for relevant physical parameters a single hole locally destroys antiferromagnetic order, creating a ferromagnetically ordered cluster (s.c. spin polaron or ferron). This process takes place in any dimension including the one-dimensional lattice. An analogous process is known to occur also in the one-band Hubbard model (and in the t-J model), but it takes place only in two- and three-dimensional cases. Another difference of one- and three-band (3BHM) Hubbard models is that in the last model a ferromagnetic cluster formation takes place at much smaller $U/T$ ratio than in the first one (in 1BHM $U/t \leq 20$, in 3BHM $U_1/T \sim 3$). This difference comes from the fact that in the three-band Hubbard model a single-hole hopping probability depends on the spin orientation, which is not the case in the one-band Hubbard model (1BHM). The highest occupied band in the 3BHM belongs to the hybridized state of a singly occupied (Cu) and a doubly occupied atom (O), while in the one-band Hubbard model it belongs only to singly occupied atoms. A hole on a doubly occupied atom (O) creates a spin-polarized state thus causing in the 3BHM the dependence of the hole hopping probability on the spin orientation. In the three-band Hubbard model the creation of ferromagnetic (spin-polaron) hole clusters is also supported by a Kondo-type O-Cu spin-spin interaction.

In real systems the creation of spin-polarons on doping is also supported by charge-compensating dopant ions (e.g. Sr, Ba, O, etc.), as they favor the localization of holes.

At higher hole concentrations hole clusters join and form larger clusters of a fractal type. The fractal clusters with many holes lose easily spin orientation, which allows the holes of both spins to move in the same cluster and by this to lower the local Fermi energy. As a result the metal insulator transition (MIT) is provided by the percolation of metallized fractal clusters. Thereby the electronic phase separation of a percolative type takes place in the system and an inhomogeneous electronic structure with a microscopic characteristic length is established. This state has an essential influence on superconductivity allowing to switch on pairing interactions via non-screened long wave optical phonons and high frequency spin excitations of the insulating subphase. Charge-compensating ions support the stability of the conducting network.

There are a number of experimental evidences that such percolative phase separation of electronic origin indeed takes place in weakly and moderately doped high-$T_c$ cuprate superconductors. For example a strong thermal treatment of $La_2CuO_4$ samples [1] shows that high $T_c$ superconductivity is connected with fractal percolative conducting (superconducting) clusters. When quenching the samples to temperatures lower than 200 K, one can "freeze in" the clusters diffusion, the formation of large percolative structures is suppressed and, consequently, a further slow cooling below $T_c$ leads to a very small diamagnetic fraction of the sample. In contrast, when the initial quenching ends up at temperatures higher than 200 K the diamagnetic fraction becomes maximal. For completeness we mention that recently the existence of spin-polarized clusters was directly observed by EPR [9] and magnetic resonance experiments [10].

## References


1. R.K. Kremer,E. Sigmund, V. Hizhnyakov, F. Hentsch, A. Simon, K.A. Müller, M. Mehring,*Z.Phys.B Cond.Matter* **86**, (1992) 319
2. P.C. Hammel, A.P. Reyes, Z. Fisk, M. Takigawa, J.D. Thompson, K.H. Heffner, S.W. Cheong, *Phys.Rev.* B**42**, (1990) 6781
3. J.D. Jorgensen, B. Dabrowski, Pei Shiyou, D.G. Hinks, L. Soderholm, B. Morosin, J.E. Schirber, E.L. Venturini, D.S. Ginley,*Phys.Rev.* B**38**, (1988) 11337
4. C. Chaillout, S.W. Cheong, Z. Fisk, M.S. Lehmann, M. Marezio, B. Morosin, J.E. Schirber, *Physica* C**158**, (1989) 183
5. V. Hizhnyakov,E. Sigmund, *Physica* C**156**, (1988) 655
6. A. Bill,V. Hizhnyakov,E. Sigmund, this volume
7. V. Hizhnyakov,E. Sigmund,M. Schneider, *Phys.Rev.* B**44**,(1991) 795
8. C. Taliani,A. Pal,G. Ruani,R. Zamboni,X. Wei and Z.V. Vardeny in *Electronic Properties of High $T_c$ Superconductors and Related Compounds* ed. by H. Kuzmany,M. Mehring and J. Fink, Springer Series on Solid State Science, Berlin 1990
9. G. Wübbeler,O. Schirmer,*Phys.Stat.Sol* B**174**, (1992) K21
10. P.G. Baranov,A.G. Badalyan, to be published
11. G. Benedeck,K.A. Müller (eds.),*Proceedings of the ERICE-Workshop on Phase Separation in Cuprate Superconductors*, World Scientific (Singapore),May 1992
12. V.J. Emery,*Phys.Rev.Lett.* **58**, (1987) 2794
13. G. Kotliar,A. Ruckenstein,*Phys.Rev.* B **57** (1968) 1362
14. A.M. Oleś,J. Zaanen,*Phys.Rev.* B**39**,(1989) 9175
15. G. Seibold,E. Sigmund,V. Hizhnyakov,*Phys.Rev.* B **48** (1993) 7537
16. J. Hubbard,*Proc.Roy.Soc.* A**276**, (1963) 238
17. L.M. Roth,*Phys.Rev.Lett.* **20**, (1968) 1431
18. H. Eskes,M.B.J. Meinders,G.A. Sawatzky,*Phys.Rev.Lett.* **67**, (1991) 1035
19. V. Drchal,*J.Phys.:Cond.Mat.* **1**, (1988) 4773
20. J. Kanamori,*Prog.Theor.Phys.* **30**, (1963) 275
21. J.R. Schrieffer,P.A. Wolff,*Phys.Rev.* **149** (1966) 491
22. P.G.de Gennes,*J.Phys.Radium* **23** 630
23. W.E. Brinkman, T.M. Rice,*Phys. Rev.* B**2**,(1976) 1324

# On the Role of Electronic and Chemical Phase Separation: Susceptibility and Conductivity Experiments on $La_{2-x}CuO_{4+\delta}$

R.K. Kremer[1], A. Simon[1], E. Sigmund[2,3], and V. Hizhnyakov[2,3*]

[1] Max-Planck-Institut für Festkörperforschung, Heisenbergstraße 1, 70506 Stuttgart, Germany
[2] Universität Stuttgart, Institut für Theoretische Physik III, Pfaffenwaldring 57, 70569 Stuttgart, Germany
[3] Technische Universität Cottbus, Lehrstuhl für Theoretische Physik, Postfach 101344, 03013 Cottbus, Germany

**Summary.** The mechanisms contributing to the metalization and the formation of the conducting and at low temperature superconducting phase in high–$T_c$ superconductors are discussed. Related experimental investigations on $La_{2-x}CuO_{4+\delta}$ are reviewed with special emphasis on our recent magnetic susceptibility and conductivity experiments on low excess oxygen doped ceramic and single crystal samples of $La_{2-x}CuO_{4+\delta}$. Particular attention was paid to the thermal history of the samples following our early observation that the superconducting phase can be significantly reduced by rapidly quenching the samples to temperatures below about 220 K. The reviewed experiments give very strong evidence that the formation of the conducting and at low temperature superconducting phase in high–$T_c$ superconductors follows the concept of percolative electronic phase separation.

## 1. Introduction

Despite the vast number of investigations and treatments covering almost all conceivable experimental and theoretical methods and techniques, high–temperature superconductivity remained an intricate phenomenon which up to now successfully withstood to unveil its origins and roots. This insight, on the other hand, leaves us with a curious discrepancy since judging from the phenomenon itself high–temperature superconductivity experimentally appears to be a comparatively simply accessible and to a wide extend *stable* phenomenon which even survives in samples being far from optimal conditions. Even more, nature seems to adjust the conditions for high–$T_c$superconductivity e.g. chemical compositions necessary to achieve or preserve this stability and to make high-$T_c$ superconductivity a widely *insensitive* and *resistant* phenomenon. For example, one very illuminating experimental observation in this respect is the formation of the 90 K and 60 K plateaus upon varying the oxygen content in $YBa_2Cu_3O_{7-y}$. Another high–$T_c$ system which exhibits a related behavior is excess oxygen doped lanthanum cuprate, $La_{2-x}CuO_{4+\delta}$.

---

* Permanent address: Tartu University, Institute of Physics of Estonian Academy of Sciences, Riia 142, 2400 Tartu, Estonia.

It was the initial discovery of high-$T_c$ superconductivity that by replacing about 8 % of the trivalent La atoms in $La_{2-x}CuO_4$ by divalent Ba (or Sr) and hence doping the Cu–O planes with holes superconducting compounds with a transition temperature of about 40 K can be produced [1]. Shortly after this finding it was shown that also by excess oxygen doping (i.e. $O_4 \rightarrow O_{4+\delta}$, $\delta \approx 1$ %) a bulk superconductor can be generated. Despite the low doping level, bulk superconductors with $T_c$ close to 40 K are generated which is very similar to the transition temperature of optimally Sr-doped $La_{2-x}CuO_4$ compounds [2]. Besides the occurrence of superconductivity, another phenomenon of high interest is that superconductivity and antiferromagnetic ordering coexist in excess oxygen doped $La_{2-x}CuO_{4+\delta}$.

The high–$T_c$ superconductor $La_{2-x}CuO_{4+\delta}$ exhibits interesting properties which were extensively discussed before [3]. For example, we reported that the formation of the superconducting phase can very effectively be suppressed by rapidly quenching $La_{2-x}CuO_{4+\delta}$ ($\delta \approx 0.02$) samples from room temperature to temperatures below about 200 K [4]. From the time scales involved we inferred that the construction of the superconducting phase is a diffusion controlled process which, as demonstrated in a subsequent experiment, can be supported by a strong polarizing magnetic field applied at 220 K. A hint on a thermal history dependence of the diamagnetic signal was first found by Yoshizaki et al. [6] but misinterpreted as due to an anomalous enhancement of the superconducting fractional volume due to field cooling the sample in the measuring field. A closer inspection of our thermal history dependent experiments and related observations [5] gave strong evidence that the process of the formation of the metallic phase ('metalization') is electronically driven and is consistent with the theoretical predictions of the model of percolative electronic phase separation proposed shortly after the discovery of high–$T_c$ superconductivity [7].

Chemically, $La_{2-x}CuO_{4+\delta}$ likewise exhibits some intricate properties which seem to be intimately connected to the occurrence of superconductivity. For $La_{2-x}CuO_{4+\delta}$ a chemical or structural *phase separation* was found by Jorgensen et al. by powder neutron diffraction experiments and subsequently studied intensively by numerous authors [8]. Samples which were loaded with excess oxygen in high pressure oxygen atmosphere on cooling below 320 K besides the reflections related to $La_2CuO_4$ displayed sharp additional reflections. This finding was interpreted as due to a segregation of the sample into two chemically different phases as a consequence of oxygen migration. The fraction of the new phase of the total sample volume was sizeably large and ranged up to 60 %. The new phase was attributed to oxygen enriched areas in the sample and found always present when superconductivity was observed, and was not detected in nonsuperconducting samples. The oxygen enriched areas were estimated to have a size of several thousand Angstroms.

From two phase refinements of neutron single crystal diffraction results the positions of the excess oxygen atoms in the $La_{2-x}CuO_4$ structure were successfully located on interstitial positions in the $La_2O_2$ sublattice, and they repel surrounding oxygen atoms [9].

Despite these clarifying and conclusive results the puzzle remains why occupation of only a tiny number of the available interstitial sites in the enriched phase ($\approx$ 2–3 %) and the local character of the phase separation [10] areas of a coherent homogeneous phases are built up which extend over several hundred lattice constants and are larger than the coherence lengths of neutrons and X–rays.

From detailed study of $La_{2-x}CuO_{4+\delta}$ samples which were oxygen loaded by the room–temperature electrochemical oxidation technique [11] it was proposed that the chemical phase separation is the consequence of a miscibility gap in the phase diagram which extends from $\delta \approx 0.01$ to 0.06 and on the temperature axis up to above room temperature [12]. A physical explanation for the occurrence of the miscibility gap in the phase diagram is not available at present. In view of this suggestion our initial observation that the superconducting fraction in excess oxygen loaded samples depends significantly on the thermal history of the samples, in particular, that superconductivity can be suppressed by rapidly quenching the samples from room to liquid helium temperatures, might superficially be interpreted as due to a suppression of chemical phase separation and the formation of an oxygen enriched phase. If we adopt the view that the chemical phase separation locally adjusts the proper oxygen concentration to provide optimal hole doping we are left with the question about the underlying *electronic* mechanisms that in such a chemically highly inhomogeneous and disordered media lead to metalization and superconductivity at low temperatures.

In this contribution we shall describe some new experimental results obtained by investigating the magnetic and conductivity properties of excess oxygen loaded $La_{2-x}CuO_{4+\delta}$ single crystals and ceramic samples. The experiments were directed towards an investigation of the recovery of superconductivity in initially rapidly quenched samples when annealing or equilibrating such samples to the temperature regime (150 K to 230 K) relevant for establishing superconductivity. Our results give additional evidence that beyond the chemical phase separation an electronic phase separation of percolative character is responsible for the metalization. Additional experimental support for the proposed electronic phase separation in high–$T_c$ superconductors has been obtained recently from photodoping experiments [13], Mössbauer spectroscopy [14], microwave absorption [15], neutron scattering [16] and EPR experiments [17, 18].

Before describing the experiments in detail we shall briefly review theoretical approaches to the physics of an electronic phase separation.

## 2. Electronic Phase Separation

As a necessary prerequisite of high–$T_c$ superconductivity an insulator–to–metal transition ('metalization') of the insulating parent compounds has to be initiated. It is brought about by hole– (also electron–) doping of the Cu–O planes and at higher doping levels completely destroys the long range antiferromagnetic correlation of the Cu moments in the Cu–O planes.

It is commonly believed that phase separation plays an important role in this metalization process. A promising way to gain more insight into the insulator–to–metal transition is to investigate the low doping regime where antiferromagnetic correlation is only locally disturbed such that one can hope to be able to prepare and to study independent quasiparticles induced by the doping process.

### 2.1 Quasiparticles

One scenario for the ground state of a hole doped antiferromagnetically coupled Cu–O plane comprises quasiparticles which as an essential ingredient have nonzero spin and magnetic momentum. It proposes that the doped holes initially generate *spin–polarized* Cu–O clusters (typically involving 5 to 10 Cu atoms) [7]. The underlying physical idea is that a hole may move more freely in a plaquette of Cu–O atoms if locally the antiferromagnetic alignment of the Cu spins in such a plaquette is broken and a spin–aligned ('spin–polarized') Cu–O cluster is formed. For example, flipping one spin creates a plaquette of five 'aligned' spins. A hole in such a plaquette gains energy by delocalization. The spin–polarized clusters themselves have a considerably lower mobility due to the small probability of spin–flip processes which have to accompany the motion of a cluster. Some direct experimental evidence for such spin–polarized clusters has emerged recently. Inelastic neutron scattering experiments on $TmBa_2Cu_3O_{6.95}$ give strong support for low energy $Cu^{2+}$ spin excitations originating from magnetic Cu–clusters of the size of a pentanuclear cluster [20]. EPR measurements on $La_{2-x}CuO_{4+\delta}$ reveal a temperature dependence of the linewidth of the spectra which has been interpreted in terms of unresolved transitions between the Zeeman levels of spin–polarized clusters [21]. Our observation that the formation of the superconducting phase is enhanced by a strong polarizing magnetic field applied at 220 K points to the fact that magnetic quasiparticles are involved in the formation of the superconducting phase [4].

Other models that treat hole doping of Cu–O planes take a different point of view [22]: In general, they consider the holes as highly mobile quasiparticles which may hop to neighboring sites provided that those are not already occupied.

## 2.2 Phase Separation and Metalization

Phase separation and metalization in the model proposing spin–polarized clusters as relevant quasiparticles comes about when by increasing the hole doping level clusters are linked together to build up a fractal clusters. The spin–polarized clusters become unstable with respect to spin depolarization when growing in size. The underlying mechanism is a lowering of the Fermi energy of the clusters by depolarization [23]. As a result in such cluster aggregates spin polarization vanishes and metalized pathways in the Cu–O planes are formed. Beyond the percolation threshold conductivity and at low temperatures superconductivity can occur. The process of metalization is essentially governed by *percolative* mechanisms and therefore depends crucially on the dimensionality of the underlying lattice structure since e.g. three dimensional percolation due to the larger coordination number appears at approximately three times smaller concentrations than two dimensional percolation.

In this model the electronic state is strongly inhomogeneous on the microscopic scale; the conducting (superconducting) subphase penetrates the insulating (AF–ordered) subsystem like 'water in a sponge'. Such an inhomogeneous electronic structure has an essential effect on pairing interactions and critical temperature because carriers of the conducting subsystem can now interact with the spin and phonon excitations of the insulating subsystem. One result of this novel situation is a strong enhancement of the interaction of the holes with long–wavelength optical phonons which leads to a strong distortive–type Cooper pairing [24].

Macroscopic electronic phase separation in models assuming the holes as (neutral) mobile band–holes comes about as a generic feature and is due to the attraction of holes embedded in an antiferromagnetic background. Recently it was suggested that if the holes are treated properly as charged particles long–range Coulomb repulsion between the holes frustrates a macroscopical phase separation and may lead to large–amplitude, low–energy fluctuations in the hole density at intermediate length scales ('clumping' of holes) [25]. The relevance of such structures for high–$T_c$ superconductivity needs to be established.

## 2.3 Counterions; chemical versus electronic phase separation

An essential but still controversial question concerns the mutual dependence of the chemical and electronic phase separation. Hole mobility requiring charge compensation viz. mobility of the counterion might support or even initiate a chemical phase separation. Mobility of the counterions, on the other hand may considerably promote the hole mobility, at least in the low doping limit, and by this support electronic phase separation.

As a first consequence of Coulombic attraction to the counterions the holes spatially are confined to the vicinity of the doping ions. However, owing to the large dielectric constant in these perovskite–like materials Coulomb attraction is significantly reduced. In fact, Chen et al. estimate the binding energy of the hole to the oxygen impurity in $La_{2-x}CuO_{4+\delta}$ to only 35–40 meV [29].

For $La_{2-x}CuO_{4+\delta}$ it is known from neutron diffraction experiments [9] that the excess oxygen atoms (O4 in the notation of ref. [9]) resides at an interstitial site close to the apical oxygen atoms (O3) belonging to the $La_2O_2$ layer which, on the other hand, is slightly displaced from its old position. The observed distance is d(O3–O4) = 1.59 Å. Distances in diatomic oxygen species are well known [26]. They range from 1.12 Å via 1.21 Å, 1.28 Å, 1.49 Å for $O_2^+$, $O_2$, $O_2^-$ and $O_2^{2-}$, respectively. These distances closely follow Pauling's bond length–bond order relationship [27] which for the distance 1.59 Å yields a bond order $n \approx 0.5$ [28]. In terms of a local bond description the highest antibonding molecular orbital $\sigma_p^*$ is half occupied. So, although each excess oxygen atom initially dopes two holes into system by bonding of the excess oxygen to an apex oxygen atom one hole becomes localized while the second hole dopes the $CuO_2$ plane.

There are several obvious consequences of this special bonding situation:

(i) The weak O–O bond allows for a high mobility of the excess O atom via bond breakage and formation (in contrast to diffusion of $O^-$). This may explain the appreciable rate of diffusion of the oxygen atoms even at or below room temperature such that e.g. quantitative and very effective anodic oxidation of $La_{2-x}CuO_4$ in a KOH electrolyte solution can be performed [30].

(ii) Each excess oxygen atom creates one hole in the adjacent $CuO_2$ plane as Sr does. In recent experiments this was in fact confirmed as long as $\delta \leq 0.08$ [31]. However, since the $\sigma_p^*$ states add to the top of the valence band the holes created have both O p–Cu d as well as $\sigma^*$ ($O_2$) character.

(iii) As a consequence, the holes are delocalized from the $CuO_2$ planes onto the apical O atoms, i.e. into the $La_2O_{2+x}$ region between adjacent Cu–O planes which allows hole transfer also between the planes.

In summary, in contrast to hole doping with Sr or Ba atoms insertion of excess oxygen into $La_{2-x}CuO_4$ not only introduces holes but also leaves mobile counterions. Holes and doping oxygen atoms are only weakly bound to each other. Owing to the special character of the excess oxygen as interstitial atom with weak oxygen–oxygen bonding a hole transfer between Cu–O planes is likely and this would distinguish $La_{2-x}CuO_{4+\delta}$ from the Sr or Ba doped systems. There the holes remain quasi two–dimensional and consequently a much higher doping level for the two–dimensional percolation is required.

## 3. Experimental

Ceramic and single crystalline samples of $La_{2-x}CuO_{4+\delta}$ were doped with excess oxygen by slowly cooling them in air after the final sintering process ('as–prepared' samples) or loaded with larger amounts of excess oxygen by annealing at 550 °C in a 200 atm high pressure oxygen atmosphere. Although we did not determine the precise excess oxygen dopinglevel, a comparison of our data with the results of other investigations indicates that our samples were in the *low doping regime* with $\delta$ approximately or less than 0.02 i.e. the oxygen content provided samples at the lower boundary value of the proposed miscibility gap in the phase diagram. Oxygen distribution in the single crystal was judged to be homogeneous from the sharpness of the Néel peak in the susceptibility which was found at 240 K (maximum temperature) with $\pm$ 22 K half width at half maximum (see. fig. 4.2b).

The temperature dependence of the electrical resistance was determined via standard 'four-probe' techniques. Contacts were applied as sputtered gold dots which were connected to gold wires by silver epoxy resin. Thermal quenching of the sample was performed by fast quenching in liquid nitrogen followed by insertion into the cryostat which was precooled to liquid helium temperature. Resistance measurements were performed while slowly (1 K/min) raising and lowering the temperature of the sample. Quenching and slow cooling/heating cycles were done successively keeping the contact geometry unchanged.

The magnetic susceptibilities (B = 100 G) were acquired with a MPMS Quantum Design magnetometer. To study the influence of thermal treatments on the phase separation two different experimental procedures were performed with special attention to the thermal history of the samples ('equilibration and annealing experiments'):

*Quenching to $T_{eq}$ and subsequent equilibration at $T_{eq}$:* The samples were always quenched from room temperature to $T_{eq}$ and kept there ('equilibrated') for a specified time interval, $\tau$. Subsequently they were further quenched to 5 K, and after applying the magnetic field the diamagnetic signal was measured at several temperatures, $T_M$, as indicated in the figures. All temperature changes and equilibration above 50 K were performed in zero field.

*Quenching to 5 K and subsequent annealing to $T_{an}$*: At first, the samples were quenched from room temperature to 5 K by lowering them into the precooled cryostat. Subsequently, the samples were annealed to $T_{an}$. They were kept at $T_{an}$ for a specified time interval, $\tau$, to promote the formation of the superconducting phase ('construction experiment') and in the following quenched from $T_{an}$ to 5 K again. After applying the magnetic field the diamagnetic signal was measured at several temperatures, $T_M$, as indicated in the figures.

A systematic variation of $T_{eq}$ and $T_{an}$ allows to find out the temperature ranges relevant for establishing superconductivity which is probed at the temperatures $T_M$ ($\leq$ 45 K).

*Isochronal annealing* experiments at a fixed $T_{an}$ were realized by performing the annealing procedure repeatedly with a fixed $\tau$ (3 min or 10 min). The diamagnetic signal was probed at 5 K.

# 4. Results and Discussion

## 4.1 Conductivity experiments

Figure 4.1 displays the results of typical thermal history dependent resistivity measurements on an 'as–prepared' and a high–pressure oxygen loaded sample.

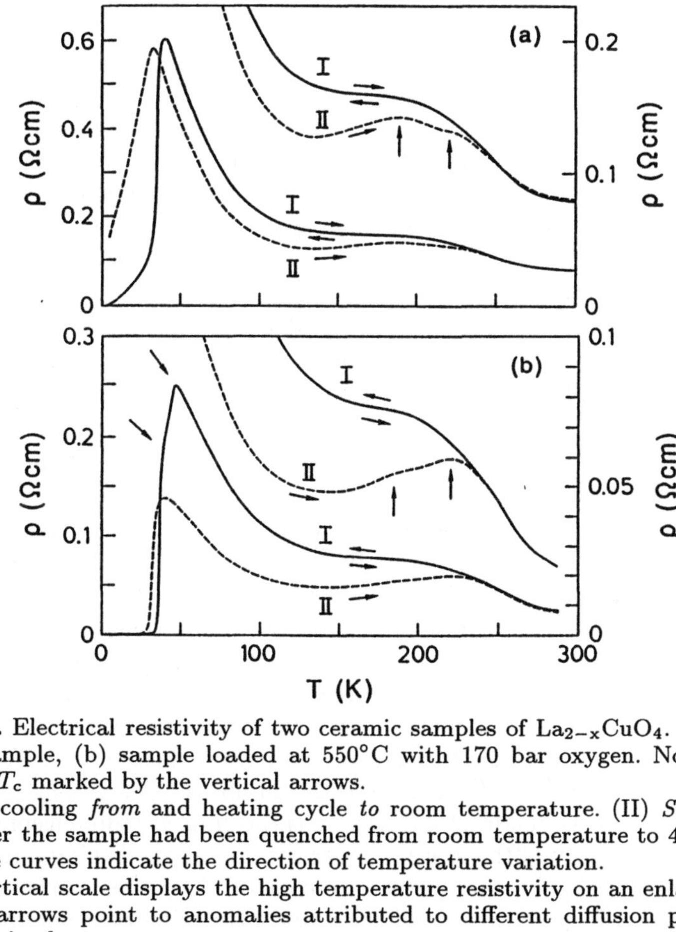


**Fig. 4.1.** Electrical resistivity of two ceramic samples of $La_{2-x}CuO_4$. (a) 'as prepared' sample, (b) sample loaded at 550°C with 170 bar oxygen. Note the two different $T_c$ marked by the vertical arrows.
(I) *Slow* cooling *from* and heating cycle *to* room temperature. (II) *Slow* heating cycle after the sample had been quenched from room temperature to 4 K. Arrows along the curves indicate the direction of temperature variation.
Right vertical scale displays the high temperature resistivity on an enlarged scale. Vertical arrows point to anomalies attributed to different diffusion processes as discussed in the text

Independent of the particular oxygen content there are a number of features in the temperature dependence common for both samples:

(i) The resistivity curves show the typical temperature dependence of excess oxygen loaded samples. Below room temperature the resistivities rise towards a maximum ('weak localization maximum') and below drop to vanish at low temperatures. However, there is a characteristic deviation between the resistivities obtained when heating the previously rapidly quenched sample and the resistivities measured during slow cooling or heating cycles [32]. The latter within experimental errors yield identical resistivities which between 220 K and 150 K exhibit a broad shoulder. The resistivities measured after first rapidly quenching the sample followed by slow heating also pass through broad, however, structured maxima between 230 and 170 K. Below 220 K resistivities are significantly smaller than those measured with slow temperature variation. This behavior was generally observed in all ceramic samples investigated so far.

(ii) The low–temperature maxima shift drastically depending on the thermal history. The maxima observed during slow cooling/heating cycles appear at significantly higher temperatures (50 K compared to 40 K in sample (b)) and zero–resistance is also shifted to higher temperatures proving the *stabilization* of the superconducting phase as a consequence of slow cooling rate.

Besides these common features there are a number of remarkable differences brought about by the increase in excess oxygen doping:

(i) Increasing the oxygen content by the high pressure treatment reduces the resistivity and also leads to a larger diamagnetic signal in the susceptibility (not shown here). Obviously due to a very low oxygen content the 'as–prepared' sample when quenched does not reach bulk superconductivity down to 4 K (fig. 4.1a, curve II).

(ii) The resistivity maxima at low temperatures change their shape and also their relative magnitude when raising the oxygen content.

(iii) The resistivity curve of the sample oxidized at displays further structure in the transition to superconductivity, apparently indicating two different critical temperatures (see fig. 4.1b, indicated by arrows).

We have discussed the different behavior of the resistivities of the quenched and slowly cooled/heated samples in detail before and given arguments that it reflects the fractal percolative character of the phase separation [5, 34]. These arguments based on the fact that the critical concentration, $c_{cr}$, for percolation and with it bulk metalization of fractal clusters with (finite) size $N$ is given by $c_{cr} = \mathrm{N}^{d/(D-d)}$ where $N$ is the number of elementary single–hole clusters (plaquettes) in a fractal cluster, $d$ is the *Hausdorf* dimensionality and $D$ the dimensionality of the system under consideration. Taking into account that the sizes of the fractal clusters are controlled by their surface energies [33] $N$ and $d$ will grow due to diffusion and attraction of elementary spin–polarized clusters leading to an increase of the superconducting fraction, with it of the $T_c$ values of the fractal coagulated clusters and of the percolation threshold $c_{cr}$. For a given concentration, $c$ ($< c_{cr}$) this leads, however, to an increase of the resistivity above $T_c$ since the total resistance in a metalized fractal network is not determined by the mean resistances of all clusters but rather by the largest resistance in a cluster pathway.

The double–peak feature in the maxima between 170 K and 230 K was most clearly resolved in a recent anisotropic conductivity investigations on single crystals [19] (fig. 4.2). However, it appears only in the resistivity measured perpendicular to the Cu–O planes of the quenched sample (d in fig. 4.2a) while the resistivity curves taken during slow cooling and heating the sample in the relevant temperature range remain flat. The double–peak feature is not induced by three–dimensional antiferromagnetic ordering as can be seen from the comparison with the susceptibility measurement (Fig. 4.2b) but rather arises from two different mechanisms of hole–cluster motion during the phase separation. It will be discussed in detail below together with related features showing up in the equilibrated and annealed magnetic susceptibilities at exactly the same temperatures.

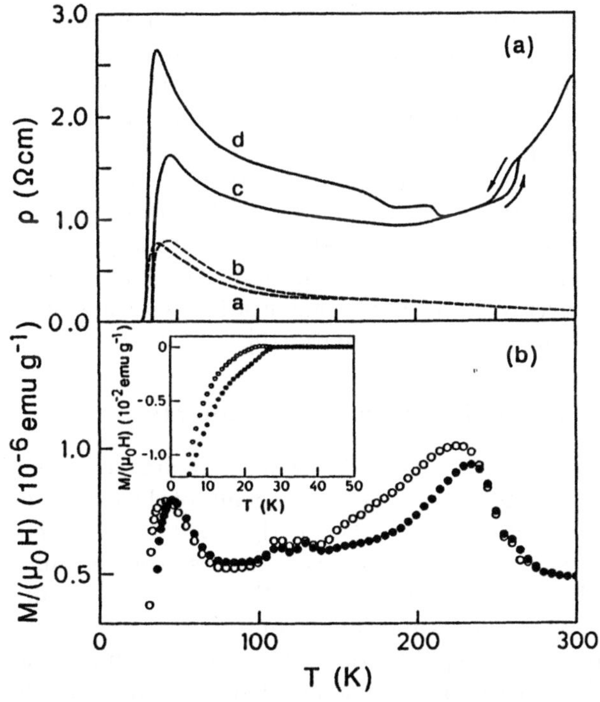


**Fig. 4.2.** (a) Resistivity of a $La_{2-x}CuO_{4+\delta}$ single crystal measured parallel (a and b) and perpendicular (c and d) to the Cu–O planes, respectively. Curves a and d were obtained on heating the rapidly *quenched* sample. Curves b and c were measured while *slowly* (1 K/min) cooling and heating the sample.
(b) Diamagnetic shielding of $La_{2-x}CuO_{4+\delta}$ measured perpendicular to the Cu–O planes. o refers to the susceptibility measured after quenching and • to the susceptibility obtained after slowly cooling the sample from room temperature to 5 K, respectively.
Notice the perfect coincidence of the characteristic features in resistivity and magnetization experiments.

Long range antiferromagnetic ordering rather causes a hysteresis loop in the resistivity when cooling and heating the sample. It appears between

245 K and 265 K at the high temperature wing of the susceptibility peak and is related to the Dzyaloshinski–Moriya tilting of the Cu moments out of the Cu–O planes discussed in detail by Thio et al. [35]. However, to the best of our knowledge the hysteresis loop associated to this process has not been detected before.

The resistivities reveal another very interesting property: As in the ceramic samples, a lowering of the resistivity of the quenched sample is seen for the component parallel to the Cu–O planes. However, the resistivity measured perpendicular to the Cu–O planes displays the opposite behavior with the resistivities of the slowly cooled and heated sample being considerably smaller. In addition, as has been reported before by Preyer et al. [36], the anisotropy of the resistivities, $\rho_\perp/\rho_\parallel$, decreases with decreasing temperature. Both features indicate an increasingly isotropic three–dimensional percolated metalized network which because of the reduced threshold concentration in three–dimensions can be improved through a diffusional growth process during the slow cooling procedure. Owing to the increased threshold concentration in two dimensions a metalized network with interconnected and conducting pathways can less well be established in the $CuO_2$ planes. There a system of larger coagulated metalized fractals with a large mean intercluster distance and high resistivity is formed initially. Quenching, on the other hand, freezes a more evenly distributed configuration with a reduced intercluster mean distance and better conductivity [19].

## 4.2 Equilibration and annealing experiments

The experiments described in the following were aimed at a more systematic investigation of the thermal history dependence of the magnetic susceptibility by *'equilibration and annealing'*. Our early result had marked temperatures around 220 K as critical for the relevant process and the time scale pointed to a diffusionally controlled mechanism [4]. The present results indicate that the process at 220 K splits into two subprocesses each characterized, when raising the annealing or equilibration temperature, at first by an increase and subsequently by a decay of the amount of superconducting phase.

When performing the measurements shown in Fig. 4.3a and b ('quenching to $T_{eq}$ and subsequent equilibration at $T_{eq}$') after probing the magnetization at $T_M$ the systems are brought to room temperature, and the samples loose the history of the preceding measuring cycle. The measured diamagnetic signal is built up during the equilibration interval at $T_{eq}$. Already at equilibration temperatures above $T \geq 150$ K superconducting phase starts to form and becomes rather pronounced above $T \geq 180$ K as marked by the strong increase in the diamagnetic response particularly for the latter process.

When comparing the curves obtained for different measuring temperatures $T_M$ (as indicated in the figures) it becomes obvious that the formation of the superconducting phase in the temperature range from 150 K to 180 K is responsible for the superconductivity at high $T_c$'s. This is demonstrated by

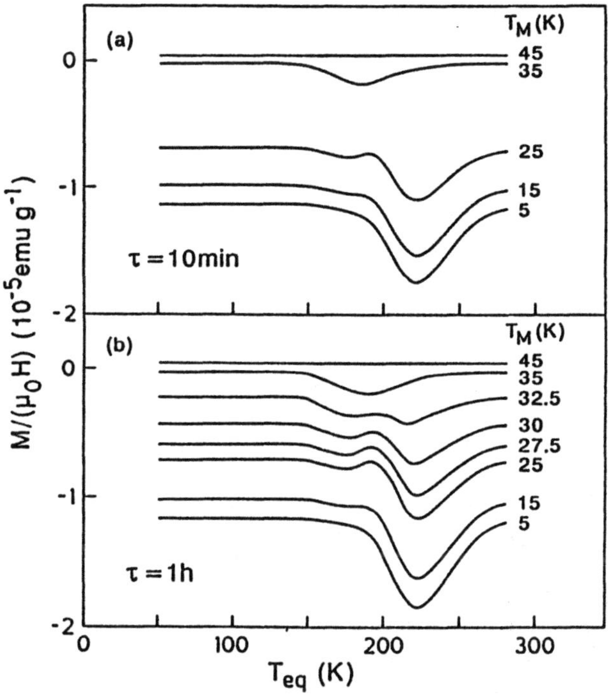


**Fig. 4.3.** Dependence of the diamagnetic signal of an 'as-prepared' $La_{2-x}CuO_{4+\delta}$ sample on the temperature $T_{eq}$ at which the sample was equilibrated after having been quenched from room temperature to $T_{eq}$. The diamagnetic signal was probed at the indicated temperatures, $T_M$. Equilibration time intervals, $\tau$, at $T_{eq}$ were 10 min (a) and 1 h (b).

the curve measured at $T_M = 35$ K showing the diamagnetic signal to built up at 150 K $\leq T_{eq} \leq$ 180 K ($T_c \geq 35$ K). The data not only reveal the creation but also the destruction of the superconducting phase with increasing temperature $T_{eq}$. The latter process prevailing the former at higher temperatures is different for systems with different $T_c$'s. Subsystems with higher $T_c$'s are thermally less stable than subsystems with lower $T_c$'s: some of the high–$T_c$ subsystems start to dissociate already at $T \approx 180$ K (cmp. fig. 4.3, $T_M = 25$ K), while low–$T_c$ subsystems decay above T $\approx$ 220 K (see fig. 4.3, $T_M = 5$ K). *'Annealing'* experiments reveal the same features and also prove two distinctly different temperature ranges as being relevant for generation and subsequent decay of superconductivity in the samples [5].

The necessary prerequisites for the occurrence of superconductivity are established not only in different temperature ranges, but also different microscopic processes prove responsible as becomes obvious from *'isochronal annealing'* experiments (fig. 4.4). Annealing $La_{2-x}CuO_{4+\delta}$ samples at different annealing temperatures ($T_{an}$= 170 K and 220 K) in the two temperature ranges reveals that the recovery of the diamagnetic shielding measured at

5 K occurs with an exponential time dependence with times constants being distinctly different for the two annealing temperatures. The diamagnetic signal due to annealing at 170 K is reestablished about an order of magnitude faster than at 220 K.

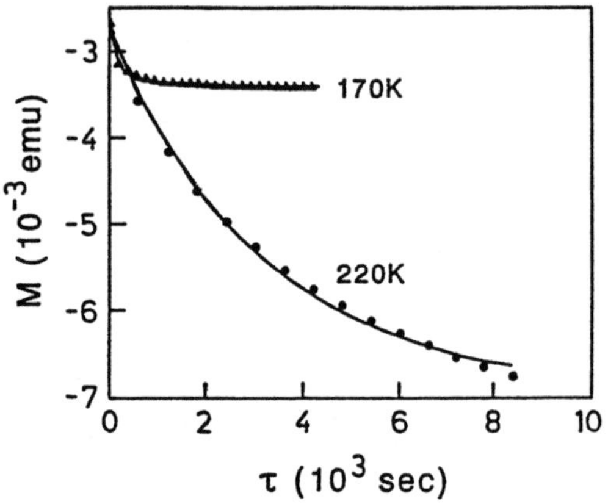


**Fig. 4.4.** Recovery of the magnetization at $T_M = 5$ K of an 'as–prepared' $La_{2-x}CuO_{4+\delta}$ sample probed in an *'isochronal annealing'* experiment at $T_{an}$= 170 K and 220 K. The magnetization (•) was measured at 100 G. Full line are fits with an exponential law.

Within the framework of percolative electronic phase separation two different hole–cluster motion processes were suggested to explain the occurrence of the two different features in the equilibration and annealing experiments [5]:

At lower temperatures hole motion and hence cluster mobility due to Coulombic attraction is spatially limited to the vicinity of the counterion. Therefore, although being a fast process, this motion quantitatively can only slightly improve the formation of a metalized network. Towards higher temperatures (T > 180 K) thermal fluctuations increasingly allow to overcome the attraction between the single hole and the charge compensating ions, and the motion of the hole is no longer space limited. This (unrestricted) hole (cluster) motion, can contribute significantly more to phase formation, and the magnitude of the diamagnetic signal is considerably intensified. However, it is a diffusionally controlled process and consequently phase formation happens much more slowly.

This interpretation is substantially supported by the conductivity results: The different hole–cluster motion processes display themselves very clearly in the resistivity curves (see e.g. fig. 4.2a, resistivity perpendicular to the Cu–O planes): 'Dips' in the resistivity occur exactly at the two temperature ranges where improved metalized pathways as prerequisites for the superconductivity are built up during the equilibration or annealing interval. Improvement of the metalized network due to the space limited motion apparently causes the resistivity to decrease at $\approx$ 180 K and space unlimited motion accounts for the resistivity dip at $\approx$ 210 K.

Most spectacular in the annealing and equilibration experiments is the dramatic decrease of the diamagnetic signal when annealing or equilibrating the samples at $T > 230$ K indicating a decomposition of the metalized phase. It is paralleled by a strong increase of the resistivity above 240 K (see fig. 4.2a. The decomposition of the metallic phase above 230 K is related to the decay of the chemical or structural phase separation induced by beginning oxygen atom mobility. The observed destruction of the superconducting phase by oxygen diffusion contradicts models according to which oxygen motion should always favor the phase formation.

## 5. Summary and Conclusions

Low excess oxygen doped $La_{2-x}CuO_{4+\delta}$ represents an outstanding example to study the microscopic mechanism leading to doping induced insulator–to–metal transition ('metalization') in high–$T_c$ superconductors. The process of 'metalization' preceding the appearance of superconductivity below about 40 K can very effectively be manipulated and influenced by a number of experimental treatments which are described in detail. The reviewed results of our conductivity and susceptibility measurements provide strong evidence that metalization is a consequence of an electronically driven phase separation comprising a linking of spin–polarized Cu clusters to establish a percolating conducting network of fractal character. Superconductivity can essentially be supported by this type of electronic inhomogeneous state allowing to have pairing interaction via phonons and spin excitations of the nonconducting subphase.

A series of related observations obtained for other high–$T_c$ systems suggest that the electronic phase separation may be a common feature of high–$T_c$ superconductors.

*Acknowledgement.* We thank E. Brücher, U. Herterich and N. Weishaupt for experimental assistance. Valuable discussions with M. Mehring, K.A. Müller and O. Schirmer are gratefully acknowledged.

## References

1. J.G. Bednorz and K.A. Müller, *Z. Phys. B* **64**, 189 (1986); J.G. Bednorz, M. Takashige, and K.A. Müller, *Europhys. Lett.* **3**, 379 (1987).
2. J. Beille, R. Cabanel, C. Chaillout, B. Chevalier, G. Demazeau, F. Deslandes, J. Etourneau, P. Lejay, C. Michel, J. Provost, B. Raveau, A. Sulpice, J.L. Tholence, and R. Tournier, C. R. Acad. Sc. Paris, t. 304, Série II, no. 18 (1987);

P.M. Grant, S.S.P. Parkin, V.Y. Lee, E.M. Engler, M.L. Ramirez, G. Lim, and R.L. Greene, *Phys. Rev. Lett.* **58**, 2482 (1987); R.L. Greene, H. Maletta, T.S. Plaskett, J.G. Bednorz, and K.A. Müller, *Solid State Comm.* **63**, 379 (1987).
3. International School of Solid State Physics, Proceedings of the 3rd Workshop: *Phase Separation in Cuprate Superconductors*, eds.: K.A. Müller and G. Benedek, World Scientific, Singapore (1993).
4. R.K. Kremer, E. Sigmund, V. Hizhnyakov, F. Hentsch, A. Simon, K.A. Müller, and M. Mehring, *Z. Phys. B — Condensed Matter* **86**, 319 (1992).
5. R.K. Kremer, V. Hizhnyakov, E. Sigmund, A. Simon, and K.A. Müller, *Z. Phys. B — Condensed Matter* **91**, 191 (1993).
6. R. Yoshizaki, H. Sawada, T. Iwazumi, and H. Ikeda, *Physica Physica C* **153–155**, 1495 (988).
7. V. Hizhnyakov, and E. Sigmund, *Physica C* **156**, 655 (1988).
8. J.D. Jorgensen, B. Dabrowski, S. Pei, D.G. Hinks, L. Soderholm, B. Morosin, J.E. Schirber, E.L. Venturini, and D.S. Ginley, *Phys. Rev. B* **38**, 11337, (1988).
9. C. Chaillout, S–W. Cheong, Z. Fisk, M.S. Lehmann, M. Marezio, B. Morosin, and J.E. Schirber, *Physica C* **158**, 183 (1989); C. Chaillout, J.Chenavas, S–W. Cheong, Z. Fisk, M. Marezio,B. Morosin, and J.E. Schirber, *Physica C* **170**, 87 (1990).
10. P. Zolliker, D.E. Cox, J.B. Parise, E.M. McCarron, and W.E. Farneth, *Phys. Rev. B* **42**, 6332 (1990).
11. A. Wattiaux, J.C. Park, J.C. Grenier, and M. Pouchard, *C.R. Acad. Sci. Paris* **310(II)**, 1047 (1990).
12. J.–.C. Grenier et al. in ref. [3], see also the lecture by D.C. Johnston et al. at this workshop.
13. G. Yu, A.J. Heeger, G. Stucky, N. Herron, and E.M. McCarron, *Solid State Commun.* **72**, 345 (1992); G. Yu, C.H. Lee, A.J. Heeger, N. Herron, and E.M. McCarron, *Phys. Rev. Lett.* **67**, 2581 (1991); G. Yu, C.H. Lee, A.J. Heeger, S.–W. Cheong, and Z. Fisk, *Physica C* **190**, 563 (1992).
14. J.A. Hodges, P. Bonville, P. Imbert, and G. Jéhanno, *Physica C* **184**, 259 (1991); J.A. Hodges, P. Bonville, P. Imbert, G. Jéhanno, and P. Debray, *Physica C* **184**, 270 (1991).
15. M. Mehring, in ref. [3].
16. J. Mesot, P. Allenspach, U. Staub, A. Furrer, H. Mutka, *Phys. Rev. Lett.* **70**, 865 (1993).
17. I.N. Kurkin, I.K. Salikhov, L.L. Sedov, M.A. Teplov, R.S. Zdanov, *Zh. Eksp. Teor. Fiz.*, to be published.
18. P.G. Baranov, and A.G. Badalyan, *Solid State Comm.* **85**, 987 (1993).
19. parts of the results can also be found in: E. Sigmund, V. Hizhnyakov, R.K. Kremer, and A. Simon, *Z. Phys. B — Condensed Matter*, in press.
20. A. Furrer, J. Mesot, P. Allenspach, U. Staub, F. Fauth, and M. Guillaume, lecture at this workshop.
21. G. Wübbeler, and O. Schirmer, *phys. stat. sol. b* **174**, K21 (1992).
22. V.J. Emery, S.A. Kivelson, and H.Q. Lin, *Phys. Rev. Lett.* **64**, 475 (1990); J.C. Phillips, *Phys. Rev. B* **39**, 7356 (1989); *Phys. Rev. B*40, 7348 (1989); *Phys. Rev. Lett.***64**, 1605 (1990); M. Marder, N. Papanicolaou, and G.C. Psaltakis, *Phys. Rev. B* **41**, 6920 (1990); C. Castellani, M. Grilli, and G. Kotliar, Phys. Rev. B, *43* **8000 (1991)**, .
23. G. Seibold, E. Sigmund, and V. Hizhnyakov, lecture at this workshop.
24. A. Bill, lecture at this workshop.
25. V.J. Emery and S.A. Kievelson, in ref. [3]; V.J. Emery and S.A. Kievelson, *Physica C* **209**, 597 (1993).

26. A.F. Wells, *Structural Inorganic Chemsitry*, Fourth Ed., Clarendon Press Oxford (1975), p.418.
27. L. Pauling, *The Nature of the Chemical Bond and the Structure of Molecules and Crystals*, Third Ed., Cornell University Press (1960).
28. The bond order is defined as half of the difference of the population of the bonding and antibonding states [27].
29. C.Y. Chen, N.W. Preyer, P.J. Picone, M.A. Kastner, H.P. Jenssen, D.R. Gabbe, A. Cassanho, and R.J. Birgeneau, *Phys. Rev. Lett.* **63**, 2307 (1989); C.Y. Chen, R.J. Birgeneau, M.A. Kastner, N.W. Preyer, and T. Thio, *Phys. Rev. B* **43**, 392 (1991).
30. J.–C. Grenier, A. Wattiaux, N. Lagueyte, J.C. Park, E. Margestaut, J. Etourneau, M. Pouchard, *Physica C* **173**, 139 (1991); J.–C. Grenier, A. Wattiaux, J.-P. Doumerc, P. Dordor, L. Fournes, J.-P. Chaminade, and M. Pouchard, *J. Solid State Chem.* **96**, 20 (1992).
31. P.G. Radaelli, J.D. Jorgensen, A.J. Schultz, B.A. Hunter, J.L. Wagner, F.C. Chou, D.C. Johnston, *Phys. Rev.* **48**, 499 (1993).
32. J. Ryder, P.A. Midgley, R. Exley, R.J. Beynon, D.L. Yates, L. Afalfiz, and J.A. Wilson, *Physica C* **173**, 9 (1991).
33. B.M. Smirnov, *Usp. Fiz. Nauk* **162**, 119 (1992): *Sov. Phys. Usp.* **35**, 37 (1992).
34. E. Sigmund, V. Hizhnyakov, R.K. Kremer, and A. Simon, *Z. Phys. B — Condensed Matter*, in press.
35. T. Thio, T.R. Thurston, N.W. Preyer, P.J. Picone, M.A. Kastner, H.P. Jenssen, D.R. Gabbe, C.Y. Chen, R.J. Birgeneau, and A. Aharony, *Phys. Rev. B* **38**, 905 (1988); T. Thio, C.Y. Chen, B.S. Freer, D.R. Gabbe, H.P. Jenssen, M.A. Kastner, P.J. Picone, N.W. Preyer, R.J. Birgeneau, *Phys. Rev. B* **41**, 231 (1990).
36. N.W. Preyer, R.J. Birgeneau, C.Y. Chen, D.R. Gabbe, H.P. Jenssen, M.A. Kastner, P.J. Picone, and T. Thio, *Phys. Rev. B* **39**, 11563 (1991).

# The Phase Diagrams and Doped-Hole Segregation in $La_2CuO_{4+\delta}$ and $La_{2-x}Sr_xCuO_{4+\delta}$ ($x \leq 0.15$, $\delta \leq 0.12$)

D.C. Johnston[1], F. Borsa[1], P.C. Canfield[1], J.H. Cho[1*], F.C. Chou[1], L.L. Miller[1], D.R. Torgeson[1], D. Vaknin[1], J. Zarestky[1], J. Ziolo[1], J.D. Jorgensen[2], P.G. Radaelli[2], A.J. Schultz[2], J.L. Wagner[2], S-W. Cheong[3], W.R. Bayless[4], J.E. Schirber[4], and Z. Fisk[5]

[1] Ames Laboratory and Department of Physics and Astronomy, Iowa State University, Ames, Iowa 50011, U.S.A.
[2] Argonne National Laboratory, Argonne, Illinois 60439, U.S.A.
[3] AT&T Bell Laboratories, Murray Hill, New Jersey 07974, U.S.A.
[4] Sandia National Laboratories, Albuquerque, New Mexico 87185, U.S.A.
[5] Los Alamos National Laboratory, Los Alamos, New Mexico 87545, U.S.A.

**Summary.** The magnetic and structural phase diagrams of the title systems are reviewed, with an emphasis on our recent results obtained from magnetic and structural neutron diffraction, thermogravimetric analysis, iodometric titration, magnetic susceptibility $\chi(T)$, and $^{139}$La nuclear quadrupole resonance (NQR) measurements. From measurements on electrochemically oxidized polycrystalline samples, the known miscibility gap in the system $La_2CuO_{4+\delta}$ is found to lie between $\delta \approx 0.01$ and 0.06 at low temperatures, with a maximum phase separation temperature $T_{ps} \approx 415$ K. Within the miscibility gap, the superconducting transition temperature $T_c$ of the oxygen-rich phase and the Néel temperature $T_N$ of the oxygen-deficient phase are constant at $\approx 32$ K and 250 K, respectively. Neutron diffraction measurements showed $T_{ps} = 260(5)$ K and $T_N = 245(3)$ K for a single crystal. Beyond the miscibility gap, two distinct superconducting phases are found in polycrystalline samples at $\delta \approx 0.06$–0.08 ($T_c \approx 32$–34 K) and 0.11–0.12 ($T_c \approx 42$–45 K), separated by another two-phase region. The doped-hole concentration in the $CuO_2$ planes is found to be $p \approx 0.08$ holes/Cu and $\approx 0.16$ holes/Cu for the two phases, respectively. These data suggest that a large fraction of the excess oxygen atoms participate in oxygen-oxygen bonding in both phases. Measurements of $T_c$ *versus* pressure suggest pressure-induced changes in the doped-hole concentration in the $CuO_2$ layers. Superstructure reflections observed in the neutron diffraction patterns of both phases suggest spatial ordering of the excess oxygen atoms. In the $La_{2-x}Sr_xCuO_{4+\delta}$ system, the phase separation disappears by $x = 0.03$ for $\delta \approx 0.03$. Bulk superconductivity is found below $T_c = 40$ K for $0.01 \leq x \leq 0.15$ for maximally oxidized samples. For $\delta = 0$ and $0 < x < 0.08$, our $^{139}$La NQR and $\chi(T)$ data indicate that the doped holes condense into walls separating undoped *nanoscopic* domains, consistent with theory based on an electronic mechanism for phase separation in these systems. In the antiferromagnetic regime ($x < 0.02$) below $T_N$, the doped-hole spins are found to freeze at a temperature $T_f = (815\text{ K})x$, whereas in the spin-glass regime $0.02 < x < 0.08$, the spin-glass transition at $T_g \propto 1/x$ is found to arise from cooperative freezing of the dynamically-ordered mesoscopic undoped domains.

* Present address: Superconductivity Technology Center, Mail Stop K 763, Los Alamos National Laboratory, Los Alamos, NM 87545, U.S.A.

# 1. Introduction

The compound $La_2CuO_4$ is an insulating (at temperature $T = 0$) spin 1/2 Heisenberg antiferromagnet on a square lattice with a Néel temperature $T_N$ of about 300 K [1]–[3]. A remarkable feature of this compound is that strong dynamic two-dimensional (2D) short-range antiferromagnetic (AF) order persists in the $CuO_2$ plane to temperatures much greater than $T_N$. Excess oxygen can be incorporated to form $La_2CuO_{4+\delta}$ by subjecting $La_2CuO_4$ to high oxygen pressure at moderate temperatures [4]. For example, an oxygen pressure of 3 kbar at 575°C yields $\delta \approx 0.03$ [5]–[11]. It is now well-established that such $La_2CuO_{4+\delta}$ compositions phase-separate below $T_{ps} \sim$ 260–320 K into nearly stoichiometric $La_2CuO_{4+\delta'}$ ($\delta' \approx$ 0.00–0.01) with $T_N \approx 250$ K and oxygen-rich $La_2CuO_{4+\delta''}$ ($\delta'' \sim$ 0.05–0.08) which becomes superconducting below about 35 K [8]–[18]. The discovery of electrochemical [19]–[23] and chemical [24] oxidation techniques to produce polycrystalline $La_2CuO_{4+\delta}$ with good control of $\delta$ up to 0.12, at room temperature, has allowed much more extensive investigation of the phase diagram. Preliminary phase diagrams were proposed by Dabrowski et al. [12] and by Grenier et al. [22, 23].

It was soon realized that the phase separation below $T_{ps}$ in $La_2CuO_{4+\delta}$ may be electronically driven [25]. One mechanism is that the increase in kinetic energy of the doped holes due to confinement in the hole-rich regions is outweighed by the decrease in the magnetic exchange energy in the hole-deficient regions due to ejection of the holes. In $La_2CuO_{4+\delta}$, the long-range Coulomb repulsion of the holes in the hole-rich regions is neutralized by the simultaneous phase separation of the negatively-charged excess oxygen ions. Recently, Emery and Kivelson have offered additional arguments that the phase separation in $La_2CuO_{4+\delta}$ is electronically driven [26].

Of course, the electronic interactions presumed responsible above for the macroscopic phase separation in $La_2CuO_{4+\delta}$ should also be present in other of the lightly-doped cuprates, including for example $La_{2-x}Sr_xCuO_4$, in which the hole doping is due to replacement of La by Sr rather than incorporation of excess oxygen. In this case, macroscopic doped-hole segregation is "frustrated" [26] by the long-range Coulomb interaction between the doped holes, because the $Sr^{+2}$ ions are evidently immobile and cannot migrate along with the doped holes to neutralize the positive charge in the hole-rich regions. In the presence of these Coulomb interactions, the tendency towards phase separation is predicted to lead to a nanoscopic and possibly dynamic inhomogeneous distribution of the doped holes [26], although the length scale of the inhomogeneities and the specific spatial distribution expected for the doped holes are not yet clear. Experimental verification of an inhomogeneous doped-hole distribution in $La_{2-x}Sr_xCuO_4$ would provide strong evidence for the electronic scenario for the known phase separation in $La_2CuO_{4+\delta}$ and for the postulated frustrated phase separation in $La_{2-x}Sr_xCuO_4$.

Herein, our recent experimental work is reviewed which addresses issues associated with the above macroscopic phase separation in $La_2CuO_{4+\delta}$ and the predicted inhomogeneous doped-hole distribution in $La_{2-x}Sr_xCuO_4$, via magnetic susceptibility $\chi(T)$, $^{139}$La nuclear quadrupole resonance (NQR), magnetic and structural neutron diffraction, thermogravimetric analysis (TGA) and iodometric titration measurements.

## 2. The $La_2CuO_{4+\delta}$ System

### 2.1 Relationship Between $T_{ps}$ and $T_N$

As noted above, phase separation has been found to occur at a temperature $T_{ps} \sim 260$–$320$ K for $\delta$ in a range estimated to be 0.01 to 0.08 into an oxygen-rich superconducting phase with $T_c \approx 35$ K and $\delta'' \sim 0.05$–$0.08$ and a nearly stoichiometric (hole-free) phase with $\delta' \approx 0.00$–$0.01$ exhibiting long-range AF order with $T_N \approx 250$ K. However, thus far, a combined study of the phase separation transition and the magnetic transition had not been carried out on the same sample to ascertain the relative values of $T_{ps}$ and $T_N$. We have carried out such a study utilizing structural and magnetic neutron diffraction measurements on a single crystal of $La_2CuO_{4+\delta}$ with $\delta \approx 0.03$ [27]. The crystal was synthesized by subjecting a $La_2CuO_4$ crystal grown in a CuO flux to 3 kbar oxygen pressure at 575°C; the specific heat and superconducting and normal state magnetization of this crystal had been previously studied [17]. These neutron measurements yielded $T_{ps} = 260 \pm 5$ K, in agreement with the value ($259 \pm 2$ K) found from the specific heat measurements [17]. The neutron measurements yielded a value for $T_N$ ($245 \pm 3$ K) about 15 K lower than $T_{ps}$. The ordered magnetic moment in the AF phase at 10 K was found to be $0.30 \pm 0.05 \mu_B$/Cu, similar to values found previously for other $La_2CuO_{4+\delta}$ powders and crystals with similar $T_N$ (for a review, see Ref. [1]).

The above study on a particular crystal of $La_2CuO_{4+\delta}$ indicated that $T_{ps}$ and $T_N$ are not identical, but rather that $T_{ps} > T_N$. Our most recent structural work on powder samples with $\delta = 0.02$ to 0.04, produced with the electrochemical oxidation technique at room temperature (see below), showed that the maximum $T_{ps}$ is about 415 K [28], far higher than the maximum $T_N$ of 328 K observed for $La_2CuO_4$ [29]. These results do not support the suggestion sometimes given that the phase separation transition is driven by long-range AF ordering, since in this case, one would expect that $T_N \geq T_{ps}$. From specific heat measurements on a single crystal of $La_2CuO_4$ with $T_N = 304(1)$ K, there is no feature above the 0.1% level associated with long-range AF ordering at $T_N$ [30]. Analysis of this result was consistent with the occurrence of strong short-range AF order at and above $T_N$. Due to the long magnetic correlation length $\xi_{2D}$ within the $CuO_2$ planes at $T_N$ [3], the difference in energy and entropy between the long-range ordered state at and below $T_N$ and the short-range ordered state at $T_N$ is very small; this explains why the long-range AF transition has negligible influence on the specific heat. Thus, the interactions giving rise to the three-dimensional

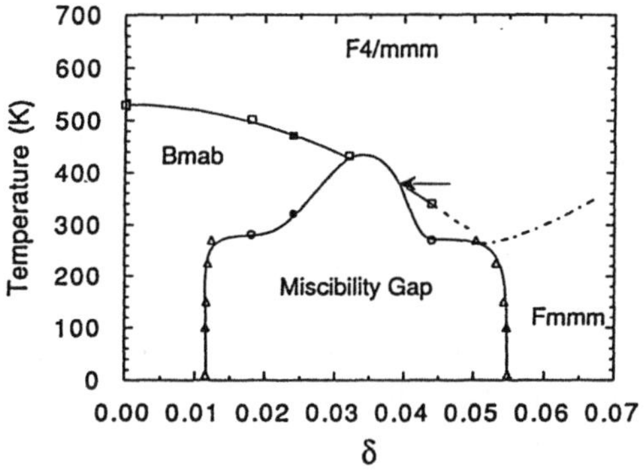


**Fig. 2.1.** Structural phase diagram of $La_2CuO_{4+\delta}$ [28]. The arrow indicates the tetragonal-to-orthorhombic phase transition temperature of the Fmmm phase in the phase-separated $\delta \approx 0.032$ sample. The dot-dashed line indicates the position of a possible additional phase line.

(3D) long-range AF ordering transition would not be expected to act as a significant driving force for phase separation. However, it is important to note that the electronic mechanism for phase separation proposed by Emery and coworkers only requires the presence of 2D short-range, not 3D long-range, AF ordering [25, 26].

## 2.2 $La_2CuO_{4+\delta}$ Phase Diagram and Structures

In the following, we summarize our studies of $La_2CuO_{4+\delta}$ samples obtained using the room-temperature electrochemical oxidation technique [19]. Upon increasing $\delta$ from 0.02 to 0.07 in polycrystalline samples, a single $T_c$ = 32–34 K is observed, where the volume fraction of superconducting material increases with increasing $\delta$ [31, 32]. At the same time, the peak in $\chi(T)$ associated with long-range AF ordering at 250 K in the oxygen-deficient phase decreases in magnitude and disappears by $\delta = 0.07$ [31, 32]. Together, these results indicate that the miscibility gap ends by $\delta = 0.07$. Our recent powder neutron diffraction study of samples with $\delta = 0.018$, 0.032 and 0.044 determined that the miscibility gap limits at low temperature are $\delta \approx 0.012$ and 0.055, close to previous estimates, as shown in Fig. 2.1 [28], where the maximum $T_{ps}$ observed is 415 K.

Further increasing $\delta$ to 0.11–0.12 results in samples with a single $T_c$ of 42–45 K [31, 32]. Samples with intermediate values of $\delta$ between 0.08 and 0.11 showed two distinct $T_c$'s of 32–34 K and 42–45 K, indicating that a two-phase region exists in this $\delta$ range. Extensive x-ray diffraction measurements of the lattice parameters *versus* $\delta$ clearly demonstrate the existence of this two-phase region [21]. Neutron diffraction structure measurements on powders with $\delta = 0.08$ ($T_c$ = 32 K) and $\delta = 0.12$ ($T_c$ = 42 K) and on a single crystal (see below) with $\delta \approx 0.1$ ($T_c$ = 40 K) showed that these compositions remain single phase down to 10–16 K, with no evidence for phase separation, confirming that these compositions are beyond the miscibility gap in the phase diagram [33]. These structural data of the metallic oxygen-rich phases also showed that the excess oxygen atoms are located between the two LaO layers in the unit cell, as found previously for the oxygen-rich phase in phase-separated $La_2CuO_{4.03}$ [10][11], and that the crystallographic space group (of the average unit cell) is Fmmm. This space group is different from that (Cmca) proposed for the oxygen-rich phase in Refs. [10] and [11]. The importance of the space group is that it reflects the manner in which the apical oxygens in the $CuO_6$ octahedra distort to accommodate the interstitial excess oxygen atoms situated between the LaO layers.

Our neutron diffraction data for $\delta$ = 0.08–0.12 also demonstrated the existence of a long-period superstructure both parallel and perpendicular to the $CuO_2$ planes [33], presumably associated with crystallographic ordering of the excess oxygen atoms. Evidence for superlattices has also been obtained from electron diffraction measurements [22]–[24]. Oxygen ordering in the metallic phases may be related [31] to the anomalous variation with $\delta$ in the orthorhombicity originally observed by Rudolf et al. [21], in which the orthorhombicity first decreases with $\delta$ (for the first oxygen-rich phase), but then increases to a value larger than in the parent $La_2CuO_4$ starting material for the highest $T_c$, highest $\delta$ phase.

To date, the available neutron and x-ray diffraction data do not distinguish the two oxygen-rich phases, apart from their compositions and lattice parameters. In particular, it is not yet clear why there should be a two-phase region separating them, or what structural features distinguish them. TGA measurements of the two phases suggest that there are two inequivalent environments for the excess oxygen atoms, where the relative proportion of the two is different in the two oxygen-rich phases [31].

### 2.3 The Oxidation State(s) of the Excess Oxygen and the Doped-Hole Concentration in the $CuO_2$ Plane

Measurements of the oxidation state(s) of the excess oxygen atoms can help to clarify their coordinations. This can be done indirectly by measuring the number of holes ($p$) donated to the $CuO_2$ layers by the intercalated oxygen. For example, the excess oxygen contents, determined by TGA measurements, and hole doping levels in the $CuO_2$ planes, determined by iodometric titration, for the two powder samples measured by neutron diffraction were found to be $\delta = 0.08(1)$, $p = 0.08(2)$ and $\delta = 0.12(1)$, $p = 0.14(2)$ [33]. These data suggest that each excess oxygen atom contributes about 1 hole in the phase with $\delta = 0.08$, and that the next 0.04 excess oxygen atoms incorporated to form the phase with $\delta = 0.12$ contribute about 2 holes per excess oxygen atom. An independent measure of the hole doping levels can be inferred from $\chi(T)$ data, which are shown in Fig. 2.2. By comparing the data for $\delta = 0.07$

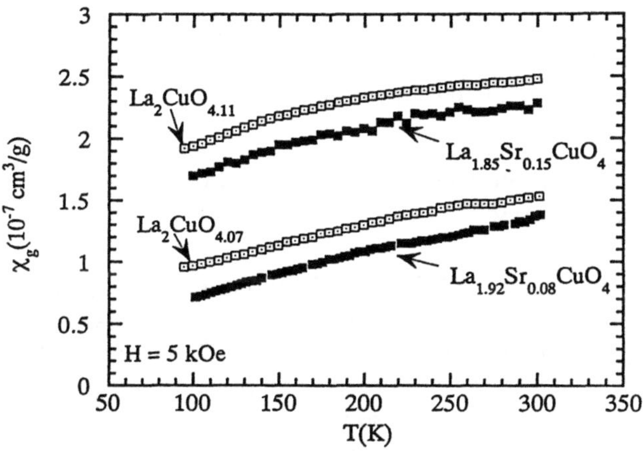


**Fig. 2.2.** Magnetic susceptibility $\chi_g$ *versus* temperature $T$ for polycrystalline $La_2CuO_{4+\delta}$ samples with $\delta = 0.07$ and 0.11 [31]. Data for conventionally prepared $La_{2-x}Sr_xCuO_4$ ceramics with $x = 0.08$ and 0.15 are shown for comparison [31].

and 0.11 with those of $La_{2-x}Sr_xCuO_4$ samples with $x = 0.08$ and 0.15, we infer that the hole doping level is $p = 0.08(1)$ for $\delta = 0.07(1)$ and $p = 0.16(1)$ for $\delta = 0.11(1)$, consistent with the above TGA and titration results. Together, these data suggest that most of the excess oxygen atoms in $La_2CuO_{4.07}$ and about 0.07 of those in $La_2CuO_{4.11}$ are partially bonded to the apical oxygen atoms of the $CuO_6$ octahedra, consistent with some proposed models for the detailed structure [10, 11, 33].

## 2.4 Single Crystal Synthesis and Characterization

We have found that oxygen can be intercalated at room temperature into single crystal $La_2CuO_4$ using the electrochemical oxidation technique, although low currents (10 $\mu$A) and long times (1-2 months) are necessary to attain homogeneous crystals of the $La_2CuO_{4+\delta}$ metallic phases [34]. Similar to the powders, two distinct $T_c$'s were found with increasing $\delta$; however, the maximum $T_c$ observed for the crystals was about 40 K, substantially lower than the maximum $T_c$ of 45 K observed for the powders. For a maximally oxidized crystal with $T_c = 40$ K, the resistivity anisotropy, $\rho_c/\rho_{ab}$, increases from about 100 at 290 K to about 300 at 50 K.

The room temperature magnetic susceptibility anisotropy $\chi_c - \chi_{ab}$ of the $La_2CuO_{4+\delta}$ phase with $T_c = 40$ K was found to be $9.7 \times 10^{-5}$ cm$^3$/mol [34]. This value is significantly larger than that of, e.g., $La_{1.9}Ba_{0.1}CuO_4$ ($6.6 \times 10^{-5}$ cm$^3$/mol) [35]. Since this anisotropy is thought to arise primarily from the orbital Van Vleck susceptibility of the Cu ions [1, 35], which in turn is strongly affected by the local crystalline electric field, one possible explanation is that the environments of the Cu atoms in the undoped and oxygen-doped crystals are significantly different [34]. These observations are clearly relevant to the validity and details of the various structural models for the metallic oxygen-rich $La_2CuO_{4+\delta}$ phases.

## 2.5 Dependence of Superconducting Properties on Temperature and Pressure History

It has been known since 1987 that the $T_c$ and apparent superconducting fraction $F$ in samples of $La_2CuO_{4+\delta}$ which show phase separation as described above are dependent on the magnetic field and/or thermal histories of the samples [18],[36]–[40]; these samples were all prepared by treating $La_2CuO_4$ in $O_2$ gas at elevated temperatures, so that $\delta$ in these samples is expected to be in the range 0.01–0.03. Rapidly quenching the samples to low temperatures (e.g., 4 K) from 300 K causes $F$ determined from magnetization measurements to be small. Subsequent annealing at higher temperatures increases both $T_c$ and $F$, where the optimum annealing temperature in several of the studies was found to be about 200–220 K. These results are qualitatively consistent with expectation. After an ideal quench, there should be no phase separation; a composition like $La_2CuO_{4.03}$ would not be expected to be superconducting, because the hole-doping level ($p \approx 0.03$ holes/Cu, see above) is too small. By annealing at temperatures high enough that the excess oxygen can diffuse and allow macroscopic phase separation to occur, regions of sufficiently high oxygen concentration are formed to support bulk superconductivity. Of course, when interpreting magnetization measurements of the superconducting properties, one must keep in mind that the superconducting regions exhibit only a partial Meissner effect if the superconducting particle size is less than a London penetration depth ($\sim 10^3 - 10^4$ Å), and/or if flux pinning is sufficiently strong. In addition, in the phase-separated mixtures, temperature and magnetic field history dependence of the ferromagnetic component of the ordered moment in the oxygen-deficient phase may play a role in the observed history-dependent magnetization data.

It is further necessary to know whether the pure bulk oxygen-rich superconducting phases show such history effects (in the absence of phase separation). The electrochemical oxidation technique for synthesizing $La_2CuO_{4+\delta}$ has enabled new insight to be gained regarding this issue. Examples of our results are shown in Fig. 2.3 [31], where $\chi$(T) data are shown for a composition $\delta \approx 0.03$ within the miscibility gap, and for a sample with $\delta \approx 0.07$ which remains single phase to low temperature. Data are shown for each sample after quenching or slow-cooling from 300 K to 6 K in a SQUID magnetometer. Whereas the superconducting properties of the highest $T_c$ phase $La_2CuO_{4.11}$ are not influenced by our different thermal treatments (not shown), those of $La_2CuO_{4.07}$ and the phase-separated mixture containing this phase are significantly affected as demonstrated in Fig. 2.3. The latter result for $La_2CuO_{4.07}$ may be due to oxygen ordering effects which depend on the thermal treat-

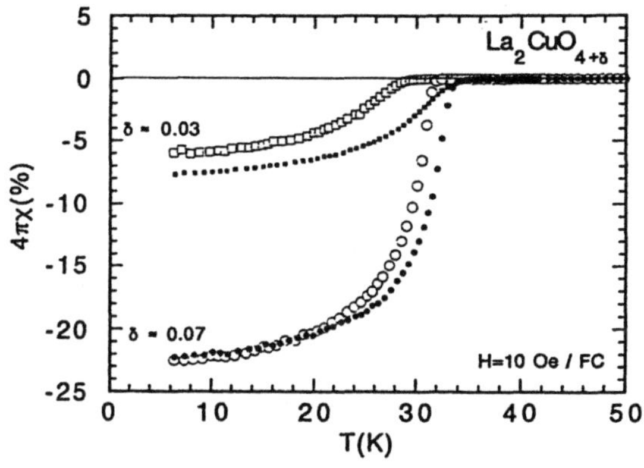


**Fig. 2.3.** Superconducting Meissner fraction $4\pi\chi$ *versus* temperature $T$ for $La_2CuO_{4+\delta}$ samples with $\delta = 0.03$ and 0.07 [31]. The open symbol data were taken after quenching from 300 K to 6 K, and the filled symbols after slow-cooling from 300 K to 6 K. The data were taken in a 10 Oe field upon warming after field-cooling to 6 K.

ment. In any case, the reported thermal history-dependent results are at least partially explained by this behavior of the pure $La_2CuO_{4.07}$ phase.

Further information regarding the superconducting properties of $La_2Cu$-$O_{4.07}$ and $La_2CuO_{4.11}$ was obtained from measurements of the variation of $T_c$ with pressure [41]. The results are strikingly different from those observed in $La_{1.85}Sr_{0.15}CuO_4$, which has a similar maximum value for $T_c$. In the latter compound, $T_c$ increases with pressure at a rate of about 0.2-0.3 K/kbar. In single phase polycrystalline $La_2CuO_{4+\delta}$ samples with $T_c = 43$ K, on the other hand, $T_c$ is found to be nearly independent of pressure up to 5 kbar. In contrast, $dT_c/dp$ for the phase with $T_c = 32$ K increases at about the same rate as in $La_{1.85}Sr_{0.15}CuO_4$. The value of $dT_c/dp$ for these samples is essentially independent of whether the pressure is changed at 300 K or at 60 K. Additional measurements were made on $La_2CuO_{4+\delta}$ single crystals oxygenated under high oxygen pressure at 575°C, and are therefore expected to be phase-separated at low temperature. For a crystal with $T_c = 28$ K, $dT_c/dp$ increases from $0.0 \pm 0.15$ K/kbar to $0.3 \pm 0.2$ K/kbar if the pressure is changed at 300 K and 60 K, respectively, whereas for a crystal with $T_c = 39$ K, $dT_c/dp$ did not depend on the temperature at which the pressure was changed. These results are not yet understood in detail, but suggest that a positive contribution to the pressure derivative ubiquitous to the copper oxide superconductors is canceled to a varying degree by a negative derivative due to pressure-induced changes in the charge transfer to the $CuO_2$ planes from the excess oxygen atoms [41].

## 3. The $La_{2-x}Sr_xCuO_{4+\delta}$ System

We carried out a study of this system with $\delta \approx 0.03$ and varying $x \leq 0.03$ to study how the phase separation observed in the $La_2CuO_{4+\delta}$ system evolves with Sr doping $x$ at fixed excess oxygen content [42]. From magnetization studies, $T_c$ decreases linearly with $x$, and is nonzero (20 K) as the superconducting fraction approaches zero at $x = 0.03$. Thus, the known phase separation for $x = 0$ and $\delta \approx 0.03$, which produces the oxygen-rich superconducting phase, no longer occurs above $x = 0.03$ for $\delta \approx 0.03$ [42]. This result is consistent with the observation that ceramic $La_{1.84}Sr_{0.16}CuO_{4.019}$, produced by high-pressure (2.3 kbar) and high temperature (575°C) oxygen annealing, does not show phase separation below room temperature [43].

By electrochemically oxidizing $La_{2-x}Sr_xCuO_4$ to the maximum extent possible using our techniques, we found bulk superconductivity with $T_c =$ 40 K for the entire range $0.01 \leq x \leq 0.15$ [31], whereas conventionally prepared ceramics show optimum superconducting properties only for $x = 0.15$–$0.20$ for which $p$ is in the same range. Iodometric titration and $\chi(T)$ measurements revealed that the maximum doping level in the $CuO_2$ planes for the oxygen-doped samples was $p \approx 0.16$ holes/Cu, independent of $x$, consistent with the composition-independent $T_c$ [31].

### 3.1 The $La_{2-x}Sr_xCuO_4$ System ($\delta = 0$)

If the interactions responsible for the *macroscopic* phase separation in $La_{2-x}$-$Sr_xCuO_{4+\delta}$ are electronic rather than elastic in nature and intrinsic to the $CuO_2$ planes, then these interactions should presumably also be present in other of the cuprate systems and should not depend on the details of how the $CuO_2$ planes become doped with holes. In the following, we review our studies of the $La_{2-x}Sr_xCuO_4$ ($\delta \approx 0$) system which reveal the presence of an inhomogeneous distribution of the doped holes on a *nanoscopic* scale, consistent with the concept of "frustrated" [26] electronically-driven phase separation.

**3.1.1 The Néel-ordered Regime: $0 \leq x \leq 0.02$.** One would expect that the detailed spatial distribution of the doped holes in the $CuO_2$ planes should have a determinative influence on the variation of $T_N$ with doped-hole concentration. To examine this issue, we carried out a study of $T_N(x)$ with high resolution in $x$ for samples of $La_{2-x}Sr_xCuO_4$ with $0 \leq x \leq 0.03$ which were annealed in 1 bar $N_2$, 1 bar $O_2$ or 230 bar $O_2$ [42]. Magnetic susceptibility $\chi(T)$ data for the series annealed in $N_2$ are shown in Fig. 3.1(a) [42]. None of these samples showed any trace of superconductivity, and hence no trace of the macroscopic phase separation discussed above. $T_N$ is the temperature of the peak in $\chi(T)$ [44]. For this series, we found $T_N(x) = T_N(0)[1 - (x/x_c)^n]$ with $T_N(0) = 301(9)$ K, $x_c = 0.0212(3)$ and $n = 1.9(2)$ [42]. A power-law dependence for $T_N(x)$ is expected from finite size scaling theory, which predicts

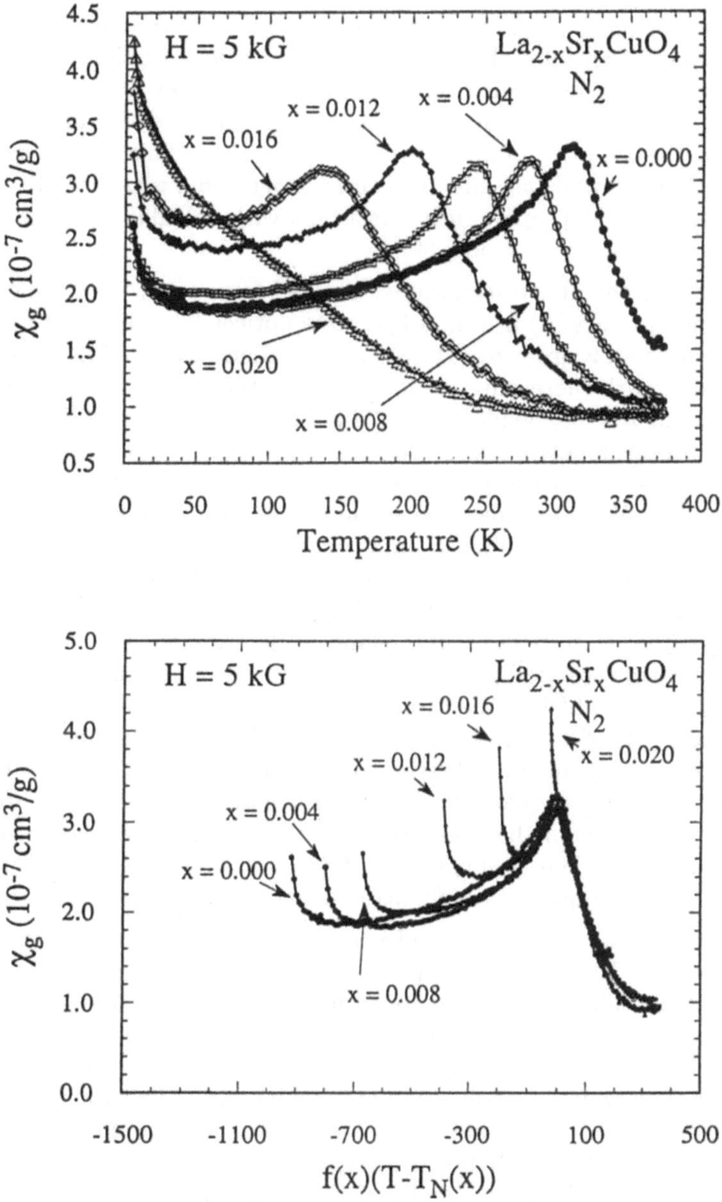


**Fig. 3.1.** (a) Magnetic susceptibility $\chi_g$ *versus* temperature for various samples of $La_{2-x}Sr_xCuO_4$ annealed in 1 bar $N_2$ [42]. (b) The $\chi_g(T)$ data in (a) replotted with a scaled temperature axis (see text) [42].

that the measured $T_N$ is limited by the finite size (linear dimension $L$) of a system. Our $T_N(x)$ data, then, suggest that the doped holes break the AF system into nearly decoupled undoped domains of size $L$; the doped holes evidently condense into walls separating these undoped domains. Utilizing mean-field theory, one obtains $L(x) \propto x^{-n/2} = 1/x$. This $L(x)$ dependence indicates that the walls are effectively one-dimensional with a width which is independent of $x$ [42], apparently consistent with calculations based on the 2D single-band Hubbard model [45].

Support for the above scenario is obtained from analysis of the composition dependence of $\chi(T)$ in Fig. 3.1(a). One notices that the peak associated with the canted AF ordering [1] becomes broadened, as well as depressed in temperature, with increasing Sr doping $x$. At the same time, the height of the peak is almost independent of $x$. This suggests that the $\chi(T)$ data for all AF compositions follow a scaling relationship $\chi(x,T) = \chi\{f(x)[T - T_N(x)]\}$, where $f(x)$ is an empirically determined scaling function. This scaling is indeed obeyed well both above and below $T_N$ as shown in Fig. 3.1(b), where the data in Fig. 3.1(a) collapse onto a single curve (except for the lowest temperature data associated with Curie-like upturns). In the analysis of the scaling function $f(x)$ above $T_N$, we identified $L(x)$ with the 2D AF correlation length $\xi_{2D}(x, T = 0)$ in the $CuO_2$ plane as determined, e.g., from inelastic magnetic neutron scattering measurements. With this identification, we found that $L(x) \propto 1/x$, which is the same dependence found from the above independent analysis of $T_N(x)$. An important conclusion from this work is that inelastic magnetic neutron scattering measurements yield a $\xi_{2D}(T = 0)$ which is not just a statistical quantity, but rather is a measure of the real-space size of undoped domains in the samples. As further support, we showed that the temperature-dependent $\xi_{2D}(x,T)$ data for $x = 0.02$–$0.04$ reported by Keimer et al. [46] could be quantitatively fitted using this real-space picture [42].

The above analysis of $T_N(x)$ and $\chi(x,T)$ were carried out for temperatures above about 50 K, where the doped holes are known to be delocalized [47]. However, below about 50 K, the doped holes become localized, possibly near the Sr sites, with a localization length of about 8 Å [47]. When the doped holes localize with decreasing $T$, they may redistribute themselves randomly. Such a redistribution would probably destroy the domain walls without destroying the long-range AF order, and cause a crossover in $L(x)$ from $L(x) \propto 1/x$ above 50 K to $L(x) \propto 1/\sqrt{x}$ below 50 K, where the latter relation is expected for a random distribution of holes if $L(x)$ measures the average distance between holes [3].

Previous $^{139}$La NQR measurements of $La_2CuO_{4+\delta}$ [48] and $La_{2-x}Ba_x$-$CuO_4$ [49] in the AF regime have revealed the presence of a strong peak in the longitudinal ($R_1$) and/or transverse relaxation rates at a temperature $T_f$ of about 10 K. We have confirmed and extended these studies via $^{139}$La pulse NQR measurements on $La_{2-x}Sr_xCuO_4$ samples with high resolution in $x$, for $x$ in the AF regime ($x < 0.02$). Representative $R_1(T)$ data are

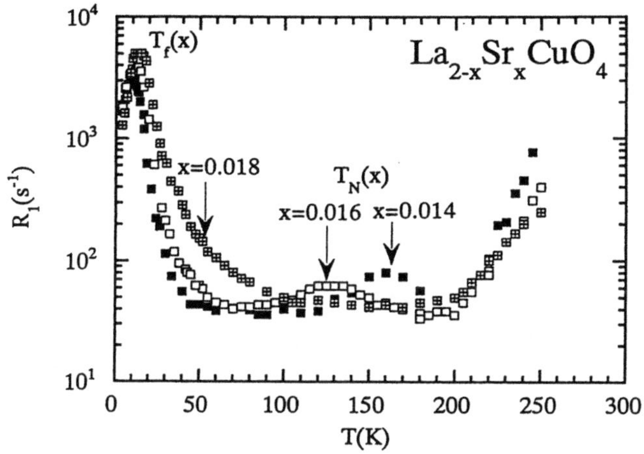


**Fig. 3.2.** $^{139}$La nuclear spin-lattice relaxation rate $R_1$ *versus* temperature $T$ for representative samples of $La_{2-x}Sr_xCuO_4$ [65]. The arrows denote the Néel temperatures for the respective samples.

shown in Fig. 3.2 [50]. A comparison of the initial nuclear magnetization decay rates at $T_f$ of the $2\nu_Q$ and $3\nu_Q$ transitions, and the lack of any features in the temperature dependences of the quadrupole frequency $\nu_Q$ or the linewidth at $T_f$, indicated that the peak in $R_1$ at $T_f$ arises from a magnetic mechanism. The peak height of $R_1$ at $T_f$ was shown to be consistent with slowing down into the MHz range of the autocorrelation time of the spins associated with the doped holes. Additional measurements showed that the static effective internal magnetic field at the La sites, due to the long-range AF order of the Cu spins, was unaffected at $T_f$. It was concluded from these results that the peaks in $R_1$ at $T_f$ arise from freezing of the effective spin degrees of freedom associated with the doped holes and that these degrees of freedom are independent of, and coexist with, the (longitudinal) Cu spin degree of freedom which orders below $T_N$ [50]. The data appear to be consistent with the transition at $T_f$ being a transverse spin (component) freezing transition [51] (see also below).

As shown in Fig. 3.3, $T_f$ is proportional to $x$ with a large slope: $T_f = (815\ K)x$. In the context of known spin glasses, both features are remarkable. In Fig. 3.3, there is no discernable threshold in $x$ for freezing of the doped hole spins, in contrast to the usual case where a percolation threshold must first be exceeded. This indicates that the effective interaction between the effective doped hole spins is of long range and that mean field theory should be applicable, consistent with $T_f \propto x$. The slope $T_f/x = 815$ K is comparable to the exchange interaction (≈1500 K) between adjacent Cu spins in the $CuO_2$ planes, suggesting that the doped hole spin interaction is indirect and mediated by the AF Cu spin system [50].

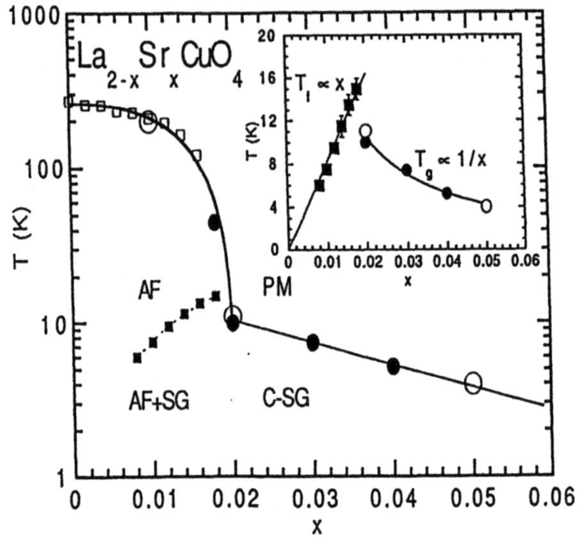


**Fig. 3.3.** Magnetic phase diagram for lightly-doped $La_{2-x}Sr_xCuO_4$ [42,50]. Inset: Linear plots of the spin-freezing temperatures $T_f$ (for $x < 0.02$) and $T_g$ (for $x > 0.02$) *versus* $x$. Abbreviations: PM (paramagnetic), AF (antiferromagnetic), SG (spin glass), CSG (cluster spin glass).

These observations provide a testing ground for microscopic theory of hole-doped $CuO_2$ planes. Recent microscopic theory [52, 53] predicts that (dilute) localized holes couple to each other via a long range perturbation induced in the AF background, leading to freezing of the holes' effective spins with $T_f \propto x$; the predicted $T_f$ values are close to our observed ones. From the work of Gooding et al. [52], the spin degrees of freedom associated with the doped holes are the transverse (to the ordering direction) components of the Cu spins near the hole which are induced by the presence of the hole on a plaquette of four oxygen ions in the $CuO_2$ layer. This is consistent with our data, which indicate that the spin degrees of freedom associated with the doped holes are distinct from the (longitudinal) Cu spin degree of freedom which orders below $T_N$. It is apparently the localization of the doped holes below about 50 K [47] which allows for subsequent observation of ordering of their spin degrees of freedom below $T_f$. It would be of interest to see whether the microscopic ferron theory [54] can account for the spin glass freezing at $T_f$.

**3.1.2 The Spin Glass Regime $0.02 \leq x \leq 0.08$.** Strong peaks in the $^{139}$La nuclear relaxation rates similar to those in Fig. 3.2 (observed in the AF regime) have been reported below a temperature $T_g$ of about 10 K for compositions of $La_{2-x}Ba_xCuO_4$ in the so-called spin glass regime with $0.02 \leq x \leq 0.08$ [49, 55]. A static internal magnetic field develops (only) below $T_g$ [56]; these authors suggested that the spin glass state should be viewed as a disordered antiferromagnet [56], as proposed by Aharony et al. [57]. In the $La_{2-x}Sr_xCuO_4$ system, evidence for such a low temperature magnetic transition was found from muon spin rotation ($\mu$SR) [58]–[61], neutron scattering [61, 62], $\chi(T)$ and electron spin resonance (ESR) [63], and $^{139}$La NQR [64] measurements.

To clarify the nature of the spin glass state, we carried out high precision $^{139}$La NQR measurements of $La_{2-x}Sr_xCuO_4$ for $0.02 \leq x \leq 0.08$ [65]. The nuclear spin-lattice relaxation rate ($R_1$) peaks were strong at the spin glass ordering temperature $T_g$, as for $x < 0.02$, but were sharper; the variation of $T_g$ with $x$ is shown in Fig. 3.3, where one sees that $T_g \propto 1/x$. Analysis of $R_1$(T) indicated that the transition at $T_g$ is associated with cooperative freezing of weakly-coupled dynamically-ordered undoped domains (which are evidently separated by walls of hole-rich material) [65]. The nature of the hole distribution is therefore the same as discussed above for the AF regime at temperatures above about 50 K, where the holes are delocalized [47]. In the present case, this distribution evidently persists to low temperatures $T < T_g$, which suggests that the doped holes may have a much longer localization length at low temperatures for $x > 0.02$ than for $x < 0.02$. These ideas appear to be consistent with the specific heat measurements of $La_{2-x}Ba_xCuO_4$ by Kumagai et al. [66], which show that the linear heat capacity coefficient is zero for $x < 0.02$ (in the AF regime), but rapidly increases with $x$ for $x > 0.02$ (in the spin-glass regime). In addition, localized holes would be expected to lead to strongly-coupled domains in the spin-glass regime [67], in contrast to the weakly-coupled domains inferred from the NQR measurements.

Our NQR analysis [65] revealed that the size of the undoped domains is $L(x) \propto 1/\sqrt{x}$ for $x > 0.02$; this is the same composition dependence as for the 2D AF correlation length $\xi_{2D}(x, T = 0)$ found from inelastic neutron scattering experiments for $x \geq 0.02$ [3], and as for $x < 0.02$ above $\sim$ 50 K, we identify $L(x)$ with $\xi_{2D}(x, T = 0)$. By using the molecular field approximation result, $T_g \approx J'(\xi_{2D}/a)^2$, where $J'$ is the effective Cu-Cu interplane coupling and $a = 3.8$ Å is the intraplane Cu-Cu distance, one expects $T_g \propto L^2 \propto 1/x$, as in Fig. 3.3, for a system where $\xi_{2D}$ is limited to a finite size $L$.

From Fig. 3.3, a distinct crossover at $x = 0.02$ occurs in the composition dependence of the spin freezing temperature: $T_f \propto x$, whereas $T_g \propto 1/x$. This novel behavior indicates that there is a crossover at $x = 0.02$ in the nature of the spin-glass transition. According to the above discussion, for $x < 0.02$ the transition at $T_f$ is due to freezing of the spin degrees of freedom associated with the doped holes, superimposed on the preexisting AF long-range order. For $x > 0.02$, on the other hand, the system undergoes a collective phase transition from a paramagnetic state, with 2D dynamical correlations in weakly-coupled undoped mesoscopic domains, to a cluster spin glass phase in which these domains freeze due to their mutual interaction.

## 4. Concluding Remarks

The issue of whether the observed phase separation in the $La_2CuO_{4+\delta}$ system is electronically or elastically driven is important to understand, because the answer is relevant to determining which interactions are most important to the observed physical properties of the cuprates, including the anomalously high superconducting transition temperatures. We have studied and characterized the phase diagram of this system using a variety of techniques. Based on these data alone, however, it is difficult to make a definitive case for one or the other mechanism for phase separation.

Extensive measurements on the lightly-doped $La_{2-x}Sr_xCuO_4$ system with $x < 0.08$, containing little or no excess oxygen, indicated that the doped holes condense into walls separating undoped domains, resulting in an electronically and magnetically inhomogeneous phase on an $x$-dependent length scale of $\sim 100$ Å. For the region $0.02 < x < 0.08$, we find a cluster-spin-glass phase at low temperatures. Both of these observations confirm the predictions of theory based on an electronic mechanism for phase separation, where in this system the phase separation is "frustrated" [26] because the $Sr^{+2}$ dopant ions cannot migrate along with the doped holes. As a result, macroscopic phase separation does not occur; rather, an inhomogeneous distribution of the doped holes occurs on a nanoscopic length scale which depends on the Sr doping level $x$. This interpretation allows a wide range of magnetic susceptibility and NQR data for this system to be understood self-consistently. As discussed in the Introduction, these results therefore strongly suggest that the phase separation transition in $La_2CuO_{4+\delta}$ is also electronically driven.

Looking to the future, we anticipate that for certain temperature and composition regimes, it will be possible to obtain more direct proof for the existence of the undoped domains separated by walls of hole-rich material via real-space images, as well as to obtain information about the domain dynamics from these images. These doped-hole inhomogeneities in the $CuO_2$ planes should also be observable in insulating members of other classes of cuprate superconductors.

Finally, one would of course like to experimentally determine whether these effects are observable in the metallic cuprate compositions exhibiting high temperature superconductivity. Several recent results appear to be relevant and important in this regard. In particular, various investigators have found evidence from Cu NMR for the existence of two types of inequivalent Cu atoms in the metallic phase(s) of $La_2CuO_{4+\delta}$ and $La_{2-x}Sr_xCuO_4$ [18, 68], which may arise from different electronic environments due to doped-hole inhomogeneity. Similarly, Mesot and coworkers have found neutron spectroscopic evidence for doped-hole inhomogeneity in both the insulating and metallic composition regimes of $ErBa_2Cu_3O_x$ [69].

*Acknowledgement.* We are grateful to V.J. Emery, R.J. Gooding, V. Hizhnyakov, R.K. Kremer, M. Mehring, P. Rudolf and E. Sigmund for helpful discussions with some of us. Ames Laboratory is operated for the U.S. Department of Energy by Iowa State University under Contract No. W-7405-Eng-82. The work at Ames was supported by the Director for Energy Research, Office of Basic Energy Sciences. The work at Argonne National Laboratory was supported by the U.S. Department of Energy under Contract No. W-31-109-Eng-38 (JDJ, AJS) and by the National Science Foundation Office of Science and Technology Centers Grant No. DMR-91-20000 (PGR, JLW). The work performed at Sandia National Laboratories was supported by the U.S. Department of Energy under Contract No. DE-AC04-76DP00789 and at Los Alamos National Laboratory under the auspices of the U.S. Department of Energy Office of Basic Energy Sciences.

## References

1. D.C. Johnston, J. Magn. Magn. Mater. 100 (1991) 218.
2. E. Manousakis, Rev. Mod. Phys. 63 (1991) 1.
3. R.J. Birgeneau and G. Shirane, in *Physical Properties of High Temperature Superconductors I*, edited by D.M. Ginsberg (World Scientific, Singapore, 1989), Ch. 4, pp. 151-211.
4. J. Beille, R. Cabanel, C. Chaillout, B. Chevalier, G. Demazeau, F. Deslandes, J. Etourneau, P. Lejay, C. Michel, J. Provost, B. Raveau, A. Sulpice, J.-L. Tholence and R. Tournier, Compt. Rend. Acad. Sci. (Paris), 304 (1987) 1097; J.L. Tholence, Physica 148B (1987) 353; P.M. Grant, S.S.P. Parkin, V.Y. Lee, E.M. Engler, M.L. Ramirez, J.E. Vazquez, G. Lim, R.D. Jacowitz and R.L. Green, Phys. Rev. Lett. 58 (1987) 2482.
5. J.E. Schirber, B. Morosin, R.M. Merrill, P.F. Hlava, E.L. Venturini, J.F. Kwak, P.J. Nigrey, R.J. Baughman and D.S. Ginley, Physica C 152 (1988) 121.
6. J. Zhou, S. Sinha and J.B. Goodenough, Phys. Rev. B 39 (1989) 12331.
7. B. Andraka, U. Ahlheim, J.S. Kim, G. Fraunberger, G.R. Stewart, B. Morosin, E.L. Venturini, D.S. Ginley and J.E. Schirber, Phys. Rev. B 42 (1990) 10016.
8. J.D. Jorgensen, B. Dabrowski, S. Pei, D.G. Hinks, L. Soderholm, B. Morosin, J.E. Schirber, E.L. Venturini and D.S. Ginley, Phys. Rev. B 38 (1988) 11337.
9. P. Zolliker, D.E. Cox, J.B. Parise, E.M. McCarron III and W.E. Farneth, Phys. Rev. B 42 (1990) 6332.
10. C. Chaillout, S.W. Cheong, Z. Fisk, M.S. Lehmann, M. Marezio, B. Morosin and J.E. Schirber, Physica C 158 (1989) 183.
11. C. Chaillout, J. Chenavas, S.W. Cheong, Z. Fisk, M. Marezio, B. Morosin and J.E. Schirber, Physica C 170 (1990) 87.
12. B. Dabrowski, J.D. Jorgensen, D.G. Hinks, S. Pei, D.R. Richards, H. B. Vanfleet and D.L. Decker, Physica C 162-164 (1989) 99.
13. E.J. Ansaldo, J.H. Brewer, T.M. Riseman, J.E. Schirber, E.L. Venturini, B. Morosin, D.S. Ginley and B. Sternlieb, Phys. Rev. B 40 (1989) 2555.
14. M.F. Hundley, J.D. Thompson, S-W. Cheong, Z. Fisk and J.E. Schirber, Phys. Rev. B 41 (1990) 4062.
15. P.C. Hammel, A.P. Reyes, Z. Fisk, M. Takigawa, J.D. Thompson, R. H. Heffner, S-W. Cheong and J.E. Schirber, Phys. Rev. B 42 (1990) 6781.
16. M.F. Hundley, R.S. Kwok, S.-W. Cheong, J.D. Thompson and Z. Fisk, Physica C 172 (1991) 455.
17. L.L. Miller, K. Sun, D.C. Johnston, J.E. Schirber and Z. Fisk, J. Alloys Compds. 183 (1992) 312.

18. P.C. Hammel, E.T. Ahrens, A.P. Reyes, J.D. Thompson, D.E. MacLaughlin, Z. Fisk, P.C. Canfield and J.E. Schirber, in *Phase Separation in Cuprate Superconductors*, edited by K.A. Müller and G. Benedek (World Scientific, Singapore, 1993), pp. 139-157.
19. A. Wattiaux, J-C. Park, J-C. Grenier and M. Pouchard, Compt. Rend. Acad. Sci. (Paris) 310 (1990) 1047.
20. J-C. Grenier, A. Wattiaux, N. Lagueyte, J.C. Park, E. Marquestaut, J. Etourneau and M. Pouchard, Physica C 173 (1991) 139.
21. P. Rudolf, W. Paulus and R. Schöllhorn, Adv. Mater. 3 (1991) 438; P. Rudolf et al., unpublished.
22. J.-C. Grenier, N. Lagueyte, A. Wattiaux, J-P. Doumerc, P. Dordor, J. Etourneau, M. Pouchard, J.B. Goodenough and J.S. Zhou, Physica C 202 (1992) 209; J.C. Grenier, A. Wattiaux, J.P. Doumerc, P. Dordor, L. Fournes, J.P. Chaminade and M. Pouchard, J. Solid State Chem. 96 (1992) 20.
23. J-C. Grenier, A. Wattiaux and M. Pouchard, in Ref. 18, pp. 187-207.
24. P. Rudolf and R. Schöllhorn, J. Chem. Soc., Chem. Commun. (1992) 1158; E. Takayama-Muromachi, T. Sasaki and Y. Matsui, Physica C 207 (1993) 97.
25. V.J. Emery, S.A. Kivelson and H.Q. Lin, Phys. Rev. Lett. 64 (1990) 475; S.A. Kivelson, V.J. Emery and H.Q. Lin, Phy. Rev. B 42 (1990) 6523; see also Ref. 54 below; for a review, see E. Dagotto, Rev. Mod. Phys. (submitted).
26. V.J. Emery and S.A. Kivelson, in Ref. 18, pp. 1-16; Physica C 209 (1993) 597.
27. D. Vaknin, J.L. Zarestky, D.C. Johnston, J.E. Schirber and Z. Fisk, Phys. Rev. B (submitted).
28. P.G. Radaelli, J.D. Jorgensen, B. Kleb, B.A. Hunter, F.C. Chou and D.C. Johnston, (unpublished).
29. S-W. Cheong, M.F. Hundley, J.D. Thompson and Z. Fisk, Phys. Rev. B 39 (1989) 6567.
30. K. Sun, J.H. Cho, F.C. Chou, W.C. Lee, L.L. Miller, D.C. Johnston, Y. Hidaka and T. Murakami, Phys. Rev. B 43 (1991) 239.
31. F.C. Chou, J.H. Cho and D.C. Johnston, Physica C 197 (1992) 303; F.C. Chou et al. (unpublished).
32. D.C. Johnston, F. Borsa, J.H. Cho, F.C. Chou, D.R. Torgeson, D. Vaknin, J.L. Zarestky, J. Ziolo, J.D. Jorgensen, P.G. Radaelli, A.J. Schultz, J.L. Wagner and S.-W. Cheong, J. Alloys Compds. (to be published).
33. P.G. Radaelli, J.D. Jorgensen, A.J. Schultz, B.A. Hunter, J.L. Wagner, F.C. Chou and D.C. Johnston, Phys. Rev. B 48 (1993) 499.
34. F.C. Chou, D.C. Johnston, S-W. Cheong and P.C. Canfield, Physica C 216 (1993) 66.
35. D.C. Johnston, T. Matsumoto, Y. Yamaguchi, Y. Hidaka and T. Murakami, in *Electronic Properties and Mechanisms of High $T_c$ Superconductors*, edited by T. Oguchi, K. Kadowaki and T. Sasaki (Elsevier, Amsterdam, 1992), pp. 301-306, and references cited.
36. R. Yoshizaki, H. Sawada, T. Iwazumi and H. Ikeda, Solid State Commun. 65 (1988) 1539; Physica C 153-155 (1988) 1495; Synthetic Metals 29 (1989) F735.
37. A. Sulpice, P. Lejay, R. Tournier, B. Chevalier, G. Demazeau and J. Etourneau, Physica B 165&166 (1990) 1157.
38. R.K. Kremer, E. Sigmund, V. Hizhnyakov, F. Hentsch, A. Simon, K.A. Müller and M. Mehring, Z. Phys. B 86 (1992) 319.
39. M. Mehring, M. Baehr, P. Gergen, J. Gross, C. Kessler and N. Winzek, in Ref. 18, pp. 67-84.
40. E.T. Ahrens, A.P. Reyes, P.C. Hammel, J.D. Thompson, P.C. Canfield, Z. Fisk and J.E. Schirber, Physica C 212 (1993) 317.

41. J.E. Schirber, W.R. Bayless, F.C. Chou, D.C. Johnston, P.C. Canfield and Z. Fisk, Phys. Rev. B 48 (1993) 6506.
42. J.H. Cho, F.C. Chou and D.C. Johnston, Phys. Rev. Lett. 70 (1993) 222; and unpublished.
43. B. Morosin, G.H. Kwei, J.E. Schirber, J.A. Voigt, E.L. Venturini and J.A. Goldstone, Phys. Rev. B 44 (1991) 7673.
44. T. Thio, T.R. Thurston, N.W. Preyer, P.J. Picone, M.A. Kastner, H. P. Jenssen, D.R. Gabbe, C.Y. Chen, R.J. Birgeneau and A. Aharony, Phys. Rev. B 38 (1988) 905.
45. J. Yang and W.P. Su, Phys. Rev. B 44 (1991) 6838, and references therein.
46. B. Keimer, R.J. Birgeneau, A. Cassanho, Y. Endoh, R.W. Erwin, M.A. Kastner and G. Shirane, Phys. Rev. Lett. 67 (1991) 1930; B. Keimer, N. Belk, R.J. Birgeneau, A. Cassanho, C.Y. Chen, M. Greven, M.A. Kastner, A. Aharony, Y. Endoh, R.W. Erwin and G. Shirane, Phys. Rev. B 46 (1992) 14034.
47. C.Y. Chen et al., Phys. Rev. B 43 (1991) 392, and references therein.
48. F. Hentsch, H. Seidel, M. Mehring, J.G. Bednorz and K.A. Müller, Physica C 153-155 (1988) 727; K. Kumagai, I. Watanabe, Y. Nakamura and H. Nakajima, J. Phys. (Paris) 49 (1988) C8-2145.
49. I. Watanabe, K. Kumagai, Y. Nakamura, T. Kimura, Y. Nakamichi and H. Nakajima, J. Phys. Soc. Jpn. 56 (1987) 3028; K. Kumagai, I. Watanabe, H. Aoki, Y. Nakamura, T. Kimura, Y. Nakamichi and H. Nakajima, Physica B 148 (1987) 480; I. Watanabe, I. Kumagai, Y. Nakamura and H. Nakajima, J. Phys. Soc. Jpn. 59 (1990) 1932.
50. F.C. Chou, F. Borsa, J.H. Cho, D.C. Johnston, A. Lascialfari, D.R. Torgeson, and J. Ziolo, Phys. Rev. Lett. 71 (1993) 2323.
51. J. Villain, Z. Phys. B 33 (1979) 31; D.H. Ryan, J.O. Strom-Olsen, R. Provencher and M. Townsend, J. Appl. Phys. 64 (1988) 5787; J.R. Thomson, H. Guo, D.H. Ryan, M.J. Zuckermann and M. Grant, Phys. Rev. B 45 (1992) 3129.
52. R.J. Gooding, Phys. Rev. Let 66 (1991) 2266; R.J. Gooding and A. Mailhot, Phys. Rev. B 44 (1991) 11852; ibid. 48 (1993) 6132; R.J. Gooding, N.M. Salem and A. Mailhot (unpublished); R.J. Gooding, K.J.E. Vos and P.W. Leung (unpublished).
53. M.A. Ivanov, V.M. Loktev and Yu.G. Pogorelov, Zh. Eksp. Teor. Fiz. 101 (1992) 596 [Sov. Phys. JETP 74 (1992) 317].
54. V. Hizhnyakov and E. Sigmund, Physica C 156 (1988) 655; E. Sigmund, V. Hizhnyakov and G. Seibold, in Ref. 18, pp. 46-66; G. Seibold, E. Sigmund and V. Hizhnyakov, Phys. Rev. B 48 (1993) 7537; E.L Nagaev, J. Magn. Magn. Mater. 110 (1992) 39, and references therein.
55. K. Kumagai, H. Aoki, I. Watanabe, Y. Nakamura and H. Nakajima, J. Phys. Soc. Jpn. 57 (1988) 1155.
56. Y. Kitaoka, K. Ishida, T. Kobayashi, K. Amaya and K. Asayama, Physica C 153-155 (1988) 733; Y. Kitaoka, H. Hiramatsu, K. Ishida, T. Kohara and K. Asayama, J. Phys. Soc. Jpn. 56 (1987) 3024.
57. A. Aharony, R.J. Birgeneau, A. Coniglio, M.A. Kastner and H.E. Stanley, Phys. Rev. Lett. 60 (1988) 1330.
58. H. Kitazawa, K. Katsumata, E. Torikai and K. Nagamine, J. Phys. (Paris) 49 (1988) C8-2149.
59. J.I. Budnick, B. Chamberland, D.P. Yang, Ch. Niedermayer, A. Golnik, E. Recknagel, M. Rossmanith and A. Weidinger, Europhys. Lett. 5 (1988) 651.
60. D.R. Harshman, G. Aeppli, G.P. Espinosa, A.S. Cooper, J.P. Remeika, E.J. Ansaldo, T.M. Riseman, K.Ll. Williams, D.R. Noakes, B. Ellman and T.F. Rosenbaum, Phys. Rev. B 38 (1988) 852.

61. B.J. Sternlieb, G.M. Luke, Y.J. Uemura, T.M. Riseman, J.H. Brewer, P.M. Gehring, K. Yamada, Y. Hidaka, T. Murakami, T.R. Thurston and R.J. Birgeneau, Phys. Rev. B 41 (1990) 8866.
62. C.J. Peters, R.J. Birgeneau, M.A. Kastner, H. Yoshizawa, Y. Endoh, J. Tranquada, G. Shirane, Y. Hidaka, M. Oda, M. Suzuki and T. Murakami, Phys. Rev. B 37 (1988) 9761.
63. M.E. Filipkowski, J.I. Budnick and Z. Tan, Physica C 167 (1990) 35.
64. F. Borsa et al., Nuovo Cimento D11 (1989) 1785; A. Rigamonti et al., in *Earlier and Recent Aspects of Superconductivity*, edited by J.G. Bednorz and K.A. Müller (Springer, Berlin, 1990).
65. J.H. Cho, F. Borsa, D.C. Johnston and D.R. Torgeson, Phys. Rev. B 46 (1992) 3179.
66. K. Kumagai, Y. Nakamichi, I. Watanabe, Y. Nakamura, H. Nakajima, N. Wada and P. Lederer, Phys. Rev. Lett. 60 (1988) 724.
67. I. Morgenstern, Z. Phys. B 80 (1990) 7.
68. P.C. Hammel, A.P. Reyes, S-W. Cheong and Z. Fisk, Phys. Rev. Lett. 71 (1993) 440, and references therein; and these proceedings.
69. J. Mesot, P. Allenspach, U. Staub, A. Furrer and H. Mutka, Phys. Rev. Lett. 70 (1993) 865; and these proceedings.

# Neutron Spectroscopy in $RBa_2Cu_3O_x$ (R=Rare Earth, 6≤x≤7) Compounds: Charge Transfer, Phase Separation, Spin Fluctuations

A. Furrer, J. Mesot, P. Allenspach, U. Staub, F. Fauth and M. Guillaume

Laboratory for Neutron Scattering, ETH Zürich and Paul Scherrer Institut CH–5232 Villigen PSI, Switzerland

**Summary.** Inelastic neutron scattering has been employed to study the perovskite-type high-$T_c$ superconducting compounds $RBa_2Cu_3O_x$ (R = rare earth; $6 \leq x \leq 7$). The variation of the energies and intensities of the observed crystalline-electric-field (CEF) transitions versus the oxygen content $x$ is shown to be predominantly related to a charge transfer process between the chains and the planes. The observed energy spectra are the result of a superposition of two different metallic components and a semiconducting one, i.e., there is clear experimental evidence for phase separation. A two-dimensional bond percolation model explains the appearance of superconductivity as well as the critical oxygen concentrations associated with the two-plateau structure of $T_c$. The line shape of some low-energy $R^{3+}$ excitations turns out to be highly asymmetric which we interpret in terms of an exchange interaction between the $R^{3+}$ spins and fluctuating $Cu^{2+}$ spins. The latter are most likely associated with low-energy spin excitations of spin-polarized polynuclear clusters of $Cu^{2+}$ ions. From a line-width analysis of the CEF transitions we derive the evolution of the fractal sizes of the clusters versus the oxygen concentration $x$. Magnetic field-dependent neutron spectroscopic experiments give further evidence for the existence of spin-polarized clusters which are suggested to be the elementary building blocks giving rise to high-temperature superconductivity above the percolation threshold.

## 1. Introduction

It has been realized that the superconducting transition temperature of the perovskite-type copper-oxide compounds is essentially unchanged upon replacing the Y and La ions by "magnetic" rare-earth (R) ions. This surprising observation is in contrast to conventional superconductors, for which paramagnetic ions usually have a large detrimental effect on superconductivity. It is therefore important to achieve a detailed understanding of the low-energy electronic properties which define the magnetic ground state of the R ions. In particular, information on the crystalline-electric-field (CEF) interaction at the $R^{3+}$ site as well as the coupling mechanism between the $R^{3+}$ ions is highly desirable.

Very detailed and direct information about the low-energy electronic properties results from inelastic neutron scattering (INS) experiments, particularly for optically opaque materials like the high-$T_c$ compounds. Today there is a complete quantitative understanding of the CEF level structure for a large number of high-$T_c$ compounds of type $RBa_2Cu_3O_x$, $RBa_2Cu_4O_8$,

$R_{2-x}Ce_xCuO_4$, and $Bi_2Sr_2Ca_{1-x}R_xCu_2O_8$ [1]–[5]. In addition, changes of the CEF interaction induced by oxygen nonstoichiometry, pressure, and doping at the copper sites have been studied in detail [6]. INS measurements of the temperature dependence of inelastic CEF transitions have brought important information about the opening of the energy gap [7]. As to the coupling between the $R^{3+}$ ions, INS measurements have only been performed for a grain–aligned sample of $HoBa_2Cu_3O_7$ [8] and for the diluted compound $Ho_{0.1}Y_{0.9}Ba_2Cu_3O_7$ [9] so far.

In this paper we present the CEF level structures obtained for $ErBa_2Cu_3O_x$ ($6 \le x \le 7$) by INS measurements. The overall variation of the spectral response versus the oxygen content $x$ is shown to be largely due to charge transfer effects between the chains and the planes. Fine structures in the observed energy spectra are interpreted as a superposition of three stable states which we attribute to local regions of semiconducting and metallic character, i.e., there is clear experimental evidence for separation into three phases. The superconductivity is shown to result from the formation of a two–dimensional percolative network. From a line–width analysis of the CEF transitions we derive information on the fractal sizes of the clusters associated with the three phases as a function of the oxygen concentration $x$. The asymmetric line shape observed for some low–energy $Ho^{3+}$ excitations in $Ho_{0.1}Y_{0.9}Ba_2Cu_3O_7$ is interpreted in terms of $Cu^{2+}$ spin fluctuations which are most likely due to low–energy spin excitations within spin–polarized polynuclear $Cu^{2+}$ clusters. Magnetic field–dependent neutron spectroscopic experiments performed for $YBa_2Cu_3O_7$ give further evidence for the existence of spin–polarized clusters. This is consistent with theoretical predictions that doping the semiconducting perovskite–type compounds with charge carriers creates a spin polarization in the antiferromagnetic $CuO_2$ planes [10], [11].

## 2. CEF excitations in $ErBa_2Cu_3O_x$ ($6 \le x \le 7$)

In $ErBa_2Cu_3O_x$ the CEF interaction splits the ground–state J–multiplet $^4I_{15/2}$ of the $Er^{3+}$ ions into eight Kramers doublets. We have been able to resolve the seven ground–state transitions as shown in Fig. 2.1: we observe three inelastic lines $A$, $B$, $C$ in a low–energy window ($\Delta E < 12$ meV) and four inelastic lines $D$, $E$, $F$, $G$ in a high–energy window ($65 < \Delta E < 82$ meV). The gap between the two windows is structureless. A detailed analysis of the temperature and momentum transfer dependence of these transitions allows us to interpret all seven lines in terms of CEF excitations out of the ground state. Thus the CEF level structure is completely determined.

Upon oxygen reduction the transitions $B$, $D$, $E$ shift slightly to higher energies, and the transitions $A$, $F$, $G$ move to lower energies, whereas the energy of the transition $C$ remains unchanged, see Fig. 2.1. Note that the line width of the transitions $B$–$G$ is a monotonic function of $x$, in contrast to the line width of the transition $A$ which shows a highly unusual behavior:

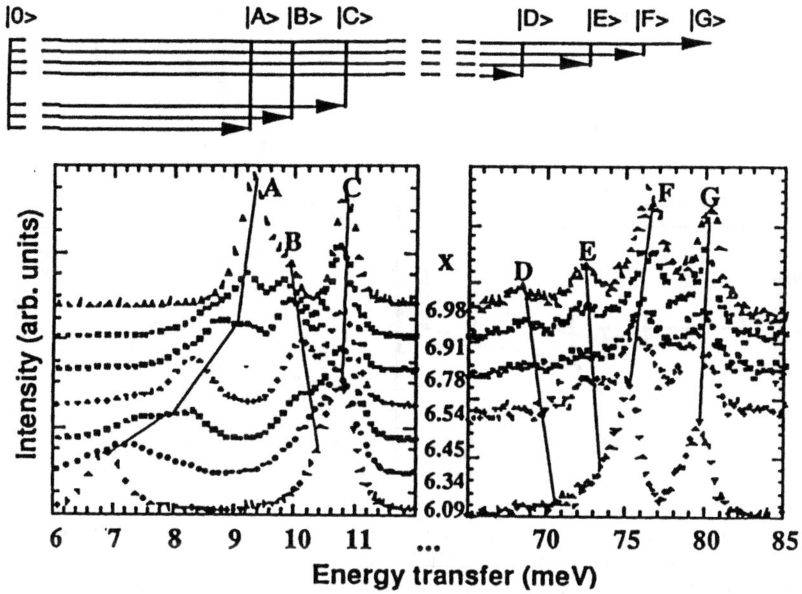


**Fig. 2.1.** Energy spectra of neutrons scattered from $ErBa_2Cu_3O_x$ at $T = 10$ K, measured with use of the time–of–flight spectrometers IN4 (left part, $E_i = 17.2$ meV) and HET (right part, $E_i = 100$ meV) installed at the Institute Laue–Langevin, Grenoble, and at the Rutherford Appleton Laboratory, Chilton, respectively. The lines indicate the $x$ dependence of the observed ground–state CEF transitions. The top of the figure shows the resulting CEF level scheme for $x = 6.98$.

it is rather narrow for $x = 6.98$, $x = 6.53$ and $x = 6.09$, but for the other concentrations $x$ it is considerably broadened and even exhibits substructures. These unusual features will be discussed in detail in Section 3.

We base the analysis of the observed energy spectra on the CEF Hamiltonian

$$H_{\mathrm{CEF}} = \sum_{n=1}^{3} \sum_{m=0}^{n} A_{2n}^{2m} (Y_{2n}^{2m} + Y_{2n}^{-2m}), \tag{2.1}$$

where $A_{2n}^{2m}$ denote the CEF parameters and $Y_{2n}^{2m}$ are spherical tensor operators. Details of the least–squares fitting procedure as well as the resulting CEF parameters have been given elsewhere [12]. The leading CEF parameters are displayed in Fig. 2.2.

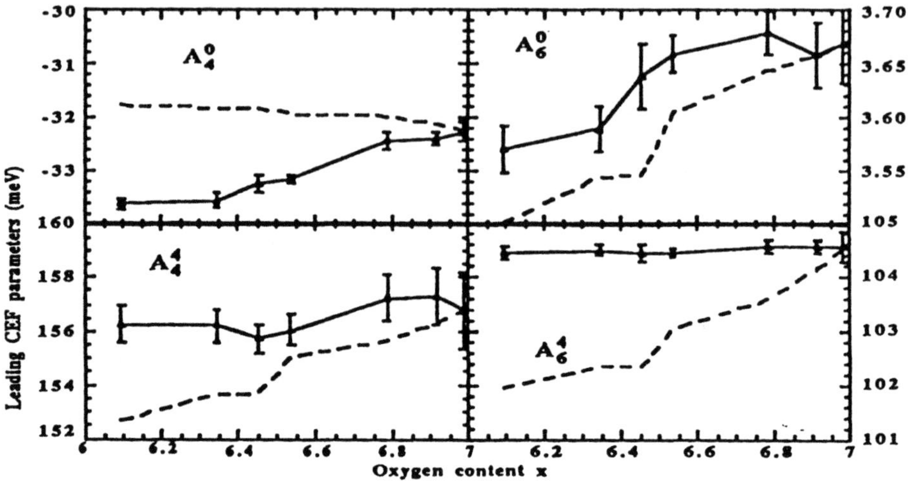


**Fig. 2.2.** Leading fourth– and sixth–order CEF parameters determined for $ErBa_2Cu_3O_x$ in the present work (triangles and solid curves). The values extrapolated from the structural changes alone are indicated by the dashed curves.

# 3. Analysis of the CEF results obtained for $ErBa_2Cu_3O_x$

## 3.1 Charge transfer

The CEF parameters are related to both the structural arrangement and the charge distribution of the ligand ions by definition of the CEF potential:

$$V_{\mathrm{CEF}}(\boldsymbol{r}) = \sum_i \int \frac{\rho(\boldsymbol{R}_i)}{|\boldsymbol{r} - \boldsymbol{R}_i|}\, d\boldsymbol{R}_i . \tag{3.1}$$

$\rho(\boldsymbol{R}_i)$ denotes the charge distribution around the ligand site $\boldsymbol{R}_{i0}$. In the point–charge approximation we have $\rho(\boldsymbol{R}_i) = Z_i e\,|e|\,\delta(\boldsymbol{R}_i - \boldsymbol{R}_{i0})$, thus the structure and charge aspects are decoupled, so that we can differentiate between them by writing the CEF parameters in the form

$$A_{2n}^{2m} = e\,|e|\,\langle r^{2n}\rangle \sum_i Z_i \gamma_{2n}^{2m}(\boldsymbol{R}_{i0}). \tag{3.2}$$

Here, $\langle r^{2n}\rangle$ is the $2n$th moment of the distribution of the 4f electrons, $Z_i$ is the charge of the ligand ion at site $\boldsymbol{R}_{i0}$ in units of the electron charge $|e|$, and $\gamma_{2n}^{2m}(\boldsymbol{R}_{i0})$ are geometrical coordination factors as defined, e.g., by Hutchings [13]. Restricting to the nearest–neighboring oxygen coordination shell for the definition of $\gamma_{2n}^{2m}(\boldsymbol{R}_{i0})$ in (3.2), we found that structural considerations alone are not sufficient to reproduce the observed variation of the CEF parameters versus the oxygen content $x$. As an example the modulus of the CEF parameter $A_4^0$ increases upon oxygen reduction, while the structure expands.

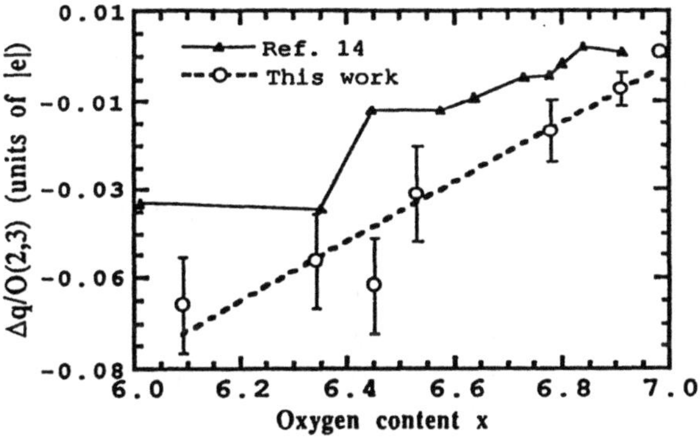


**Fig. 3.1.** Charge transfer versus oxygen content $x$ derived for $ErBa_2Cu_3O_x$ in the present work, in comparison with the results obtained for $YBa_2Cu_3O_x$ from bond valence sum considerations [14].

An increase of the plane–oxygen charge, however, can reproduce the experimental findings. This is consistent with the mechanism of electronic charge transfer from the chains to the planes [14]. A quantitative analysis yields a charge transfer of 0.07 $|e|$ /O when going from $x = 6$ to $x = 7$, see Fig. 3.1, which means that about 28% of the created holes go into the planes. This is slightly less than the value of 40% derived from resistivity measurements [15] and slightly more than the charge transfer 0.04 $|e|$ /O (0.08 $|e|$ /Cu) derived from diffraction data by means of the bond valence sum formalism [14] for the related $YBa_2Cu_3O_x$ compound. Our results suggest a linear dependence of the charge transferred as a function of $x$, in contrast to Cava et al. [14] who deduce a close relation between the two–plateau structure of $T_c$ and the charge transfer with a pronounced discontinuity at $x \approx 6.4$. In the bond valence sum formalism a (non–)linear charge transfer is a direct consequence of a (non–)linear decrease of the cell parameter c with increasing $x$. In this respect the data by Cava et al. [14] are at variance with diffraction work published by Radaelli et al. [16] who observed a linear relation between the lattice parameter $c$ and the oxygen concentration $x$ for $ErBa_2Cu_3O_x$ ($6 < x < 7$), which nicely supports the results of our CEF analysis.

## 3.2 Phase separation

As mentioned in section 2. and shown in Fig. 3.2, there is evidence for substructures associated with the CEF transition $A$. More specifically, the CEF transition $A$ appears to be decomposed into three individual transitions $A_1$, $A_2$ and $A_3$ as indicated by the dashed lines in Fig. 3.2 and discussed in detail in Ref. [17]. The main characteristics can be summarized as follows:

*Intensities:* The transitions $A_1$, $A_2$ and $A_3$ have maximum weight close to $x = 7.0$, $x = 6.5$ and $x = 6.0$, respectively. With the CEF interaction being a local probe, there is no doubt that the above substructures originate from different local environments of the $Er^{3+}$ ions which obviously coexist in the compound $ErBa_2Cu_3O_x$.

*Energies:* Whereas all the CEF transitions are independent of energy for oxygen contents $x \geq 6.5$ within the experimental error, they shift slightly when going from $x \approx 6.5$ to $x \approx 6.0$. This may be due to the structural discontinuities at the orthorhombic to tetragonal phase transition at $x \approx 6.4$.

*Line–widths:* As visualized in Fig. 3.3 the intrinsic line widths of the transitions $A_i$ are much smaller for oxygen contents where these transitions individually reach their maximum weight, namely for $x \approx 6.0$, 6.5 and 7.0. The line widths obviously reflect the differences in the local structure and charge distribution around the $Er^{3+}$ ions. A high degree of homogeneity is expected when only one particular cluster type dominates (i.e., for $x = 6$ and $x = 7$). The rather high degree of homogeneity around $x \approx 6.5$ may also be related to superlattice ordering of the CuO chains.

These findings provide clear experimental evidence for cluster formation or phase separation. It is tempting to identify the three clusters associated with the transitions $A_1$, $A_2$ and $A_3$ by two local regions of metallic ($T_c \approx 90$ K, $T_c \approx 60$ K) and a local region of semiconducting character, respectively. Fig. 3.4 shows the fractional proportions of the three cluster types which correspond to the observed intensities of the transitions $A_i$ corrected by the corresponding matrix element [17]. They exhibit a continuous behavior versus the oxygen content $x$, consistent with our earlier findings that the transfer of holes into the $CuO_2$ planes is linearly related to the oxygenation process.

Combined statistical and geometrical considerations may be useful to understand the $x$–dependent profiles of the fractional proportions of the three cluster types visualized in Fig. 3.4. In an earlier publication [18] we have developed a local symmetry model and defined the following probabilities $P_k(y)$ to have, for a given oxygen content $x = 6 + y$, $k$ of the four oxygen chain sites (0,1/2,0), (1,1/2,0), (0,1/2,1) and (1,1/2,1) nearest to the $Er^{3+}$ ion occupied:

$$P_k(y) = \binom{4}{k}(1-y)^{4-k} , \ (0 \leq k \leq 4). \tag{3.3}$$

The fractional proportion of the cluster type $A_1$ exhibits the behavior predicted by the probability function $P_4(y)$ (i.e., all the oxygen chain sites being occupied). Similarly, the fractional proportions of the cluster types $A_2$ and $A_3$ follow the sum of the probability functions $P_3(y) + P_2(y)$ (i.e., one or two empty oxygen chain sites) and $P_1(y) + P_0(y)$ (i.e., one or none oxygen chain site being occupied), respectively. The above probability functions are shown in Fig. 3.4 by solid lines which excellently reproduce the experimental data.

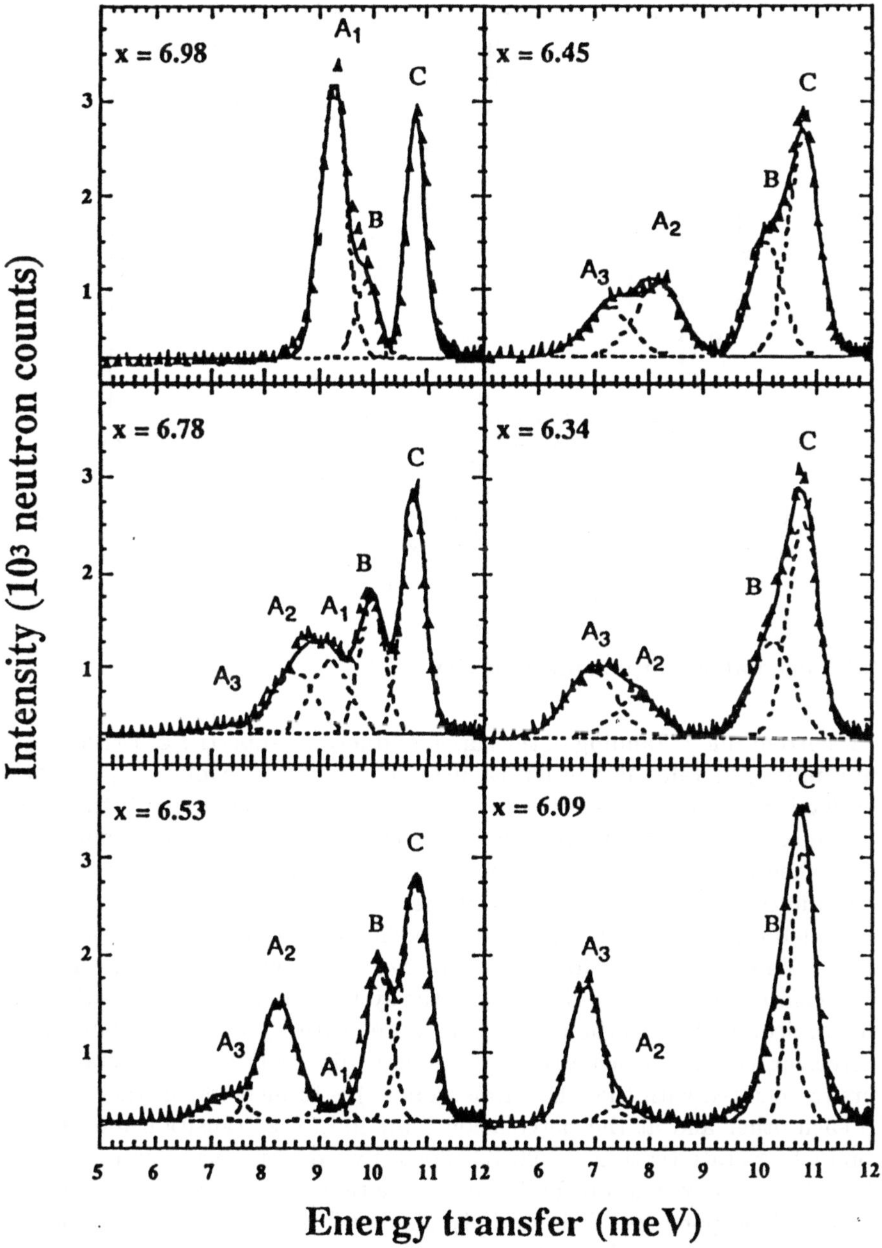


**Fig. 3.2.** Low-energy part of the neutron spectra observed for $ErBa_2Cu_3O_x$ at $T = 10$ K. The lines are the result of a decomposition into individual Gaussian peaks as explained in detail in [17].

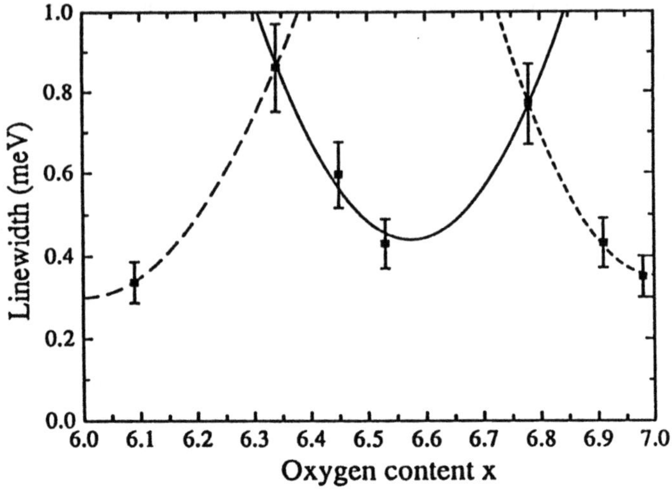


**Fig. 3.3.** Intrinsic line widths of the CEF transitions $A_i$ determined for $ErBa_2Cu_3O_x$ at $T = 10$ K. The lines are explained in the text.

### 3.3 Percolative superconductivity

Our current understanding of the superconducting properties of $ErBa_2Cu_3O_x$ (and more generally all the $RBa_2Cu_3O_x$ compounds) involves a percolation mechanism of electric conductivity as recently discussed in both theoretical [19] and experimental [20] work. For $x = 6$ the system is a perfect semiconductor. When adding oxygen ions into the chains, holes are continuously transferred into the $CuO_2$ planes. By this mechanism the number of local regions with metallic character (associated with the CEF transition $A_2$) rises, which can partially combine to form larger regions. For some critical concentration a percolative network is built up, and the system undergoes a transition from the semiconducting to the conducting state (with $T_c \approx 60$ K). Upon further increasing the hole concentration a second (different) type of metallic clusters (associated with the CEF transition $A_1$) is formed which start to attach to each other and at the percolation limit induce a transition into another conducting state (with $T_c \approx 90$ K). For a two–dimensional square structure the critical concentration for bond percolation is $p_c = 50\%$ [21]. From the fractional proportions of $A_2$ and $A_1$ displayed in Fig. 3.4 we can then immediately determine the critical oxygen concentrations for the transitions from the semiconducting to the $T_c \approx 60$ K–superconducting state and to the $T_c \approx 90$ K–superconducting state to be $x_2$=6.42 and $x_1$=6.86, respectively, which is in excellent agreement with the observed two–plateau structure of $T_c$ [14], [22].

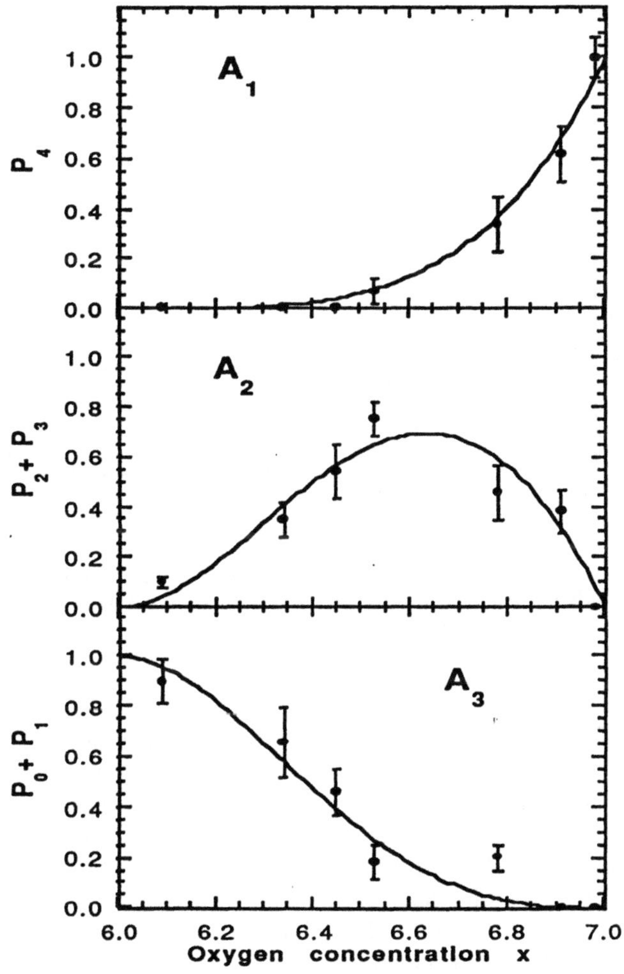


**Fig. 3.4.** Proportions of the lowest-lying CEF transitions $A_i$ of $ErBa_2Cu_3O_x$ as a function of the oxygen content $x$. The lines refer to geometrical probability functions as explained in the text.

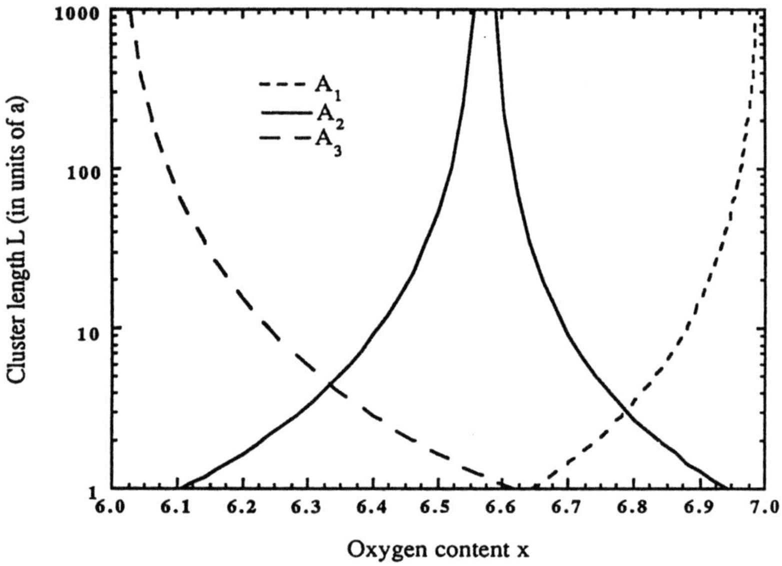


**Fig. 3.5.** Mean cluster lengths L for the three phases $A_1$, $A_2$ and $A_3$ derived from the CEF line widths observed for $ErBa_2Cu_3O_x$ (see Fig. 3.3).

# 4. Fractal sizes of the clusters

The $x$–dependent line width of the CEF transition A (see Fig. 3.3) allows to estimate the size of the above mentioned clusters. Line broadening is due to structural inhomogeneities and gradients in the charge distribution which occur predominantly at the border lines between different cluster types. For a cluster of mean length $L$ (in units of the lattice parameter a) in the (a,b)–plane the fraction of unit cells at the border of the cluster and inside the cluster is $4(L-1)/L2$ and $[(L-2)/L]2$, respectively. In a first approximation the total line width is then given by

$$\Gamma = \frac{1}{L}\sqrt{(L-2)^2\Gamma_0^2 + 4(L-1)\left[\left(\frac{\Delta E}{2}\right)^2 + \Gamma_0^2\right]}\,, \tag{4.1}$$

where $\Gamma_0$ is the intrinsic line width for infinite cluster size (i.e., for $x = 6$ and $x = 7$ exactly realized) and $\Delta E$ corresponds to the energetic separation of the CEF transitions $A_i$ for two coexisting cluster types. More specifically, from Fig. 3.3 we derive $\Delta E = E(A_1) - E(A_2) \approx 1.0$ meV and $\Delta E = E(A_2) - E(A_3) \approx 1.5$ meV at high and low oxygen concentrations $x$, respectively, and $0.3 \leq \Gamma_0 \leq 0.4$ meV. In order to arrive at numerical values of the mean cluster size versus the oxygen concentration $x$, we have interpolated the observed line widths by second–order polynomials, see Fig. 3.3. Based on eq. (4.1) we derive for the three phases mean cluster lengths $L$ versus the oxygen content $x$ as shown in Fig. 3.5. Of particular interest are the fractal sizes of the clusters at percolation which turn out to be $L_1 \approx 5$ and $L_2 \approx 10$ for $x_1 = 6.86$ and $x_2 = 6.42$, respectively. These numbers compare favourably with theoretical estimates by Hizhnyakov and Sigmund [10].

## 5. Low–energy $Cu^{2+}$ spin fluctuations

### 5.1 Introductory remarks

In the preceding section we have presented experimental evidence for the percolative nature of the superconductivity in the $RBa_2Cu_3O_x$ compounds. Moreover, based on the observed CEF transition line widths we have discussed the evolution of the clusters versus the oxygen content $x$. The immediate question is then: what is the nature of these clusters? So far there is no unambiguous answer, however, there are several theoretical concepts and experimental facts which shed light on this important question as summarized below.

Doping or oxidation of the antiferromagnetic parent materials for high–$T_c$ superconductors such as $La_2CuO_4$ or $YBa_2Cu_3O_6$ leads to the creation of holes in the $CuO_2$ planes. In analogy to electron–induced antiferromagnetic semiconductors a hole in the $CuO_2$ plane rearranges the antiferromagnetic order in its nearest vicinity forming a spin–polarized cluster for which various models have been suggested [10], [11]. A particular property of spin–polarized clusters is their magnetic ground state which is expected to give rise to low–energy spin excitations (e.g., under the action of an external magnetic field or due to exchange anisotropy). Indeed, EPR experiments on $La_2CuO_{4+\delta}$ reveal large linewidths of the spectra which have been interpreted in terms of unresolved transitions between Zeeman levels of spin–polarized clusters [23]. EPR measurements in aligned powders of $Y_{0.99}R_{0.01}Ba_2Cu_3O_{6.85}$ (R = Er, Yb) show at the impurity R sites the presence of fluctuating magnetic fields due to copper ions [24]. Recent infrared spectroscopic measurements using linearly polarized light performed for $YBa_2Cu_4O_8$ revealed a depolarization of the reflected beam below 200 K which has been interpreted as experimental evidence for the formation of ferromagnetic microregions [25].

INS is clearly the experimental technique of choice to search for the existence of spin–polarized clusters. The present situation, however, is rather controversial. Very early neutron scattering experiments using three–dimensional polarization analysis on $YBa_2Cu_3O_7$ by Brückel et al. [26] gave no evidence at all for magnetic scattering below 25 meV (for $T$ = 300 K) and below 12 meV (for $T$ = 70 K). Extensive studies of the $YBa_2Cu_3O_x$ system by Rossat–Mignod et al. [27] showed an energy gap $E_g$ in the spin excitation spectra at low temperatures in the superconducting samples, namely $E_g$ = 16 meV for $x = 6.69$ and $E_g$ = 28 meV for $x = 6.92$. This result is at variance with recent polarized neutron experiments on $YBa_2Cu_3O_7$ by Mook et al. [28] where magnetic scattering has been observed below the above mentioned energy

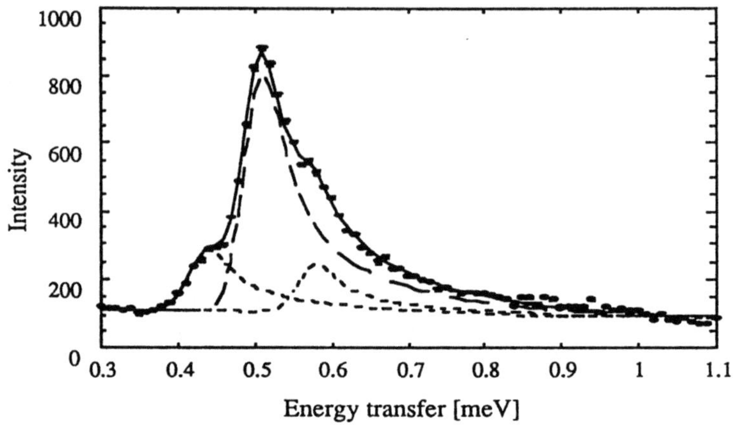


**Fig. 5.1.** Energy spectrum of neutrons scattered from $Ho_{0.1}Y_{0.9}Ba_2Cu_3O_7$ at $T = 1.5$ K and $Q \approx 1.2$ Å$^{-1}$, measured with use of the high–resolution spectrometer IRIS installed at the Rutherford Appleton Laboratory, Chilton. The dashed and dotted lines refer to monomer and dimer excitations, respectively.

gap. Similarly, Mezei et al. [29] found evidence for diffuse magnetic scattering in $YBa_2Cu_3O_{6.59}$ at low energy transfers. In the following paragraphs we provide further neutron spectroscopic evidence in favour of spin–polarized clusters and we discuss possible INS experiments to unambiguously establish their existence.

## 5.2 Asymmetric line shape of CEF transitions in $Ho_{0.1}Y_{0.9}Ba_2Cu_3O_7$

In several INS experiments on $RBa_2Cu_3O_7$ compounds we have observed an asymmetric line shape of CEF transitions, particularly at low energy transfers. Fig. 5.1 shows a typical example obtained for $Ho_{0.1}Y_{0.9}Ba_2Cu_3O_7$ which, according to statistics, contains 66% monomers, 29% dimers, and less than 5% trimers, tetramers, etc. of $Ho^{3+}$ ions in the (a,b)–plane. The spectral response of $Ho^{3+}$ monomers corresponds to the CEF energy level sequence which has been established in detail [5]. The magnetic coupling of the $Ho^{3+}$ ions in the dimers produces a splitting of the CEF states. Trimers, tetramers, etc. will produce a more complicated splitting pattern, however, we can neglect them due to their small statistical probability. Fig. 5.1 shows that our expectations are clearly supported by the experimental data. The energy splitting associated with the pair transitions amounts to about 0.15 meV which is directly related to the nearest–neighbor $Ho^{3+}$ coupling [9].

A characteristic of the observed spectrum displayed in Fig. 5.1 is the asymmetric line shape with a pronounced tail on the high–energy side. An obvious mechanism for the line shape is the exchange interaction between the $Ho^{3+}$ spins and fluctuating $Cu^{2+}$ spins in the superconducting $CuO_2$ planes.

We have treated the $Cu^{2+}$ spin fluctuations in the dynamic effective–field approximation [30] which indeed produces a line shape compatible with the experimental findings, see Fig. 5.1. The variance of the (Gaussian) probability distribution for the fluctuating field, left free as a fitting parameter in the least–squares procedure, turned out to be of the order of 1 T, which nicely confirms the results obtained by EPR measurements in aligned powders of $Y_{0.99}R_{0.01}Ba_2Cu_3O_{6.85}$ (R = Er, Yb) [24]. The above explanation of the asymmetric line shape is clearly compatible with the presence of low–energy $Cu^{2+}$ spin excitations.

## 5.3 Neutron scattering from spin–polarized clusters

Of course it would be highly desirable to prove the existence of spin–polarized clusters in a more direct way. In principle this can be achieved by a detailed analysis of the observed magnetic scattering in terms of cluster excitations which is governed by the neutron cross–section in the dipole approximation [31]:

$$\begin{aligned}\frac{d^2\sigma}{d\Omega d\omega} \sim\ & \frac{1}{Z}\exp\left\{-\frac{E_n}{k_B T}\right\} F^2(\boldsymbol{Q}) \cdot \\ & \cdot \sum_{\alpha,\beta}\left[\delta_{\alpha,\beta} - \frac{Q_\alpha Q_\beta}{Q^2}\right]\sum_{i,j}\exp\{i\boldsymbol{Q}\cdot(\boldsymbol{R}_i - \boldsymbol{R}_j)\} \cdot \\ & \cdot \langle\psi_n \mid \hat{S}_i^\alpha \mid \psi_m\rangle\langle\psi_m \mid \hat{S}_j^\beta \mid \psi_n\rangle\delta(E_n - E_m \pm \hbar\omega),\end{aligned} \tag{5.1}$$

where Z is the partition function, $\psi_n$ and $\psi_m$ are the eigenfunctions of the initial and final cluster states with energies $E_n$ and $E_m$, respectively, $F(\boldsymbol{Q})$ is the magnetic form factor, $\alpha$ and $\beta$ stand for the Cartesian coordinates x, y, and z, the summation indices i and j run through all the ions positioned at $\boldsymbol{R}_i$ and $\boldsymbol{R}_j$ in the cluster, respectively, and $\hat{S}_i^\alpha$ is a spin operator. Eq. (5.2) has been successfully applied to describe neutron scattering from various types of magnetic clusters in the past, e.g., dimers [32], trimers [33], tetramers [34], septamers [35], etc. Of particular importance in eq. (5.2) is the structure factor whose oscillatory behaviour reflects in detail the cluster geometry and therefore readily allows to identify the particular cluster type. This is visualized for a five–fold $Cu^{2+}$ cluster in Fig. 5.2 for excitations within the ferromagnetic ground state (i.e., $\Delta S = 0$ with $\boldsymbol{S} = \sum_i \boldsymbol{S}_i$). In addition, the value of the structure factor for $Q = 0$ is proportional to the number of $Cu^{2+}$ ions in the cluster.

Detailed measurements of the $Q$ dependence of the intensities are therefore essential to unambiguously prove the existence of spin–polarized clusters as well as to identify their nature. The experiments, however, will be extremely difficult. Firstly, a high energy resolution is required in order to resolve the cluster excitations from the elastic line. Secondly, kinematical

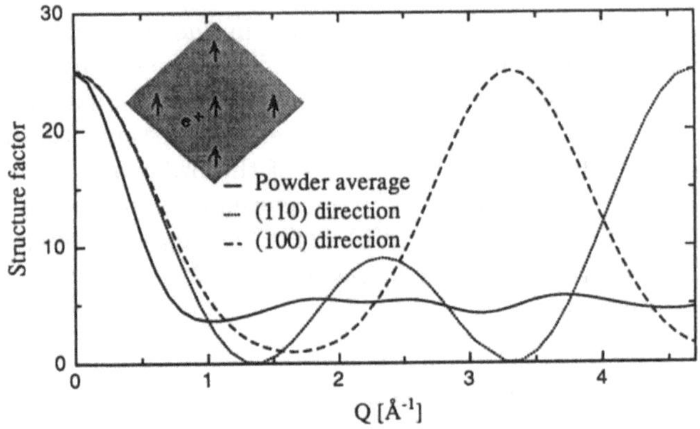


**Fig. 5.2.** Calculated structure factor for Zeeman–like ground–state transitions for a spin–polarized pentanuclear $Cu^{2+}$ cluster (schematically shown in the insert).

constraints will limit the accessible range of scattering vectors $\boldsymbol{Q}$. Thirdly, and most importantly, meaningful data in terms of the cluster picture outlined above are only expected for systems where the cluster size is reasonably small. According to Figs. 3.4 and 3.5 a small cluster size is intimately connected with a small volume fraction of the corresponding phase, thus the intensity of the scattering will be at the limit of detectability.

The above experimental difficulties can be partly overcome by the following strategy. From theoretical work [10] it is known that in the percolation process the spin orientation of the clusters will be important, i.e., only clusters with the same spin orientation contribute to form a conducting pathway for the carriers in the percolative network. Moreover, the spin polarization of the cluster is decreasing with increasing cluster size and vanishes when an infinite network is realized [36]. By the application of an external magnetic field the spins will tend to align along the field direction which increases the number of clusters with the same spin orientation. In turn the overall cluster connectivity (with respect to electric conductivity) is improved and at the same time the overall spin polarization is reduced. As a result the intensity of the magnetic scattering from cluster excitations will be decreasing with increasing magnetic field strength. We have actually performed such an experiment for $YBa_2Cu_3O_{7.0}$ as shown in Fig. 5.3 where we plot the difference of the energy spectra observed in zero field and in a field of 4 T. The peak centered at zero–energy transfer is indeed compatible with the expected field–induced decrease of the magnetic scattering from low–energy spin excitations.

## 6. Conclusion

The neutron scattering experiments described for $ErBa_2Cu_3O_x$ have given further support for the concept of phase separation as a generic feature of

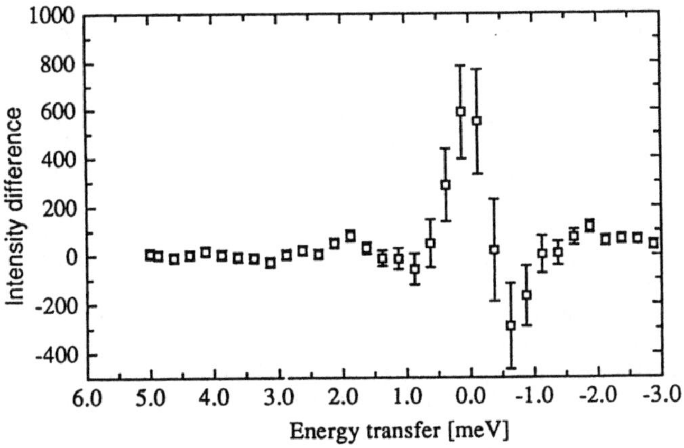


**Fig. 5.3.** Energy spectrum of neutrons scattered from $YBa_2Cu_3O_7$ at $T = 4.2$ K and Q = 1.3 $Å^{-1}$, measured with use of the triple–axis spectrometer IN2 (constant analyzer energy 13.7 meV) installed at the Paul Scherrer Institut, Würenlingen. The data correspond to the difference $I(H = 0) - I(H = 4T)$.

high–$T_c$ superconductivity in the $RBa_2Cu_3O_x$ compound series. The line widths of the CEF transitions yield information on the $x$–dependent fractal sizes of the clusters associated with the coexisting phases. The observation of an asymmetric line shape in the magnetic excitation spectrum of $Ho_{0.1}Y_{0.9}Ba_2Cu_3O_7$ as well as magnetic field–dependent neutron spectroscopic measurements of $YBa_2Cu_3O_7$ give evidence for the existence of spin–polarized clusters of $Cu^{2+}$ ions which may be considered to be the basic elements contributing to the formation of a percolative network and above a critical carrier concentration to the realization of a metallic state.

*Acknowledgement.* We have greatly benefited from discussions with K.A. Müller, E. Sigmund and R.K. Kremer. Financial support by the Swiss National Science Foundation is gratefully acknowledged. We are indebted to H. Mutka and C.J. Carlile for the assistance with the neutron experiments performed at the ILL Grenoble and at the RAL Didcot, respectively.

## References

1. L. Soderholm, C.K. Loong, G.L. Goodman, and B.D. Dabrowski, *Phys. Rev. B* **43**, 7923 (1991).
2. A.T. Boothroyd, S.M. Doyle, D. McK. Paul, and R. Osborn, *Phys. Rev. B* **45**, 10075 (1992).
3. A.I. Goldman, Y. Gao, S.T. Ting, J.E. Crow, W.H. Li, and J.W. Lynn, *J. Magn. Magn. Mater.* **76** & **77**, 607 (1988).

4. U. Staub, P. Allenspach, J. Mesot, A. Furrer, R. Müller, T. Schweizer, L.J. Gauckler, H. Blank, and H. Mutka, *Z. Phys. B—Condensed Matter* **85**, 35 (1991).
5. A. Furrer, P. Brüesch, and P. Unternährer, *Phys. Rev. B* **38**, 4616 (1988).
6. A. Furrer, P. Allenspach, J. Mesot, U. Staub, H. Blank, H. Mutka, C. Vettier, E. Kaldis, J. Karpinski, S. Rusiecki, and A. Mirmelstein, *Eur. J. Solid State Inorg. Chem.* **28**, 627 (1991).
7. E.A. Goremychkin, R. Osborn, and A.D. Taylor, *JETP Lett.* **50**, 380 (1989).
8. U. Staub, F. Fauth, M. Guillaume, J. Mesot, A. Furrer, P. Dosanjh, and H. Zhou, *Europhys. Lett.* **21**, 845 (1993).
9. A. Furrer, J. Mesot, U. Staub, F. Fauth, and M. Guillaume, to be published in *J. Alloys and Compounds.*
10. V. Hizhnyakov and E. Sigmund, *Physica C* **165**, 655 (1988).
11. N.F. Mott, *J. Phys.: Condens. Matter* **5**, 3487 (1993), and references cited therein.
12. J. Mesot, P. Allenspach, U. Staub, A. Furrer, H. Mutka, R. Osborn, and A. Taylor, *Phys. Rev. B* **47**, 6027 (1993).
13. M.T. Hutchings, in F. Seitz and D. Turnbull (eds.), *Solid State Physics*, Vol 16, Academic, New York, 1964, p. 227.
14. R.J. Cava, A.W. Hewat, E.A. Hewat, B. Battlog, M. Marezio, K.M. Rabe, J.J. Krajewski, W.F. Peck Jr., and L.W. Rupp Jr., *Physica C* **165**, 419 (1990).
15. U. Welp, S. Flescher, W.K. Kwok, J. Downey, Y. Fang, G.W. Crabtree, and J.Z. Liu, *Phys. Rev. B* **42**, 10189 (1990).
16. P.G. Radaelli, C.U. Segre, D.G. Hinks, and J.D. Jorgensen, *Phys. Rev. B* **45**, 4923 (1992).
17. J. Mesot, P. Allenspach, U. Staub, A. Furrer, and H. Mutka, *Phys. Rev. Lett.* **70**, 865 (1993).
18. P. Allenspach, A. Furrer, and B. Rupp, in V.L. Aksenov, N.N. Bogolubov, and N.M. Plakida (eds.), *Progress in High-Temperature Superconductivity*, Vol.21, World Scientific, Singapore, 1990, p. 318.
19. V. Hizhnyakov, N. Kristoffel, and E. Sigmund, *Physica C* **160**, 119 (1989).
20. R.K. Kremer, E. Sigmund, V. Hizhnyakov, F. Hentsch, A. Simon, K.A. Müller, and M. Mehring, *Z. Phys. B-Condensed Matter* **86**, 319 (1992).
21. S. Kirkpartick, *Rev. Mod. Phys.* **45**, 574 (1973).
22. B. Rupp, E. Pörschke, P. Meuffels, P. Fischer, and P. Allenspach, *Phys. Rev. B* **40**, 4472 (1989).
23. G. Wübbeler and O.F. Schirmer, *phys. stat. sol. (b)* **174**, K21 (1992).
24. I.N. Kurkin, I.Kh. Salikhov, L.L. Sedov, M.A. Teplov, and R.Sh. Zdanov, *Zh. Eksp. Teor. Fiz.* **103**, 1342 (1992) & *JETP* **76**, 657 (1993).
25. P. Wachter and R. Pittini, Proc. ICMAS-93, *Superconducting Materials* (Paris, December 13-15, 1993), to be published.
26. T. Brückel, H. Capellmann, W. Just, O. Schärpf, S. Kemmler-Sack, R. Kiemel, and W. Schaefer, *Europhys. Lett.* **4**, 1189 (1987).
27. J. Rossat-Mignod, L.P. Regnault, C. Vettier, P. Bourges, P. Burlet, J. Bossy, J.Y. Henry, and G. Lapertot, *Physica C* **185-189**, 86 (1991).
28. H.A. Mook, M. Yethiraj, G. Aeppli, T.E. Mason, and T. Armstrong, *Phys. Rev. Lett.* **70**, 3490 (1993).
29. F. Mezei, B. Faragó, C. Pappas, Gy. Hutiray, L. Rosta, and L. Mihály, *Physica C* **153-155**, 1669 (1988).
30. A. Furrer and H. Heer, *Phys. Rev. Lett.* **31**, 1350 (1973).
31. S.W. Lovesey, *Theory of neutron scattering from condensed matter*, Vol. 2 (Clarendon, Oxford, 1984), p. 6.
32. A. Furrer and H.U. Güdel,, *Phys. Rev. Lett.* **39**, 657 (1977).

33. U. Falk, A. Furrer, H.U. Güdel, and J.K. Kjems, *Phys. Rev. Lett.* **56**, 1956 (1986).
34. U. Falk, A. Furrer, N. Furer, H.U. Güdel, and J.K. Kjems, *Phys. Rev. B* **35**, 4893 (1987).
35. M.A. Aebersold, H. Blank, B. Briat, A. Furrer, and H.U. Güdel, *Inorg. Chem.* **30**, 3556 (1991).
36. E. Sigmund and G.U. Seibold, private communication.

# Phase Separation in Cuprate Superconductors: Magnetic Resonance Studies

P.G. Baranov and A.G. Badalyan

A.F. Ioffe Physical–Technical Institute, Russian Academy of Sciences, 194021 St. Petersburg, Russia

**Summary.** Intense magnetic resonance (MR) signals have been observed below 40 K in cuprate superconductors R-Ba-Cu-O (123, R = Y, Nd, Gd, Sm-ceramics and single crystals) after fast quenching from temperature 800–1050 K. MR signal intensity decreases after slow cooling from the temperatures 300–700 for 123 compounds, and at the same time the superconducting phase recovers. Strong temperature and orientation dependences of the positions of MR lines (g-factors) and a totally different behaviour of MR signals at X- and Q-bands have been observed. The appearing of the MR signal was concluded to be common phenomenon for cuprate superconductors and seems to be connected with formation of the phase built up by small localized hole rich clusters with magnetic ordering in $CuO_2$ plane. These magnetic clusters have a quasi one-dimensional structure and MR fields shift is a result of short-range order effects. During annealing the magnetic clusters diffuse through the sample and start to aggregate above 200 K in La-Cu-O, and 300 K in R-Ba-Cu-O cuprate superconductors, forming a percolative network of superconducting paths. Rotation of R-Ba-Cu-O crystals which contain a big fraction of the superconducting phase with $T_c \sim 80$ K results in broadening of the MR line followed by temperature dependent relaxation of the signal to the original linewidth. The effect of flux motion on MR signals of spin-polarized clusters might be a direct evidence of coexistence of spin-polarized clusters and bulk superconductivity or phase separation in cuprate superconductors and also of the fact that the magnetic clusters are pinning centers for flux motion.

## 1. Introduction

One of the most remarkable features of the cuprate superconductors is the fact that the starting compounds are antiferromagnetic insulators, which after "self-doping" with oxygen become superconducting. The superconductivity apparently originates from holes on oxygen ions, which are mainly confined to the $CuO_2$ layers and move in the presence of a concentrated spin lattice of $Cu^{2+}$-ions. The negative results in the observation of the magnetic resonance of $Cu^{2+}$-ions in copper-oxides high-temperature superconductors and their insulating relatives has stimulated many discussions [1] because the present theories assume that copper ions in these compounds are divalent. Our basic idea was to observe magnetic resonance in cuprate superconductors in the transition region of oxygen concentration (e.g., when in 123 $O_x$ compounds x changes between 6.7 and 6.4). In cuprate superconductors the hole concentration may be varied simply by doping with oxygen. The holes locally destroy the antiferromagnetic coupling between the $Cu^{2+}$ spins. In this case a "spin glass" regime seems to be observed where there is a local magnetic order which is quite limited in extent.

The doping with holes has been suggested in [2]–[4] to lead to the creation of ferromagnetically ordered clusters. The cluster size depends on the Coulomb repulsion on the Cu-sites. Thus doping of the $CuO_2$ planes results in an electronic phase separation of the system into hole-rich spin clusters and hole-free antiferromagnetically ordered areas. The holes inside of the spin clusters can move freely whereas the clusters have low mobility. For sufficiently large hole concentration and as a result of cluster diffusion, the clusters start to overlap, and a percolation network is built up. The percolation of single-hole clusters gives rise to the insulator-metal (superconducting below $T_c$) transition.

Strong anisotropic magnetic resonance signals have been observed in oxygen-deficient $RBa_2Cu_3O_x$ (R = Y, Gd) crystals [5]–[8] at X-band below 40 K. The intensities of the signals were shown to correspond to a concentration of paramagnetic centers larger than $10^{20}$ $cm^{-3}$ and the magnetic resonance signals must be associated with ions of the host lattice copper and oxygen. Analogous magnetic resonance signals were detected in ceramics [9],[10].

In this paper an investigation of strong magnetic resonance signals in oxygen-deficient $RBa_2Cu_3O_x$ (R-Ba-Cu-O) compounds with R = Y, Nd, Gd, Sm at X- and Q-bands is reported. The oxygen deficiency has been created by quenching of the R-Ba-Cu-O compounds with $x = 6.9$–$7.0$ and $T_c$ about 90 K from 800–1050 K.

The existence of spin-polarized clusters with quasi one-dimensional structure and the formation of a percolative network of superconducting paths as a result of temperature controlled diffusion of the clusters in cuprate superconductors has been shown with direct magnetic resonance measurements of annealing processes in these systems. The effect of flux motion on magnetic resonance signals of spin-polarized clusters which was observed [6]–[8] might be direct evidence of coexistence of spin-polarized clusters and bulk superconductivity and also of the fact that the magnetic clusters are pinning centers for flux motion.

## 2. Experimental details

Ceramics and single crystals of the high-temperature cuprate superconductors R-Ba-Cu-O (123, R = Y, Nd, Gd, Sm; both ceramics and single crystals), which were prepared in several laboratories, have been used. Before the heat-treatments all samples were pure well-annealed superconductors with $T_c$ about 90 K. The microwave absorption at low magnetic fields, characteristic for the superconducting state, has only been observed in these samples. The samples have been heated up to 770–1050 K and then rapidly quenched to

300 K or 77 K by dropping into water or liquid nitrogen. After the quenching from temperature above ~ 1000 K the 123 systems transform from an orthorhombic to a tetragonal crystal structure as oxygen is depleted. The tetragonal phase does not show superconductivity.

JEOL X- and Q-band spectrometers have been employed for the present measurements. A home-made set up with microwave power modulation has been used for Q-band experiments at 1.6 K.

## 3. Results and discussion

The quenching of the 123 high-temperature superconductors from temperature higher than 750 K is known [11] to result in oxygen deficiency and to decrease $T_c$. In the superconducting region, transition temperatures $T_c$ plotted versus oxygen stoichiometry usually exhibit two plateaus, one near 90 K and another near 50 K.

Fig. 3.1 shows the magnetic resonance spectra typical for the 1020 K quenched 123 crystals (Gd-Ba-Cu-O with a small amount of superconducting phase in the Fig. 3.1) which were registrated at X- (a) and Q-band (b) for two orientations of magnetic field, H||c and H⊥c. The registration temperatures were 10 K at X-band and 8 K at Q-band. We see that MR spectra are anisotropic, and the striking result is that the positions (g-factors) and linewidths of the signals at X- and Q-bands are totally different. This result cannot be explained for a paramagnetic system with one electron spin $S = 1/2$. The orientation dependences for g-factors also do not correspond to the usual equation with $g_{\parallel}$ and $g_{\perp}$.

Fig. 3.2 shows magnetic resonance signals (absorption) at Q-band in Nd-Ba-Cu-O (a) and Gd-Ba-Cu-O (b) crystals which were measured at 1.6 K using microwave power modulation at 200 Hz. The crystals were quenched from 1020 K. Before the quenching no MR signals in these crystals were observed. In comparison with the MR spectra in Fig. 3.1 the linewidths are larger, and this result seems to be connected with lower registration temperature. The magnetic resonance signals and the temperature dependences of linewidth and g-factor do not depend qualitatively on R (Y, Gd, Nd), but some difference in the MR spectra parameters was observed.

The MR signal appears below 40 K and its intensity increases more rapidly than it would be expected according to the Curie law. We measured the g-factor temperature dependences and plotted these values as a function of temperature in Fig. 3.3. Preliminary data about g-factor and MR linewidth temperature dependences were published previously in [5]–[8].

Fig. 3.3 shows these dependences for a Gd-Ba-Cu-O crystal, but a similar temperature dependences were observed for the other 123 compounds. It should be noted that the g-factors depend very strongly on the temperature, and the temperature dependences are different at X- and Q-bands.

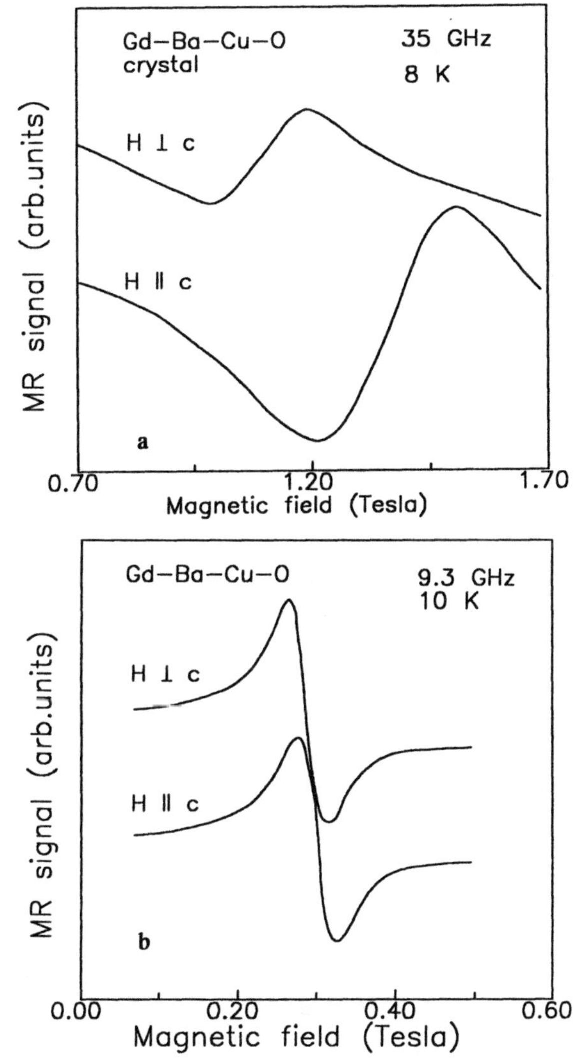


**Fig. 3.1.** Magnetic resonance spectra in 1020 K quenched Gd-Ba-Cu-O crystals (with small amount of superconducting phase) recorded for two orientations of magnetic field H$\|$c and H$\perp$c
a) $T = 8$ K, 35 GHz (Q-band)
b) $T = 10$ K, 9.3 GHz (X-band)

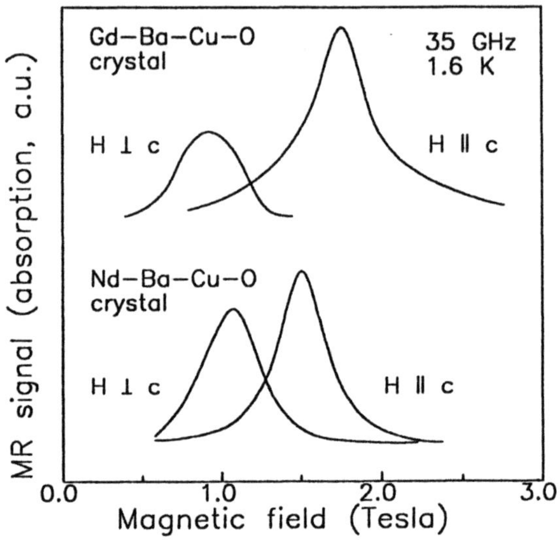


**Fig. 3.2.** Magnetic resonance signal (absorption) recorded at 1.6 K with modulation of microwave power (Q-band, modulation frequency 200 Hz) in 1020 K quenched crystals (small amount of superconducting phase) Gd-Ba-Cu-O (a) and Nd-Ba-Cu-O (b).

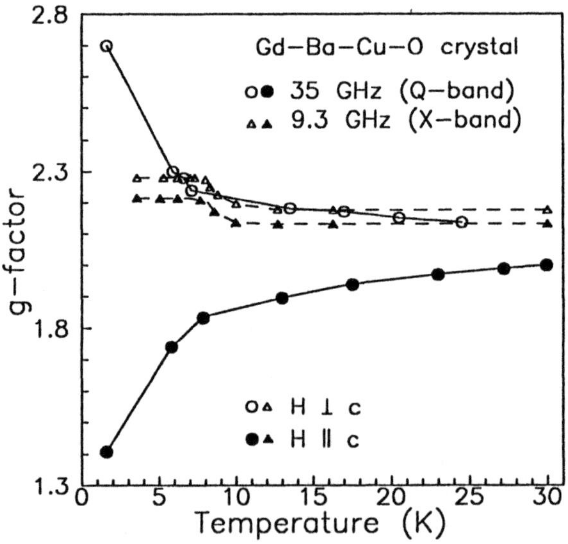


**Fig. 3.3.** g-factor temperature dependences of the MR signals in crystal Gd-Ba-Cu-O quenched from 1020 K (with small amount of superconducting phase) measured at Q-band (circles) and X-band (triangles) for magnetic field perpendicular to c-axis (open circles and triangles) and parallel to c-axis (filled circles and triangles).

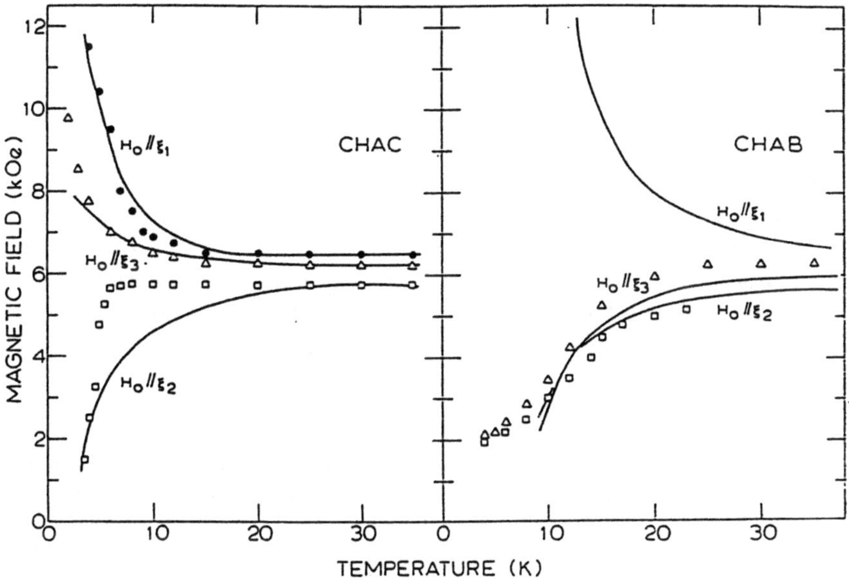


**Fig. 3.4.** Resonance fields as a function of temperature for three orientations of the external magnetic field in a quasi one-dimensional Heisenberg ferromagnet $[C_6H_{11}NH_3]CuCl_3$ (CHAC) and CHAB (Br instead of Cl) at 18.3 GHz. In both compounds the c-axis is the chain axis along which the copper ions are coupled through a strong intrachain exchange interaction. $\xi_3$ is the chain axis, $\xi_1$ is the hard axis; this hard axis direction and perpendicular $\xi_2$ are found to be in a-b plane.
The drawn lines are the calculated values (after [14]).

At X-band the g-factors are approximately constant or decrease slowly above 10 K. They increase abruptly as the temperature is lowered. The g-factor for the crystal orientation in the magnetic field H⊥c is larger as compared with the orientation H||c, and in both cases g-factors exceed 2.1.

At Q-band the g-factor temperature dependences are essentially different. In this case we observe a big increase of the g-factor in the orientation H⊥c as the temperature is lowered and the opposite behaviour in the orientation H||c. In Fig. 3.3 one can see the two characteristic parts of the g-factor temperature dependences. The g-factor changes slowly above 10 K and abruptly below this temperature. At 1.6 K the Q-band g-factors are equal 2.7 and 1.4 in the orientations H⊥c and H||c, respectively! The MR linewidth at Q-band is much bigger in comparison with it at X-band.

The shifts of the resonance fields (g-factors) as a function of temperature for different orientations of the external magnetic field (Fig. 3.3) is qualitatively coinciding with analogous dependences, which have been observed in the quasi one-dimensional (1D) antiferromagnets and ferromagnets[12]–[14]. Fig. 3.4 (after [14]) shows resonance fields as a function of temperature for three orientations of the external magnetic field in a quasi 1D Heisenberg ferromagnet $[C_6H_{11}NH_3]CuCl_3$ (CHAC) and CHAB (Br instead of Cl) at 18.3 GHz. In both compounds the c-axis is the chain axis along which the copper ions are coupled through a strong intrachain exchange interaction. The resultant magnetization vector spontaneously orients into a so-called "easy" direction, whereas energy must be spent to turn it into a "hard" direction. Phenomenologically this behaviour can be described by energy minimization with respect to the orientation within the coordinate system of the crystal axes. In Fig. 3.4 $\xi_3$ is a chain axis and an easy axis for magnetization, $\xi_1$ is a hard axis; both the hard axis direction, $\xi_1$, and perpendicular one, $\xi_2$, are found to be in the a-b plane.

In the paramagnetic range the resonance fields for quasi 1D systems with short-range ordered spins was shown in [12],[13] to shift as a function of temperature. The resonance field in a 1D ferromagnet is expected to increase (g-factor to decrease) upon lowering the temperature when the external magnetic field is applied along a hard axis ($\xi_1$ in Fig. 3.4, for $T = 5$ K g-factor along hard axis is about 1.3 for CHAC), whereas the resonance field should decrease (g-factor increases) when the external magnetic field is oriented along the easy axis or is lying in the easy plane. This theory gives a relation between the resonance field and the one-dimensional susceptibilities. The lines drawn in Fig. 3.4 are the calculated values for intrachain exchange interactions which were reported to be 35 K and 25 K for CHAC and CHAB respectively. The anisotropic exchange parameters (roughly along hard axis) were much smaller.

We described the experiments on quasi 1D ferromagnets (similar results have been obtained for 1D antiferromagnets) in detail because the g-factor temperature dependences which are presented in Fig. 3.3 (for 123 superconductors at Q-band) and Fig. 3.4 (for 1D ferromagnet) are qualitatively the same. From these results it is clear that MR signals which we observed in 123 systems belong to quasi one-dimensional magnetic clusters where the resonance fields shift is a result of short-range order effects and of the anisotropy in the intrachain exchange interaction. The hard axis for magnetic clusters seems to be the c-axis of the 123 crystals. In this orientation the maximum resonance field (the smallest g-factor) has been observed (Fig. 3.3). The spin-polarized magnetic clusters are probably lying in the $CuO_2$ plane which is also an easy plane.

The striking result is that the temperature dependence of the g-factor (or the resonance line position) (Fig. 3.3) is strongly connected with the microwave frequency: it is weak enough for 9.3 GHz and drastically changes for 35 GHz. The measurements of frequency dependences of the resonance line positions are now in progress. However, the results of X- and Q-band studies give a preliminary estimate of the value of the exchange interaction energy for spin-polarized clusters about 1 K.

The fact that g-factor for H⊥c (H in the $CuO_2$ plane) is greater than for H||c (H is perpendicular to the $CuO_2$ plane) can give information about the wave function of unpaired electron in the magnetic cluster. It should be noted that an unpaired electron is connected not only with copper but also with oxygen. For an $O^-$-ion in an ionic crystal (see, e.g. [15]) $g_{\|} = 1.983$ and $g_{\perp} = 2.174$ (in RbCl) where the parallel axis is the oxygen-metal direction (or in-plane direction for $CuO_2$ plane).

To prove that the magnetic resonance signals which we observed belong to isolated spin-polarized clusters or magnetic polarons [2]–[4] we made special heat-treatment experiments with 123 ceramics like it was made for La-Cu-O [3],[16]. The striking result was that the amount of superconducting phase in La-Cu-O samples could be varied drastically for constant oxygen content just by slow and fast cooling. Fast quenching from 300 K gives rise to a small fraction of the superconducting phase. The observations were interpreted with the scenario that hole rich clusters diffuse through the sample and start to aggregate slightly above 200 K. If the cooling from a temperature above 200 K is too fast, only small hole rich aggregates (spin-ordered clusters) are formed, whereas under slow cooling a percolative network of metallic (superconducting below $T_c$) paths will form due to overlapping of spin-polarized clusters. MR signal which were observed in [16] correlated with the results of susceptibility measurements [3].

This phenomenon qualitatively should be common for all cuprate superconductors. Therefore we decided to look again at our previous MR results with 123 compounds [5]–[8],[10]. It was shown that after quenching the MR decayed during 2–3 months at room-temperature, while the superconducting phase appeared again, as correlated with susceptibility measurements [17].

It should be noted that up to now we observed MR signal in 123 single crystals which were like “sandwich” consisting of thin plates with $\sim 1\ \mu$ thickness (one can see it through the microscope). The total thickness of the samples was about 50–100 $\mu$. Such structure seems to create the best conditions for penetration of the microwave field.

Fig. 3.5 shows magnetic resonance spectra of the Y-Ba-Cu-O ceramic, fast quenched from 1050 K to 77 K, measured at X-band as quenched (curve 1) and after 5 min annealing at 650 K (2), 710 K (3) and 730 K (4). The spectra have been recorded at 5 K.

The fluctuations of microwave absorption (it is not noise!) which were observed after annealing (Fig. 3.5, curves 3–4) is direct indication of appearing of superconducting phase and are caused by weak links or Josephson junctions. Also the strong increase of the intensity of the low field signal which indicates the recovery of the superconducting phase was observed after these thermal treatments.

In some 123 samples the form of the MR signal was also changed after the annealing treatment. In the as quenched sample the magnetic resonance signal had a complicated structure, and a asymmetrical line shape was ob-

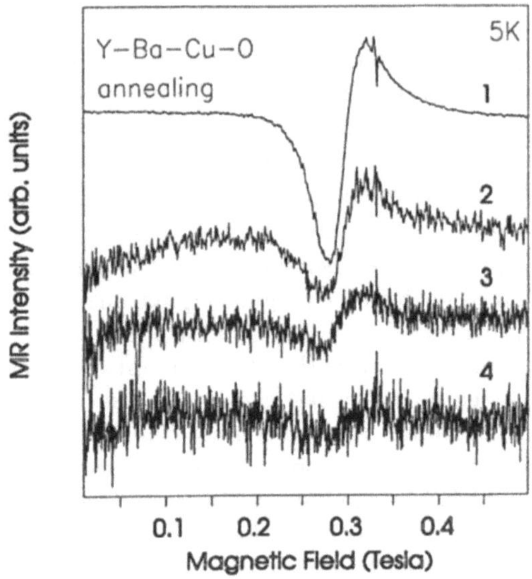


**Fig. 3.5.** Magnetic resonance spectra of the Y-Ba-Cu-O ceramic, fast quenched from 1050 K to 77 K, measured at X-band as quenched (curve 1) and after 5 min annealing at 650 K (2), 710 K (3) and 730 K (4). The spectra have been recorded at 5 K. The sharp line belongs to DPPH. The fluctuations of microwave absorption for curves 2–4 indicate a recovery of superconducting phase with $T_c$ about 50 K.

served. At the temperature below 10 K the cavity quality factor $Q$ abruptly decreased (the losses on the sample increased). After annealing at temperatures above 300 K the losses decreased and the signal became symmetrical. The deviation from the normal signal form can possibly be attributed to the appearing of spin dependent microwave conductivity which changes under magnetic resonance conditions.

Thus, the MR signal is not a single line but the superposition of several lines which seems to belong to magnetic clusters with different structures and with different annealing temperatures.

For some samples of 123 compounds a similar behaviour of decreasing of MR intensity versus annealing temperature and the recovery of the superconducting phase with $Tc \sim 50$ K have been observed at lower temperatures.

The results of annealing experiments for three different samples of 123 compounds number 2, 3, and 4 are displayed in Fig. 3.6. The sample number 2 is a Nd-Ba-Cu-O crystal, the samples 3 and 4 correspond to Y-Ba-Cu-O ceramics; the ceramic density of the sample 4 was larger compared with sample 3. Fig. 3.6 shows magnetic resonance intensities of fast quenched samples of 123 compounds (curves 2, 3, and 4) and of La-Cu-O samples (1) as depending on the 5 min annealing temperature; For La-Cu-O ceramic the result after [16] have been used. For samples 2, 3, and 4 the MR intensity

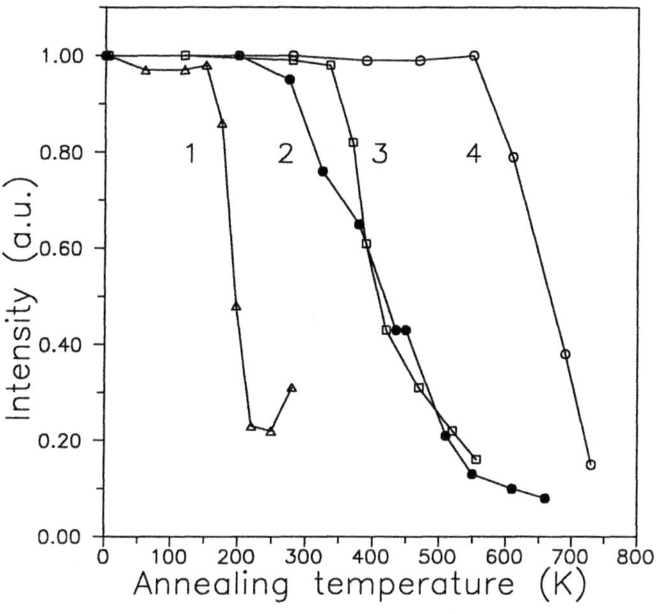


**Fig. 3.6.** Magnetic resonance signal intensities of fast quenched R-Ba-Cu-O samples: Nd-Ba-Cu-O crystal (curve 2), Y-Ba-Cu-O ceramics (3, 4) and La-Cu-O ceramic (1) as a function of 5 min annealing temperature. For La-Cu-O ceramic the results after [16] have been used.

remain nearly constant up to temperatures of 250 K, 350 K, and 550 K, respectively. Then an abrupt decrease of the intensities follows.

Thus, when the samples were quenched from temperatures above 1000 K only a very small fraction of the superconducting phase could be detected and the biggest MR signal was observed. The amount of superconducting phase in these samples with $T_c$ about 50 K for 123 compounds increases drastically and at the same time, the magnetic resonance signal intensity decreases just after slow cooling from the temperatures 300–700 K for 123 compounds. The ratio of temperature of the abrupt decrease of the magnetic resonance signal (and the recovery of the superconducting phase) to the quenching temperature is about 0.4 for the samples number 2 and 3, i.e. the same for La-Cu-O and some samples of 123 compounds.

These processes of magnetic clusters diffusion are thermally activated and our estimates of the activation energy obtained from dependences in Fig. 3.6 result in a values more than 1 eV for 123 systems.

The MR signal appears after the quenching at temperatures higher than 800 K. At the interval of the quenching temperatures 800–950 K the MR lines at X-band are about two times broader than those after quenching at higher temperatures. The same effect has also been observed in the samples with MR signals after annealing at $\sim$ 750 K in oxygen (the effect studied up to

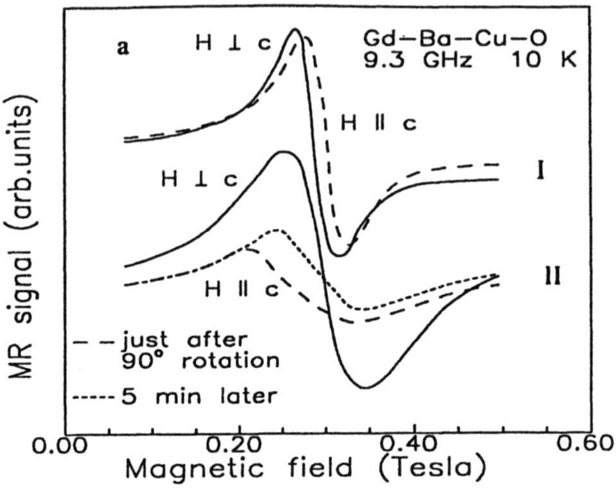


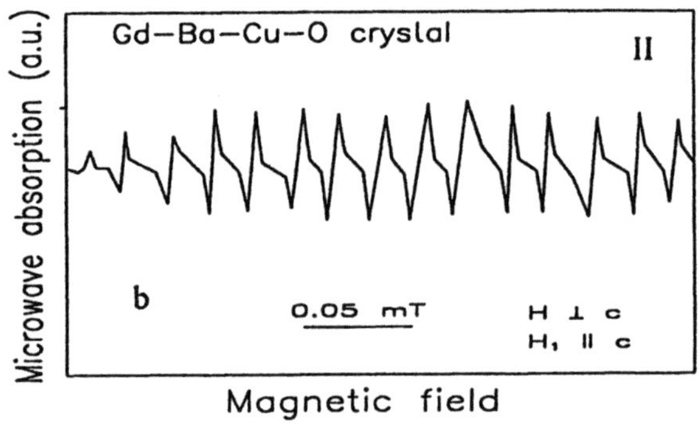


**Fig. 3.7.** (a) MR signal in a Gd-Ba-Cu-O crystal after fast quenching from 1020 K (with a small fraction of superconducting phase) measured at 10 K for two orientations (crystal type I). MR signal in a Gd-Ba-Cu-O crystal which contain a large fraction of superconducting phase with $T_c$ about 80 K (crystal type II). Spectra have been measured first in the orientation H⊥c (solid line) and then in the orientation H∥c (dashed lines) immediately after $\pi/2$ rotation and 5 min later. (b) Low field microwave absorption in the crystal with a large fraction of superconducting phase (type II) where MR signal type II (Fig. 3.7a) was observed.

now only for 123 systems). For these samples, the decrease of the MR signal amplitude and the line broadening have been observed after the rotation of the crystal in magnetic field.

Fig. 3.7a shows MR signal in a Gd-Ba-Cu-O crystal after fast quenching from 1020 K (with a small fraction of superconducting phase) measured at 10 K for two orientations H⊥c and H∥c (crystal type I). The MR signal in a Gd-Ba-Cu-O crystal which contain a big fraction of superconducting phase with a $T_c$ of about 80 K is presented in Fig. 3.7a (crystal type II). Spectra have been measured first in the orientation H⊥c (solid line) and then in the orientation H||c (dashed lines) immediately after $\pi/2$ rotation and 5 min later.

In the crystal with a big fraction of superconducting phase where the MR signal (crystal type II) was measured we observed low field microwave absorption which is characteristic for single superconducting crystals [18]. The principal difference between the microwave absorption spectrum of ceramics and single crystals is the appearance of line spectra in the latter like the one shown in Fig. 3.7b. Their origin seems to be microwave current induced creation and annihilation of fluxons within section of a Josephson junction in a superconductor. Thus there are fluxons in this crystal and as a result the broadening and relaxation of the MR lines which is caused by flux motion.

Fig. 3.8a (line 1) shows a MR signal recorded at 18 K in the orientation H⊥c for a sample Gd-Ba-Cu-O type II. The crystal was cooled for the measurements in this orientation. After the crystal was rotated by 360° in a magnetic field of 0.5 T, i.e., again in the orientation H⊥c, we first observed line 2 (dashed) and then the signal amplitude increased. After 18 minutes, it assumed shape 4.

Fig. 3.8b illustrates the relaxation of the MR signal after the crystal was rotated by 360° in 20 s in an external magnetic field of 0.5 T, the orientation before and after the rotation was H⊥c. Two rotations at 18 K (solid line) and one rotation of a new sample at 9.5 K were made. The points on solid curve (Fig. 3.8b) correspond to the line numbers in Fig. 3.8a.

The relaxation time depends on the temperature and decreases with increasing temperature. A rapid and complete restoration of the original resonance signal required that the crystal was heated to a temperature of the order of 80 K and then recooled.

The vertical local fields seem to broaden the MR signal, and the pinning and depinning of these vortex fields give rise to relaxation and hysteresis effects [6]–[8]. For the samples quenched from temperatures higher than 1000 K the MR linewidth is mainly determined by spin-spin interactions. The local magnetic fields which are created by spin-spin interaction depend on temperature and external magnetic field. Variations of temperature, value, and direction of the magnetic field (magnetic fields are different at X- and Q-bands) change the local magnetic fields and as a result change the positions and linewidths of the MR signals.

## 4. Summary

The intense magnetic resonance signals have been observed in cuprate superconductors R-Ba-Cu-O both in ceramics and single crystals after fast quenching from temperature 800–1050 K. These signals appear below 40 K and its intensity increases more rapidly than it would be expected according to the Curie law.

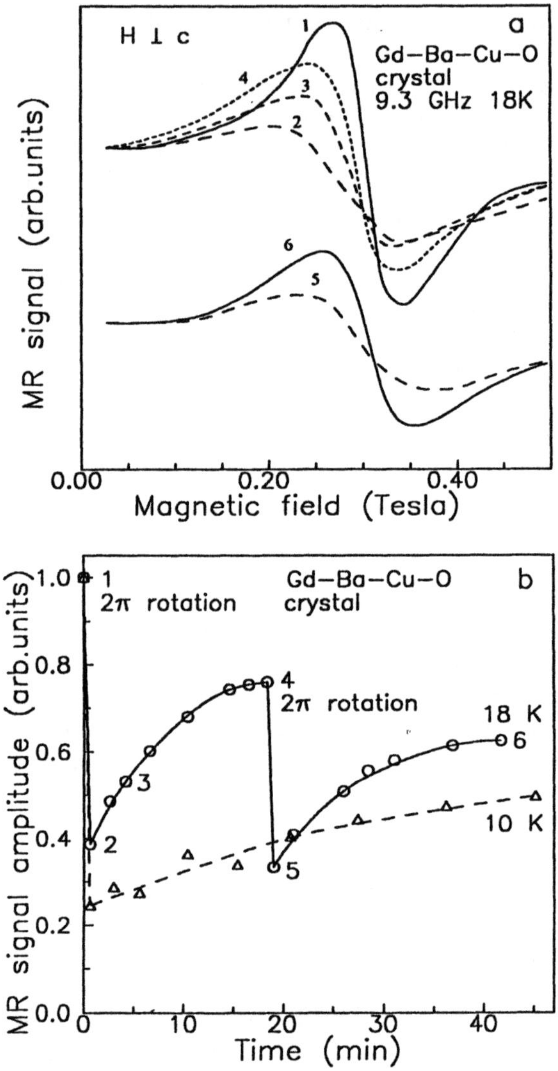


**Fig. 3.8.** (a) Magnetic resonance signal recorded at 18 K in the orientation H⊥c for a sample Gd-Ba-Cu-O which was cooled in this orientation (curve 1), was quickly rotated 360° (2), measured in different moment of the time after the beginning of the first rotation (curves 3–6). (b) Relaxation of the MR signal amplitude as a function of time after two rotations 360° at 18 K (solid line) and after a rotation at 9.5 K (dashed line). Figures on the solid line correspond to the time when the lines in Fig. 3.8a were recorded.

When the samples were quenched from a temperature above 1000 K only a very small fraction of the superconducting phase could be detected. The amount of superconducting phase in these samples with $T_c$ about 50 K for 123 compounds increases drastically and at the same time, the magnetic resonance signal intensity decreases immediately after slow cooling from the temperatures 300–700 K for 123 compounds. The ratio of the temperature of the abrupt decrease of the magnetic resonance signal (and recovery of the superconducting phase) to the quenching temperature is about 0.4, i.e. the same for La-Cu-O and some samples of 123 compounds.

Anisotropic magnetic resonance signals have been measured in R-Ba-Cu-O crystals at X- and Q-bands. Strong temperature dependences of the positions of magnetic resonance lines (g-factors) and linewidths and a totally different behaviour of the magnetic resonance signals at X- and Q-bands have been observed.

We conclude that the magnetic resonance signal is connected with the formation of the phase built up by small localized hole rich clusters with magnetic ordering in $CuO_2$ planes in cuprate superconductors. During the annealing process the magnetic clusters diffuse through the sample and start to aggregate effectively above 200 K in La-Cu-O and 300 K in R-Ba-Cu-O cuprate superconductors, forming a percolative network of metallic (or below $T_c$ — superconducting) paths due to overlapping clusters with magnetic ordering.

These magnetic clusters (magnetic polarons [2],[4]) have a quasi one-dimensional structure in 123 compounds and the resonance fields shift is a result of short-range order effects and of the anisotropy in the intrachain exchange interaction. The hard axis for magnetic clusters seems to be the c-axis of the 123 crystals. In this orientation the maximum resonance field (the smallest g-factor) has been observed (Fig. 3.3). The spin-polarized magnetic clusters are probably lying in a $CuO_2$ plane which also is an easy plane.

The rotation of the R-Ba-Cu-O crystals which contain a large fraction of the superconducting phase (with $T_c$ about 80 K) gives rise to a broadening of the magnetic resonance line and to the temperature dependent relaxation of the signal to the original linewidth [8]. The effect of flux motion on magnetic resonance signals of spin-polarized clusters might be the direct evidence of coexistence of spin-polarized clusters and bulk superconductivity or phase separation in cuprate superconductors and also of the fact that the magnetic clusters are pinning centers for flux motion.

All these phenomena qualitatively should be common for all cuprate superconductors. Now Bi-Sr-Ca-Cu-O systems are under studies.

*Acknowledgement.* We are deeply grateful to K.A. Müller, M. Mehring, E. Sigmund, V. Hizhnyakov, and O. Schirmer for the very fruitful discussions during the 2nd Workshop on Phase Separation in Cuprate Superconductors, and to Ivan Il'in for the annealing experiments with Nd-Ba-Cu-O crystals.

## References


1. F. Mehran and P.M. Anderson, *Solid State Commun.* **71**, 29 (1991).
2. V. Hizhnyakov and E. Sigmund, *Physica C* **156**, 655 (1988); V. Hizhnyakov and E. Sigmund, *Mat. Lett.* **9**, 425 (1990).
3. R.K. Kremer, E. Sigmund, V. Hizhnyakov, F. Hentsch, A. Simon, K.A. Müller and M. Mehring: *Z. Phys B — Condensed Matter* **86**, 319 (1992).
4. G. Seibold, E. Sigmund and V. Hizhnyakov, *Phys. Rev. B* **48**, 7537 (1993–II).
5. V.I. Aleksandrov, A.G. Badalyan, P.G. Baranov, V.S. Vikhnin, V.V. Osiko and V.T. Udovenchik, *JETP Lett.* **47**, 207 (1988).
6. A.G. Badalyan, P.G. Baranov, V.I. Aleksandrov, M.A. Borik and V.V. Osiko, *JETP Lett.* **49**, 697 (1989).
7. A.G. Badalyan and P.G. Baranov, in: *Extended Abstracts 25 Congress AMPERE on Magnetic Resonance*, M. Mehring, J.H. von Schütz and H.C. Wolf (eds.) p. 255, Stuttgart (1990).
8. P.G. Baranov and A.G. Badalyan, *Solid State Commun.* **85**, 987 (1993).
9. D. Shaltiel, J. Genossar, A. Grayevscy, Z.H. Kalman, B. Fisher and N. Kaplan, *Solid State Commun.* **63**, 987 (1987).
10. V.I. Aleksandrov, A.G. Badalyan, P.G. Baranov, V.S. Vikhnin, V.V. Osiko and V.T. Udovenchic, *Sov. Phys. Solid State* **29**, 2123 (1987).
11. K. Nakahigashi, S. Nakanishi, M. Kogachi, R. Kawano, J. Inoue, S. Noguchi and K. Okuda, *International Journal of Modern Physics B* **2**, 1431 (1988).
12. K. Nagata and Y. Tazuke, *J. Phys. Soc. Jpn.* **32**, 337 (1972).
13. T. Karasudani and H. Okamoto, *J. Phys. Soc. Jpn.* **43**, 1131 (1977).
14. R. Hoogerbeet and A.J. van Duyneveldt, *Physica B* **121**, 233 (1983).
15. A.G. Badalyan and P.G. Baranov, *Sov. Phys. Solid State* **22**, 2020 (1980).
16. G. Wübbeler and O.F. Schirmer, *phys. stat. sol. B* **174**, K21 (1992).
17. B.W. Veal, A.P. Paulikas, Hoydoo You, Hao Shi, Y. Fang and J.W. Downey, *Phys. Rev. B* **42**, 6305 (1990).
18. K.W. Blazey, A.M. Portis, K.A. Müller and F.H.Holtzber, *Europhys. Lett.* **6**, 457 (1988).

# Paramagnetic Resonance Studies of Phase Separation in Cuprate Superconductors

O.F. Schirmer and G. Wübbeler and Th. Wahlbrink

University of Osnabrück, D–49069 Osnabrück

**Summary.** Using paramagnetic resonance techniques phenomena are identified in $La_2CuO_{4+\delta}$ ceramics which support the phase separation model. A signal near $B = 0$, proportional to the superconducting volume fraction, decreases after heating to $T_i \geq 270$ K and fast quenching to 4.2 K and a new signal, related to $Cu^{2+}$, appears. The initial intensities are approached under annealing at $T_{an} \geq 200$ K. In single crystal $La_2CuO_{4+\delta}$ the $Cu^{2+}$ related signal is not observed. $La_{2-x}Sr_xCuO_{4+\delta}$ ceramic does not show any of the indicated responses. The effects can be attributed to evaporation of spin polarized clusters from a percolating network of metallic – and below $T_c$ superconducting – paths under heating and subsequent quenching and condensation into such paths under annealing. The origin of the $Cu^{2+}$ related signal is discussed.

## 1. Introduction

The energy of a hole in a $CuO_2$ plane of a cuprate superconductor will be lowered by delocalization if the $Cu^{2+}$ ions encompassed by the mobile hole, mainly having $O^-$ character, are ferromagnetically spin polarized [1, 2]. This defines a cluster of polarized $Cu^{2+}$ spins in the antiferromagnetic background. In the model of phase separation in high–$T_c$ superconductors [2] such clusters have the tendency to condense into a percolating network of metallic – and below $T_c$ superconducting – paths embedded into the antiferromagnetic phase, if the hole density increases upon doping. The spin polarization decreases with increasing size of the metallic clusters [3].

Clear evidence for phase separation into metallic and insulating regions is observed in the system $La_2CuO_{4+\delta}$ [4, 5, 6, 7]. Also in other high–$T_c$ compounds hints to such processes were found [8, 9, 10]. Results of previous investigations by Kremer et al. [4] on $La_2CuO_{4+\delta}$ suggest that such clusters can preferentially be isolated in materials prepared close to the metal/insulator transition by adjusting $\delta$. Above $T \geq 270$ K the metallic network is thought to be destroyed and evaporated into separated clusters. This state is frozen in by fast quenching to temperatures below 200 K. A subsequent heating to $T > 150$ K then allows the clusters to diffuse towards each other, thus reestablishing the metallic network, which causes superconductivity and diamagnetism, if studied below $T_c$. The question arises whether this phase separation is driven electronically, by the motion of the holes, and/or chemically, by diffusion of excess oxygen, incorporated on interstitial sites [11].

Isolated spin polarized clusters, for instance five $Cu^{2+}$ spins in a $CuO_2$ plane sharing one oxygen hole, have a magnetic moment by which they can couple to an external magnetic field. The resulting Zeeman splittings can possibly be observed by paramagnetic resonance techniques.

We have found [7] and investigated paramagnetic resonance like signals in such compounds and studied their dependence on the thermal history of the material. The results support the above scenario. Additional experiments and the obtained findings will be described in the following. The results will be discussed in the framework of the phase separation model. A number of questions, which remain unanswered at the present stage, will be pointed out.

## 2. Investigated materials and experimental details

Our studies began with $La_2CuO_{4+\delta}$ ceramic probes, which had been made available by Dr. R.K. Kremer. These specimens had been prepared in parallel to those investigated by Kremer et al. [4] by magnetic susceptibility measurements. Later we observed identical results from $La_2CuO_{4+\delta}$ ceramics produced by ourselves. The oxygen excess $\delta$ of the samples was adjusted by equilibration in oxygen rich ($P_{O_2} \approx 10^5$ Pa, 770 K) and poor ($P_{O_2} \approx 10^{-2}$ Pa, 540 K) atmospheres for 18 h. No high pressure or electrochemical oxidation was applied, because the effects reported here all occur in the low doping regime near to the metal/insulator transition. In addition $La_2CuO_{4+\delta}$ single crystals and $La_{2-x}Sr_xCuO_{4+\delta}$ ceramics were investigated.

All measurements were performed like typical EPR studies with a Bruker 200 D EPR spectrometer, with frequency near 9 GHz, equipped with an Oxford cryostat system for low temperature measurements.

In order to investigate the influence of different thermal treatments on the sample properties the following heating and cooling procedures were employed. After heating to $T_i > 250$ K (see below) the material was fast (20 K/sec) quenched to 4.2 K, where EPR was measured. In order to study the reestablishment of the metallic network at higher temperatures, $T_{an}$, the samples were heated to such $T_{an}$, kept there for five minutes and cooled again slowly to 4.2 K, where a new EPR measurement was performed. The same procedure was used for successively higher $T_{an}$ up to 270 K. In this method, known under the name "isochronal anneal", all EPR measurements are performed at the same temperature in order to monitor the changes in EPR intensity caused by annealing unimpeded by the intrinsic temperature dependence of the EPR signals.

## 3. Experimental results

### 3.1 $La_2CuO_{4+\delta}$ ceramics

Two types of signals were observed in this material. Before any heat treatment a microwave absorption near magnetic field $B = 0$ is found below $T_c$ .

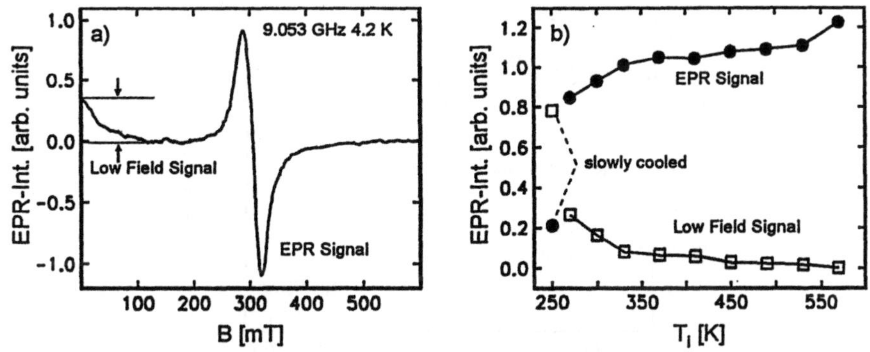


**Fig. 3.1. a)** EPR spectrum and low field signal of an oxygen treated $La_2CuO_{4+\delta}$ sample, fast quenched from 470 to 4.2 K. **b)** Signal heights of the low field and the $B = 300$ mT signal after quenching as depending on the initial temperature $T_i$.

This is usually attributed [12] to the microwave induced resistive movement of fluxoids and thus indicates the presence of superconducting regions in the investigated material. The signal height is proportional to the superconducting volume fraction [13]. It thus increases with rising hole concentration, induced by raising the oxygen excess, $\delta$. We shall first deal with specimens with $\delta$ close to the metal/insulator transition.

After heating to $T > 270$ K and fast quenching to 4.2 K the low field signal was found to decrease and a signal near 300 mT to appear (Fig. 3.1a). As shown below the latter is certainly related to $Cu^{2+}$. Here we shall concentrate first on the dependence of both signals on the thermal history of the samples.

Fig. 3.1b demonstrates the dependence of the signal heights on the initial temperature $T_i$ and the quenching speed. After slow cooling the low field signal is high, corresponding to fully developed superconductivity, and the 300 mT signal is low. This intensity ratio is inverted after fast quenching, indicating that the heat treatment had partially destroyed superconductivity. The symmetric dependence of the signal heights on this procedure clearly demonstrates the intimate connection of both values.

The signal at 300 mT is not thermally stable. Fig. 3.2 shows its dependence on the isochronal anneal temperature $T_{an}$. Near 150 K it is seen to decrease, while the low field signal (indicated by SC) rises again.

The dependence of the diamagnetic susceptibility on temperature, determined by Kremer et al. [4], is included in Fig. 3.2. It is found to be strongest where the SC signal appears, as expected. Furthermore the increase of the diamagnetic volume fraction is accompanied by a nearly parallel decrease of the $Cu^{2+}$ related signal, indicating that two complementary phenomena have been identified.

A synopsis of the results obtained with a ceramic specimen, prepared and loaded with excess oxygen in our laboratory, is exhibited in Fig. 3.3. In Fig. 3.3a the increase of diamagnetism [4] and of the low field signal are com-

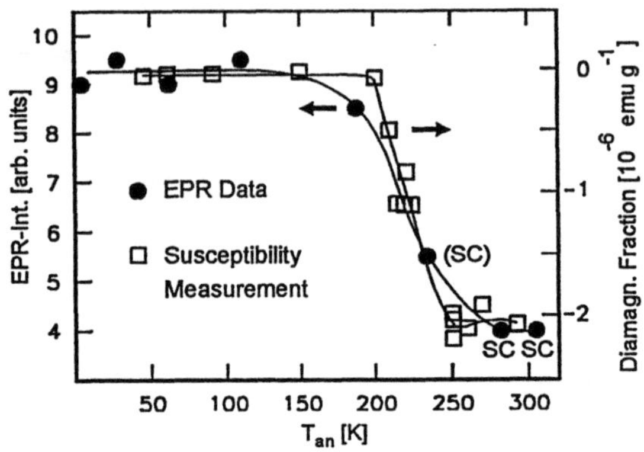


**Fig. 3.2.** Result of the isochronal anneal experiment compared to the diamagnetic susceptibility measured by Kremer et al. [4]. SC indicates the observation of the low field signal. Lines are guides to the eye.

pared to each other. A close parallelity is seen, although the measurements were performed on samples prepared in different laboratories. The decrease of the 300 mT signal is exhibited in Fig. 3.3b. It starts decaying already at about 150 K, lower than the temperatures where the changes in Fig. 3.3a occur. This will be discussed below.

Information on the microscopic structure responsible for the 300 mT signal can possibly be obtained from an analysis of the resonance features of the latter. The signal, Fig. 3.4, can be reproduced by assuming a powder-like distribution of crystallites in the ceramic, each containing $Cu^{2+}$ related paramagnetic objects of axial symmetry. Their splitting in the magnetic field $\boldsymbol{B}$ can be described by the spin hamiltonian $\mathcal{H} = \mu_B(g_\perp B_x S_x + g_\perp B_y S_y + g_\| B_z S_z)$ (1) with S=1/2, hyperfine interaction not being resolved. Here $\mu_B$ is the Bohr magneton, $g_\| \approx 2.33 \pm 0.01$ and $g_\perp \approx 2.08 \pm 0.01$, and the lineshape has to be assumed to be Lorentzian. The values of the g–tensor are typical for $Cu^{2+}$ paramagnetic ions with $(x^2 - y^2)$ type orbitals as groundstates [14]. Neither the g–tensor components nor the linewidth show a dependence on temperature (Fig. 3.5). The convoluted linewidth was constant at a value of about 17.5 mT. The product $I * T$ (signal intensity times temperature), however, decreases. The multiplication by $T$ eliminates the intrinsic temperature dependence of the EPR lines, caused by Boltzmann population of the Zeeman levels involved.

Such a decline is expected for a groundstate level separated by a few meV from a band with high degeneracy, as it is typical for shallow donors in semiconductors, e.g. for $P^0$ in silicon. In the present case the decrease of $I * T$ can be explained if the ratio of the degeneracies of ground and excited states is 1:52 and the energy distance, about 4 meV (solid line in Fig. 3.5b). Presently it is open whether there are further models to explain the temperature dependence of the $I * T$ product.

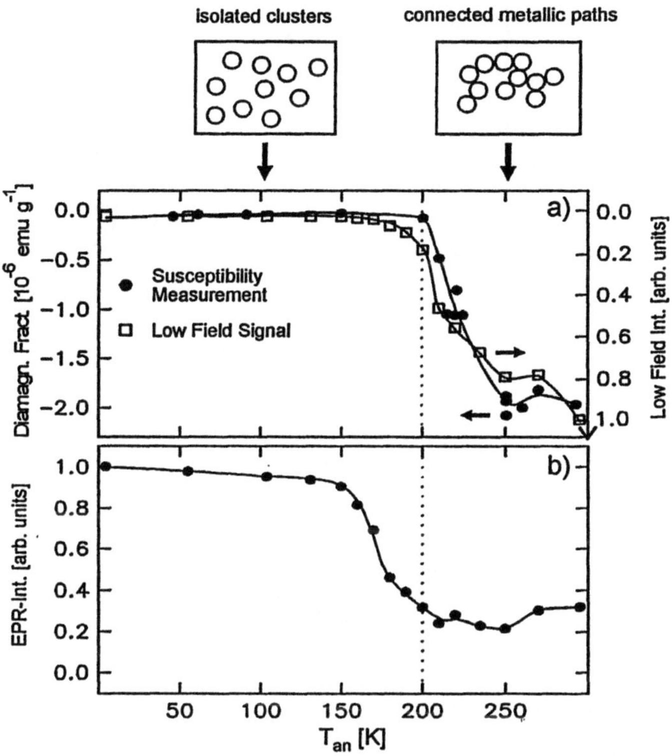


**Fig. 3.3.** Results of the investigation of another $La_2CuO_{4+\delta}$ ceramic under isochronal annealing. **a)** Increase of the low field signal (scale at right with downward arrow direction) compared to that of diamagnetic susceptibility (left scale), as measured by Kremer et al. [4] with a different sample. **b)** Decline of the 300 mT signal under isochronal annealing. Models attributed to the cluster aggregate states in both temperature regions are indicated at the top.

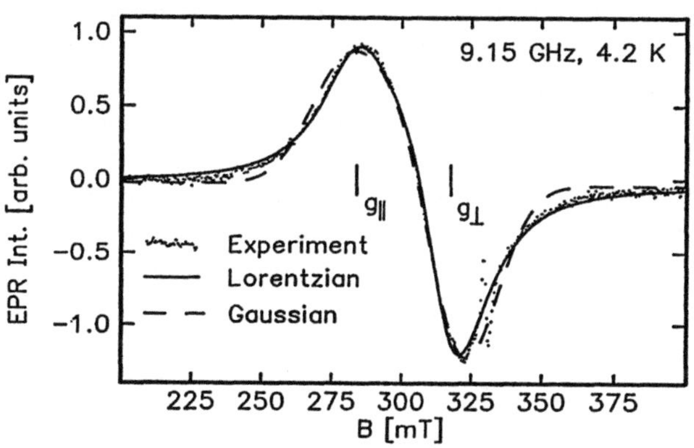


**Fig. 3.4.** Lineshape analysis of the observed EPR spectrum, based on the assumption of a powderlike distribution of grains.

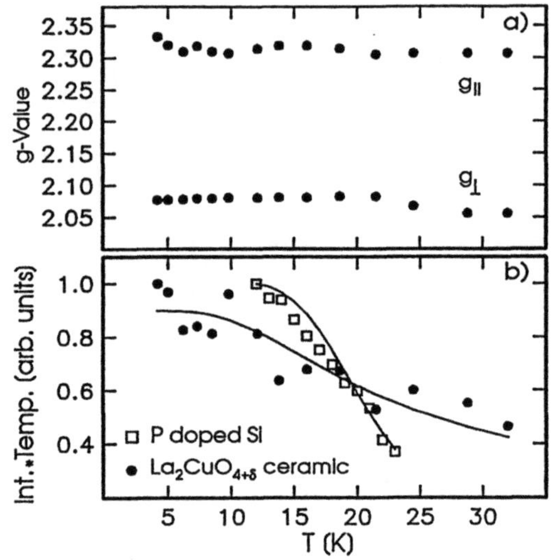


**Fig. 3.5.** Temperature dependence of the EPR parameters. **a)** g–values. **b)** Temperature normalized intensity. The solid line was calculated assuming a model of a 52–fold degenerate state about 4 meV above the ground state. For comparison the $I * T$ product for $P^0$ in silicon is plotted.

## 3.2 $La_2CuO_{4+\delta}$ single crystal

The decrease of the superconducting volume fraction after quenching, as monitored by the low field signal, is observed also in this specimen (Fig. 3.6). However no corresponding resonance at 300 mT was found, irrespective of the thermal treatments of the crystal in oxidizing or reducing atmospheres and of the initial temperature 270 K $< T_i <$ 970 K. The regeneration of superconductivity at about 200 K appears to be much sharper in single crystals than in ceramics. Perhaps the hopping energies of the diffusing particles (holes or excess oxygen ions) are distributed over a larger range in ceramics due to lattice perturbations.

The comparison of these observations with those made with $La_2CuO_{4+\delta}$ ceramics leads to the conclusion that the 300 mT signal in ceramics arises because of influences connected with structural perturbations in this material. The signal is not related to the surface area of the individual crystallites because a sample with decreased grain sizes, leading to larger internal surfaces, shows the same specific signal heights.

## 3.3 $La_{2-x}Sr_xCuO_{4+\delta}$ ceramics

Samples of this type were again partly made available by R.K. Kremer and were partly prepared by ourselves. The Sr content ranges from 2% to about 6%. We also investigated samples treated in argon atmosphere at about 770 K

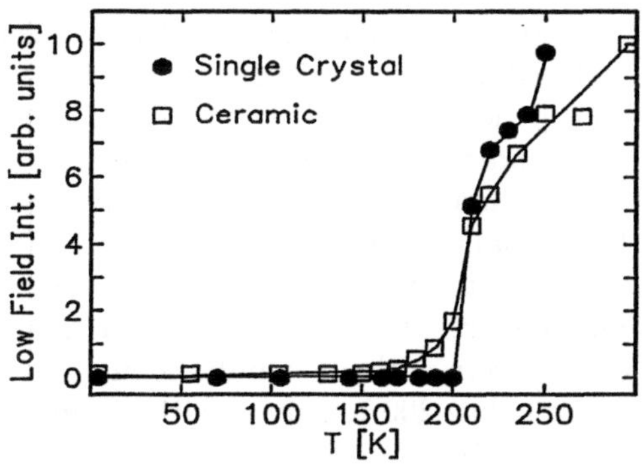


**Fig. 3.6.** Isochronal annealing of a quenched $La_2CuO_{4+\delta}$ single crystal and comparison of results obtained with ceramics.

in order to prevent effects connected with excess oxygen. Superconductivity, as indicated by the low field absorption, starts to appear at doping levels of about 3%–4% of Sr.

Neither a decrease of the low field signal nor an appearance of $Cu^{2+}$ related signals was observed for any of these specimens for 270 K $< T_i <$ 970 K.

This result apparently has to be explained by the different mobilities of the dopant ions. Whereas the excess oxygen, being incorporated as interstitial $O_i$, remains mobile down to temperatures of about 200 K [15], the Sr ion is expected to be much more fixed in the lattice.

## 4. Discussion

The above results show: The superconducting volume fraction is decreased by fast quenching in ceramic $La_2CuO_{4+\delta}$ as well as in crystalline material. Under this treatment a $Cu^{2+}$ related signal near 300 mT appears in ceramics, not in single crystalline $La_2CuO_{4+\delta}$. Both features behave contragrediently when annealed. In $La_{2-x}Sr_xCuO_{4+\delta}$ ceramics neither a decrease of superconductivity nor a rise of the 300 mT signal is observed.

If interpreted on the basis of the phase separation model these detailed findings mean: The metallic — and below $T_c$ superconducting — network is partly thermally dissociated into clusters, possibly leading to the 300 mT signal, which is observable by EPR techniques in ceramics but not in single crystals. The formation of isolated clusters is triggered by the diffusion of interstitial oxygen, the effect being absent in $La_{2-x}Sr_xCuO_{4+\delta}$, where Sr is tightly bound. – It is unlikely that any other model can consistently account for all observations.

Also the finding, Fig. 3.3, that superconductivity, as detected by diamagnetic susceptibility or low field absorption, attributed to fluxoid resistive movement, starts to increase at higher temperature, $\sim$ 200 K, than the decrease of the $Cu^{2+}$ related signal, $\sim$ 150 K, can be discussed on the basis of the above model. Flux penetration can be supported only by large metallic regions, formed by the condensation of many isolated clusters. The paramagnetic signal, on the other hand, will decrease already when isolated clusters are coagulating to form smaller metallic inclusions. As was shown by Kremer et al. [5], only above 200 K interstitial oxygen, $O_i$, and holes will be mobile enough to form larger clusters. Diffusion below 200 K is limited to those few holes which can form clusters without simultaneous movement of the compensating $O_i$. Therefore below 200 K only small clusters will form, unable to allow magnetic flux penetration.

While these observations apparently are consistent with the phase separation model, the question arises, whether the 300 mT signal can be attributed to spin polarized clusters [1], or whether it is connected with a nonsuperconducting, $Cu^{2+}$ containing, paramagnetic compound of different composition embedded into the $La_2CuO_{4+\delta}$ material.

In dealing with this question one has to consider that there is a complementary interconversion between the signals attributed to $Cu^{2+}$ paramagnetism and superconductivity. The latter, of course, can exist only in a cuprate compound identified to allow superconductivity, here $La_2CuO_{4+\delta}$ . It is difficult to see how superconductivity can be replaced by $Cu^{2+}$ paramagnetism if both features do not depend on each other. Since only short distance diffusion will occur during quenching and annealing near 200 K, where the most decisive intensity changes occur, one has to conclude that both phenomena arise from locally identical or closely spaced regions. According to this argument the $Cu^{2+}$ related resonance then would occur in regions potentially becoming superconducting and not in a chemical compound of another Cu containing structure.

But can the 300 mT signal be explained on the basis of the model [1] of spin polarized clusters? It is expected that ideally developed clusters of this kind will preferentially exist in the $CuO_2$ planes of single crystals. Here strong anisotropic exchange interactions, reflecting the two dimensional structure of the planes, will determine the spin levels and their splitting. These will be larger than the microwave quanta applied in our experiments and will not decisively be modified by an external field. The possibility to describe the 300 mT signal by the spin hamiltonian (1) means, on the other hand, that here the dominant anisotropic term is the Zeeman interaction aligning the spins along the magnetic field $\boldsymbol{B}$. These considerations can explain why in single crystals no signals corresponding to the 300 mT resonance in ceramics was observed.

Continuing this argumentation one expects that in ceramics at most a variant of such clusters has been observed, modified in such a way by the microstructure of the grains that the large anisotropic exchange splittings are quenched. It is open presently, how this can occur. Since it cannot be assumed that this will be the case for all clusters one has to conclude that also in ceramics the majority of the clusters might have escaped detection. Further experiments are being planned to identify the microstructure causing the 300 mT signal in $La_2CuO_{4+\delta}$ ceramics.

Finally we want to point out again that the absence of signal changes under quenching and therefore also under annealing in $La_{2-x}Sr_xCuO_4$ ceramics, where the $Sr^{2+}_{La}$ acceptor is tightly bound, demonstrates that the dependence of superconductivity in $La_2CuO_{4+\delta}$ ceramics and single crystals on the thermal treatment is connected to diffusion of highly mobile $O_i^{2-}$ in the latter.

*Acknowledgement.* We thank R.K. Kremer for kindly furnishing some of the samples used. This investigation was initiated by a discussion with E. Sigmund. Fruitful and critical discussions with V. Hizhnyakov, K.A. Müller, R.K. Kremer, M. Mehring and E. Sigmund are gratefully acknowledged. — The work was performed within Graduiertenkolleg "Mikrostruktur oxidischer Kristalle", funded by DFG.

## References

1. V. Hizhnyakov, E. Sigmund, Physica C **156**, 655 (1988)
2. V. Hizhnyakov, E. Sigmund, Materials Letters **9**, 425 (1990)
3. E. Sigmund, V. Hizhnyakov, G. Seibold, Preprint
4. R.K. Kremer, E. Sigmund, V. Hizhnyakov, F. Hentsch, A. Simon, K.A. Müller, M. Mehring, Z. Phys. B – Condensed Matter **86**, 319 (1992)
5. R.K. Kremer, V. Hizhnyakov, E. Sigmund, A. Simon, K.A. Müller Z. Phys. B – Condensed Matter **91**, 169 (1993)
6. J.D. Jorgensen , B. Dabrowski, Shiyou Pei, D.G. Hinks, L. Soderholm, B. Morosin, J.E. Schirber, E.L. Venturini and D.S. Ginley, Phys. Rev. **B38**, 11337 (1988)
7. G. Wübbeler and O.F. Schirmer, Phys. Status Solidi b**174**, K21 (1992)
8. J. Mesot, P. Allenspach, U. Staub, and A. Furrer, Phys. Rev. Lett. **70** ,865 (1993)
9. G. Yu, C.H. Lee, A.J. Heeger, Phys. Rev. B **45**, 4964 (1992)
10. P.G. Baranov and A.G. Badalyan, Submitted to Solid State Commun.
11. C. Chaillout, S.W. Cheong, Z. Fisk, M.S. Lehmann, M. Marezio, B. Morosin and J.E. Schirber, Physica Scripta Vol. **T29**, 97 (1989)
12. K.W. Blazey, K.A. Müller, J.G. Bednorz, W. Berlinger, Phys. Rev. B **36**, 7241 (1987)
13. R. Janes and P.P. Edwards, Bruker Report **138**, 14 (1992)
14. S.A. Al'tshuler, B.M. Kozyrev, "Electron Paramagnetic Resonance in Compounds of Transition Elements", John Wiley & Sons, New York, (1974)
15. J. Ryder, P.A. Midgley, R. Exley, R.J. Beynon, D.L. Yates, L. Afalfiz, J.A. Wilson, Physica C**173**, 9 (1991)

# Fermi Glass State in Partially Doped Insulating Cuprates

G. Yu, C.H. Lee and A.J. Heeger

Institute for Polymers and Organic Solids and Department of Physics, University of California, Santa Barbara, CA 93106-5090, USA

**Summary.** The properties of the Fermi glass state in partially doped insulating cuprates are reviewed. The localized electronic states near the Fermi energy ($E_F$) are characterized through dark transport and transient photodoping experiments. For insulating crystals close to the metal-insulator (M-I) transition, the density of localized states at the Fermi energy, $N(E_F)$, inferred from various experiments, is of the order of $10^{21} cm^{-3} eV^{-1}$. The experimental results suggest an Anderson-type M-I transition in the doped cuprates; i.e., the transition from metal to insulator is dominated by disorder-induced localization. Supporting evidence from angular-resolved photoemission, NQR and other experiments are also discussed. The Fermi glass state has also been studied spectroscopically. In partially doped $YBa_2Cu_3O_{6.3}$, the optical absorption observed throughout the infrared has no counterpart in photoconductive response $\sigma_{ph}(\omega)$; transitions below 2 eV do not create mobile carriers. The spectral gap in $\sigma_{ph}(\omega)$ corresponds to the mobility gap in this material.

## 1. Introduction

Studies of the changes in electronic structure of the high temperature superconducting (HTSC) cuprates as a function of chemical doping are of fundamental importance for understanding the mechanism of high-$T_c$ superconductivity. The electronic structure of stoichiometric (undoped) cuprates has been described with considerable success in terms of a generalized Hubbard model with half-filled Hubbard bands split by the on-site Coulomb interaction. In spectroscopic studies, a charge transfer (CT) gap of 1.5-2 eV is commonly observed between the filled O 2p band (in reality, this band is hybridized with the Cu 3d lower Hubbard band) and the empty Cu 3d upper Hubbard band (UHB) [1]–[6]. In non-stoichiometric insulating crystals (materials which have been "doped" either by hetero-valent substitution or by the addition of interstitial oxygen), substantial oscillator strength grows within the CT gap, spreading over the entire mid-infrared region [4],[7],[8]. Carrier transport in these crystals is thermally-activated or follows the behavior predicted for variable range hopping (see Section 2.). Thus, free carriers (the Drude peak centered at $\omega = 0$) are not observed in the far-infrared at low temperatures [7]. These results are inconsistent with the simplified Hubbard model which would predict a metallic phase in the case of nonstoichiometric doping.

These experimental results are reminiscent of those found in heavily doped semiconductors [7],[9],[10], suggesting that the electronic structure in partially-doped insulating cuprates is that of a Fermi-glass [6],[9]–[12]. The term "Fermi-glass" denotes an insulator in which the density of states at Fermi energy, $E_F$, is finite, but the electronic wave functions at energies near $E_F$ are localized [13]. The localization can be the result of either random positions of the atomic centers in space (so called lateral disorder) as in amorphous Si, or disorder in the crystal potential at different lattice sites, as in crystalline alloys or compounds. In cuprate crystals, the latter seems the dominating factor. When the disorder strength is moderate (the magnitude of the disorder potential is smaller than the band width), only states in the band tails are localized, with a sharp energy boundaries (the mobility edges, $E_c$) [13]–[15] separating the localized states from extended states. If the band filling is such that electronic states at $E_F$ are localized, the conduction at zero temperature vanishes. Thus, an insulating state with a finite density of states at $E_F$ can exist. The electron (or hole) density and thus the Fermi energy can be altered by chemical doping (in this text, the term of chemical doping refers generally to hetero-valent substitution, or inclusion of oxygen vacancies or interstitials). With enough carriers added to the system, one can effectively shift $E_F$ across $E_c$ into energy region where the electronic function are extended, corresponding to the formation of a metallic state.

To check whether this type of electronic structure is general to the cuprates, we examine the experimental results from various transport studies in the dark and under photoexcitation for partially doped insulating cuprates. Attention is paid to data from high quality, single crystal samples, to rule out the ambiguities due to grain boundaries and sample morphology. The results from dark transport experiments such as dc, ac conductivity, Hall coefficient and specific heat are discussed in Section 2.. In Section 3., we discuss the transport results from transient, and steady state photoexcitation experiments, in which the carrier density and thus the $E_F$ can be effectively tuned by means of photon-absorption. The Fermi-glass electronic structure can also be studied spectroscopically, an example in $YBa_2Cu_3O_{6.3}$ is given in Section 4.. In Section 5., we discuss supporting evidence for the Fermi glass state and the disordered crystal field from other types of experiments. We conclude our discussion in Section 6. with remarks on the electronic structure of "normal state" and on the effects of phase separation.

## 2. Dark transport in partially doped insulating cuprates

The electronic structure of the partially doped insulating cuprates inferred from dark transport experiments is consistent with the Fermi glass structure. In a Fermi glass, the electronic states are localized at the Fermi energy, and the conductivity vanishes at zero temperature. Carrier transport at finite temperatures can be realized via two processes: by thermal excitation from $E_F$ to the mobility edge $E_c$, or by variable-range-hopping (VRH) among the

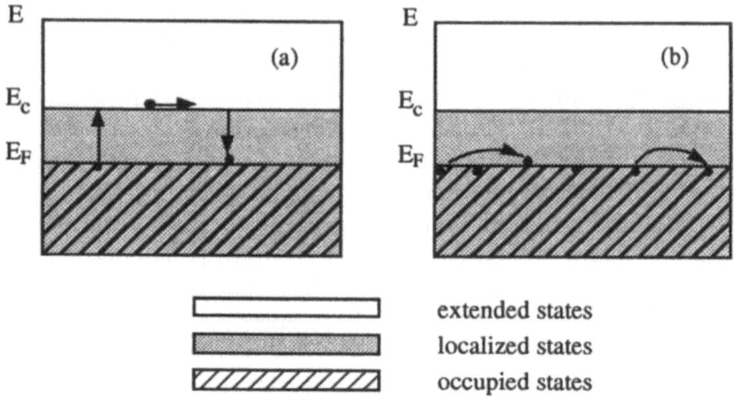


**Fig. 2.1.** Conduction in Fermi-glass insulators; (a) by thermal activation of carriers from $E_F$ to $E_c$, (b) by thermally-assisted hopping near the $E_F$.

localized states near the $E_F$. These two situations are illustrated in Fig. 2.1a and 2.1b respectively.

The first process requires a thermal activation energy of $E_a = |E_F - E_c|$, and the conductivity has the following temperature dependence:

$$\sigma = \sigma_0 \exp(-E_a/k_B T)\,. \tag{2.1}$$

When $E_F$ lies at $E_c$, the conductivity is equal to the pre-exponential factor $\sigma_0$, which is called the "minimum metallic conductivity", $\sigma_{min}$ [13].

A second process, first described by Miller and Abrahams [16], and further developed by Mott and others [14],[17], dominates the low temperature transport. This is a process in which an electron at an occupied localized state near $E_F$ receives energy from a phonon that enables it to "hop" to a nearby empty state at an energy near $E_F$. This VRH conductivity is of the form [14]

$$\sigma \sim \exp(-(T_0/T)^{1/4})\,. \tag{2.2}$$

In 2-dimension (2D), the exponent 1/4 in this expression is replaced by 1/3 [15],[18].

These two parallel conduction processes are commonly observed in insulating cuprates. In $La_2CuO_{4+\delta}$, the temperature ($T$) dependence of the in-plane resistivity $\rho(T)$ follows the thermal activated form for $T$ >50–100 K, and the $E_a$ reduced with the Néel temperature ($T_N$) when increasing oxygen concentration [19]. In the same temperature range, the Hall coefficient is also thermally activated, with a similar $E_a$ and a temperature independent Hall mobility of 5–1 $cm^2V^{-1}s^{-1}$ [19]. At temperatures below 50 K, VRH conduction has been observed in single crystal $La_2CuO_{4+\delta}$ by several groups [20],[21] This type of transport has also been observed in crystals of insulating 123 systems [9],[22] and in single crystals of $Bi_2Sr_2Ca_{1-x}Y_xCu_2O_8$ [23]. In Fig. 2.2, we plot a typical result from an insulating $YBa_2Cu_3O_{6+x}$ crystal (with

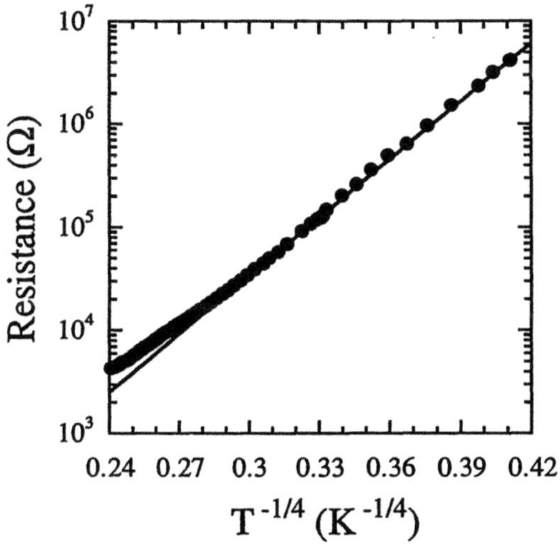


**Fig. 2.2.** Resistance vs. temperature for a $YBa_2Cu_3O_{6+x}$ crystal with $E_a \approx$ 55 meV; the data are plotted in the form appropriate to VRH in 3D.

$E_a \approx$ 55 meV) in which the temperature dependence of the sample resistance is plotted in the 3D VRH form. The deviation from 3D hopping at high temperature region results from the effect of thermal activation; examples of thermally-activated conduction will be discussed in the next Section.

VRH conduction provides direct evidence of the existence of empty localized states near $E_F$. The density of such localized electronic states can be inferred from the VRH conduction by the following relation

$$T_0 = \frac{16}{k_B\, N(E_F)\, L_{loc}^3}\,, \tag{2.3}$$

where $L_{loc}$ defines the localization length [13],[24]. This is a 3D expression; the corresponding 2D expression is similar, which can be found in literatures (for instance, in ref. [21]). For the data shown in Fig. 2.2, $T_0$ is $\sim 10^6$ K, comparable to the $T_0$ observed in insulating $La_2CuO_{4+\delta}$ with $E_a \approx 40$ meV [19],[21]. Assuming the $L_{loc}$ is of the order of the lattice size, $L_{loc} \approx 4$ Å, $N(E_F)$ obtained from Eq. 2.3 is $\sim 10^{21}$ $(cm^3 eV)^{-1}$.

The Fermi glass structure and the existence of a finite $N(E_F)$ can also be inferred from other types of transport experiments. The ac conductivity in $La_2CuO_{4+\delta}$ crystals has been measured for crystals with $E_a \approx$40–100 meV [21],[25]. A power law dependence ($\sigma(\omega) \sim \omega^\beta$) was observed over a broad frequency range, consistent with the VRH conduction. The $N(E_F)$ inferred from the $\sigma(\omega)$ is $\sim 10^{21}$ $(cm^3 eV)^{-1}$ which is of the same order as that obtained from the $\rho(T)$measurement. $N(E_F)$ can also be evaluated independently from specific heat experiments. A measurement on insulating $La_{2-x}Sr_xCuO_4$ (ceramic samples) with $x = 0.012$, 0.024, and 0.036 has been

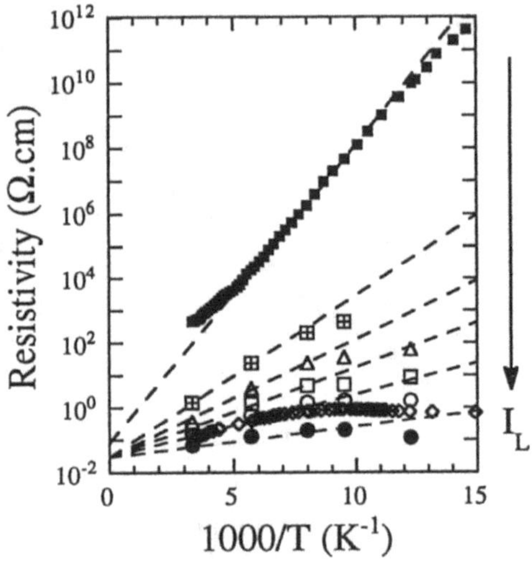


**Fig. 3.1.** The photoresistivity in a $YBa_2Cu_3O_{6+x}$ crystal at several excitation levels are plotted vs $1/T$; (■) dark, (⊞) $1 \cdot 10^{13}$, (△) $1 \cdot 10^{14}$, (□) $1 \cdot 10^{15}$, (○) $7 \cdot 10^{15}$, (◇) $1.1 \cdot 10^{16}$ and (•) $2.3 \cdot 10^{16}$ photons/cm$^2$. The dashed lines are obtained by fitting the high temperature data to Eq. 2.1.

reported by Aliev and co-workers [26]. The data were analyzed in the form of $C/T = g + bT^2$, where $g$ is proportional to $N(E_F)$, $g \sim p\,2k_B^2\,N(E_F)/3$ (assuming the term $g$ is fully contributed by the carriers at $E_F$ which sets up an up-limit). The data show that the $N(E_F)$ at three doping levels are similar, of the order of $1 \pm 0.4$ ($eV^{-1}cell^{-1}$) which is equivalent to $5 \pm 2 \cdot 10^{21}$ ($cm^{-3}eV^{-1}$).

## 3. Conductivity under photoexcitation

Photoexcitation, especially transient photoconductivity, provides an effective method to shift $E_F$ in a given crystal and to study the continuous distribution of the localized states above $E_F$. "Photodoping" offers advantages over chemical doping, since it allows one to vary the carrier concentration within a defined chemical environment. It also provides an unique method to investigate the carrier dynamics (by means of transient excitation technique). Several types of transient photoexcitation experiments have been conducted on insulating cuprates [9],[10],[22],[27]. It has been demonstrated that the $E_F$ can be effectively shifted over a relatively broad energy range by using picosecond laser pulses. In fact, photoinduced M-I transition has been achieved by shifting $E_F$ across $E_c$.

Typical data on the change of the transient peak conductivity, $\sigma_{ph}$, after photoexcitation from a single crystal of $YBa_2Cu_3O_{6.3}$ are given in Fig. 3.1 [9],[27]. In this plot, the $\sigma_{ph}$ data at several excitation levels ($I_L$)

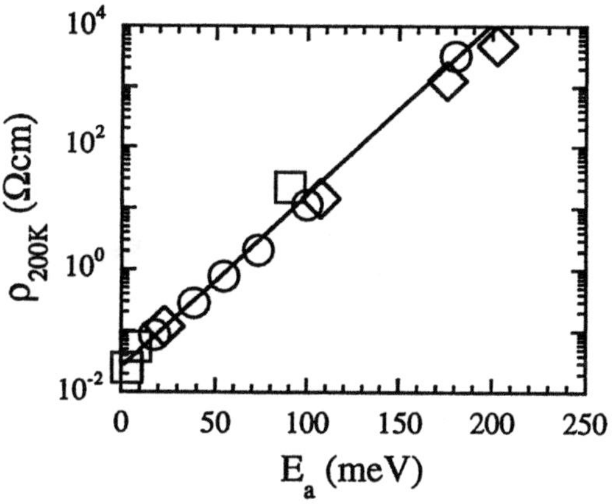


**Fig. 3.2.** Resistivity at 200 K as a function of thermal activation energy; from a crystal of $YBa_2Cu_3O_{6.3}$ at several photodoping levels (o), from a film of $DyBa_2Cu_3O_{6+x}$ at several oxygen concentrations (◇), and from a crystal of $Bi_2Sr_2Ca_{1-x}Y_xCu_2O_8$ at several Y concentrations (□).

are plotted as function of $1/T$. Also plotted in this figure is the dark conductivity in the same crystal. A change in resistivity by more than ten orders of magnitude was observed after photoexcitation. $\sigma_{ph}(T)$ is thermally activated, as is the dark conductivity, and the activation energy, $E_a$, varies with $I_L$; $E_a$ decreases from 180 meV in the dark to near zero when $I_L > 5 \cdot 10^{15}$ photons/cm$^2$. The monotonic decrease of $E_a$ ($=E_F$-$E_c$) with increasing $I_L$ reveals a continuous, monotonic distribution of shallow traps (unoccupied localized states above $E_F$). The reduction of $E_a$ toward zero at $I_L \geq 5 \cdot 10^{15}$ photons/cm$^2$ is associated with $\sigma_{ph}$ approaching $\sigma_{min}$ and with the anomaly in $\sigma_{ph}(T)$ at temperatures below 100 K, indicating a photoinduced M-I transition and the formation of photoinduced metallic/superconducting droplets (charge density fluctuations) [9],[27].

The density of localized states near $E_F$ can also be estimated from this photodoping experiment. From the data in Fig. 3.1 (through $\mathrm{d}\,E_a/\mathrm{d}\,I_L$), of the order of $10^{22}$ photons/cm$^3$eV can be obtained at $I_L \approx 10^{15}$ photons/cm$^2$ with $E_a \approx 50$ meV. Assuming a quantum efficiency of ~ 10% (as suggested by the photoinduced absorption experiments [28]), gives a density of states of ~ $10^{21}$ (cm$^3$eV)$^{-1}$, the same level as inferred from dark transport experiments (see the previous Section).

It is interesting to plot the resistivities from various insulating cuprates as a function of $E_a$ at a given temperature. Data from several cuprates are collected in Fig. 3.2, as a function of $E_a$. Note that the three sets of in-plane resistivities fall basically on a single straight line which intersects with the y axis at ~ $0.03\Omega$ cm (the point corresponds to the $\rho_0$ in Eq. 2.1). This fact not only proves the validity of the Eq. 2.1, but also suggests that a general, material and carrier concentration independent $\rho_{min}$ (~ $0.03\Omega$ cm) exists. This fact is consistent with the quasi-2D nature of the conduction,

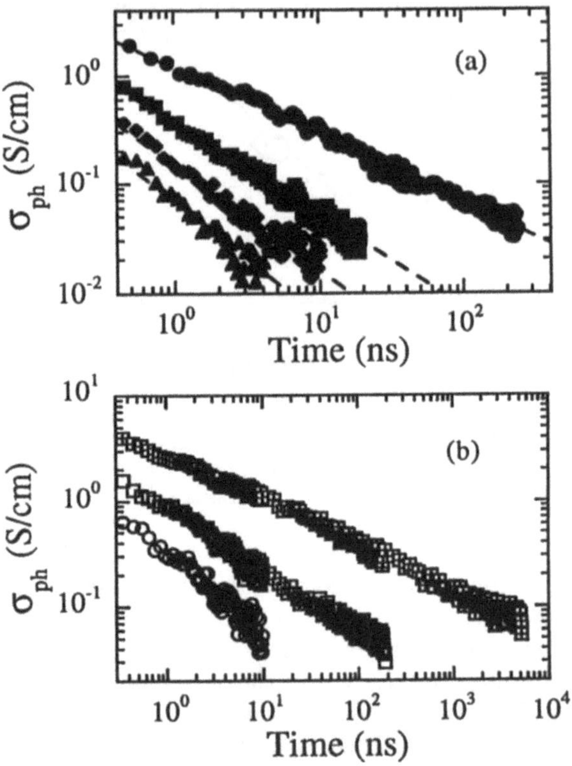


**Fig. 3.3.** (a) Time dependence of the transient $\sigma_{ph}(t)$ in a $La_2CuO_{4+\delta}$ crystal ($E_a \approx 20$ meV) at 25 K; (●) $6.3 \cdot 10^{14}$, (■) $2.8 \cdot 10^{14}$, (♦) $1.3 \cdot 10^{14}$ and (▲) $7.2 \cdot 10^{13}$ photons/cm$^2$. (b) Time dependence of the transient $\sigma_{ph}(t)$ in a $YBa_2Cu_3O_{6+x}$ crystal ($x \approx 0.4$ and $E_a \approx 33$ meV) at 35 K; (⊞) $2.1 \cdot 10^{15}$, (□) $8.6 \cdot 10^{14}$ and (○) $4.5 \cdot 10^{14}$ photons/cm$^2$.

since in 3D the "minimum metallic conductivity" is not universal. In 2D, the sheet resistivity is an universal constant ($\sim e^2/\hbar$), independent of materials and independent of carrier concentrations [13]. Taking the spacing between the $CuO_2$ planes ($\sim$ 12 Å), this universal constant is consistent with the experimentally observed sheet resistivity.

As noted earlier, picosecond transient photoconductivity enables one to study the dynamics of the photoinduced carriers. In Fig. 3.3a, we show the time dependence (log-log scale) of the transient $\sigma_{ph}(t)$ obtained from $La_2CuO_{4+\delta}$ ($E_a \approx 20$ meV) at 25 K for several different $I_L$ [11]. A similar data set from $YBa_2Cu_3O_{6+x}$ at 35 K ($x \approx 0.4$ and $E_a \approx 33$ meV) are given in Fig. 3.3b [9],[10]. For photon densities less than $10^{15}$ photons/cm$^2$, where the initial photocurrent is lower than $\sigma_{min}$, the transient photocurrent decay $\sigma_{ph}(t)$ characterizes the nature of the insulating cuprates. The data show that $\sigma_{ph}$ follows a power law decay, $t^{-\alpha}$, with the exponent $\alpha$ decreasing in magnitude as $I_L$ increases. More specifically, for the lowest curve in Fig. 3.3a, the power law is characterized by $\alpha \approx 1$, while for the highest curve, $\alpha \approx 0.6$.

This power law decay of $\sigma_{ph}(t)$ has been observed in many disordered systems, and has been understood in terms of dispersive transport of the photoinduced charge carriers among localized electronic states [29]. Since the unoccupied localized states act as traps, charge transport takes place via a sequence of multiple trapping-and-release steps among localized states in the presence of an applied electric field [13],[30]. Due to disorder, the transfer time is a random variable, and the accumulated sequence of all transfer events in the motion of a charge carrier can be viewed as a continuous-time random walk [31]–[33]. As a result, the $\sigma_{ph}(t)$ follows a power-law time dependence, $t^{-(1-\beta)}$, where the reduced temperature $\beta = T/T_0$ is related to the exponent, $\alpha$, by

$$\alpha = 1 - \beta\,. \tag{3.1}$$

In this expression, $k_B T_0$ specifies the energy range over which the localized states (traps) can be filled [29], which corresponds to the energy difference between $E_F$ and $E_c$, i.e., the thermal activation energy $\varepsilon$ in resistivity measurements (see Eq. 2.1). Thus,

$$\beta = \frac{k_B T}{E_a}\,. \tag{3.2}$$

The monotonic reduction of $\alpha$ with increasing $I_L$ is consistent with the electronic structure of a Fermi glass (as shown in Fig. 2.1). Consider, for example, the $YBa_2Cu_3O_{6.4}$ data in Fig. 3.3b: At $T = 35$ K, $k_B T = 3$ meV that $\beta \approx k_B T/E_a \approx 0.1$. Thus the exponent in the time decay of $\sigma_{ph}(t)$ at the lowest $I_L$ is $\alpha \approx 0.9$, consistent with Eqs. 3.1 and 3.2. Since the data in Fig. 3.3 were measured at a fixed temperature, the reduction of $\alpha$ with increasing $I_L$ implies a shift in $E_F$ toward $E_c$, caused by filling the localized states with photoexcited charge carriers (and thus the M-I transition at higher $I_L$) [9]–[11]. This is consistent with the fact that the initial $\sigma_{ph}$ approaches $\sigma_{min}$.

The density of localized states near $E_c$ can also be estimated from the data shown in Fig. 3.3. A value of the order of $10^{22}$ photons/cm$^3$eV is obtained from $\Delta\beta/\Delta I_L$. Again, using the quantum efficiency of 0.1 as inferred from the photoinduced absorption experiments [28], this number agrees with the $N(E_F) \approx 10^{21}$ (cm$^3$eV)$^{-1}$ inferred from other experiments.

The photoconductivity measured under steady state photoexcitation in these crystals is also consistent with dispersive transport [5],[6]. After prolonged photoexcitation, the photocurrent relaxes non-exponentially, typically fitting the stretched-exponential form $\exp(-(t/\tau)^\gamma)$, which correlates with the power law dependence of the photocurrents on chopping frequency observed with light-modulation techniques [5].

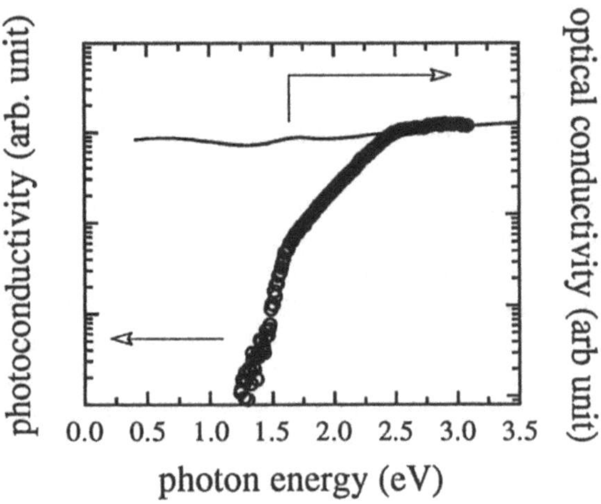


**Fig. 4.1.** $\sigma_{ph}(\omega)$ and $\sigma(\omega)$ in a single $YBa_2Cu_3O_{6.3}$ crystal at 300 K.

## 4. Spectral response of the photoconductivity in partially doped insulating cuprates

For stoichiometric (undoped) $YBa_2Cu_3O_6$, the optical conductivity, $\sigma(\omega)$, shows an energy gap of > 1.5 eV with interband transitions at 1.75 eV, 2.6 eV and 4.2 eV. Upon doping with oxygen, a broad absorption band grows across the near and middle infrared (ir), accompanied by a reduction of oscillation strength above the charge transfer (CT) gap. No discontinuous changes in the ir and visible absorption are found in association with the metal-insulator (M-I) transition, which occurs at $x \approx 0.4$. Rather, the sub-gap absorption is observed at very low doping concentrations (as soon as carriers are added to the $CuO_2$ layers) and increases in strength monotonically across the M-I transition boundary into the metallic/superconducting phase [4],[8].

The localized electronic states and the electronic structure of the Fermi-glass can also be studied spectroscopically [5],[6],[34]. With optical data alone (e.g., $\sigma(\omega)$), however, one cannot distinguish whether the initial and/or the final states in an optical transition are extended or localized, nor can one distinguish whether the resulting excitations are charged or neutral. Such distinctions require a comparison of $\sigma_{ph}(\omega)$ with $\sigma(\omega)$.

Fig. 4.1 shows a comparison of the $\sigma_{ph}(\omega)$ with $\sigma(\omega)$ in a $YBa_2Cu_3O_{6.3}$ crystal with $E_a \approx 230$ meV [5],[35]. Both data sets were measured at 300 K. Although $YBa_2Cu_3O_{6.3}$ is an insulator, *no energy gap can be seen in* $\sigma(\omega)$. Oscillator strength is distributed over the entire ir. In contrast, $\sigma_{ph}(\omega)$ decreases rapidly for $\hbar\omega < 2.1$ eV. In addition, the spectral profile of $\sigma_{ph}(\omega)$ becomes strongly temperature dependent between 1.4 eV and 2.1 eV: the lower $\hbar\omega$, the stronger the $T$ dependence. This is shown in Fig. 4.2, in which $\sigma_{ph}(\omega)$ is plotted for data obtained at 340 K, 300 K, 240 K, 190 K; the data are taken under photon flux of $10^{15}$ $cm^{-2}sec^{-1}$ and normalized to $\sigma_{ph}(2.5eV)$.

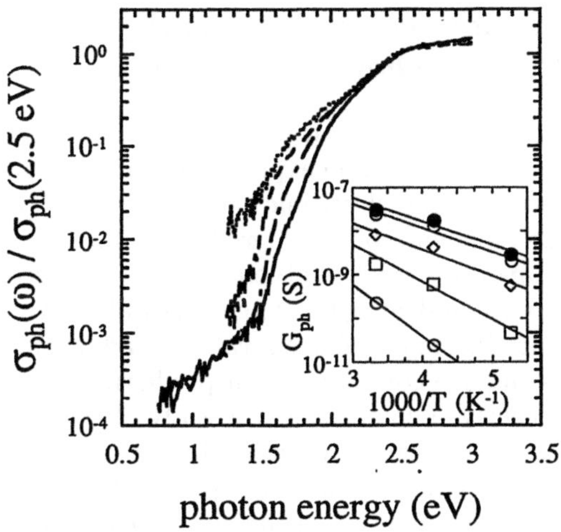


**Fig. 4.2.** $\sigma_{ph}(\omega)$ in $YBa_2Cu_3O_{6.3}$ at 340 K (top), 300 K, 240 K and 190 K (bottom). In the inset, the $\sigma_{ph}$ is plotted vs $1/T$ at 3 eV (top), 2.5 eV, 1.8 eV and 1.4 eV (bottom).

At 300 K, the ratio between $\sigma_{ph}(1.7eV)$ and $\sigma_{ph}(2.6eV)$ is about 10%, and the 1.7 eV feature shows up only as a shoulder. This ratio drops by a factor of $\sim$ 10 when the sample is cooled from 300 K to 190 K, and the shoulder disappears. This strong temperature dependence is observed only in $\sigma_{ph}(\omega)$; there is no corresponding temperature effect in $\sigma(\omega)$. Above 2.1 eV, the profile of $\sigma_{ph}(\omega)$ is insensitive to temperature; i.e., the photocarriers excited with $\hbar\omega > 2.1$ eV have a common thermal activation energy for transport. Moreover, the value of this thermal activation energy is similar to that measured in the d.c. conductivity (which measures the energy interval between $E_F$ and the mobility edge $E_c$; see section 2.). The detailed temperature dependence of $\sigma_{ph}(\omega)$ (plotted for several photon energies) is shown in the inset of Fig. 4.2; the increase in thermal activation energy with decreasing photon energy for $\hbar\omega < 2.1$ eV is clearly evident.

The data in Figs. 4.1 and 4.2 are consistent with the electronic structure of a Fermi glass. The weaker response and the stronger temperature dependence in $\sigma_{ph}(\omega)$ imply that additional thermal energy is required for the photocarriers excited with low energy photons ($\hbar\omega < 2.1$ eV) to contribute to conduction; evidence that the electronic states involved in these transitions are localized. Thus the spectral gap in the $\sigma_{ph}(\omega)$ in Figs. 4.1 and 4.2 corresponds to a mobility gap.

## 5. Supporting evidence of the Fermi-glass state from ARPES, NQR and other experiments

In addition to various transport and spectroscopic effects discussed in previous sections, there is considerable experimental evidence in support of the Fermi-glass description of the electronic structure of the partially-doped insulating cuprates and evidence of disordered crystal fields in these materials.

The results of angle-resolved photoemission spectroscopy (ARPES) on $YBa_2Cu_3O_{7-\delta}$, and $Bi_2Sr_2CaCu_2O_8$ single crystals have proven the existence of Fermi surface in the metallic cuprates; in addition, the details of the Fermi surface mapping are consistent with predictions by LDA band calculations [36],[37]. Recently, a comprehensive study of well-characterized single crystals of $YBa_2Cu_3O_{7-\delta}$ (with $\delta$ covering broad range from 0 to 0.8) was carried out at Argonne National Laboratory [38],[39]. The Fermi surface exists in metallic cuprates doped to optimized superconductivity ($\delta \approx 0.06$, $T_c \approx 92$ K), and also exists in metallic samples which are close to the M-I transition (e.g., $\delta \approx 0.5$–$0.6$). Most importantly, the Fermi-surface crossing is also observed in partially-doped insulating crystals ($\delta \approx 0.65$–$0.7$) [38],[39]. The main effect of crossing over to the insulating side of the M-I transition (by reduction of oxygens at O(1) sites) is a continuous decrease in the spectral weight near $E_F$. As pointed out by Liu et al. [38]: when the oxygen is reduced to convert the crystal into the insulating phase, the spectral features continue to show dispersion with no indication of an energy gap (within the limit of instrumental resolution). The conclusion is that *the insulator has a Fermi surface* with no gap in the density of states; the definition of a Fermi-glass. These results provide strong supporting evidence of the Fermi-glass electronic structure, implying that the M-I transition at $\delta \approx 0.6$ is due to localization of the carriers at the $E_F$ (rather than due to the opening up of an energy gap).

Nuclear magnetic resonance (NMR) and nuclear quadrupolar resonance (NQR) are effective local probes to investigate the crystal field surrounding each nucleus. Recent NQR experiments of $^{63}Cu$ on powder ceramic samples of undoped $La_2CuO_4$ and on doped samples of $La_{2-x}Sr_xCuO_4$ show that the $^{63}Cu$ NQR line broadens quickly at low levels of Sr substitution (in the insulating phase with $x$ only a few percents). The observed linewidth in carefully reduced $La_2CuO_4$ is $55 \pm 5$ KHz [40], which is twenty times narrower than that for Sr doped samples [41],[42]. As pointed out by the authors [40], this indicates that the Sr doping causes substantial disorder of the local field in the $CuO_2$ sheets. Note that this broadened NQR linewidth (and the implied disorder) persists when doping the crystal into the metallic phase and even to the doping level corresponding to optimized superconductivity ($x \approx 0.15$) [41],[42]. These experimental results provide clear evidence of disorder in the crystal potential in the $CuO_2$ planes due to random Sr substitution, consistent with the existence of localized electronic states observed in various transport experiments.

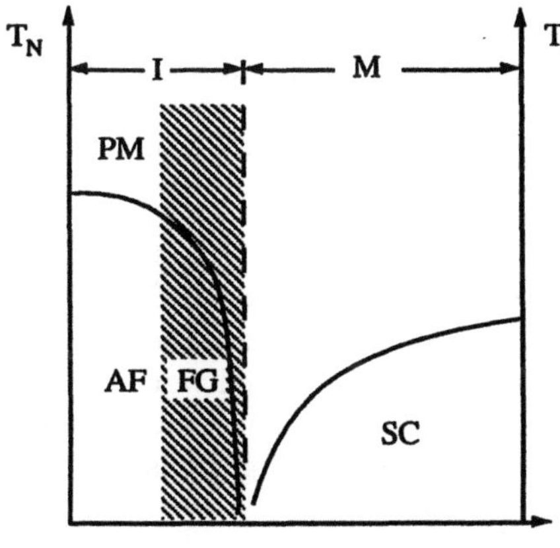


**Fig. 6.1.** Suggested phase diagram of HTSC cuprates; the Fermi glass phase is shaded. (I: insulating; M: metallic; PM: paramagnetic; AF: antiferromagnetic; FG: Fermi glass; SC: superconducting; $T_N$: Néel temperature; $T_c$: superconducting transition temperature).

It is well known that the diffusion of oxygen persists in the cuprates even at low temperatures. For this reason, for example, the YBCO crystals used for the ARPES experiments must be cleaved at 20 K before measurement to obtain a fresh surface with oxygen concentration similar to that in the bulk [43]. This type of thermally induced random oxygen motion will also contribute to the disorder of the crystal potential, as in conventional ionic conductors.

## 6. Discussions and Remarks

The experimental facts discussed in this brief review suggest that a region of Fermi glass state exists in phase diagram of the HTSC cuprates. Thermally-activated hopping behaviour (in dark conduction and the dispersive transport of the photoinduced carriers) exists not only in crystals doped near the boundary of M-I transition, but also in crystals in the antiferromagnetic phase. For example, the crystals of $La_2CuO_{4+\delta}$ used for the transport and photodoping studies have $T_N > 260$ K, higher than the temperatures used for the transport studies (25 K in the photodoping experiment). Even in carefully reduced single crystal $La_2CuO_{4+\delta}$ (with $\delta$ much smaller than 0.01 and $T_N \approx 315$ K), the $E_a$ inferred from $\rho(T)$ is only 100 meV [19],[25], much less than the charge-transfer gap energy. Thus, the reduction of the antiferromagnetic ordering (as indicated by $T_N$) of the Cu spins and the dispersive transport of the localized carriers are dual-features in partially-doped insulating cuprates. The significant reduction of $T_N$ is correlated with a significant growth of the density of localized electronic states in $CuO_2$ planes. Both phenomena are correlated with the random distribution of the dopants (oxygen atoms) and the disorder of the crystal field in the lattice. These facts are summarized in the phase diagram of Fig. 6.1. The Fermi glass state covers the shaded region in which $T_N$ drops quickly and co-exists (over some range of concentrations) with the spin glass state.

A correlation between the carrier hopping transport and paramagnetism in a conventional Fermi glass, hydrogen-doped $CaV_2O_6$, was studied a decade ago by Müller and co-workers [44]. Paramagnetism has also been observed in partially doped cuprates (see the related work in this Workshop) where it could be related with phase inhomogeneity and Fermi glass insulating components in the specimens. Further investigations on high quality crystals may be helpful to solve this puzzle.

Ordered magnetic structures have been found in partially-doped cuprate crystals through neutron scattering experiments and discussed extensively in terms of spin-clusters in this Workshop; the size of these clusters are of the order of $\sim$ 10 Å(see the related articles in this Proceedings). On the other hand, on a larger scale (for example, over hundreds of Ångstroms), the magnetic data suggest crystal field inhomogeneity and incoherence. This is consistent with the disorder and localization effects observed in the various transport experiments. This may also provide a hint to the relation between disorder and phase separation (or phase fluctuation).

In conclusion, various experimental results from high quality single crystals of partially doped insulating cuprates reveal a Fermi-glass type of electronic structure. The M-I transition is dominated by localization of the electronic wave functions at the Fermi energy as a result of local disorder in the crystal field. The density of states at $E_F$, inferred from various experiments, is of the order of $10^{21}$ $cm^{-3}eV^{-1}$ for insulating crystals near the metal-insulator transition.

The M-I transition can be induced by chemical doping, by photoexcitation, (and other physical parameters such as magnetic field or pressure etc.) to cause a relative shift in $E_F$ with respect to $E_c$. Therefore, the electronic structure of the metallic/super-conducting cuprates with optimized transition temperature can be characterized as disordered metals with the Fermi energy relatively close to the mobility edge [45]. This is in good agreement with the low Fermi temperatures observed in related experiments (100 meV in LSCO with $T_c \approx 40$ K and 200 meV in 123, 2212 and 2223 series with $T_c \approx$90–120 K) [46]. Since these materials are disordered metals with the Fermi energy relatively close to the mobility edge, the interaction between the quasi-particles at the $E_F$ and the localized electronic states near $E_c$ might be of importance. We have shown recently that the polarizability of these localized states can provide an attractive interaction to pair the quasi-particles; the same interaction provides a mechanism for the linear temperature dependence of the normal state resistivity [45].

*Acknowledgement.* This work was supported by the National Science Foundation under DMR 91-21946.

## References

1. T. Timusk and D.B. Tanner, in *Physical Properties of High Temperature Superconductors I* D.M. Ginsberg, Eds. (World Scientific, Singapore, 1989), p. 339.
2. S.L. Cooper, G.A. Thomas, A.J. Millis, P.E. Sulewski, J. Orenstein, D.H. Rapkine, S.-W. Cheong, and P.L. Trevor, Phys. Rev. B **42**, 10785 (1990).
3. J. Orenstein, G.A. Thomas, A.J. Millis, S.L. Cooper, D.H. Rapkine, T. Timusk, L.F. Schneemeyer, and J.V. Waszczak, Phys. Rev. B **42**, 6342 (1990).
4. D.B. Tanner and T. Timusk, in *Physical Properties of High Temperature Superconductors III* D.M. Ginsberg, Eds. (World Scientific, Singapore, 1992), p. 363.
5. G. Yu, C.H. Lee, D. Mehalivok, A.J. Heeger, C. Fincher, N. Herron, and E.M. McCarron, Phys. Rev. B **48**, 7545 (1993).
6. G. Yu and A.J. Heeger, Inter. J. Rev. Mod. Phys. **7**, 3751 (1993).
7. G.A. Thomas, in *Proceedings of the Scottish Universities Summer School on Physics of High-Temperature Superconductors*, St. Andrews, D.P. Dunstall, Eds. (1991).
8. G.A. Thomas, D.H. Rapkine, S.L. Cooper, S.-W. Cheong, and A.S. Cooper, Phys. Rev. Lett. **67**, 2906 (1991).
9. G. Yu, C.H. Lee, A.J. Heeger, N. Herron, and E.M. McCarron, Phys. Rev. Lett. **67**, 2581 (1991).
10. G. Yu, C.H. Lee, A.J. Heeger, N. Herron, E.M. McCarron, L. Cong, G.C. Spalding, C.A. Nordman, and A.M. Goldman, Phys. Rev. B **45**, 4964 (1992).
11. G. Yu, C.H. Lee, A.J. Heeger, S.-W. Cheong, and Z. Fisk, Physica C **190**, 563 (1992).
12. G. Yu, PhD dissertation (University of California at Santa Barbara, 1992).
13. N.F. Mott and E.A. Davis, *Electronic Processes in Non-Crystalline Materials* (Clarendon Press, Oxford, 1979).
14. N. Mott, *Conduction in Non-crystalline Materials* (Clarendon Press, Oxford, 1987).
15. N.F. Mott, *Metal-insulator Transitions* (Taylor & Francis, London, 1990).
16. A. Miller and S. Abrahams, Phys. Rev. **120**, 745 (1960).
17. N.F. Mott, J. Non-cryst. Solids **1**, 1 (1968).
18. N.F. Mott, *Metal-Insulator Transitions* (Tylor & Francis, London, 1974), Chapter 2.
19. N.W. Preyer, R.J. Birgeneau, C.Y. Chen, D.R. Gabbe, H.P. Jessen, M.A. Kastner, P.J. Picone, and T. Thio, Phys. Rev. B **39**, 11563 (1989).
20. M.A. Kastner, R.J. Birgeneau, C.Y. Chen, Y.M. Chiang, D.R. Gabbe, H.P. Jenssen, T. Junk, C.J. Peters, P.J. Picone, T. Thio, T.R. Thurton and H.L. Tuller, Phys. Rev. B **37**, 111 (1988).
21. P. Lukenheimer, M. Resch, A. Loidl, and Y. Hidaka, Phys. Rev. Lett. **69**, 498 (1992).
22. G. Yu, A.J. Heeger, G. Stuky, N. Herron, and E.M. McCarron, Sol. St. Comm. **72**, 345 (1989).
23. D. Mandrus, L. Forro, C. Kendziora, and L. Mihaly, Phys. Rev. B **44**, 2418 (1991).

24. S. Roth, in *Hopping Transport in Solids* M. Pollack, B. Shklovskii, Eds. (North-Holand, Amsterdam, 1991), p. 377.
25. C.Y. Chen, R.J. Birgeneau, M.A. Kastner, N.W. Preyer, and T. Thio, Phys. Rev. B **43**, 392 (1991).
26. F.G. Aliev, V. Kovacik, S.R. Li, V.V. Moshchalkov, N.N. Oleinikov, N.A. Samarin, J. Sebek, and L. Skrbek, J. Mag. Mag. Mat. **90-91**, 641 (1990).
27. G. Yu, C.H. Lee, and A.J. Heeger, in *Phase Separation in Cuprate Superconductors*, G. Benedek and K.A. Muller Eds. (World Scientific, Singapore, 1993).
28. S.G. Han, Z.V. Vardeny, K.S. Wong, and O.G. Symko, Phys. Rev. Lett. **65**, 2708 (1990).
29. H. Sher, M.F. Shlesinger, and J.T. Bendler, Phys. Today **44**, 26 (1991).
30. N.F. Mott, Adv. Phys. **34**, 329 (1985).
31. E.W. Montroll and H. Sher, J. Stat. Phys. **9**, 101 (1973).
32. G. Pfister and H. Sher, Adv. Phys. **27**, 747 (1978).
33. H. Sher and E.W. Montroll, Phys. Rev. B **12**, 2455 (1975).
34. G. Yu, C.H. Lee, A.J. Heeger, and S.-W. Cheong, Physica C **203**, 419 (1992).
35. G. Yu, C.H. Lee, D. Mihailovic, C. Fincher, N. Herron, E.M. McCarron, and A.J. Heeger, J. Phys. Chem. Solids **54**, 1161 (1993).
36. W.E. Pickett, Rev. Mod. Phys. **61**, 433 (1989).
37. W.E. Pickett, H. Krakauer, R.E. Cohen, and D.J. Singh, Science **255**, 46 (1992).
38. R. Liu, B.W. Veal, A.P. Paulikas, J.W. Downey, P. J. Kostic, S. Fleshler, U. Welp, C.G. Olson, X. Wu, A.J. Arko and J.J. Joyce, Phys. Rev. B **46**, 11056 (1992).
39. R. Liu, B.W. Veal, A.P. Paulikas, J.W. Downey, H. Shi, C. G. Olson, C. Gu, A.J. Arko, and J.J. Joyce, Phys. Rev. B **45**, 5614 (1992).
40. T. Imai, C.P. Slichter, K. Yoshimura, and K. Kosuge, Phys. Rev. Lett. **70**, 1002 (1993).
41. T. Imai, K.Y. T., T. Uemura, H. Yasuoka, and K. Kosuge, J. Phys. Soc. Jpn. **59**, 3846 (1990).
42. C.H. Pennington and C.P. Slichter, in *Physical Properties of High Temperature Superconductors II* D.M. Ginsberg, Eds. (World Scientific, Singapore, 1991).
43. R.S. List, A.J. Arko, Z. Fisk, S.-W. Cheong, S.D. Conradson, J.D. Thompson, C.B. Pierce, D.E. Peterson, and R.J. Bartlett, Phys. Rev. B **38**, 11966 (1988).
44. K.A. Muller, T. Penney, M.W. Shafer, and W.J. Fitzpatrick, Phys. Rev. Lett. **47**, 138 (1981).
45. A.J. Heeger and G. Yu, Phys. Rev. B **48**, 6492 (1993).
46. D.R. Harshman and J.A.P. Mills, Phys. Rev. B **45**, 10684 (1992).

# Two-Magnon Raman Scattering in Doped $YBa_2Cu_3O_x$ with Respect to Confined Hole Clusters

P. Knoll, S. Lo, U. Hohenester, and M. Pressl

Institut für Experimentalphysik, Universität Graz, Universitätsplatz 5, A-8010 Graz

**Summary.** The 2-Magnon-Ramanscattering has been measured on single crystals of $YBa_2Cu_3O_x$ with different oxygen concentration. The decrease of scattering intensity upon doping can be interpreted as a reduction of antiferromagnetic ordered spin pairs and is in good agreement with recent neutron scattering results. On the other hand the change of complex zone-edge magnon energy indicates an itinerant character of the charge carriers which can be understood as an hoping tendency which is also responsible for the electrical conductivity.

## 1. Introduction

The interplay between antiferromagnetism and superconductivity is one of the interesting phenomena in the high $T_c$ research. Especially the behavior of a charge carrier in the antiferromagnetic background raises a lot of open questions. Two extreme pictures are possible: The carriers localize and form some bounded (or confined) state with the copper spins, or they tend to delocalize over the whole crystal forming band-like states. One of the first models actually including the interplay of oxygen holes and copper spins [1] predicted a confined hole in a spin polarized background. Such a model naturally supports a tendency to phase separation into hole rich and antiferromagnetic, carrier-less regions. So far, there exist measurements which seem to support such a behavior [2, 3]. But on the other hand, a lot of experimental data are published [4] which are interpreted in a homogeneous distributed charge carrier picture. However, each experimentalist who works with HTc-materials has to deal with sample inhomogeneity, defects etc. even in single crystals. The critical question is, if such effects are of intrinsic nature of these materials and if they are essential for the superconducting properties.

In this contribution we interpret the results of Raman scattering of spin-fluctuations upon doping within the picture of localized or confined charge carriers. As a result we find good quantitative agreement with recent neutron scattering data [3]. On the other hand a consistent picture can only be given, if also a mobility of the charge carrier — as e.g. hopping — is assumed.

## 2. Light scattering of spin fluctuations

The interaction of light with localized spins antiferromagnetically ordered has been studied on several materials in the past [5]. The pioneering work of Fleury and Loudon [6] had given the explanation why the 2-magnon scattering intensity can become so strong in antiferromagnets. Usually, light does not interact strongly with spin systems. The electric field component can only interact via spin-orbit coupling which makes this interaction channel very weak. The energy density of the magnetic field component of light is very low; therefore, only very weak scattering intensity is expected for direct magnetic coupling. These two possibilities of interaction would both predict a scattering intensity which strongly decreases with higher order processes; two magnon scattering would be less intense than one magnon scattering. In antiferromagnetically ordered spin systems light can excite the spin at site i from its ground state to a virtual intermediate state via usual momentum or dipole matrix elements. In this intermediate state the spin can change by an exchange matrix element with its antiferromagnetic neighbor at site j which results in a spin flip on each sublattice of the antiferromagnet. The system relaxes to its groundstate by emitting scattered radiation depending on a usual momentum or dipole matrix element. Because of energy conservation for the whole process, the scattered radiation is reduced in energy (frequency) by at least two spin excitations each on a sublattice. Because of momentum conservation and the negligible momentum transfer of light to the crystal the two involved spin excitations must have opposite momentum. This just described scattering mechanism naturally explains Raman intensity in lowest order due to 2-magnon excitations. The scattering intensity scales with the exchange energy and can become rather strong.

In HTc-materials it is widely accepted that the magnetic moments on the Cu site arise from the S=1/2 hole of the $3d^9$ configuration. These S=1/2 spins are antiferromagnetically ordered via a superexchange involving the oxygen 2p orbitals. The polarization selection rules of the 2-magnon scattering indicate strong optical matrix elements for transitions within the Cu-O plane (a-b direction) but no significant intensity is observed for transitions along z direction [7]. Assuming such inplane matrix elements and taking into account the square plane configuration of the Cu sites scattering intensity is predicted in $B_{1g}$ symmetry in agreement with the experiment [7, 8]. Although intensity is also observed in other symmetries [7, 9] this intensity is much lower and the shape and frequency range is not in agreement with the here discussed lowest order 2-magnon process. The resonance of the 2-magnon scattering has been reported recently [10] and indicates optical transitions of the charge transfer type between Cu and O. The scattering intensity may be written to be proportional to the square of the expectation value of a lattice sum of intersite scattering operators $M_{ij}$ [7].

$$I \approx \left( \sum_{ij} M_{ij} \right)^2 = \left( \sum_{ij} \boldsymbol{E}_l T_{ij} \boldsymbol{E}_s S_i^- S_j^+ \right)^2 \tag{2.1}$$

The sum runs over all antiferromagnetically coupled spin pairs and can be splitted into a sum over next neighbors and a sum over all sites $i$. In that formulation the scattering intensity is proportional to the square of the number of antiferromagnetically ordered spin pairs within the scattering volume and the intersite scattering tensor $T_{ij}$ which connects the scattered field $\boldsymbol{E}_s$ with the laser field $\boldsymbol{E}_l$. The intersite scattering tensor includes the optical inplane matrix elements and the exchange matrix element and can be obtained by third order perturbation theory [6]. If the phase-coherence of the spin excitations is very short not the square of the sum has to be taken but the sum of the square. In that case the intensity is directly proportional to the number of spin pairs. (The Fourier-transform of the on site spin operators $S_{i,j}^{-;+}$ represent the spin excitations.) Based on that formulation it was possible to give a description of the 2-magnon scattering process in HTc-materials if additional damping of the spin excitations via magnon-phonon interaction is taken into account [11].

Upon doping the 2-magnon scattering vanishes and relaxes into a broad featureless background [12]. This behavior has been quantitatively studied in the $YBa_2Cu_3O_x$ system [13]. The exchange energy (or realpart of the one magnon energy) does decrease only by 3% if the oxygen content is reduced from $O_7$ to $O_{6.5}$. On the other hand the one magnon damping increases by a factor 2 in the same doping range. In this paper we concentrate our interest to the intensity dependence of the 2-magnon scattering.

## 3. Results

The 2-magnon Raman scattering signal has been measured with the 488nm laser excitation at room temperature for different doping concentrations [13] and are shown in Fig. 2.1. The spectra are corrected for the spectral sensitivity of the spectrometer by comparison with a standard lamp. The intensity of the 2-magnon excitation has been worked out by fitting the experimental curve with a damped spin wave model [11, 7]. The integrated intensity of the fitcurve has been taken as the scattering intensity of the 2-magnon Raman signal. This procedure is necessary because of the more complicated lineshape of the spin fluctuation signal. Even the pure 2-magnon signal is no simple lorentzian but has a strong asymmetric contribution and an intrinsic background which increases to lower frequencies. The experimentally observed spin-fluctuation signal contains not only the 2-magnon process but has also additional contributions from higher order processes [11, 7] which may have other interaction channels than that described in the previous section. Therefore, it is very important to determine only the 2-magnon contribution

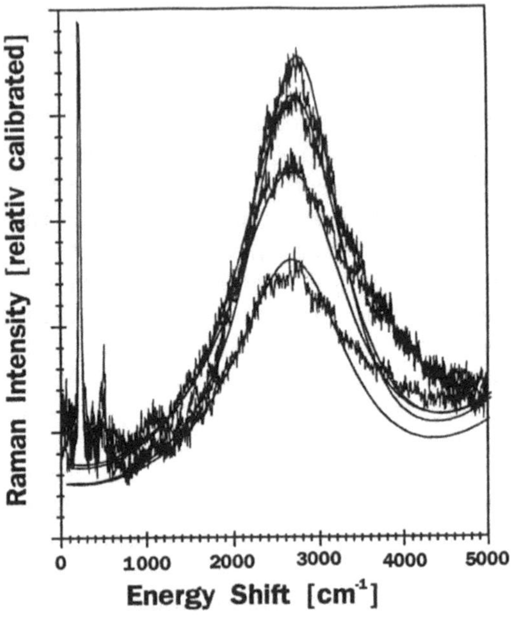


**Fig. 2.1.** Raman spectra in $B_{1g}$-symmetry of $YBa_2Cu_3O_x$ in the doping range of $6.0 < x < 6.5$. Each spectrum is calibrated in relative intensities. The broad feature at 2700 $cm^{-1}$ corresponds to spin fluctuations. The sharp peak at 335 $cm^{-1}$ is the $B_{1g}$ phonon. The thin solid lines are 2-magnon fits to the spectra with an additional luminescence background peaking around 12000 $cm^{-1}$.

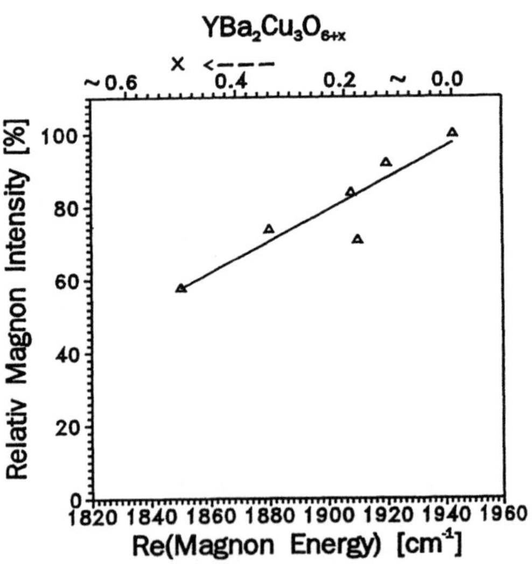


**Fig. 2.2.** The integrated intensities of the 2-magnon-fits of Fig. refknoll.fig1 plotted against the real part of the zone-edge magnon energy obtained from the same fit. The intensity of the spectrum for an oxygen content of 6.0 is set to 100%.

including its background in order to discuss the results with respect to the assumed scattering mechanism. Fig. 2.2 shows these integrated 2-magnon intensities scaled to the intensity of the undoped material by comparison to the $B_{1g}$ phonon. The change of the Raman intensity of this phonon upon doping is known and roughly has a linear behavior. Within the laser spot it is very difficult to know the exact doping concentration. Therefore, in Fig. 2.2 the intensity is plotted against the real part of the one-magnon energy at the zone boundary (corresponds directly to the exchange energy) because both values (intensity and magnon energy) are obtained by the same fit and give more accurate the behavior of the spin system. The scale of doping concentration is more inaccurate with an estimated error of 0.1. The interesting feature of Fig. 2.2 is that the intensity of the 2-magnon excitation does not decrease so strongly upon doping as one would expect from the experimental data of Fig. 2.1. The reason is that a lot of intensity is going from the magnon peak into the background because of the broadening of the signal. Following the scattering mechanism as described in the previous section this decrease of intensity can be explained by an reduction of antiferromagnetic pairs in the scattering volume. If one assumes a short coherence length of the zone edge magnons the simple linear dependence suggests a reduction of antiferromagnetic spin pairs less than a factor of 2 upon doping to 6.5. Recent neutron scattering experiments done independently have quantitatively determined the amount of different phases in the $ErBa_2Cu_3O_x$ samples upon doping [3]. Within the experimental error good agreement exists between our decrease in 2-magnon intensity and the reduction of $A_3$ phase (no oxygen sites occupied on the chains) determined by neutron scattering. On the other side one also has to take into account that the change of Raman intensity can occur from a change in optical matrix elements. Optical measurements [4] show a broadening of optical transitions which smear out the gap at higher doping concentrations in a way that spectral weight from the blue is shifted to the red. From that behavior one could expect a similar tendency of 2-magnon scattering; excitation with blue laser light will become less intense and intensity at green and red laser excitation will become stronger. For such a similarity to optical reflectivity and transmission experiments one has to consider that for 2-magnon scattering only some particular matrix elements which enable the exchange coupling to the neighboring spin can contribute. This is different to optical experiments where all optical matrix elements are relevant and one can follow the shift of states from the band to the gap. As long as the special 2-magnon scattering mechanism is valid only some selected matrix elements are probed by the Raman experiment. These matrix elements are not expected to change dramatically with charge carrier concentration. That idea is supported by preliminary investigations of the resonance Raman profile in the intermediate doped material which shows no significant change compared to the undoped material. So far, it seems that a rather quantitative picture of the doped material can be derived from Raman scattering based on destroying antiferromagnetic spin pairs. Although this interpretation of 2-magnon scattering requires only localized moments on the Cu site one can learn a lot about charge carriers from the intensity behavior upon doping. It is clear that a hole localized on the oxygen between two antiferromagnetically coupled spins will destroy the superexchange and the 2-magnon scattering mechanism. If the hole forms a bound state with one local Cu moment the 2-magnon scattering mechanism is prohibited by at least this binding energy. If the hole forms a more confined state on the oxygen site the scattering and exchange mechanism is prohibited by at least the on site Hubbard U for the oxygen 2p orbital. This is also true if the hole is delocalized in an oxygen band if there is a probability that within the timescale of the experiment (which in our case is larger than a few femtoseconds) the hole is present in order to prohibit the 2-magnon process. If the hole additional interacts with the copper spins and forms a spin polarized background [1] even at small Hubbard U on the oxygen 2p orbital the 2-magnon process will not take place because of the lack of antiferromagnetically ordered spin pairs. If the intensity values from Fig. 2.2 are taken one obtains a reduction of antiferromagnetic spin pairs to 60% or 77% (depending on the phase coherence of the spin excitations) by doping to $O_{6.5}$. For such a doping level a hole concentration of around 0.07 holes per $CuO_2$ unit or 0.14 holes per possible antiferromagnetic spin pair is estimated. That means at this doping concentration a reduction on spin pairs to 86% if one hole only destroys one antiferromagnetic pair. The experimentally determined value of 60% to 77% is obtained if one hole destroys three to two antiferromagnetic spin pairs. This is a clear result of the intensity analysis of 2-magnon scattering if one considers a time scale of longer than a few femtoseconds. Within that timescale the doped hole influences at least 2 or 3 antiferromagnetic bonds. The picture of a hole confined to a cluster with the size of around 3 copper pairs can give a simple and natural explanation of the observed intensity behavior. However, the 2-magnon experiment also shows a strong increase in one magnon damping upon doping as shown in Fig. 3.1. Such a behavior one would not expect for an antiferromagnet with some restricted regions (hole clusters) where no undisturbed antiferromagnetic order exists. Size effects of the antiferromagnet especially will dominate above the percolation limit of the hole clusters. The zone edge magnons are rather insensitive to size effects and coherence length effects because of their local spin flip nature. Therefore, at such low doping levels size effects can not explain this strong increase in zone edge magnon damping which is shown in Fig. 3.1. We assume that the hole is not completely confined to a cluster but has some probability to travel through the antiferromagnet. Because of interaction of the hole with the spin excitations the increase in damping can be explained. In the same way some mobility of the hole is required because of the strong increase in electrical conductivity upon doping. This itinerancy of the hole is not in contradiction with the assumed hole clusters because these clusters have only to exist for times longer than a few femtoseconds.

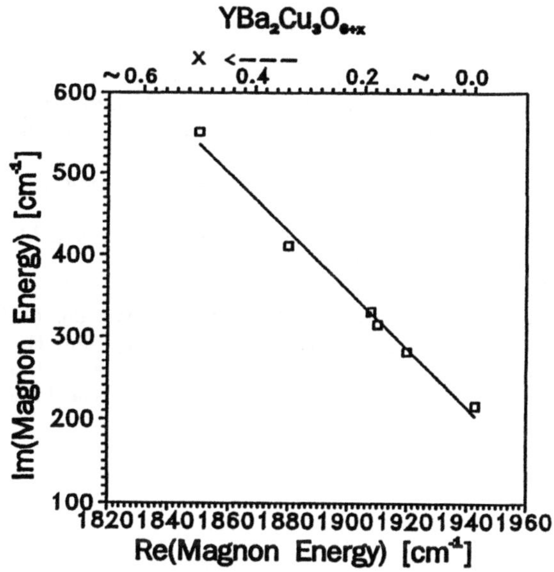


**Fig. 3.1.** Imaginary part of the zone-edge magnon energy (damping) plotted against the real part as obtained from the fits of Fig. 2.1. The accuracy of the oxygen content scale is around 0.1.

So far, the best common explanation of the 2-magnon Ramanscattering can be given by assuming confined hole areas with disturbed antiferromagnetism which do not take part in the 2-magnon scattering mechanism. The hole can not be completely restricted to its cluster but has some probability to travel through the antiferromagnet like a hopping mechanism. More details of the hole behavior, the coherence length of the spin excitations, exact size and shape of the possible hole clusters can not be given with the experimental data available so far.

*Acknowledgement.* This work was supported by the Fonds zur Förderung der wissenschaftlichen Forschung in Österreich., Projekt Nr.8181.

## References

1. V.Hizhnyakov and E.Sigmund, Physica C**165** (1988) 655 and this volume
2. R.K. Kremer, E. Sigmund, V. Hizhnyakov, F. Hentsch, A. Simon, K.A. Müller, and M. Mehring, Z. Phys. B-Condensed Matter **86** (1992) 319 and this volume
3. J. Mesot, P. Allenspach, U. Staub, A. Furrer, and H. Mutka, Phys. Rev. Lett. **70** (1993) 865 and this volume
4. S.L. Cooper, A.L. Kotz, M.A. Karlow, M.V. Klein, W.C. Lee, J. Giapintzakis, and D.M. Ginsberg, Phys. Rev. B**45**, (1992) 2549

5. see e.g. M.G. Cottam and D.J. Lockwood, Light Scattering in Magnetic Solids, Ed. J. Wiley & Sons (1986)
6. P.A. Fleury and R. Loudon, Phys. Rev. **166**, (1967) 514
7. P. Knoll, Springer Series in Solid-State Sciences, Vol.**99** (1990) 318
8. R.R.P. Singh, P.A. Fleury, K.B. Lyons, and P.E. Sulewski, Phys. Rev. Lett. **62**, (1989) 2736
9. P.E. Sulewski, P.A. Fleury, K.B. Lyons, and S-W. Cheong, Phys. Rev. Lett. **67**, (1991) 3864
10. P. Knoll, U. Hohenester, and S. Lo, Physica B (in press), M.V. Klein, R. Liu et al., Journal of Superconductivity (in press)
11. P. Knoll, C. Thomsen, M. Cardona, P. Murugaraj, Phys. Rev. B**42**, (1990) 4842
12. see e.g. K.B. Lyons, P.A. Fleury et al., J. Appl. Phys. **64**, (1988) 6075
13. P. Knoll, S. Lo, M. Mayer, M. Pressl, P. Murugaraj, and E. Schönherr, Proc. of the XIIIth Int. Conf. on Raman Spectr. (1992) 886, Ed. J. Wiley & Sons, P. Knoll, S. Lo, P. Murugaraj, E. Schönherr, and M. Cardona, Springer Series in Solid State Sciences, Vol. **113**, (1993) 286

# Electron Hole Ordering in $La_2CuO_{4+\delta}$

M. Mehring, P. Gergen, C. Kessler, and S. Krämer

2. Physikalisches Institut, Universität Stuttgart, D-70550 Stuttgart, Germany

**Summary.** Phase separation in La cuprate superconductors is studied by NMR, NQR and microwave (mw) absorption measurements. The comparison of all these data leads to a unique phase separation temperature ($T = 230$ K) at which the separation into a hole rich (metallic and superconducting) and a hole poor (insulating and antiferromagnetic) phase occurs. There is indication that this phase separation is accompanied by oxygen diffusion. Numerical simulations of the Monte Carlo type and a stochastic Langevin equation approach are applied in order to model the experimental data.

## 1. Introduction

Phase separation into an electron hole rich (conducting/superconducting) and a hole poor (insulating/antiferromagnetic) phase in the cuprate superconductors has become a fascinating subject in solid state physics. A number of theoretical as well as experimental contributions to this area of research have appeared recently [1]–[8].

Starting from insulating antiferromagnetic (undoped) cuprates, where the Cu atoms in the $CuO_2$ plane carry the localized electron spins, it has been well established that already small concentrations of hole doping reduces the antiferromagnetic ordering temperature rapidly. The electronic structure of the hole is still a matter of discussion. In the low doping regime the hole is considered immobile, however, delocalized over a number $N_{Cu}$ of about 15 – 30 Cu atoms. This number can be derived from the percolation threshold for the onset of metallic conductivity. Very interesting $^{139}$La NQR features have been observed in this low doping regime. In particular the appearance of relaxation peaks in the longitudinal ($T_1$) and transversal spin lattice relaxation ($T_2$) well below the Néel peak in the range of 5 K and 60 K are indicative of strong magnetic fluctuations at those temperatures [9]–[13]. Fig. 1.1 and Fig. 1.2 represent typical examples obtained in our laboratory by Grosshans some time ago [13]. It was demonstrated that Sr doping enhances the 5 K peak with a maximum at 3% doping [14]. This special relaxation peak disappears above 5% where metallic conductivity sets in. We have attributed this relaxation peak to the local spin dynamics of the hole doped area. It supports the view of strong spin fluctuations in those areas.

Because of the confinement of the holes in the antiferromagnetic lattice variable range hopping conductivity is observed below the percolation threshold. Above the percolation threshold metallic conductivity and superconductivity appear as has been observed in Sr doped $La_{2-x}Sr_xCuO_4$. Since the 5 K relaxation peak disappears above this threshold it supports the view of

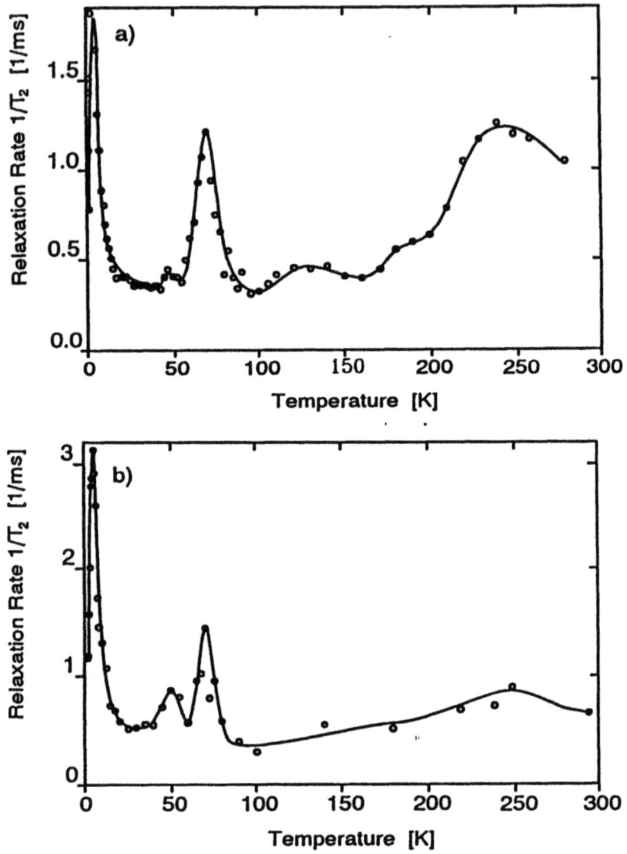


**Fig. 1.1.** Transversal relaxation rate $1/T_2$ of $^{139}$La NQR in as-prepared $La_2CuO_{4+\delta}$ for the NQR transitions a) $5/2 \leftrightarrow 7/2$ and b) $3/2 \leftrightarrow 5/2$. Besides the Néel peak at 250 K additional peaks due to hole doping are visible.

a metallic state with delocalized holes. From spin lattice relaxation measurements on Cu we know that antiferromagnetic correlations, however, are still present in this metallic state.

A different scenario is observed in oxygen doped $La_2CuO_{4+\delta}$ where $\delta$ represents the additional oxygen. It is well established that even as-prepared samples of $La_2CuO_4$ contain excess oxygen. Under high pressure or by using electrochemical oxidation, however, the oxygen content can be increased. A metallic and superconducting phase appears with optimum $T_c$. Even in the low doping regime a phase separation has been observed by dc susceptibility [7] and low field mw absorption, as well as NQR measurements [15]. It is interesting to note that all three measurements give about the same characteristic temperature, namely 230 – 240 K, where phase separation occurs. We believe it to be important that indeed a specific *phase separation tempera-*

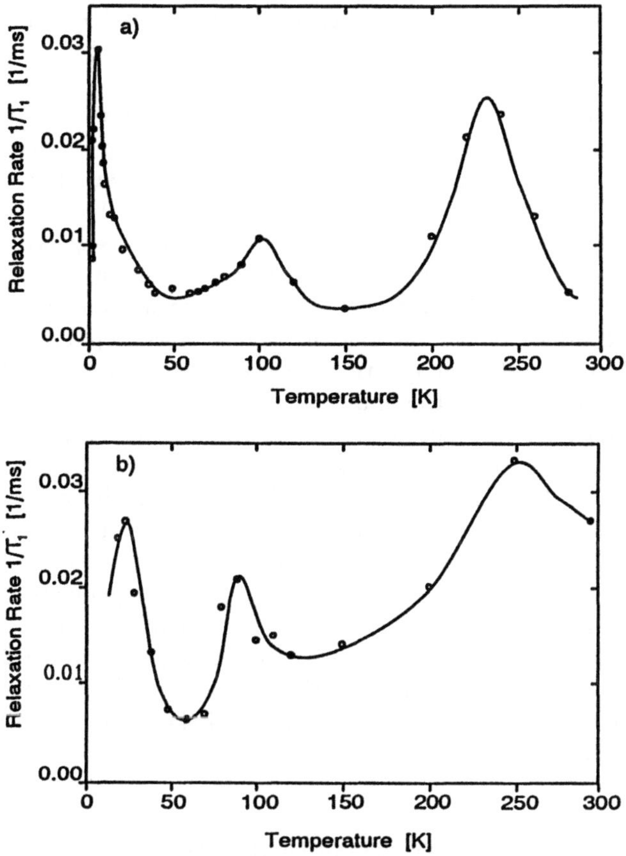


**Fig. 1.2.** $^{139}La$ $1/T_1$ relaxation rate for the a) $5/2 \leftrightarrow 7/2$ transition and b) $3/2 \leftrightarrow 5/2$. Similar features as compared to the $1/T_2$ relaxation are encountered.

*ture* is observed which allows the distinction between phase separation and all sorts of local distortions and inhomogeneities which are to be expected in this kind of material and which should not be confused with the phase separation phenomenon.

In this contribution we want to address the problem of electronic phase separation by comparing nuclear quadrupole resonance (NQR) relaxation measurements with low field microwave absorption (LFS) and dc susceptibility measurements in as-grown and electrochemically doped $La_2CuO_4$. In particular we will be concerned with the phase separation temperature and the spin dynamics around this transition. In order to support our observations we have started percolation calculations on a discrete lattice as well as a Ginsburg-Landau-Wilson continuum model. A brief report on preliminary results will be presented.

## 2. Spin dynamics in as-prepared and Sr-doped $La_{2-x}Sr_xCuO_4$

It came as a surprise that even in as-prepared samples of $La_2CuO_4$ superconductivity was observed. Conductivity measurements showed that charge carriers are in fact present in nominally undoped samples. The sensitivity of the Néel temperature to doping and the observation that pristine samples show a much higher Néel temperature as compared to as-grown samples give hints towards excess oxygen as a possible cause for the small number of holes in as-prepared samples. From the susceptibility measurements of Kremer et al. [7, 17] and others it became evident that a phase separation into metallic and insulating antiferromagnetic phases occurs and is connected with the high mobility of oxygen in this material.

From $^{139}$La NQR measurements it was indeed realized quite early that certain relaxation features could not be explained by the antiferromagnetic Néel state alone. A relaxation peak in $1/T_1$, $1/T_2$ has been observed at the Néel temperature indicating the onset of antiferromagnetic order. However, additional peaks in the relaxation rates were observed below the Néel temperature which were not supposed to be there in a homogeneous antiferromagnetic state. An example of two sets of measurements at the $3/2 \leftrightarrow 5/2$ and $5/2 \leftrightarrow 7/2$ NQR transition of $^{139}$La are shown in Figs. 1.1 and 1.2 [11, 12, 13]. These additional peaks are caused by the as-prepared doping of the antiferromagnetic $CuO_2$ layers. Ordering phenomena of the doped holes into clusters may be responsible for this.

## 3. Spin dynamics in electrochemically oxidized $La_2CuO_{4+\delta}$

By applying the electrochemical doping procedure as introduced by Grenier and co-workers [16] we have oxidized $La_2CuO_4$ to varying degree. A drastic change in the $^{139}$La NQR spectra as well as in the spin lattice relaxation rates is observed with increasing oxygen content. Fig. 3.1 represents different $^{139}$La NQR spectra at 19 MHz ($5/2 \leftrightarrow 7/2$ transition) and $T = 130\,K$ for different oxygen doping. Note the double peak feature characteristic of the antiferromagnetic ordering field at the La nuclear site. This feature is only observed below the Néel temperature and is therefore indicative of antiferromagnetic order in the sample. With increasing oxygen doping a broad spectral feature appears with the antiferromagnetic phase still present. The broad spectral feature was attributed by Hammel et al. [18] to the hole rich part of the sample. In the highly oxygenated sample the antiferromagnetic double peak has disappeared (Fig. 3.1).

DC susceptibility measurements as well as low field mw absorption measurements in as-prepared samples revealed the appearance of a metallic/superconducting phase if slow cooling is applied in the range of 200 –

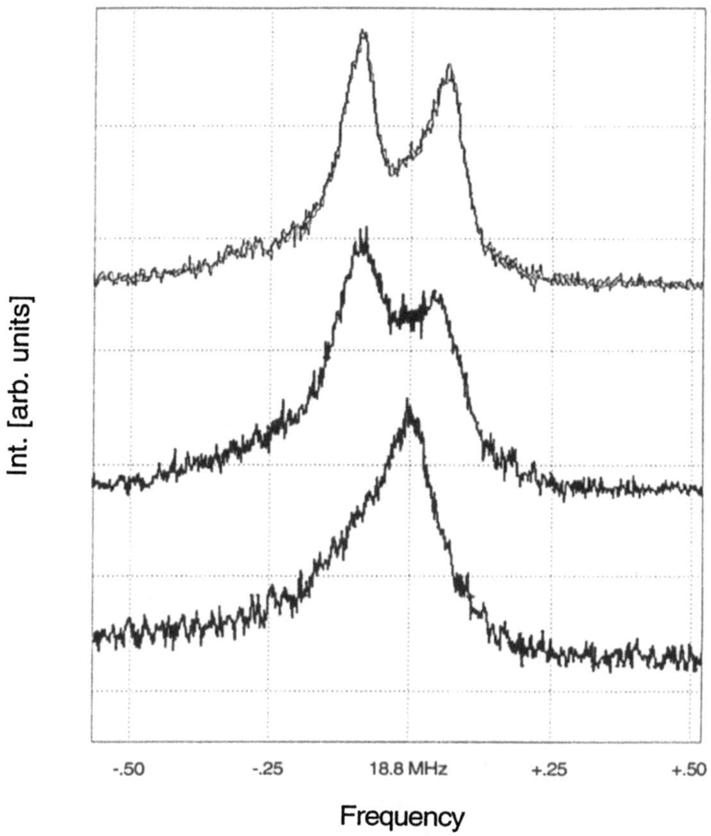


**Fig. 3.1.** $^{139}$La NQR spectrum at 19 MHz. Starting from the undoped sample (upper spectrum) higher doping levels reduce the antiferromagnetic splitting of the NQR line.

250 K. This was attributed to a phase separation taking place around this temperature. Hammel et al. [19] did observe a similar behaviour in the NQR/NMR spectra of high pressure oxygenated samples. The interesting question arises if characteristic dynamic processes appear at the phase separation temperature. In order to address this question we have performed different spin lattice relaxation measurements ($T_1$, $T_1^*$, $T_2$) around the suspected phase separation temperature. The different rates are plotted in Fig. 3.2 versus temperature.

Before we enter the discussion of these results we note that a pronounced relaxation peak occurs at the suspected phase separation temperature only in the stimulated echo relaxation rate $1/T_1^*$ whereas the longitudinal relaxation rate $1/T_1$ is rather small and the transversal relaxation rate $1/T_2$ is rather large and where both increase monotonically with temperature. In order to

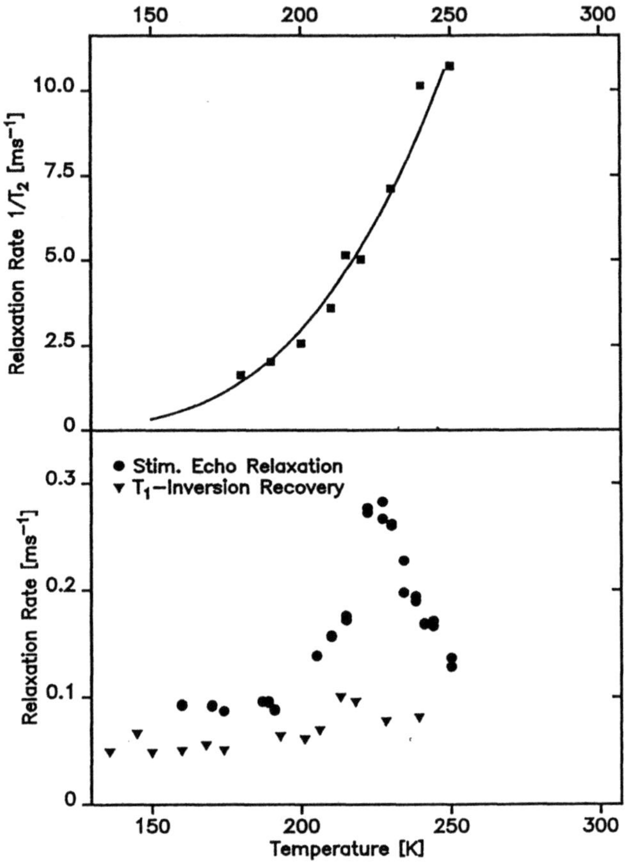


**Fig. 3.2.** Relaxation rates $1/T_1$ (triangles), $1/T_2$ (squares) and $1/T_1^*$ (circles) for $La_2CuO_{4.06}$. Note the different scales for the $T_2$ relaxation. A pronounced peak is only seen for the stimulated echo relaxation rate $1/T_1^*$.

appreciate these observations we summarize the different contributions to these relaxation rates.

(i) **transversal (spin echo) relaxation rate** $1/T_2$

The transversal spin echo relaxation rate $1/T_2$ represents the decay of coherence of the spin system ($x, y$-component) and is measured by applying a spin echo sequence, consisting of two rf pulses separated by a delay time $\tau$. With increasing $\tau$ the spin echo decays towards zero with a decay function $g_2(2\tau)$ which in the most simple case can be expressed as

$$g_2(2\tau) = M_0 \exp(-2\tau/T_2) \tag{3.1}$$

where $M_0$ is the initial magnetization. In general there is no reason for the decay function to be exponential. The echo decay is caused by a fluctuation of the energy level splitting of the $m \rightarrow m \pm 1$ transition represented by the second moment $< \Delta\omega_z^2 >$ which is either a measure of the fluctuation of the z-component of the local magnetic field or an electric field gradient $V_{zz}$ if quadrupolar interaction is involved. In the weak collision case, under the condition $\Delta\omega_z\, J(0) < 1$, where $J(0)$ is the spectral density of the fluctuation at zero frequency the transversal relaxation rate $1/T_2$ can be expressed as

$$\frac{1}{T_2} = < \Delta\omega_z^2 > J(0) + \frac{1}{2T_1} \tag{3.2}$$

with

$$J(0) = \int_0^\infty f(t)\, dt$$

where $f(t)$ is the correlation function of the fluctuating field. Both quantities, namely the strength of the fluctuation $< \Delta\omega_z^2 >$ and the spectral density contribute to the relaxation rate. In the strong collision case ($\Delta\omega_z\, J(0) \gg 1$), the transversal relaxation rate can be expressed as

$$\frac{1}{T_2} = \frac{a}{\tau_c}$$

where $a$ is a parameter of the order unity and $\tau_c$ is the correlation time. The disadvantage of $T_2$ for measuring dynamic processes is the fact that it is in most cases dominated by the slow spin dynamics of the coupled nuclear spins. It is therefore usually much shorter than any other relaxation time and does very often not allow to investigate physically relevant processes. Here, however, the situation is different as will be demonstrated below.

(ii) **longitudinal relaxation rate** $1/T_1$

The longitudinal relaxation rate represents the approach of the population difference of spin levels $m$ and $m \pm 1$ to Boltzmann equilibrium after a perturbation. The perturbation can be either a saturation sequence of rf pulses or an inversion (180°) pulse. The recovery curve depends on the way the experiment is performed and leads to a decay function $g_1(t)$ which need not be exponential, but can be non-exponential or – as in this case – a sum of three exponentials. The relaxation rate could e. g. be defined as the initial slope and can be expressed in similar terms as the transversal relaxation rate in the weak collision case:

$$\frac{1}{T_1} = < \Delta\omega_{xy}^2 > (J(\omega_1) + J(\omega_2)) \tag{3.3}$$

with

$$J(\omega) = \int_0^\infty f(t) e^{-i\omega t}\, dt$$

The second moment of the fluctuation $<\Delta\omega_{xy}^2>$ corresponds to fluctuations perpendicular to the quantization axis ($x, y$-plane) which can cause transitions between the levels $m$, $m \pm 1$ at frequencies $\omega_1$ and $\omega_2$. It is therefore obvious that here the spectral densities at the frequencies $\omega_1$ and $\omega_2$ are responsible for the relaxation.
In case of rapid fluctuations ($\omega_0\tau_c \ll 1$), where $\omega_0$ is the transition frequency, $T_1$ and $T_2$ are about equal, whereas in the case of slow motion ($\omega_0\tau_c \gg 1$) $1/T_2 \gg 1/T_1$.

(iii) **stimulated echo relaxation rate $1/T_1^*$**

The stimulated echo sequence is an extension of the spin echo sequence discussed in (i). A third pulse is applied at time $t$ after the second pulse. This results in a spin echo at time $\tau$ after the third pulse. The decay of this "stimulated" echo defines the relaxation rate $1/T_1^*$. The important aspect here is that its decay function is directly given by the correlation for slow spin dynamics

$$g_{se}(t) = < e^{-\tau\delta\omega(t)} \cdot e^{-\tau\delta\omega(0)} > \tag{3.4}$$

where $\delta\omega(t)$ is the frequency shift of the resonance frequency at time $t$ and which can be combined with the longitudinal relaxation function to the total decay function for the stimulated echo as

$$g_3(t) = g_{se}(t) \cdot g_1(t) \tag{3.5}$$

The decay function $g_1(t)$ for the longitudinal magnetization has been included in order to allow for the additional decay due to $T_1$ processes. The relaxation rate $1/T_1^*$ could again be represented by the initial slope if $g_3(t)$ is non-exponential. In this approximation it will be a sum of two contributions, namely

$$1/T_1^* = 1/T^* + 1/T_1 \tag{3.6}$$

In the more general case the correlation function needs to be known. In the case of slow motion the $g_{se}(t)$ or $1/T^*$ contribution is dominant and allows to investigate slow motion with high sensitivity. The stimulated echo relaxation rate $1/T^*$ does not suffer from the slow nuclear spin dynamics of the spin echo. It is therefore the appropriate sequence for the investigation of slow motions in solids. We apply the sequence here in particular to investigate the slow oxygen diffusion and the slow phase separation dynamics.

In Fig. 3.2 all three relaxation rates are plotted versus temperature for an electrochemically oxygenated $La_2CuO_{4+\delta}$ sample with $\delta$ about 0.06 determined from the electrochemical potential. The precise oxygen content is not known unfortunately. Note the pronounced difference in the absolute values of the relaxation rates. The longitudinal relaxation rate $1/T_1$ is the smallest and is monotonically increasing with temperature as is the transversal relaxation rate $1/T_2$ which, however, is much larger. The huge difference in these relaxation rates points at a slow motion process which governs both relaxation rates and which is due to the oxygen diffusion. This diffusion is expected to be thermally activated and can expressed by a diffusion rate

$$D = D_0 e^{-E_a/kT} \tag{3.7}$$

where $E_a$ is the activation energy. Activation energies for oxygen diffusion in $La_2CuO_4$ have not been measured yet but have been calculated to be on the order of 0.48 eV [20]. Oxygen diffusion could affect the $^{139}La$ spin relaxation in two different ways:

- fluctuation of the local *magnetic field* due to a rearrangement of the Cu spin structure in the neighbourhood of the hole accompanying the interstitial oxygen or
- a fluctuating *electric field* gradient due to the electron holes and/or the interstitial oxygen.

Which process dominates can be decided by comparing the different rates at the different transitions $3/2 \leftrightarrow 5/2$ and $5/2 \leftrightarrow 7/2$, respectively. This has been performed and is presented in Figs. 1.1 and 1.2. All three relaxation rates are expected to be affected by either hole and/or oxygen diffusion in the following way:

$$\begin{aligned} 1/T_1 &= cA_1 D_0 e^{-E_a/kT} \\ 1/T_2 &= cA_2 D_0 e^{-E_a/kT} \\ 1/T_1^* &= cA_3 D_0 e^{-E_a/kT} \end{aligned}$$

where $c$ is the concentration of the holes and/or the oxygen ions. It should be kept in mind, however, that in $c$ the "area of influence" of the oxygen or hole which is larger than a single lattice site has to be considered. The different scaling factors $A_1$, $A_2$ and $A_3$ are due to the different relaxation mechanisms discussed above. Here the following relation applies $A_1 \ll A_3 < A_2$.

Note that only $1/T_1$ and $1/T_2$ show the expected behaviour, whereas $1/T_1^*$ deviates significantly in that a narrow peak occurs at $T = 230\,K$. It is reminiscent of the relaxation peaks observed at magnetic and structural phase transitions. Both $1/T_1$ and $1/T_2$ show a weak peak at the paramagnetic to antiferromagnetic phase transition in undoped $La_2CuO_4$. This peak is, however, much weaker and is not observed in the oxygenated sample. Moreover the quadrupole spectrum of this sample does not show any antiferromagnetic splitting anymore. We therefore conclude that the relaxation peak in $1/T_1^*$ results from the cooperative modes of the phase separation with a *phase separation temperature* $T_p = 230\,K$. If the increase of $1/T_2$ with temperature is interpreted in terms of an activated process as discussed above an activation energy of 112 meV is found. Currently it cannot be distinguished if this is due to oxygen or hole diffusion or both. Details will be published elsewhere.

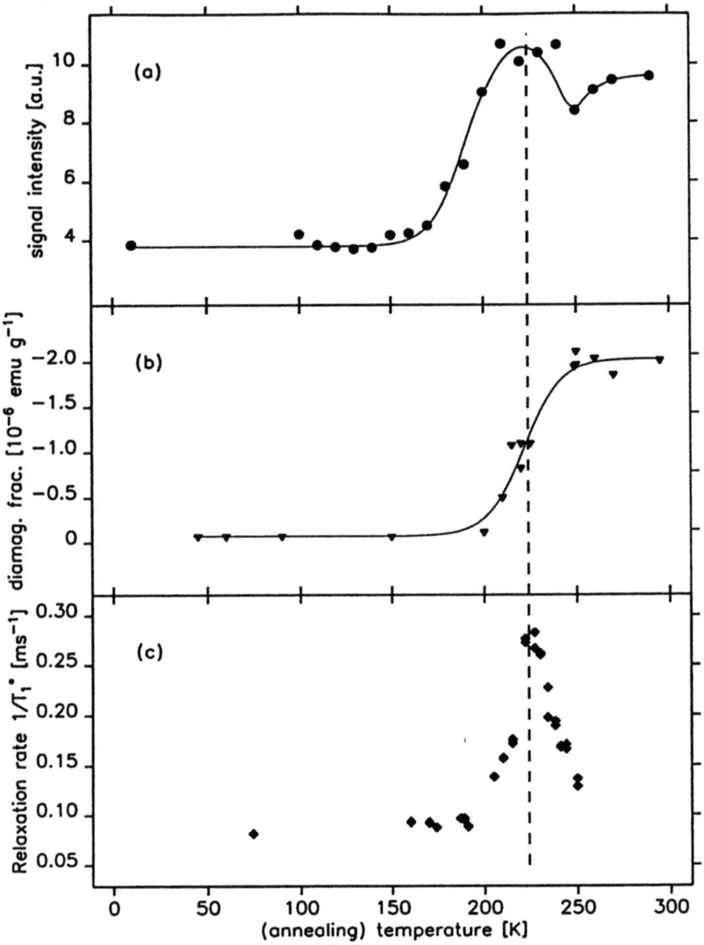


**Fig. 4.1.** Comparison of three different experimental techniques on low doped $La_2CuO_{4+\delta}$: a) low field microwave absorption annealing [15]; b) susceptibility [7]; c) Stimulated echo. All of them give a transition region around 230 K though sample and experimental procedure are different.

# 4. Comparison of $1/T_1^*$, LFS and dc susceptibility measurements

We have demonstrated before that low field microwave absorption measurements (LFS: low field signal) show the phase separation into electron rich (superconducting) and insulating antiferromagnetic domains in as-grown samples of $La_2CuO_4$ [15]. Since the LFS is sensitive to the superconducting domains it was possible to investigate their magnitude depending on the annealing temperature. We have shown that the annealing curve observed by

LFS parallels exactly the curve observed before by magnetic susceptibility measurements, with a *phase separation temperature* of $T_p = 230\,\mathrm{K}$ [15].

In the following we want to compare the results of LFS, susceptibility measurements and stimulated echo relaxation. Fig. 4.1 represents all three data sets in order to point out the similarity between these data. Note that a characteristic temperature, which we label the *phase separation temperature* $T_p$ can be defined by (a) the halfpoint temperature in the susceptibility and LFS measurements and (b) the peak position in the stimulated echo relaxation rate measurements. All three temperatures agree within $\pm 15\,\mathrm{K}$. This clearly supports the proposal that a unique phase separation temperature exists. Moreover, the dynamic relaxation measurements demonstrate that large fluctuations occur at this critical temperature.

It might be surprising, however, that the phase separation temperature for the as-grown samples and the highly oxygenated sample is very similar. This implies that the phase separation temperature is not very sensitive to the hole concentration. In fact Dichtel and co-workers [21] find from theoretical calculations that indeed the phase separation temperature is fairly insensitive to the doping. Moreover, it follows from their calculations that the phase separation temperature is near the Néel temperature of the undoped material, which in fact is observed here.

To extract an estimate of the effective activation energy involved we performed isothermal annealing measurements of the LFS at different temperatures in the range of 160 K – 250 K. The results point towards timescales of the order of minutes to hours. Assuming a singly activated process for the elementary step controlling the height of the absorption signal one can fit the experimental data by a single exponential of the form

$$S(t) = S(0) + A_0(1 - exp(-t/\tau)) \tag{4.1}$$

with

$$\tau(T) = \tau_0 e^{E_a/kT}, \tag{4.2}$$

where $\tau_0$ is the mean lifetime of a percolating cluster. An Arrhenius-plot gives an activation energy on the order of 140 meV (Fig. 4.2). This is consistent with the $T_2$ relaxation experiments reported in the previous section. It is in clear contrast, however, to the theoretical value of 480 meV as derived in [20] assuming mobile $O_2^-$ on interstitials. One might want to speculate that the lower activation energy observed in our experiments could be due to hole diffusion. Further studies are necessary to clarify this discrepancy.

## 5. The percolation model

In order to develop a suitable model for the interpretation of the experimental data we start from a lattice gas model. We have chosen a 2D-square lattice as the basic lattice where a lattice site is occupied by either one or no hole.

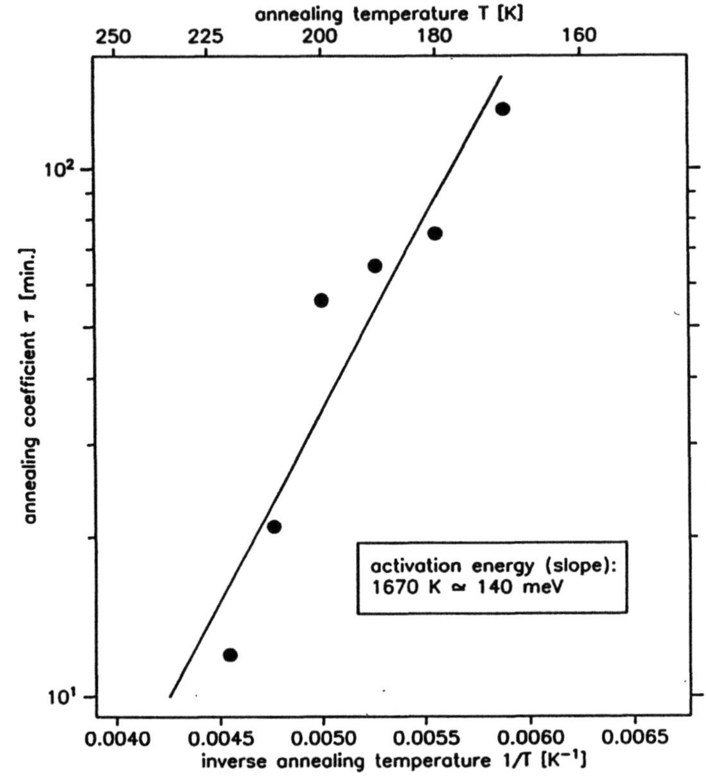


**Fig. 4.2.** Isothermal annealing coefficient $\tau$ versus inverse temperature. Data points are obtained by a single exponential fit according to equ. (4.1) to the isothermal annealing curve of the LFS. An activation energy of 140 meV is obtained.

This creates a two-state or Ising type lattice model with conserved order parameter (fixed magnetization in the Ising-model) [22]. Our model assumes the LFS to arise from Josephson coupling across domain boundaries that are connected by percolating and therefore conducting paths.

It is now well known that doping creates holes whose wavefunctions extend over many lattice sites, typically 15 – 20. The discrete lattice used for the Ising model here could be considered as representing 15 – 20 $CuO_2$ lattice sites of the real lattice. Every parameter must therefore be scaled appropriately. Clustering and percolation of holes is the important aspect to be studied and how it relates to experiments. From percolation theory it is known that the critical concentration for site percolation is $p_c = 0.59275$ for an infinite system when spin and other interactions of the particles are neglected. In order to model microwave absorption experiments we must include finite size effects. This can be seen as follows: The Josephson coupling across grain boundaries is the dominant contribution to the LFS. Typical grain sizes are on the order of $\mu$m$\approx 10^4$ Å in granular samples. The lattice constants in the $\{a, b\}$-plane

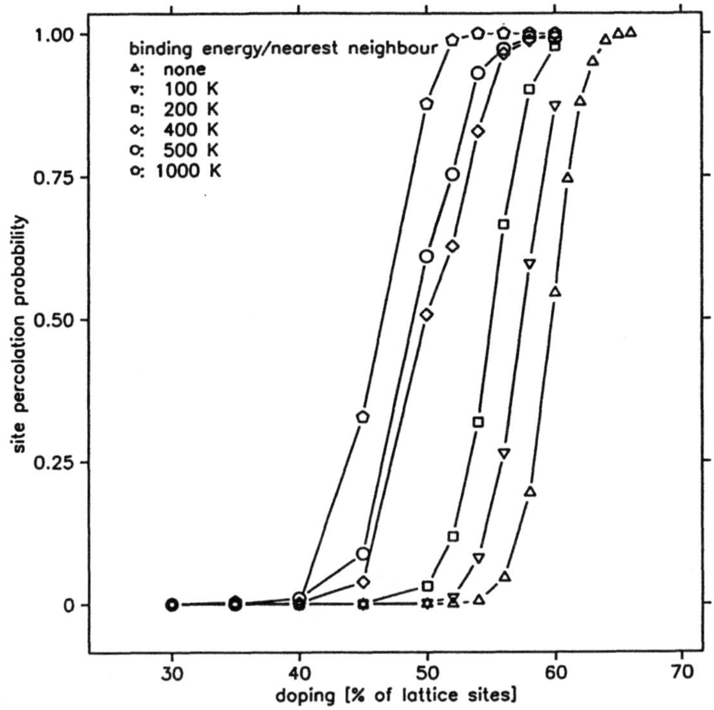


**Fig. 5.1.** Site percolation probability vs effective doping for different binding energies as calculated by Monte Carlo simulations. The system size is 50 × 50 and the simulation temperature is set to 300 K.

are about 5 Å, resulting in a square lattice of 2000 × 2000 unit cells for a single $CuO_2$ layer. This is an upper limit and even smaller lattice sizes may be considered if intragranular boundaries are taken into account reducing the size to something smaller than 1000×1000. If the extension of the hole is taken into account a further reduction factor of 4 – 5 in the linear scale must be included. Besides the finite size effects we have also investigated the influence of the binding energy between the holes on the percolation threshold.

In Fig. 5.1 we demonstrate the well known fact that the finite size and the introduction of a binding energy cause a shift to lower values of $p_c$ and a smoothening of the percolation transition. We assume that the LFS in microwave absorption measurements requires the grain to be above the percolation threshold. The annealing curve of the LFS shown in Fig. 4.1 is therefore an indication for the increasing number of percolating hole rich clusters around 230 K. It is an interesting feature that introducing a binding energy on a finite lattice renders the percolation threshold (or equivalently the percolation probability) temperature dependent. First results show a peak in the percolation probability about 100 K below the binding energy. This exactly meets the expectations that for higher temperatures large clusters are unstable and at lower temperatures more compact but large clusters are thermodynamically favoured. A more detailed discussion including the non-equilibrium properties will be presented elsewhere.

## 6. A Ginsburg-Landau-Wilson model for phase separation

In this section we introduce a numerical simulation of phase separation processes in high temperature cuprate superconductors which is based on the model of spinodal decomposition in binary alloys [22]. Our main purpose is an explanation of microwave absorption measurements in $La_2CuO_{4+\delta}$. The observed signal depends on the structural arrangement of weak links or Josephson junctions between superconducting domains or grains as discussed in the previous section. Moreover nuclear spin lattice relaxation measurements require the modeling of the fluctuation spectral density function for their interpretation.

In order to visualize the corresponding inhomogeneous topology of the superconducting layers and the temporal evolution in a phase separated system we have studied the time and temperature dependence of pattern formation in the $CuO_2$ plane. The dynamic processes involved are described in this model by the time evolution of a continuous order parameter $\phi(\boldsymbol{r},t)$ in two spatial dimensions which corresponds to the effective hole concentration in the $CuO_2$ plane. Since we consider a binary system the concentration of the antiferromagnetic part is given by $1-\phi(\boldsymbol{r},t)$. For symmetry reasons $\phi(\boldsymbol{r},t)=0$ is equivalent to a concentration of 50 %. By use of common thermodynamics we obtain a differential equation for $\phi(\boldsymbol{r},t)$ [23].

The local flux $\boldsymbol{j}(\boldsymbol{r},t)$ of the order parameter is given by

$$\boldsymbol{j}(\boldsymbol{r},t) = -M\,\boldsymbol{\nabla}(\Delta\mu) \tag{6.1}$$

The mobility term $M$ describes the hole diffusion which is very likely controlled by oxygen diffusion because of local charge neutrality. This term depends in general on all thermodynamic variables. $\Delta\mu$ is the chemical potential difference between the holes and the antiferromagnetic part which can be derived from the total free energy functional $\mathcal{F}(\phi(\boldsymbol{r},t),T)$ of the system:

$$\Delta\mu = \frac{\delta\mathcal{F}(\phi(\boldsymbol{r},t),T)}{\delta\phi(\boldsymbol{r},t)} \tag{6.2}$$

Since the order parameter $\phi(\boldsymbol{r},t)$ is conserved a continuous equation is valid:

$$\frac{\partial\phi(\boldsymbol{r},t)}{\partial t} = -\boldsymbol{\nabla}\cdot\boldsymbol{j}(\boldsymbol{r},t) \tag{6.3}$$

Combination of equations (6.1),(6.2) and (6.3) leads to

$$\frac{\partial\phi(\boldsymbol{r},t)}{\partial t} = -\boldsymbol{\nabla}\cdot\left(-M\,\boldsymbol{\nabla}\frac{\delta\mathcal{F}(\phi(\boldsymbol{r},t),T)}{\delta\phi(\boldsymbol{r},t)}\right) \tag{6.4}$$

The free energy functional is chosen in a Ginsburg Landau Wilson form with a transition temperature $T_c$

$$\mathcal{F} = \frac{1}{2}\int_V \mathrm{d}V\left(A\,(T-T_c)\,\phi^2 + \frac{1}{2}\,u\,\phi^4 + \kappa\,(\boldsymbol{\nabla}\phi)^2\right). \tag{6.5}$$

The oxygen and hole diffusion are considered as thermally activated processes. The activation energy is set to 480 meV as suggested by theoretical calculations [20]. Furthermore the influence of thermal fluctuations has to be included. A noise term $\eta(\boldsymbol{r},t)$ with

$$\begin{aligned} <\eta(\boldsymbol{r},t)> &= 0 \\ <\eta(\boldsymbol{r},t),\eta(\boldsymbol{r}\,',t')> &= -2\,k_{\mathrm{B}}\,T\,M\,\delta(t-t')\,\boldsymbol{\nabla}^2\delta(\boldsymbol{r}-\boldsymbol{r}\,') \end{aligned}$$

is therefore added. Finally we obtain the following diffusion like Langevin equation which is the basis for our numerical simulation:

$$\frac{\partial\,\phi}{\partial\,t} = \frac{M_0}{T}\exp\left(-\frac{E_a}{k_{\mathrm{B}}\,T}\right)\,\boldsymbol{\nabla}^2\left(A\,(T-T_c)\,\phi + u\,\phi^3 - \kappa\boldsymbol{\nabla}^2\phi\right) + \eta \tag{6.6}$$

In the high temperature limit equation (6.6) becomes an ordinary diffusion equation if only linear terms in lowest order are considered. Equation (6.6) can be scaled in order to reduce the number of parameters. For numerical simulations a finite difference approximation with discrete space and time coordinates is used [24].

$$\begin{aligned} t &\mathrel{\hat{=}} n\,\Delta t \\ \boldsymbol{r} &\mathrel{\hat{=}} i\,\Delta x\;\boldsymbol{e}_x + j\;\Delta y\;\boldsymbol{e}_y \\ \phi(\boldsymbol{r},t) &\mathrel{\hat{=}} \phi_{i,j}(n) \end{aligned}$$

In all simulations the average hole concentration is set to 50 % (i. e. $<\phi(\boldsymbol{r},t)>= 0$).

Fig. 6.1 displays the variation of the order parameter in the two-dimensional lattice. The characteristic transition temperature for phase separation $T_c$ was set to 250 K, whereas the lattice temperature in the simulation shown in Fig. 6.1 is 160 K. The system was prepared initially in a homogeneous equilibrium state with small fluctuations. After some evolution times percolating domains are observed. The growth of phase separated regions can be viewed as an increasing array of superconducting domains with Josephson junctions which give rise to a microwave absorption signal. The spatial extension of phase separated regions is estimated by integration of the radial correlation function in its discrete form

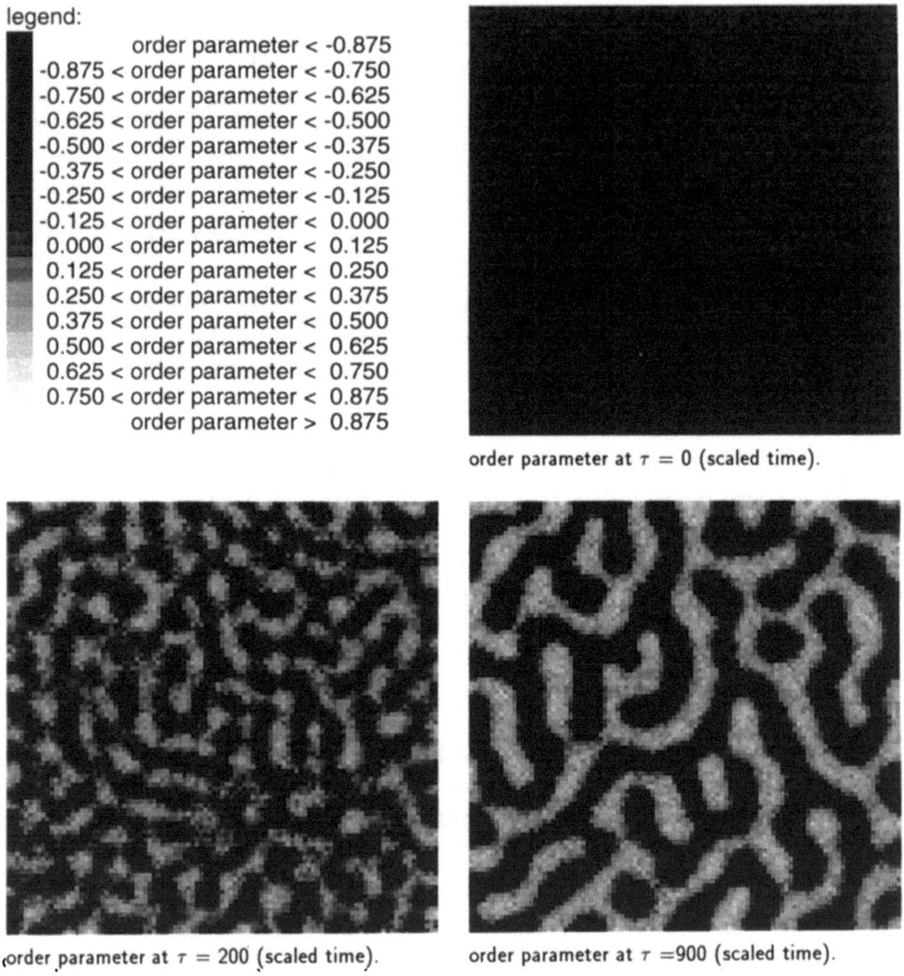


**Fig. 6.1.** Results of a Ginsburg-Landau-Wilson phase separation calculation. Shown is the temporal evolution of the order parameter on a discrete lattice. Cluster formation and percolation are displayed. The figures were simulated at a temperature of 160 K whereas the phase transition temperature was $T_c = 250\,\mathrm{K}$.

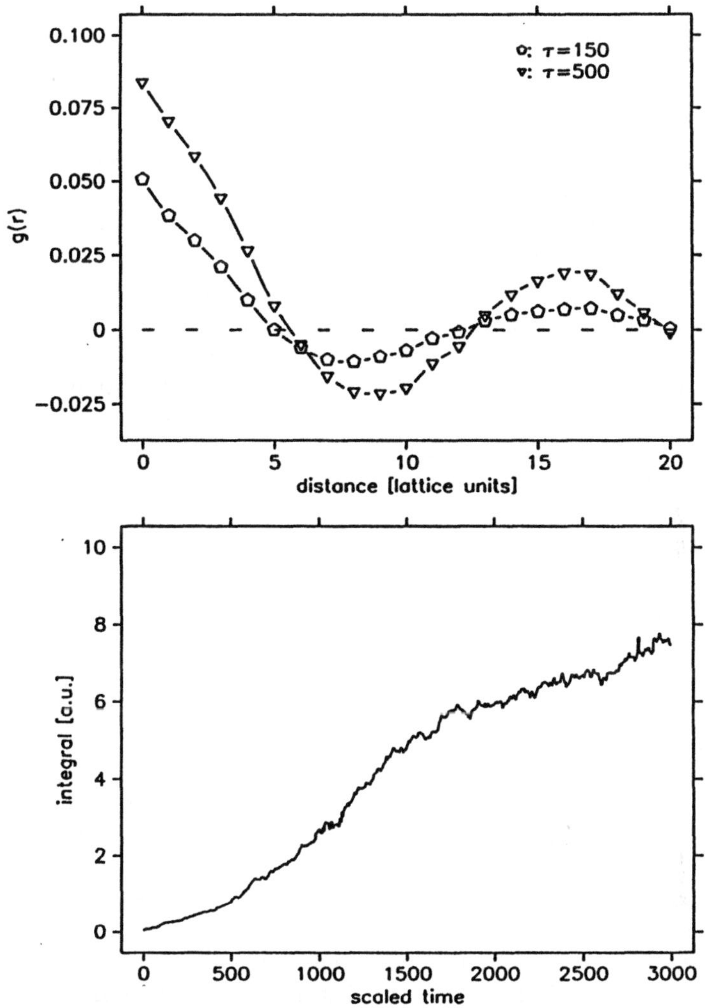


**Fig. 6.2.** *Top:* Spatial correlation function as a measure of cluster size at different times. The area under $g(r)$ from the origin to the first zero crossing serves as a measure of the size of the phase separated clusters. *Bottom:* Cluster growth as determined from the integrated spatial correlation function. Note the increase and saturation of the mean cluster size.

$$g(r,t) = \frac{1}{N_r} \sum_{i,j} \left( \sum_{k,l} \phi_{i,j}(t)\, \phi_{k,l}(t) \right) \tag{6.7}$$

$$\text{with} \quad ((i-k)\,\Delta x)^2 + ((j-l)\,\Delta y)^2 = r^2$$

between r=0 and its first zero crossing. Fig. 6.2 shows a typical correlation function and the mean cluster size.

In order to model the microwave absorption and susceptibility experiments one set of simulations was performed in the following way: the system, initially in a high temperature equilibrium state, was rapidly cooled to an annealing temperature and was kept there for about 10 minutes. At present some fundamental results obtained by these simulations shall be discussed.

One important feature is that the extent of phase separation and its onset at high temperatures depends on the transition temperature $T_c$ which is closely related to the binding energy of a single hole (oxygen) in a hole-rich cluster. The behaviour at low temperatures shows the influence of the activation energy $E_a$ which sets the time scale for phase separation processes. For $T \to 0$ the time scale in which phase separation is observed diverges. The recent simulations suggest that the introduced continuous model is appropriate for an explanation of the microwave absorption signal in $La_2CuO_{4+\delta}$. More details will be presented in a forthcoming publication.

## 7. Summary

We have demonstrated by NQR, microwave absorption and susceptibility measurements that a unique phase separation temperature exists in the range of 230 K. The long time scale seen by these experiments is controlled by oxygen diffusion. Longitudinal ($1/T_1$, $1/T_1^*$) and transversal relaxation rate ($1/T_2$) measurements were performed at the NQR transition $5/2 \leftrightarrow 7/2$ of $^{139}$La. Their different sensitivity to different types of interactions and spectral densities was utilized in order to distinguish rapid from slow motion. In particular "stimulated echo relaxation" $1/T_1^*$ was applied to demonstrate the slow motion character of the dynamics around the phase separation temperature. Since $1/T_1^*$ is much more sensitive to slow motion than $1/T_1$ it is proposed here to be the relaxation rate of choice in the investigation of phase separation.

Computer simulations of two different kinds were presented, namely a lattice gas model of the Ising type and a stochastic Langevin equation simulation with Ginsburg-Landau-Wilson Ansatz. A preliminary goal of these simulations was to provide a visualization of the percolation process involved in the phase separation and to model certain experimental parameters. Although the qualitative features of the phase separation and the corresponding annealing process are very well represented by the simulation, no quantitative comparison with experiment can be given at the current state.

*Acknowledgement.* We greatfully acknowledge financial support by the Bundesministerium für Forschung und Technologie (BMFT) under grant 13N5845.

# References

1. Proceedings of the Workshop on *Phase Separation in Cuprate Superconductors*, Erice, Italy, 6.–12.5.1992, eds. K. A. Müller and G. Benedek, World Scientific, Singapore, London (1993) and references therein.
2. P. B. Visscher, *Phys. Rev. B* **10**, 943 (1974).
3. V. J. Emery, S. A. Kivelson, H. Q. Lin, *Phys. Rev. Lett.* **64**, 475 (1990).
4. M. Marder, N. Papanicolaou, G. C. Psaltakis, *Phys. Rev. B* **41**, 6920 (1990).
5. E. Sigmund, V. Hizhnyakov, G. Seibold, in "Proceedings of the Workshop on *Phase Separation in Cuprate Superconductors*", Erice, Italy, 6.–12.5.1992, eds. K. A. Müller and G. Benedek, World Scientific, Singapore, London (1993); V. Hizhnyakov, E. Sigmund, *Physica C* **156**, 655 (1988); V. Hizhnyakov, E. Sigmund, M. M. Schneider, *Phys. Rev. B* **44**, 795 (1991).
6. J. H. Cho, F. Borsa, D. C. Johnston, D. R. Torgenson, *Phys. Rev. B* **46**, 3179 (1992).
7. R. K. Kremer, E. Sigmund, V. Hizhnyakov, F. Hentsch, A. Simon, K. A. Müller, M. Mehring, *Z. Phys. B* **86**, 319, (1992).
8. J. H. Cho, F. C. Chou, D. C. Johnston *Phys. Rev. Lett.* **70**, 222 (1993).
9. I. Watanabe, K. Kumagai, Y. Nakamura, T. Kimura, Y. Nakamichi, H. Nakajima, *J. Phys. Soc. Jpn.* **56**, 3028 (1987); Y. Kitaoka, S. Hiramatsu, K. Ishidam T. Kohara, K. Asayama, *J. Phys. Soc. Jpn.* **56**, 3024 (1987).
10. H. Lütgemeier, M. W. Pieper, *Sol. Stat. Com.* **64**, 267 (1987).
11. F. Hentsch, H. Seidel, M. Mehring, J. G. Bednorz, K. A. Müller, Physica C **153-155**, 727 (1989).
12. M. Mehring, IBM Journal of Research and Development **33**, 342 (1989).
13. U. Grosshans, Diploma thesis, 2. Physikalisches Institut, Universität Stuttgart, Germany, January 1990.
14. M. Baehr, Diploma thesis, 2. Physikalisches Institut, Universität Stuttgart, Germany, June 1992.
15. M. Mehring, M. Baer, P. Gergen, J. Gross, C. Kessler, N. Winzek in "Proceedings of the Workshop on *Phase Separation in Cuprate Superconductors*", Erice, Italy, 6.–12.5.1992, eds. K. A. Müller and G. Benedek, World Scientific, Singapore, London, p. 67–84, (1993).
16. J. C. Grenier, A. Wattiaux, N. Lagueyte, J. C. Park, E. Marquestaut, J. Etourneau, M. Pouchard, *Physica C* **173**, 139 (1991).
17. R. K. Kremer, V. Hizhnyakov, E. Sigmund, A. Simon, K. A. Müller, *Z. Phys. B* **91**, 169, (1993).
18. P. C. Hammel, A. P. Reyes, S.-W. Cheong, Z. Fisk, J. E. Schirber, *Phys. Rev. Lett.* **71**, 440 (1993).
19. P. C. Hammel, E. T. Ahrens, A. P. Reyes, R. H. Heffner, P. C. Canfield, S.-W. Cheong, Z. Fisk, J. E. Schirber, *Physica C* **185-189**, 1095 (1991).
20. N. L. Allan, W. C. Mackrodt, *Phil. Mag.* **64**, 1129 (1991).
21. J. Carstensen, K. Dichtel, *Phys. Rev. B* **46**, 11040 (1992).
22. J. D. Gunton, M. San Miguel, Paramdeep S. Sahni in *Phase Transitions and Critical Phenomena*, eds. C. Domb and J. L. Lebowitz, Academic Press, London, Vol. 8, p. 267–466 and references therein.

23. K. Elder, *Computers in Physics* **7**, 27 (1993).
24. T. M. Rogers, K. R. Elder, Rashmi C. Desai,*Phys. Rev. B* **37**, 9638 (1988).

# Microscopic Study of Local Structure and Charge Distribution in Metallic $La_2CuO_{4+\delta}$

P.C. Hammel[1], A.P. Reyes[1], E.T. Ahrens[1], Z. Fisk[1], P.C. Canfield[1], J.D. Thompson[1], and J.E. Schirber[2]

[1] Condensed Matter and Thermal Physics Group, Los Alamos National Laboratory, Los Alamos, NM 87545, U.S.A.
[2] Sandia National Labs, Albuquerque, NM 87185, U.S.A.

**Summary.** We employ NMR and NQR spectroscopy as probes of local structure and charge environments in metallic $La_2CuO_{4+\delta}$ ($T_c = 38$ K). We discuss the effect of annealing the sample at various temperatures $T_a$ ($T_c < T_a < 300$ K) on the superconducting $T_c$. The dependence of $T_c$ on $T_a$ indicates that annealing allows the development of structural order which is important for $T_c$. The $^{139}$La quadrupole frequency $\nu_Q$ in the in metallic phase is smaller than in the undoped material. This is unexpected and may indicate a smaller charge on the apex oxygen in the doped material and thus a different distribution of charge between the La–O layer and the planes. We find that the presence of doped holes induces a *distribution of displacements* of the apex oxygen off of the vertical La–Cu bond axis. The distribution in displacements extends from zero to the value observed in lightly doped (antiferromagnetic) $La_2CuO_{4+\delta}$. This directly demonstrates considerable inhomogeneity in the crystal structure of the La–O layer. Copper NQR spectroscopy shows that there are *two distinct copper sites* in the $CuO_2$ planes and thus that either the structure or the charge distribution in the planes is inhomogeneous as well. These inhomogeneities are the intrinsic response of the crystal to doped holes; they are not the result of distortions of the lattice due to the presence of interstitial oxygen atoms.

## 1. Introduction

Structure is an underlying determinant of electronic behavior in all materials. A thorough understanding of structure of the high $T_c$ cuprates is essential to unraveling the puzzle of the normal state electronic structure of these materials and thus superconductivity. It has been established that the superconductivity in some of the cuprates, if not all, is exceptionally sensitive to subtle details of the structure. It is becoming evident that the local structure of these materials deviates considerably from structures deduced from crystallographic determinations. There is growing recognition that these little understood structural properties may be at the root of the novel and potentially technologically important properties of these materials. We have undertaken an extensive *microscopic* study of local structural and charge environments with the goal of characterizing the nature of these local environments and gaining insight into the relationships between these and the novel electronic and magnetic behaviors observed in the cuprates.

For several reasons we have chosen to focus on metallic, oxygen-doped $La_2CuO_{4+\delta}$. A remarkably strong sensitivity of superconductivity to doping near $x = 0.125$ has been observed in $La_{2-x}Ba_xCuO_{4+\delta}$ [1]. This has been attributed to a subtle change in structure from the LTO to the LTT structure [2]. There is recent evidence that the suppression of superconductivity may involve more than the simple structural transition [3]. Büchner et al. [4] have shown that superconductivity is suppressed in the LTT phase only when the octahedral tilt exceeds a critical angle. Such sensitivity of electronic properties to structure indicates that the lanthanum cuprates might be good place to begin a study of the effects of local inhomogeneities, be they structural or electronic. Of this class of materials, the oxygen-doped variety is attractive for magnetic resonance studies. The observed La NMR line does not exhibit the line shift nor the broadening which results from the oxygen displacements around an interstitial oxygen. This supports the result obtained earlier [5] that the structural distortion associated with the interstitial is very short range. The $^{139}$La NMR lineshape is extremely sensitive to small structural changes in the majority of the crystal which is not affected by the distortion around the interstitial oxygen. Samples with uniform doping of $0.01 < \delta < 0.06$ such as the one studied here undergo macroscopic phase separation [6]–[8]. Although this represents a complication for many probes, this is much less so for magnetic resonance techniques which readily distinguish the oxygen-rich and oxygen-poor phases [8].

We will address the question of oxygen mobility in $La_2CuO_{4+\delta}$ and show evidence that oxygen mobility is lost below 200 K, thus explaining the vertical phase lines in the phase diagram below this temperature [8]. Several experiments have shown there to be a dependence of the superconducting properties, including critical temperature $T_c$ on the rate at which the sample is cooled from room temperature [10]. We will focus on the sensitivity of $T_c$ to cooling rate and determine the temperature at which the sample must be annealed in order to achieve a higher $T_c$. Given the sensitivity of $T_c$ to doping it is attractive to attribute the reduction in $T_c$ to incomplete phase separation which would decrease the doping of the metallic phase. We have performed studies in which the sample is held at an annealing temperature $T_a$ for varying times then quenched to low temperatures to measure $T_c$ [12]. These experiments reveal the dependence of $T_c$ on $T_a$ and suggest that the reduction on $T_c$ after quenching is not due to a reduction of the oxygen content of the metallic phase. Instead annealing may allow development of an optimum local structure or charge environment.

The remainder of the paper will be devoted to a microscopic study of local structure and charge environments by means of magnetic resonance techniques. These studies have shown that the addition of doped holes alters the local structure of $La_2CuO_{4+\delta}$ and that these changes are the intrinsic response of the crystal to the presence of doped holes — they are not manifestations of the distortion associated with the presence of the oxygen dopant. La NQR spectroscopy probes the electric field gradient at the La site, and this is determined by the location and charge of the neighbors. The unusual

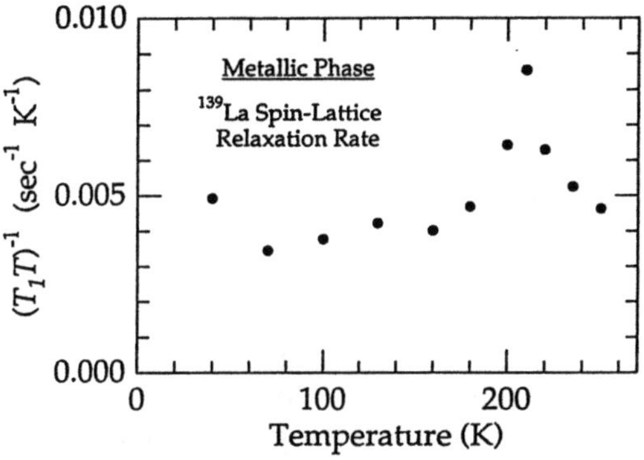


**Fig. 2.1.** Temperature dependence of the La relaxation rate divided by temperature is shown. Note the sharp peak in $(T_1T)^{-1}$ at 210 K. La nuclear spin relaxation is due to fluctuations of the EFG caused by motions of neighboring oxygen atoms. A likely interpretation of the peak is that at this temperature the frequency of fluctuations become equal to the nuclear Larmor frequency.

temperature dependence of the NQR frequency $\nu_Q$ and comparison of the experimental results with results expected on the basis of point charge calculations suggests that there is temperature dependent redistribution of charge from the La–O layer to the planes.

## 2. Macroscopic phase separation

The phase diagram of oxygen doped $La_2CuO_{4+\delta}$ at doping levels in the phase separation regime shows some unusual features. Among these are the very rapid change in the doping of both phases at the phase separation temperature $T_{PS}$ and the vertical sides of the phase diagram below about 200 K. In Fig. 2.1 we show the temperature dependence of La $(T_1T)^{-1}$ as a function of temperature. Such a sharp peak as occurs at 210 K often occurs when the correlation time for oxygen motion relevant for La relaxation equals the Larmor frequency. We plan to confirm this hypothesis by determining the frequency dependence of $1/T_1$. The Larmor frequency is very low ($\sim$ 50 MHz) so oxygen motion is essentially frozen out at this temperature. This result is consistent with the observation that below this temperature the doping of both the oxygen–rich and oxygen–poor phases remains constant also showing the loss of oxygen motion.

## 3. Cooling rate dependence of $T_c$

Many workers [10] have observed a cooling rate dependence of the superconducting properties in oxygen doped $La_2CuO_{4+\delta}$. In Fig. 3.1 we have plotted

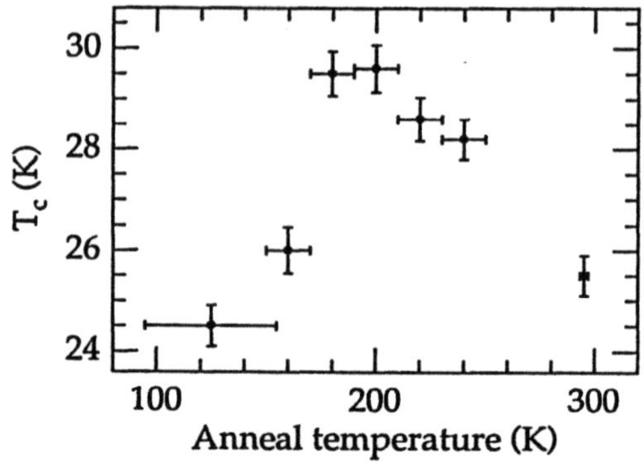


**Fig. 3.1.** The dependence of $T_c$ on the temperature $T_a$ at which the sample is annealed during cooldown is shown. $T_c$ increases with decreasing anneal temperature from room temperature down to 200 K. It is unlikely that this phenomenon can be attributed simply to reduction in metallic phase oxygen content $\delta$. Although one might expect rapid cooling to prevent full phase separation, the oxygen mobility changes by over four orders of magnitude between 250 K and 200 K, much more than the factor of four difference between fast and slow cooling rates important for $T_c$.

the $T_c$ which results when the sample is held at various annealing temperatures $T_a$. Annealing at 195 K for times as short as 20 minutes produces the higher $T_c$. We observe that $T_c$ increases steadily with decreasing $T_a$ down to approximately 200 K. Annealing below this temperature results in a low $T_c$ similar to that achieved by rapid cooling from room temperature.

One possible explanation of this behavior is that rapidly cooling the sample leaves the metallic phase oxygen content $\delta_M$ at a lower value than is achieved by slow cooling and thus decreases $T_c$. While the magnitude of the change in $T_c$ is consistent with this explanation [11], the weak dependence of $T_c$ on $T_a$ shown in Fig. 3.1 is not. Studies of oxygen mobility in the cuprates find [12] a conventional Arrhenius behavior. In $La_{2-x}Sr_xCuO_4$ the activation energy U which falls in the range of 0.7 eV to 1.1 eV depending on $x$. In $La_2CuO_{4+\delta}$ we find that a slow cooling rate $R_s$ of about 1 K/minute is sufficient to produce a higher $T_c$; while a fast rate $R_f$ of less than 4 K/minute will result in lower $T_c$. The difference in these rates is small compared to the several orders of magnitude change in oxygen mobility between 250 K and 200 K. Let us consider the possibility that the reduction in $T_c$ is attributable to inadequate oxygen mobility which in turn causes the metallic phase to have reduced oxygen content. In this case this small change in cooling rate would not be expected to affect the low temperature value of $\delta_M$ and thus outside a narrow range of annealing temperatures near a temperature $T_f$ ($\sim 200$ K) where oxygen mobility is lost. Assuming Arrhenius behavior ($U = 0.7$ eV) we estimate the width of this range to be $\Delta T \approx (T_f^2/U)\ln(R_f/R_s) \approx 7$ K. The result in Fig. 3.1 is inconsistent with this scenario since $T_c$ is changing with annealing temperature from room temperature down to 200 K. Thus the observed sensitivity of $T_c$ to cooling rate is more complex. One possibility is that the oxygen mobility in the oxygen doped material has a very different temperature dependence than is observed in the Sr doped material. In order to explain the observed $T_c$ dependence on annealing temperature the mobility would have remain nearly constant between room temperature and 200 K instead of decreasing with the strong activated behavior. This would imply that the kinetics of oxygen motion is very different in this material. This is not implausible if the interstitial oxygen plays a key role in determining oxygen mobility. Perhaps once the interstitial site is occupied the activation barrier for motion is greatly decreased leading to a much weaker temperature dependence. If the interstitial oxygen were then to become ordered in the lattice at 200 K this could reintroduce a (larger) activation barrier and thus initiate a strong temperature dependence to the oxygen mobility below this. A more likely possibility is that this is a conventional annealing effect and that the slow cooling allows the local structural environments to achieve a configuration more favorable for superconductivity. It has been proposed that the observation that $T_c$ increases in $YBa_2Cu_3O_{7-y}$ with annealing [13] results from the optimization of the local configuration of oxygen in the chains which allows improved charge transfer to the planes and thus increases $T_c$. As we will discuss shortly we have observed that the local structure of the La–O layer is quite temperature dependent, opening the possibility that an analogous effects may be important in the La–O layer. It may also be that residual local disorder is pair–breaking and that this can be reduced by annealing [14].

## 4. Determination of local structure and local charge environments

We have employed NQR [15] and NMR [16] as local probes of the La–O layer and the $CuO_2$ planes. In both cases the key information we have obtained reflects the interaction of the nuclear quadrupole moment with the electric field gradient (EFG) at the nuclear site. The EFG is described by a symmetric traceless tensor [17]. In its principal axis frame this tensor is diagonal and can be described by two parameters: the quadrupole frequency $\nu_Q$ which specifies the magnitude of the largest component of the diagonal EFG tensor $V_{zz}$, and the asymmetry parameter $\eta$ which specifies the degree of departure from axial symmetry. We determine both $\nu_Q$ and $\eta$ from zero-field (NQR) measurements.

### 4.1 La NQR

At high temperature we find that $\nu_Q$ is the same for both the metallic and AF phases. However, below 200 K we observe the shift of the spectral weight associated with the metallic phase to lower frequency as well as substantial broadening of the spectrum. Fig. 4.1 depicts the La NQR spectrum at low temperature. The spectrum from the metallic phase has shifted to substantially lower frequency and broadened substantially. While the broadening is not unexpected in view of the distribution of structures we observe in NMR experiments (see below and [16]), the shift to lower frequency is. The addition of interstitial oxygen has the effect of compressing the lattice. The EFG at the nuclear site is the sum of the contributions from the distribution of charge on the ion as well as from the lattice of ions surrounding the site. In the case of a closed shell ion such as La, these two contributions are essentially proportional to one another and so $\nu_Q$ is proportional to the lattice contribution to $V_{zz}$. Thus $\nu_Q \propto \sum_i eq_i(3z_i^2 - r_i^2)/r_i^5$ where $i$ labels the various ions having relative location $r_i$ with respect to the La site and charge $q_i$. One expects that this compression should lead to an increase in $\nu_Q$. We have used point charge calculations [18] to estimate the EFG in the presence of the interstitial dopant oxygen. They confirm this picture and predict an *increase in* of approximately 10% (for ≈ 50% of the sites, roughly, another 10% of La sites will have $\Delta\nu_Q > 100\%$).

In order to understand the observed changes in the metallic phase spectrum we must assume that below 200 K there is a decrease in $q_i$ with decreasing temperature. This would be consistent with a redistribution of charge from the La–O layer into the planes. It should be noted that this is a transfer of electrons, not holes, i.e. the magnitude of the (negative) oxygen valence is decreasing. The smaller value of $\nu_Q$ for the broad metallic line may be related to its unusual temperature dependence which is shown in Fig. 4.1(b). The behavior of the average frequency observed in the AF phase is indicative of the behavior of $\nu_Q$ with temperature in $La_2CuO_{4.01}$. The contrast with the metallic phase is clear: the center of gravity of the La NQR resonance is non-monotonic in temperature, decreasing at low temperatures. The more rapid decrease just above $T_c$ is interesting, indicating that charge redistribution accelerates in this temperature regime.

## 4.2 La NMR

In the NMR experiment we measure the quadrupole interaction at the La site in the presence of a large applied field. In this case the resonance frequency is sensitive to the orientation of the EFG principal axis frame with respect to the field which was applied along the c-axis (perpendicular to the planes) in our experiments. The second order quadrupole shift $\nu^{(2)}$ of the $m = 1/2 \leftrightarrow -1/2$ transition frequency by an axially symmetric EFG depends on the angle $\theta$ between the applied magnetic field and $V_{zz}$. If we define $\mu \equiv \cos\theta$, then $\nu^{(2)} = (15/16)(\nu_Q^2/\nu_z)(1-\mu^2)(1-9\mu^2)$ where the Zeeman frequency is $\nu_z = (1+K)\gamma H_0$, $K$ is the Knight shift, $\gamma$ the $^{139}$La gyromagnetic ratio and $H_0$ the magnitude of the applied magnetic field. The

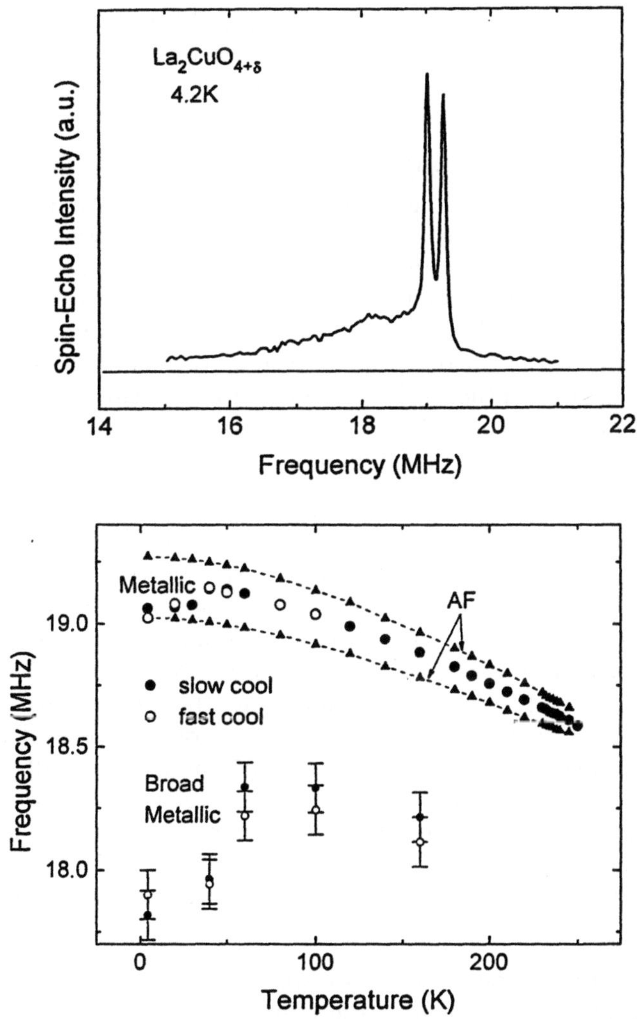


**Fig. 4.1.** Variation of the $^{139}$La spectrum with temperature. In panel (a) the two sharp peaks originate in the oxygen–poor, antiferromagnetic (AF) phase, the splitting arising from the hyperfine field of the ordered copper moments. The broad line at lower frequency arises from metallic phase of the crystal. The temperature dependence of the four lines is shown in (b). At high temperatures the metallic line falls between the two (AF) lines. Around 200 K intensity begins to shift from this line into the broad metallic line at lower frequency leaving a small fraction at low temperatures. The reduced frequency for the broad line is unexpected as discussed in the text as is the temperature dependence which has a broad maximum. The broadening of the line at low temperatures is consistent with the observation (through NMR) of a distribution of tilts of the oxygen octahedra which develop with decreasing temperature.

field dependence of the quadrupole shift ($\propto 1/H_0$) allows it to be easily distinguished from magnetic shifts ($\propto H_0$). Spectra were taken at 52.5900 MHz and 30.3629 MHz at all temperatures and occasionally at 39.1983 MHz as well to establish that the shifts were linear in $1/H_0$.

We have also modeled the relationship between the configuration of the oxygen ions surrounding the La and the orientation of $V_{zz}$. We find that this is primarily determined by the orientation of the bond between the La ion and the apex oxygen located vertically (along the c–axis) above it. Thus the development of a finite tilt of the EFG away from the c–axis reflects a lateral shift of the apex oxygen off the La–Cu bond axis. Point charge calculations indicate that, in fact, the tilt of the EFG exceeds the tilt of the La–O bond by about 60%. The tilts of the EFG we observe at the La site in the AF phase ($\approx 19°$) exceed the predictions of our simple point charge calculations ($\approx 7°$) by a factor of 2.5. This is not understood.

It is worthwhile to consider the effects of fluctuations of the EFG on the NMR spectrum. For small tilts we can make the approximation $\nu^{(2)} \propto \theta^2$. If these fluctuations are rapid compared to the Larmor frequency then $\nu^{(2)} \propto \langle\theta^2\rangle$ where $\langle f\rangle \equiv$ time average of $f(t)$. One might imagine a situation in which the EFG is tilted by a small angle away from the c–axis but rotating about it in such a way that $\langle\theta\rangle = 0$ on the NMR time scale. In this case $\langle\theta^2\rangle$ is not zero and $\langle\nu^{(2)}\rangle$ will not vanish, and the spectrum will be sharp whether the EFG is fluctuating or not. Another possibility is that the EFG at a given site is occasionally parallel to the field and occasionally tilted. If this fluctuation is slow (compared to the Larmor frequency) we will observe a broad line representing the distribution of tilts, while rapid fluctuations will produce a narrow line where the shift is given by $\langle\theta^2\rangle$. Thus, in this case rapid fluctuations can narrow the line, but they will not cause the shift to vanish.

We turn now to the results which we wish to emphasize: the determination of the off–axis displacement of the apex oxygen in the metallic phase and its description in terms of a distribution of orientations of the EFG at the La site. In Fig. 4.2 we show that above and just below $T_{PS}$ the $^{139}$La NMR line is sharp and unshifted. Thus although we expect that the material is orthorhombic [6], there is no lateral displacement of the apex oxygen. That we observe a sharp, unshifted line observed at high temperature is also very interesting because it shows that the La signal *we observe* is *not* affected by the structural distortion in the immediate vicinity of an interstitial oxygen. Note that oxygen motion cannot be responsible for the unshifted, narrow line. Our point charge modeling [18] confirms that the signal from the fraction ($\sim 10\%$) of La nuclei neighboring an interstitial oxygen is unobservable due to the very large values of $\nu_Q$ and large tilts. The behavior of the metallic phase should be compared with that of the oxygen–poor ($\delta \cong 0.01$) antiferromagnetic (AF) phase which displays a finite and well defined shift

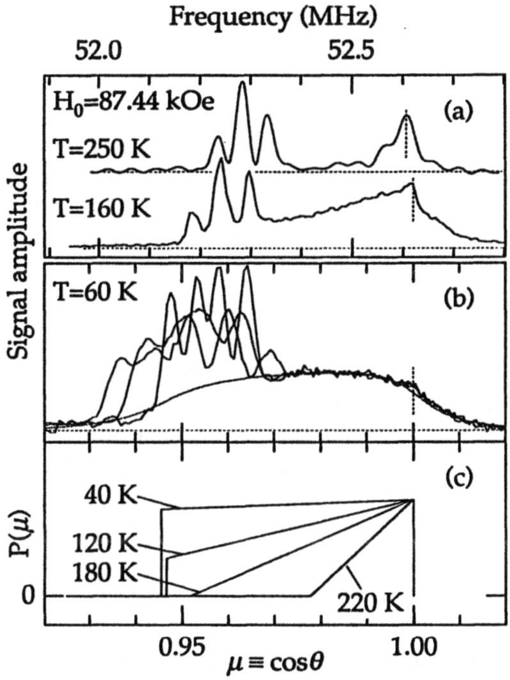


**Fig. 4.2.** Variation of the octahedral tilts with temperature. The shifts of the several lines seen in the $^{139}$La NMR spectrum in (a) reflect different tilts of the oxygen octahedra in both metallic and AF phases. The NMR frequencies are very sensitive to the off–axis displacement of the apex oxygen and thus the tilts of the oxygen octahedra. Analysis of the NMR lineshapes such as those shown in (a) and (b) reveals a distribution of tilts ($\mu \equiv \cos\theta$ where $\theta$ is roughly the angle the La–O bond makes with the c–axis) which evolves with decreasing temperature. Panel (c) shows the measured distribution of tilts obtained at several temperatures.

at all temperatures consistent with the tilt of the EFG (angle $\theta_{\mathrm{AF}}$) due to the off–axis displacement of the apex oxygen present in the orthorhombic phase.

With decreasing temperature the metallic $^{139}$La NMR line becomes broadened as a distribution of apex oxygen off-axis displacements develop. We have [16] analyzed the lineshape as a function of temperature to obtain a distribution of tilts $P(\mu)$. In Fig. 4.2(b) we show $P(\mu)$ at four temperatures to indicate its the evolution with temperature. Note that there is never any spectral weight at shifts greater than are observed in the AF ($\delta \cong 0.01$) phase, thus, in the metallic phase, there is a cutoff in the tilt distribution at $\theta_{\max} = \theta_{\mathrm{AF}}$. At low temperature we find a simple but remarkable distribution of tilts: $P(\mu) \cong$ constant, i.e. the distribution is *uniform* (any point on the unit sphere with $\theta < \theta_{\max}$ is occupied with equal probability). Note that the sharply peaked tilt distribution characteristic of the distortion associated with a point defect is *not* observed. Thus, even at low temperatures, NMR is unable to observe the signal from La atoms affected by the distortion due to the interstitial oxygen. We conclude that the distributed structure we observe is the intrinsic response of the crystal to the presence of doped holes. Note that for a low ($\delta \cong 0.01$) doping of the AF phase, there is a single, well defined lateral displacement of the apex-oxygen at all temperatures. Thus there is some threshold doping necessary to induce the distributed structure. Unfortunately the macroscopic phase separation prevents one from varying the doping of the metallic phase in the region $0.01 < \delta < 0.06$ in order to determine this threshold. It is tempting to speculate that it might coincide with the metal–insulator transition.

Recently neutron scattering measurements in powders and single crystals of electrochemically prepared $La_2CuO_{4+\delta}$ with sufficient oxygen content to be in the single phase region of the phase diagram have been performed [19]. They find the apex oxygen located directly above the copper (Fmmm structure) *on average*, however superstructure peaks indicate a long wavelength, commensurate modulation of the oxygen position. Given evidence of long range order, we conclude that the varying lateral displacements of the apex oxygens we observe locally are described by a long wavelength modulation commensurate with the lattice. Similar superlattices have been observed in BSSCO. Bianconi et al. have found [20] the superlattice structure to be associated with variations in the length of the Cu–apex oxygen bond and have hypothesized the existence of an ordered polaron superlattice.

### 4.3 Cu NQR

It is important to understand the effect this distribution of octahedral tilts has on the electronic properties of the $CuO_2$ planes, since this is the location of the itinerant holes which become superconducting at low temperature. Copper NQR is a local probe of the charge environment of the copper. Earlier copper NQR work in this compound is discussed in [22]. In Fig. 4.3 we show the copper NQR spectrum obtained in oxygen–annealed $La_2CuO_{4+\delta}$ at 40 K (it is very little different at 250 K). The important feature of this spectrum is that when one takes into account the effect of the two copper isotopes, the appearance of three lines demonstrates unambiguously the existence of two distinct copper sites in the planes. Note that the signal from the antiferromagnetic oxygen–poor phase [22] occurs at a frequency of approximately 100 MHz due the very large hyperfine field of the ordered copper moments in the Néel state. Thus all of the spectral features shown here arise from the metallic phase of the sample alone. It is natural to attribute the second site to the proximity of an interstitial oxygen. This explanation is rendered implausible by the fact that this spectrum is essentially identical to the copper NQR spectrum observed in $La_{2-x}Sr_xCuO_4$. The dependence of this spectrum on doping has been extensively studied [23]–[26] and it has been found that the frequencies of the two $^{63}Cu$ spectral features depend on doping as shown in the inset to Fig. 4.3. Both features increase in frequency with doping, the minority feature increasing more slowly.

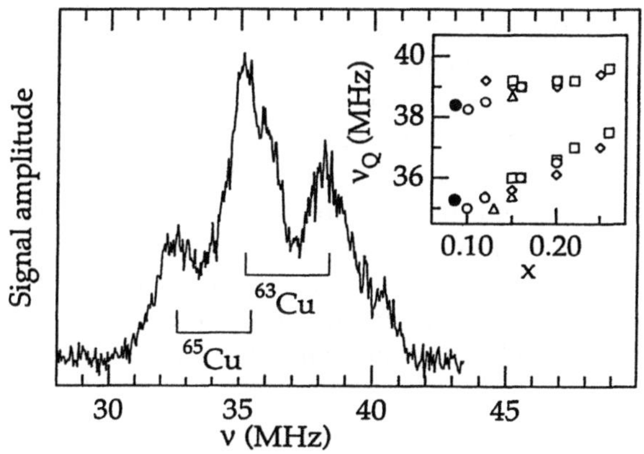


**Fig. 4.3.** Copper NQR spectrum in $La_2CuO_{4.06}$ is shown. The spectrum is the sum of spectra from $^{63}$Cu and $^{65}$Cu isotopes. The existence of three peaks unambiguously demonstrates that there are two distinct copper sites. The plot in the inset compares the frequencies of these peaks observed in oxygen doped material (present work) to those seen in Sr doped material. The frequencies are essentially identical. This clearly demonstrates that the second peak does not arise as a consequence of the proximity of the dopant impurity. Note that the copper signal arising from the oxygen–poor phase occurs at approximately 100 MHz due to the very large hyperfine field of the antiferromagnetically ordered copper moments, and so is not visible here.

The magnetic susceptibility of most metallic cuprates including lanthanum cuprate is very sensitive to the level of hole doping in the planes. These measurements in metallic $La_2CuO_{4+\delta}$ have allowed a determination of the hole density arising from oxygen annealing [8], [27]. In particular, the hole density in the metallic phase of phase separated $La_2CuO_{4+\delta}$ is approximately 0.09 holes/copper. When we compare (see inset to Fig. 4.3) the frequencies of the $^{63}$Cu features with those observed in $La_{2-x}Sr_xCuO_4$ at this doping we find that they both coincide perfectly.

The location of the dopant impurities in $La_2CuO_{4+\delta}$ and $La_{2-x}Sr_xCuO_4$ are very different: the excess oxygen is interstitial and located between the La–O planes at the center of four tetrahedrally coordinated La atoms whereas the strontium substitutes for a La atom. Furthermore the valence of the two dopants relative to the $La_2CuO_{4.0}$ background is different. The charge of the oxygen dopant is $-1.4$ to $-2.0$ while the charge of the strontium relative to the La it replaces is $-1.0$. It is quite implausible that these two very different lattice charge arrangements could be directly responsible for the two sets of spectral features having identical frequencies. Thus it is clear that the second copper site is not differentiated simply by the proximity of the dopant impurity. The agent which distinguishes the second site must be common to both systems and independent of the specific nature of the dopant. The most likely explanation is that the second site is a manifestation of the response of the system to the presence of doped holes themselves.

Identification of the specific origin of the second site is difficult. Ultimately, the quadrupole frequency $\nu_Q$ is determined by the distribution of charge surrounding the copper nucleus. Thus the larger value of $\nu_Q$ may be due either to a change in the local lattice structure around the copper site or due to a redistribution of charge in the lattice surrounding the copper or within the copper ion itself: one possibility is a transfer of charge from the $d_{x^2-y^2}$ to the $d_{3z^2-r^2}$ orbital. It is important to point out that the feature which distinguishes the second site must be static on the time scale of at least the inverse of $\nu_Q$. Thus whatever local variation in charge density is responsible for the second copper site must be essentially stationary or very slowly moving.

## 5. Interpretation and discussion

At this point we would like to propose two mechanisms which might explain the inhomogeneous properties of the crystal structure and charge distribution. We must reconcile two seemingly inconsistent facts: the tilt of the oxygen octahedra are distributed *continuously* between zero tilt and the large tilt associated with very lightly doped material ($\delta \cong 0.01$) whereas we observe *two distinct* charge environments in the $CuO_2$ planes.

If we focus first on the copper NQR results it is natural to think of a local region of increased hole density. The association of the feature at higher frequency with a region of increased hole density would be consistent with the fact that $\nu_Q$ increases with doping in $La_{2-x}Sr_xCuO_4$. The observation of two distinct sites would imply well defined regions having different hole densities. Thus this is, in effect, a *microscopic* phase separation of doped holes occurring within the metallic phase (which is unrelated to the macroscopic phase separation). The tilt of the oxygen octahedra decreases as hole density is increased so we might expect that the region of high hole density in the planes would produce a region of reduced tilts. In order to understand the observation of a continuous distribution of tilts, we might hypothesize that some feature of the cuprates prevents the magnitude of this tilt angle from changing rapidly with distance. If this were the case we might expect a suppression of the octahedral tilt at the location of the high hole density in the planes, and that this tilt would increase with distance from the hole–rich drop. The observation of superlattice points in the neutron scattering [19] would suggest that the locations of the hole rich regions in the planes exhibit long range order. One might imagine that in this case the interstitial oxygen would attempt to order as well.

On the other hand, the continuous distribution of octahedral tilts suggests a modulation of this tilt having a wavelength long compared to the lattice parameter. Such a modulation of local properties is reminiscent of materials which exhibit charge density waves (CDW). In $TaSe_2$ a CDW driven by Fermi surface nesting develops at low temperature [28]. This mechanism favors an incommensurate CDW because the nesting wave vector is incommensurate. However, lattice energies favor a commensurate CDW. The system resolves this competition by forming domains in which the CDW is commensurate. Phase slips in the domain walls allows the long range variation of the phase of the CDW to be incommensurate. If the tilt of the octahedra were correlated with the phase of the CDW a similar phenomenon could explain the observed distribution of tilts. In this scenario the interior of the domains and the domain walls could provide two distinct environments which would explain the two features in the copper NQR spectrum. The CDW and the domain structure would have to be static as previously mentioned, possibly pinned to the lattice by the presence of the interstitial oxygen atoms in the La–O layer.

## 6. Conclusions

We have found in $La_2CuO_{4+\delta}$ clear evidence that the addition of doped holes induces inhomogeneity into the octahedral tilt pattern and the charge environment of copper in the $CuO_2$ planes. We have argued that these inhomogeneities are not manifestations of the expected local distortion associated with the interstitial oxygen, but are, instead, the response of the material to the presence of doped holes. This observation is important in itself because it demonstrates a previously unrecognized tendency toward inhomogeneity in a cuprate when doped holes are added. The recognition that the inhomogeneity in the planes in both strontium– and oxygen–doped material is identical and therefore cannot be the direct result of the proximity of a dopant reinforces the belief that this is a general feature of doped lanthanum cuprate. The sensitivity of $T_c$ to annealing suggests that these local structures play an important role in the superconductivity in these materials.

## References


1. A.R. Moodenbaugh, Y. Xu, M. Suenaga, T.J. Folkerts and R.N. Shelton, *Phys. Rev. B* **38**, 4596 (1988).
2. J.D. Axe, A.H. Moudden, D. Hohlwein, D.E. Cox, K.M. Mohanty, A.R. Moodenbaugh and Y. Xu, *Phys. Rev. Lett.* **62**, 2751 (1989); M.K. Crawford, R.L. Harlow, E.M. McCarron, W.E. Farneth, J.D. Axe, H. Chouand and Q. Huang, *Phys. Rev. B* **44**, 7749 (1991).
3. S.J.L. Billinge, G.H. Kwei, A.C. Lawson, J.D. Thompson and H. Takagi, *Phys. Rev. Lett.* **71**, 1903 (1993).
4. B. Büchner, M. Cramm, M. Braden, W. Braunisch, O. Hoffels, W. Schnelle, R. Müller, A. Freimuth, W. Schlabitz, G. Heger, D.I. Khomskii and D. Wohlleben, *Europhys. Lett.* **21**, 953 (1993).
5. C. Chaillout, S–W. Cheong, Z. Fisk, M.S. Lehmann, M. Marezio, B. Morosin and J.E. Schirber, *Physica C* **158**, 183 (1989).

6. J.D. Jorgensen, B. Dabrowski, S. Pei, D.G. Hinks, L. Soderholm, B. Morosin, J.E. Schirber, L. Venturini and D.S. Ginley, *Phys. Rev. B* **38**, 11337 (1988).
7. P.C. Hammel, E.T. Ahrens, A.P. Reyes, J.D. Thompson, D.E. MacLaughlin, Z. Fisk, P.C. Canfield and J.E. Schirber, in: *Phase Separation in Cuprate Superconductors*, eds.: K.A. Müller and G. Benedek, World Scientific Publishing (1993), p. 139.
8. P.C. Hammel, A.P. Reyes, E.T. Ahrens, D.E. MacLaughlin, J.D. Thompson, Z. Fisk, P.C. Canfield, S-W. Cheong and J.E. Schirber, *Physica B*, in press.
9. P.C. Hammel, A.P. Reyes, Z. Fisk, M. Takigawa, J.D. Thompson, R.H. Heffner, S–W. Cheong and J.E. Schirber, *Phys. Rev. B* **42**, 6781 (1990).
10. R.K. Kremer, E. Sigmund, V. Hishnyakov, F. Hentsch, A. Simon, K.A. Müller and M. Mehring, *Z. Phys. B — Condensed Matter* **86**, 319 (1992); Y. Oda and M Yamada,, *Physica C* **178**, 158 (1991); J. Ryder, P.A. Midgley, R. Exley, R.J. Beynon, D.L. Yates, L. Afalfiz and J.A. Wilson, *Physica C* **173**, 9 (1991).
11. E.T. Ahrens, A.P. Reyes, P.C. Hammel, J.D. Thompson, P.C. Canfield, Z. Fisk and J.E. Schirber, *Physica C* **212**, 317 (1993).
12. See e.g. J.L. Routhbort et al., *J. Mater. Res.* **3**, 116 (1988).
13. J.D. Jorgensen, D.G. Hinks, B.A. Hunter, R.L. Hitterman, A.W. Mitchell, P.G. Radaelli, B. Dabrowski, J.L. Wagner, H.Takahashi and E.C. Larson, in: *Lattice Effects in High–$T_c$-Superconductors*, eds.: Y. Bar–Yam, T. Egami, J. Mustre–de Leon and A.R. Bishop, World Scientific Publishing (1992), p. 84.
14. A. Kampf, private communication.
15. A.P. Reyes, E.T. Ahrens, P.C. Hammel, J.D. Thompson, P.C. Canfield, Z. Fisk, R.H. Heffner and J.E. Schirber, *J. Appl. Phys.* **73**, 6323 (1993); A.P. Reyes, P.C. Hammel, E.T. Ahrens, J.D. Thompson, P.C. Canfield, Z. Fisk and J.E. Schirber, to appear in: *Spectroscopies in Novel Superconductors*, eds.: F. Mueller, A. Arko and A. Bansil, Pergamon, New York.
16. P.C. Hammel, A.P. Reyes, S–W. Cheong, Z. Fisk and J.E. Schirber, *Phys. Rev. Lett.* **71**, 440 (1993).
17. For a review of nuclear quadrupole resonance (NQR) see: T.P. Das and E.L. Hahn, "Nuclear Quadrupole Resonance Spectroscopy", in: *Solid State Physics*, eds.: F. Seitz and D. Turnbull, Academic Press, New York (1958), suppl. 1, chap. I.
18. The point charge modeling was based on the structural determination reported by C. Chaillout, S–W. Cheong, Z. Fisk, M. S. Lehmann, M. Marezio, B. Morosin and J.E. Schirber, *Physica C* **158**, 183 (1989).
19. P.G. Radaelli, J.D. Jorgensen, A.J. Schultz, B.A. Hunter, H.L. Wagner, F.C Chou and D.C. Johnston, *Phys. Rev. B* **48**, 499 (1993).
20. see article by A. Bianconi in this volume and A. Bianconi, in: *Phase Separation in Cuprate Superconductors*, eds.: K.A. Müller and G. Benedek, World Scientific Publishing (1993), p. 125.
21. K. Ueda, T. Sugata, Y. Kohori, T. Kohara, Y. Oda, M. Yamada, S. Kashiwai and M. Motoyama, *Solid State Commun.* **73**, 49 (1990); T. Kohara, K. Ueda, Y. Kohori and Y. Oda, *Physica C* **185–189**, 1189 (1991).
22. T. Tsuda, T. Ohno and H. Yasuoka, *J. Phys. Soc. Jpn.* **61**, 2109 (1991).
23. Y.–Q. Song, M.A. Kennard, M. Lee, K.R. Poeppelmeier and W.P. Halperin, *Phys. Rev. B* **44**, 7159 (1991).
24. S. Ohsugi et al., *J. Phys. Soc. Jpn.* **60**, 2351 (1991).
25. K. Kumagai et al., *Z. Naturforschung A* **45**, 433 (1990).
26. H. Lütgemeier and M.W. Pieper, *Solid State Commun.* **64**, 267 (1987).
27. F.C. Chou, J.H. Cho and D.C. Johnston, *Physica C* **197**, 303 (1992).
28. B.H. Suits, S. Couturie and C.P. Slichter, *Phys. Rev. Lett.* **45**, 194 (1980) and references therein.

# Mössbauer Probe Measurements in $La_{2-x}Sr_xCuO_4$ and $YBa_2Cu_3O_x$

P. Imbert, J.A. Hodges, C. Garcin, P. Bonville, and G. Jéhanno

Service de Physique de l'Etat Condensé, Centre d'Etude de Saclay, 91191 Gif-sur-Yvette, France

**Summary.** We present some results concerning the local magnetic properties of doped superconducting cuprates obtained using Mössbauer spectroscopy. These results which follow on from those presented at the Erice Workshop [1] concern the $La_{2-x}Sr_xCuO_4$ system studied using the $^{57}Fe$ probe and the $YBa_2Cu_3O_x$ system studied using the $^{170}Yb$ probe.

## 1. Mössbauer emission measurements on highly dilute $^{57}Fe$ impurities in $La_{2-x}Sr_xCuO_4$

The initial Mössbauer absorption measurements were made on $^{57}Fe$ doped into the $La_{2-x}Sr_xCuO_4$ series when the substitution level (Fe/Cu) was 0.005 [1, 2]. We have repeated the measurements with a considerably lower substitution level by doping with the radioactive parent $^{57}Co$. The $^{57}Fe$ atoms, which are then generated by the decay of $^{57}Co$, can be considered as isolated impurities in the matrix, and were studied by Mössbauer emission spectroscopy. In the superconducting samples, we continue to observe the Cu spin-glass like environment for low $x$ values, and probe induced magnetic freezing for high $x$ values. However we no longer observe any clear evidence of the coexistence of the two behaviours. The new results provide information concerning the nature of the local perturbation induced by the Fe substitution at the Cu site for both $x$ ranges, and they led us to partially revise our previous interpretation of the absorption measurements.

### 1.1 $^{57}Co$ doping process

$La_{2-x}Sr_xCuO_4$ pellets were wetted with a solution of $^{57}CoCl_2$ in 0.1 N HCl neutralized with a convenient amount of $NH_4OH$-alcohol solution. The $^{57}Co$ diffusion was performed by annealing in air at 1100°C for 12 h, by regrinding the sample and by reannealing. The samples were then quenched and finally annealed in air at 700°C for 3 days.

### 1.2 $^{57}Fe$ room temperature emission measurements

Fig. 1.1 shows the variation of the isomer shift $\varepsilon$ and of the quadrupole splitting $\Delta$ on the $^{57}Fe^{3+}$ probe versus $x$ at room temperature. These variations

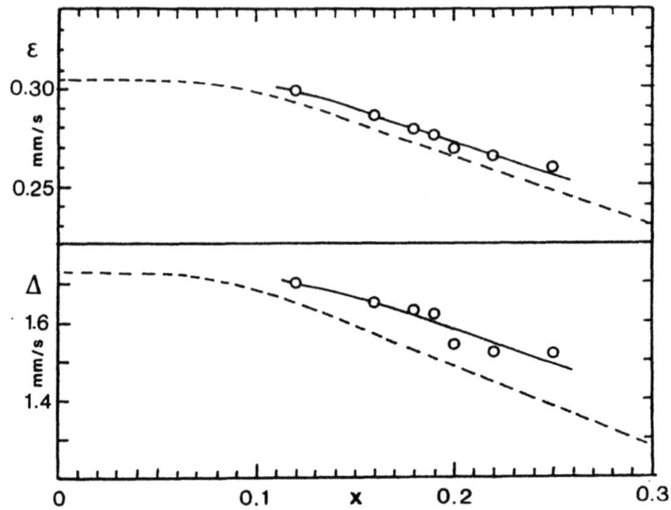


**Fig. 1.1.** $^{57}$Fe hyperfine parameters versus $x$ at 295 K in $La_{2-x}Sr_xCuO_4$. $\varepsilon$: isomer shift relative to iron metal. $\Delta$: quadrupole splitting. Experimental values and full line: from emission measurements. Dotted line: from absorption experiments.

(full line curves) follow the same trend as in the previous absorption measurements (dashed line curves). The decrease of both $\varepsilon$ and $\Delta$ in the metallic region of the phase diagram reflects the increase of the number of delocalized holes in the Cu-$O_2$ planes versus $x$. However a shift is observed between the emission and the absorption curves, and it appears as if the effective $x$ values in the source samples are slightly lower (by $\delta x = 0.02$ to 0.04) compared to the values in the absorber samples. This shift could be due in part to a small reduction of the actual Sr concentration level produced when the samples were wetted with the $^{57}$Co / HCl (and $NH_4OH$) solution and annealed.

### 1.3 $^{57}$Fe low temperature emission measurements

A magnetic hyperfine structure appears at low temperature on the $^{57}$Fe probe whatever the value of $x$. The Fig. 1.2 compares the low temperature measurements in the emission experiments (right side) and in the absorption experiments (left side). The figure displays 3 parameters versus $x$: first, the local spin-freezing temperature $T_m$, defined by the onset of the magnetic hyperfine structure; second, the saturated magnetic hyperfine field $H_o$ observed at 1.4 K; third, the average value of the orientation angle $\theta = (\overline{OZ}, \overline{H_o})$ of $\overline{H_o}$ relative to the main axis $\overline{OZ}$ of the electric field gradient which, to a rough approximation, remains near the crystal $\bar{c}$ axis for all values of $x$ [3]. Let us recall first that the previous $^{57}$Fe absorption measurements together with resistivity measurements led us to divide the $x$ composition range into successive regions, with the following features [1, 2].

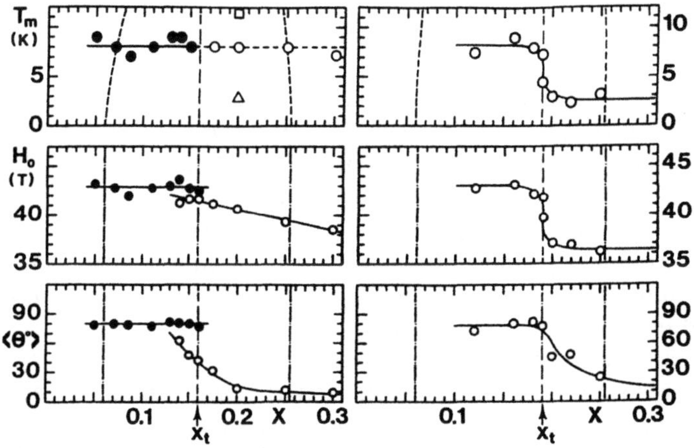


**Fig. 1.2.** $^{57}Fe$ low temperature Mössbauer data versus $x$ in $La_{2-x}Sr_xCuO_4$. $T_m$: spin freezing temperature. $H_o$: saturated magnetic hyperfine field. $<\theta>$: average value of the angle $\theta = (\overline{OZ}, \overline{H_o})$ (note that large distributions of $\theta$ are observed for $x > x_t$). *Left part:* data from absorption experiments [ standard Fe concentration: c=0.5% (circles); for x=0.2, $T_m$ has also been measured using c=0.2% (triangle) and c=1% (square)]. *Right part:* data from emission experiments (extremely low $^{57}Co$ or $^{57}Fe$ concentration).

- The semiconducting compounds in the series show long range Cu antiferromagnetic ordering in zone a ($0 < x < 0.02$) and low temperature Cu spin-glass ordering in zone b ($0.02 < x < 0.06$). In both zones a and b the Cu spin are frozen along the basal plane ($<\theta>= 80$ to $90°$).
- The superconducting compounds show two different local spin freezing behaviours depending on the $x$ values: in zone c ($0.06 < x < 0.16$) the majority of the $^{57}Fe$ probes still experience the same type of local Cu spin-glass ordering as in zone b; in zone d ($0.16 < x < 0.25$) the local spin freezing is different and it is clearly induced by the probe itself: $T_m$ becomes Fe-concentration dependent and with increasing $x$ the probe magnetization rotates from the basal plane towards the $\bar{c}$ direction. In the transition region between zones c and d, the absorption spectra are composite and the relative weight of the Cu spin-glass fraction (with $<\theta>\sim 80°$) decreases when $x$ increases.
- Finally in zone e ($x > 0.25$), where the thermal dependence of the resistivity evidences a metallic state, the local magnetic behaviour remains the same as in zone d.

The emission experiments were performed in the superconducting composition range ($0.06 < x < 0.25$). The results, presented on the right part of Fig. 1.2, show again the two local magnetic behaviours in zones c and d, but the limiting Sr concentration $x_t$ between these zones is about 0.19 instead of 0.16 in absorption experiments. We notice that the shift between the two values of $x_t$ is roughly equivalent to the mean shift $\delta x$ observed between the absorption and emission measurements of $\varepsilon$ and $\Delta$ at room temperature, and we conjecture that it has the same origin.

- In zone c, the spin-freezing temperature $T_m$ ($8 \pm 1$K) and the hyperfine parameters $H_o$ ($\sim 43$ T) and $< \theta >$ ($80 - 90°$) are essentially the same as in absorption experiments, as expected because the behaviour of the probe is governed by the local Cu spin-glass ordering.
- In zone d where the spin-freezing is induced by the Fe, $T_m$ is Fe-concentration dependent in the absorption experiments, and it reaches a low limit value of about 3 K for the non-interacting Fe impurities observed in emission experiments.
- The transition between zones c and d occurs in a narrower $x$ range in the source experiments than in the absorber experiments (see $T_m$ and $H_o$ variations near $x_t$ in Fig. 1.2. Besides, there is no clear evidence of superimposition of c-type and d-type subspectra in the transition region contrary to the absorption experiments, and we rather observe a continuous variation of the parameters $H_o$ and $< \theta >$ for increasing $x$ values.

## 1.4 Discussion and conclusions

Two main questions have to be answered: first, what is the nature of the coexistence of the Cu spin-glass character and of the superconducting properties within a microscopic scale in zone c? And second, what is the nature of the magnetic perturbation induced by the Fe probe on the matrix in zone d?

- The persistence of a Cu spin-glass environment around the Fe probe in zone c has to be compared with the observation of a spin-glass fraction in the same composition range by other techniques, such as $\mu$SR [4] or $^{139}$La NQR measurements [5]. It seems that Cu spin-glass domains coexist with percolating superconducting regions in the samples below $x_t$. The fact that the very dilute $^{57}$Fe probes (source experiments) all experience a Cu spin-glass behaviour in zone c, proves that the $^{57}$Fe impurity cannot be considered as a purely passive probe. We think that the $^{57}Fe^{3+}$ impurity induces a pinning of a Cu spin-glass cluster for $x < x_t$, which prevents from observing the response of the probe in regions where the Cu has no intrinsic magnetic character (presumably superconducting fraction). Let us mention that our observation of Cu spin-glass areas surrounding Fe impurities in superconducting $La_{2-x}Sr_xCuO_4$ for $x < x_t$ seems to support the model of "frustrated phase separation" developed by Emery and Kivelson [6]. This model predicts hole density fluctuations for moderate charge doping levels in high $T_c$ compounds, with metallic regions where the hole density is greater than average and antiferromagnetic droplets (or cluster spin-glass) where it is smaller than average. In addition, the model implies that substitution of the Cu should have the effect of pinning the fluctuations and inducing spin-glass behaviour.

A remark has to be made on our previous interpretation of the composite absorption spectra near $x_t$. Since this inhomogeneous behaviour has disappeared in the emission experiments, we must conclude that it was due not so much to the intrinsic properties of the matrix, but rather to vicinity effects between iron impurities. Therefore the relative weight of the minor subspectrum with the lower $< \theta >$ value which was present below $x_t$ in the absorption experiments cannot be considered as representative of the relative weight of the superconducting fraction, as we thought at first.

- Concerning the response of the probe for $x > x_t$ (zones d and e), we have to discard first the hypothesis according to which the magnetic reorientation of the probe towards the $\bar{c}$ direction is due to the weak single-ion $Fe^{3+}$ anisotropy. In fact, in an octahedral site with elongated tetragonal symmetry for which the electric field gradient $V_{zz}$ due to the neighbouring lattice charges is positive (as is confirmed experimentally), it can be shown that the second-order spin hamiltonian parameter $D$ of the $Fe^{3+}$ ion is positive [7]. That is, the ground electronic level is the $S_z = \pm 1/2$ Kramers doublet and the $Fe^{3+}$ ion presents an easy plane of magnetization. Therefore, it seems that the observed orientation of the $Fe^{3+}$ magnetization near the $\bar{c}$ axis is linked with the anisotropy of the local spin polarization induced in the Cu-$O_2$ medium by the substitution of Fe at the Cu site. This conclusion is based on the following considerations. First, it has been observed that non-magnetic dopants such as $Zn^{2+}$, $Ga^{3+}$, $Al^{3+}$ in superconducting $La_{2-x}Sr_xCuO_4$ break the moment compensation in the dynamic antiferromagnetic state and create the equivalent of approximately one paramagnetic $Cu^{2+}$ moment per doped ion, by locally removing one $Cu^{2+}$ (S = 1/2) spin [8, 9]. Second, it can be shown that in its elongated oxygen octahedron an isolated $Cu^{2+}$ ion presents an anisotropic $g$ tensor with $g_c > g_\perp$ [10]. We think that a similar conclusion probably also applies to the origin of the magnetic orientation of the $^{57}Fe^{3+}$ probe along the $\bar{c}$ direction in superconducting Bi-HTSC [11].

As a conclusion, the emission Mössbauer measurements on highly dilute $^{57}Fe$ impurities in superconducting $La_{2-x}Sr_xCuO_4$ show that Cu spin-glass regions are pinned around the probe for moderate carrier levels, whereas for higher carrier levels the $Fe^{3+}$ impurity induces a localized Cu moment, which freezes near the $\bar{c}$ direction at low temperature. The threshold value $x_t$ probably corresponds to the highest carrier density where spin-glass correlations can still exist.

## 2. Local properties of substituted $YBa_2Cu_3O_x$ examined by Mössbauer Spectroscopy on $Yb^{3+}$ ion probes

We previously presented results concerning the evolution of the local properties of $YBa_2Cu_3O_7$ as the density of doped holes in the the $CuO_2$ planes was reduced by lowering the average occupancy of the oxygen chain sites [1, 11, 12]. We recall that these measurements, made on non-perturbing $^{170}Yb^{3+}$ probes, relate to the detection of a molecular field which is present on an isolated $Yb^{3+}$ ion when the Cu(2) in its immediate neighbourhood are magnetically correlated. This method can distinguish the regions of the sample where the Cu(2) are magnetically correlated (including the regions showing cluster spin-glass order) from the regions where there are no visible magnetic correlations (these regions are presumably superconducting). At intermediate oxygen levels where magnetic and non-magnetic (spin-glass) regions coexist, we were able to follow the evolution, as a function of the average oxygen content, of the amount of the sample showing respectively magnetic and non-magnetic local behaviour. It was also possible to follow the dynamics of the magnetic fluctuations of the Cu(2) spin-glass state over the range from $2.0 \times 10^8$ to $10^{10}$ $s^{-1}$.

### 2.1 Influence of the spatial distribution of oxygen on the magnetic properties of $YBa_2Cu_3O_{6.55}$

The superconducting properties for intermediate oxygen levels depend not only on the average oxygen content but also on the spatial arrangement around the Cu(1) (chain) sites [13] and this arrangement can be modified by an appropriate heat treatment. For example at intermediate oxygen levels, a heat treatment tending to accentuate the orthorhombic character and to accentuate the local ordering amongst the oxygens of the chains, is found to lead to more pronounced superconducting properties. An example of how the heat treatment modifies the relative amounts of the magnetic and non-magnetic fractions as evidenced by the present technique, is shown by the data on Fig. 2.1. Here a pellet of $Y_{0.97}Yb_{0.03}Ba_2Cu_3O_7$ was studied after having been partially de-oxidized in two different ways. For one treatment the de-oxidization was made at relatively a low temperature inside a thermobalance whilst the second was made by quenching from a relatively high temperature into liquid nitrogen. Both treatments led to orthorhombic samples with the same average oxygen content ($x \sim 6.55$) as deduced from the common value of the room temperature c-axis lattice parameter (1.1734 nm) but with different levels of orthorhombicity. The relative weights of the magnetic and non-magnetic subspectra are also quite different for the two cases. The ratio of the weight of the non-magnetic fraction to that of the magnetic fraction is 55/45 in one case and 75/25 in the other and it is the sample which has the higher orthorhombicity which has the higher non-magnetic fraction.

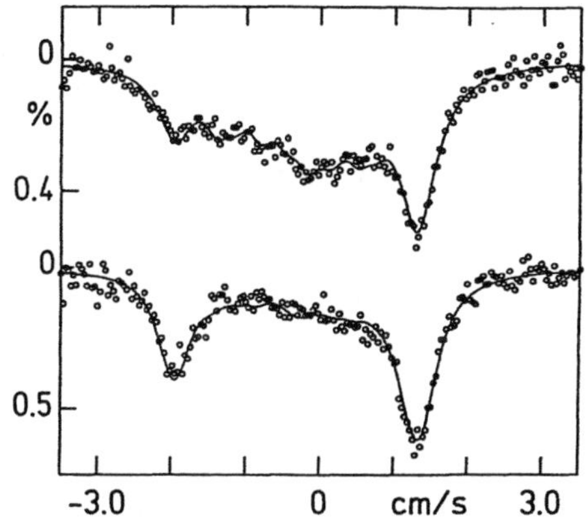


**Fig. 2.1.** $^{170}Yb^{3+}$ Mössbauer absorption spectra at 1.4 K for two samples of $Y_{0.97}Yb_{0.03}Ba_2Cu_3O_x$ having the same average oxygen content ($x = 6.55$) but with different spatial distributions in the oxygen occupancies around the chain sites. Each spectra is fitted in terms of two subspectra, one characteristic of a neighbourhood of magnetically correlated (spin-glass) Cu(2) and one characteristic of a neighbourhood which is non-magnetic (and presumably superconducting). The ratio of the relative weight of the subspectrum associated with the superconducting fraction to that due to the spin-glass fraction is 55/45% (upper) and 75/25% (lower). The lower sample has a higher orthorhombicity.

The observed evolution is consistent with the overall tendency for superconductivity at intermediate oxygen levels to be more developed when the orthorhombicity is accentuated [14]. In the present case it was the quenched sample that showed the higher orthorhombicity.

## 2.2 Influence of cationic substitutions

We have examined the evolution of the local properties, as sampled by the $^{170}Yb^{3+}$ probe, when the Cu is partially replaced by $Zn^{2+}$ and when the $Y^{3+}$ is partially replaced by hybridized $Pr^{3+}$. Both substitutions lead to a considerable weakening of the superconductivity properties. The substitutions could give rise to different local environments due to the introduction of statistical variations in the local site occupancies.

## 2.3 $Zn^{2+}$ substitution [15]

Measurements on $Yb^{3+}$ probes in a sample of $YBa_2(Cu_{0.94}Zn_{0.06})_3O_7$ that is fully oxidized show that the introduction of $Zn^{2+}$ does not lead to the creation of correlated Cu(2) magnetic moments in any part of the sample even though $T_c$ is reduced from 92 to 32 K. However if $T_c$ is further reduced by lowering the oxygen content (for example to $x = 6.88$ where $T_c = 13$ K), correlated Cu(2) moments do appear in a part of the sample. The amount by which the oxygen content has to be lowered in order to introduce correlated Cu(2) moments is lower for the sample containing zinc than for the zinc-free sample. Although the influence on the $T_c$ due to the lowering of the oxygen content and to the substitution of zinc are cumulative, the mechanisms controlling the lowerings are different in the two cases: when $T_c$ is lowered by reducing the oxygen occupancy of the chains so as to reduce the carrier density in the planes, this leads to the introduction of correlated Cu(2) magnetic moments, whereas no such moments are introduced when $T_c$ is reduced by substituting $Zn^{2+}$.

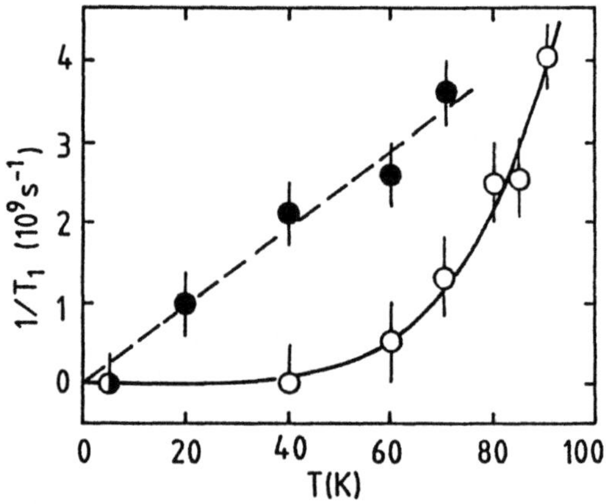


**Fig. 2.2.** From $^{170}Yb^{3+}$ Mössbauer measurements: paramagnetic relaxation rate within the sublevels of the $Yb^{3+}$ Kramers doublet ground state in $YBa_2Cu_3O_7$ with $T_c = 91$ K (open circles) and in $YBa_2(Cu_{0.94}Zn_{0.06})_3O_7$ with $T_c = 32$ K (full circles) with in each case 3% of the $Y^{3+}$ replaced by $Yb^{3+}$. A Korringa dependence is present for the $Zn^{2+}$ substituted sample.

We have also examined the thermal evolution of the relaxation rate within the Kramers doublet ground state of the $Yb^{3+}$ ion in fully oxidized zinc-free and zinc substituted samples. The results are shown on Fig. 2.2. The variation in the zinc-free sample was initially attributed to coupling with phonons [2] but part of the decrease which is observed as the temperature is lowered may, in fact, be due to the opening of a gap within the excitations to which the paramagnetic $Yb^{3+}$ ion is coupled. The substitution of $Zn^{2+}$ leads to a marked change in the evolution of the $Yb^{3+}$ relaxation rate which now follows a Korringa dependence. This behaviour suggests the relaxation is driven by the exchange interaction ($J_{sf}$) between the $Yb^{3+}$ electrons and a conduction band so implying that there are now well established states at the Fermi level. From the slope of the thermal dependence of the relaxation rate shown on Fig. 2.2 ($4.8 \times 10^7$ $s^{-1}$/K), we obtain the modest value $|J_{sf}N(E_F)| = 0.012$ where $N(E_F)$ is the density of states at the Fermi level. The so detected density of states is still present below $T_c$ which is evidence in favour of gapless (d - wave) superconductivity. Evidence that the superconductivity becomes gapless in $Zn^{2+}$ doped samples of $YBa_2Cu_3O_7$ has previously been obtained from specific heat [16], point spectroscopy [17] and Cu NMR measurements [18]. With the present measurement accuracy, it is not possible to say whether or not different local $Yb^{3+}$ paramagnetic relaxation rates coexist, that is whether or not there are spatial inhomogeneities in the local band structure.

### 2.4 $Pr^{3+}$ substitution [19]

Superconductivity is suppressed in fully oxidized $Pr_{1-x}Y_xBa_2Cu_3O_7$ when $x$ becomes lower than about 0.45 [20]. In $PrBa_2Cu_3O_7$ long range antiferromagnetic order exists both within the Cu(2) sublattice ($T_N(Cu)$ = 285 K [21]) and within the $Pr^{3+}$ sublattice ($T_N(Pr)$ = 17 K [22]). The suppression of superconductivity (and the introduction of correlated Cu(2) magnetic moments) is probably related to the influence of the hybridization of the $Pr^{3+}$ 4f states with the $CuO_2$ valence band states which leads to the localization of the previously mobile holes.

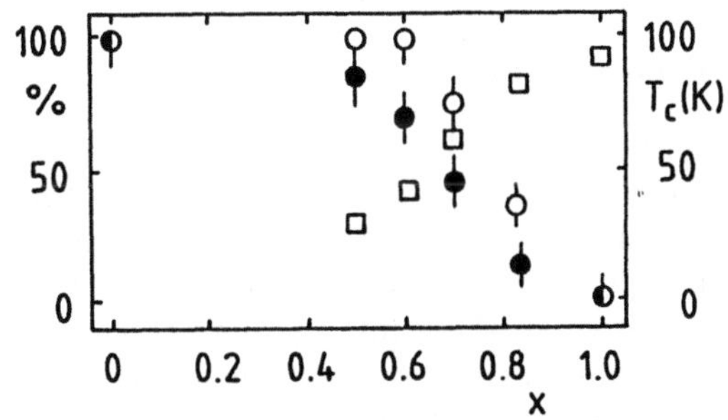


**Fig. 2.3.** For $Pr_{1-x}Y_xBa_2Cu_3O_7$ where 2.5% of the rare earth sites are occupied by $^{170}Yb^{3+}$: $x$ dependence of $T_c$, the superconducting transition temperature (squares), of the $^{170}Yb^{3+}$ subspectrum weight (fraction of the sample volume) which, above $T_N(Pr)$, evidences collective fluctuations within the $Pr^{3+}$ subsystem (open circles) and of the subspectrum weight which, at 1.4 K, evidences $Pr^{3+}$ magnetic order (full circles).

Measurements on $Yb^{3+}$ probes on samples of $Pr_{1-x}Y_xBa_2Cu_3O_7$ which show superconductivity, evidence heterogeneous local behaviour stemming from the statistical variations in the local Pr/Y occupancies. It is also possible that some further heterogeneity could be introduced by the presence of $Pr^{3+}$ ions at the Ba sites. Although some evidence of local Cu(2) magnetic order was obtained, it was not possible to make any quantitative estimates of the $x$ dependence of the amount of the sample showing this order. From measurements at $T < T_N(Pr)$ it was possible to obtain quantitative estimates of the $x$ dependence of the amount of the sample showing $Pr^{3+}$ magnetic order and from measurements at $T > T_N(Pr)$ it was possible to estimate the amount where the $Pr^{3+}$ subsystem in the neighbourhood of the probe undergoes collective magnetic excitations (the $Yb^{3+}$ ion undergoes paramagnetic relaxation driven by interaction with these excitations). The $x$ dependence of these two amounts as well as that of $T_c$, are shown in Fig. 2.3. We note that superconductivity is well developed in the $x = 0.6$ sample ($T_c = 42$ K) but all of the $Yb^{3+}$ are coupled to collective magnetic excitations within the $Pr^{3+}$ subsystem. It thus appears that superconductivity coexists with these excitations. However, there is an indication of an anti-correlation between superconductivity and $Pr^{3+}$ magnetic order : superconductivity tends to become completely quenched near the value of $x$ for which all of the $Pr^{3+}$ become magnetically ordered at low temperature. The suppression of superconductivity is obviously not directly due to the $Pr^{3+}$ magnetic ordering (the magnetic ordering of normal rare earth coexists with superconductivity) but it is rather due to the changes produced by the $Pr^{3+}$ on the $CuO_2$ system. As superconductivity is suppressed when the concentration of $Pr^{3+}$ approximately reaches that needed to support $Pr^{3+}$ magnetic order throughout the sample, this probably means that at this concentration all the Cu(2) become magnetically ordered.

## 2.5 Summary and Discussion

In the $YBa_2Cu_3O_x$ series, it is well known that the reduction of $T_c$ produced by a reduction of the average oxygen content leads to the introduction of correlated magnetic moments within the Cu(2) sublattice and that superconducting samples with intermediate oxygen levels contain regions which are magnetically correlated. In these intermediate oxygen level samples, the $^{170}Yb^{3+}$ Mössbauer measurements show that the magnetically correlated moments have spin glass character with the average magnetic fluctuation rate slowing down as the temperature is lowered. The amount of the sample showing magnetic character depends not only on the average oxygen content but on how the oxygens are distributed about the chain sites. To date, local magnetic character has been identified in samples with an average oxygen content of $x = 6.55$ which is near the lower end of the $T_c = 60$ K plateau region. It is planned to compare the influence of heat treatment on the amount of magnetic and non-magnetic fractions for oxygen levels higher than 6.55 and in particular to examine how much of the sample retains local magnetic character when the oxygens in the chains are ordered in a well developed ortho - II structure.

The substitution of $Zn^{2+}$ into fully oxidized $YBa_2Cu_3O_7$, which leads to a marked decrease in $T_c$, does not lead to the introduction of correlated Cu(2) moments in any part of the sample. The substitution of $Zn^{2+}$ does however, introduce a sizeable paramagnetic susceptibility with a Curie-Weiss like temperature dependence. Superconducting samples for which the $T_c$ has been further reduced by lowering the oxygen content contain regions of magnetically correlated Cu(2) showing spin glass behaviour.

Finally, as evidenced by the results on mixed samples containing both $Y^{3+}$ and $Pr^{3+}$, we recall that chemical disorder can lead to the coexistence of regions with quite different local properties.

# References


1. P. Imbert, J. A. Hodges, G. Jéhanno, P. Debray, P. Bonville, C.Garcin. Phase Separation in Cuprate Superconductors, Edited by K.A. Müller and G. Benedek, World Scientific Publishing (Singapore), (1993) 158.
2. P. Imbert, G. Jéhanno, P. Debray, C. Garcin, J.A. Hodges. J. Phys. 1 France 2 (1992) 1405.
3. P. Imbert, G. Jéhanno, C. Garcin, J.A. Hodges, M. Bahout-Mouallem. Physica C190 (1992) 316.
4. A. Weidinger, Ch. Niedermayer, A. Goldnik, R. Simon, E. Recknagel, J.I. Budnick, B. Chamberland, C. Baines. Phys. Rev. Lett. 62 (1989) 102.
5. T. Kobayashi, S. Wada, Y. Kitaoka, K. Asayama, K. Amaya. J. Phys. Soc. Jpn 58 (1989) 779 and ibid. 2662.
6. V.J. Emery and S.A. Kivelson. Physica C209 (1993) 597.
7. G. Harris. J. Chem. Phys. 48 (1968) 2191.
8. G. Xiao, M.Z. Cieplak, J.Q. Xiao, C.L. Chien. Phys. Rev. B42 (1990) 8752.
9. J. Arai, S. Ogawa, K. Umezawa. Physica C206 (1993) 257.
10. A.M. Finkel'stein, V.E. Kataev, E.F. Kukovitskii, G.B. Teitel'baum. Physica C168 (1990) 370.
11. J.A. Hodges, P. Bonville, P. Imbert, G. Jéhanno. Physica C 184 (1991) 259.
12. J.A. Hodges, P. Bonville, P. Imbert, G. Jéhanno, P. Debray. Physica C 184 (1991) 270.
13. G.V. Uimin, V.F. Gantmakher, A.M. Neminsky, L.A. Novomlinsky, D.V. Shovkum, P. Brüll. Physica C 192 (1992) 481.
14. Y. Nakazawa, M. Ishikawa. Physica C 158 (1989) 381.
15. J.A. Hodges, P. Bonville, P. Imbert, D. Colson, N. Genand-Riondet, A. Pinatel-Philippot. Physica C (1993). To be published.
16. J.W. Loram, K.A. Mirza, P.F. Freeman. Physica C 171 (1990) 243.
17. A.I. Akimenko, G. Goll, I.K. Yanson, H.v. Löhneysen, R. Ahrens, T. Wolf, H. Wühl. Z. Phys. B 82 (1991) 5.
18. K. Ishida, Y. Kitaoka, Y. Yoshitomi, N. Ogata, T. Kamino, K. Asayama. Physica C 179 (1991) 29.
19. J.A. Hodges, G. le Bras, P. Bonville, P. Imbert, G. Jéhanno. Physica C (1993). To be published.
20. L. Sonderholm, K. Zhang, D.G. Hinks, M.B. Beno, J.D. Jorgenson, C.U. Segré, I.K. Schuller. Nature 328 (1987) 604.
21. D.W. Cooke, R.S. Kwok, R.L. Lichti, T.R. Adams, C. Boelema, W.K. Dawson, A. Kebede, J.Schwegler, J.E.Crow, T.Mihalisin. Phys. Rev. B 41 (1990) 4801.
22. G. Wortmann, I. Felner. Solid State Comm. 75 (1990) 981.

# Indications for a Phase Separation in $YBa_2Cu_3O_{7-x}$

K. Conder[1], D. Zech[1], Ch. Krüger[1], E. Kaldis[1], H. Keller[2], A.W. Hewat[3], and E. Jilek[1]

[1] Laboratorium für Festkörperphysik ETH, CH-8093 Zürich, Switzerland
[2] Institute of Physics, University of Zürich, CH-8001 Zürich, Switzerland
[3] ILL-Grenoble, France

**Summary.** Slope changes of the lattice constants and dc magnetization measurements as a function of the oxygen content of powder 123 samples, indicate the existence of three phases with phase boundaries at $x \cong 6.75$ and $x \cong 6.91$. In addition, a splitting of the derivative of the magnetization curve appears at $x \geq 6.95$. These results are discussed in terms of a three-phase model. It is shown that oxygen inhomogeneities of the grains cannot be the reason for the double peaks in specific heat and magnetization.

## 1. Introduction

As already shown in the past [1], small lattice constant anomalies appear in near "equilibrium" 123 samples at high oxygen concentrations (e.g $x \cong 6.90 - 6.95$), which coincide with anomalies of the physical properties: maximum of $T_c$ and strong increase of the muon depolarization rate at $x \cong 6.90$ [1, 2]. Later, anomalies were measured in the specific heat and resistivity [3] indicating at the same $x$ values the existence of phase separation. We have consistently measured splitting of the derivatives of the dc magnetization curves at $x \geq 6.95$ [4]. Very recent specific heat measurements [5], confirmed these results showing two peaks.

In view of the numerous investigations of the lattice constants dependence on the oxygen content, it was amazing that no other laboratory has reported these anomalies. We were, therefore, very cautious to make sure that this was not an artifact. To this end, two kinds of investigations were started: neutron diffraction to determine the bond lengths as a function of the oxygen content; and variation of the conditions of synthesis to find out how far the anomalies of the lattice constants depend on the chemical or thermal history of the samples. The samples were characterized by magnetization measurements, as well as with x-ray lattice constants and chemical analysis for oxygen. Our work is based on ceramic samples, because for a composition dependent phenomenon the exact determination of the oxygen stoichiometry is very important. In the past, we have developed a very sensitive volumetric determination of oxygen in cuprates [6] which, unfortunately needs samples of at least 100mg weight. This can be fulfilled mainly by ceramic samples and only very seldom by high quality single crystals. Large crystals of 123 grown from the flux tend to have defects due to impurities and Cu1-vacancies in the single chains. For the single chains of 247 this has been shown with the structure refinement of 15 crystals [7]. It seems very likely that this is the case, also for many of the up to now investigated 123 single crystals in the literature. Recently, work on single crystals supported the maximum $T_c$ at $x = 6.90$ [8] which we have reported in the past [1] and showed a reversible splitting of the c-axis [9]. These anomalies are attributed to a phase separation which appears in the overdoped range ($x > 6.90$). On the other hand, concentration inhomogeneities in sample grains caused by insufficient oxygen diffusion rate is believed to be a reason for the observed splitting in susceptibility and specific heat measurements [5]. The existence of a highly oxidized outer layer on the surface of each grain acting as a oxygen diffusion barrier and showing different $T_c$ as a core has been proposed [5].

In this work, the oxygen diffusion rate in 123 phase will be estimated and a model assuming oxygen concentration inhomogeneities will be critically discussed. We present the dependence of the lattice constants on the oxygen content (x-ray), structure refinements by neutron diffraction for 14 samples with different oxygen stoichiometry, as well as dc susceptibility at various oxygen concentrations. These measurements indicate the existence of a phase separation in 123. A possible model for three phases is given.

## 2. Synthesis of the samples

To find out a possible influence of the chemical and thermal history of the samples on the lattice constant anomalies, two methods of synthesis, both different from that used in the past [1], have been investigated.

In the past, we have used $O_2$ and Ar mixtures at various temperatures and reaction times to reduce the oxygen content in an open flow system. $BaCO_3$ has been decomposed (40h, 950°C) in oxygen atmosphere and annealed with the other metal oxides at 950°C (20h). Three grindings and annealings were performed, the last in pellet form. Samples of high oxygen content were first synthesized with this procedure and then subjected to a controlled reduction with $O_2$-Ar mixtures. Our previous work [1] and the neutron diffraction data reported here (Sect. 4.), are based on these samples.

The additional methods of synthesis used in the present work are:

- direct oxidation of the pure metals in a closed system followed by reaction of the binary oxides (DO method). This method is suitable also for the synthesis of isotope samples with high isotope content [10]. Unfortunately, the experimental setup did not allow simultaneous processing of the whole batch, and therefore a certain scattering due to unavoidable slight thermal history deviations is observed. Starting with high oxygen content samples, lower oxygen contents have been reached by equilibration with a known mass of copper in a small sealed ampoule (900°C).

- In the second method, the classical approach of direct reaction of $BaCO_3$ with the metal oxides has been used, but the reduction of the samples has been performed by equilibration with $YBa_2Cu_3O_{6.05}$ (CAR method) in a small sealed ampoule (600°C). In this method, the experimental setup allowed the processing of the whole batch, decreasing strongly the scattering of the lattice constants. Unfortunately, fewer samples could be synthesized with this method. In both methods, a slow cooling rate (0.1°C/min CAR, 0.25°C/min DO) was used to obtain homogeneous samples under near equilibrium conditions.

## 3. X-ray lattice constants

Figures 3.1 and 3.2 show the dependence of the c-, a-, and b-lattice constants, orthorhombicity, and cell volume, on the oxygen concentration of the directly oxidized samples (method DO). The lattice constants have been determined with a Guinier Nonius camera, using NIST silicon as an internal standard. The accuracy of the a- and b-axes is 0.001Å and that of the c-axis 0.002Å. Characteristic for the lattice constants of samples from the DO method (Fig. 3.1 and 3.2), as performed here, is a) some scattering of the values at higher oxygen concentration, and b) a rather clear change of slopes in two concentrations: $x \cong 6.75$ and $x \cong 6.91$. On the other hand, the lattice constants of the CAR method (Fig. 3.3 and 3.4) show less clearly these changes of slopes, and particularly that at $x \cong 6.75$. Also it seems that in these samples the anomaly appears at $x \cong 6.9$.

It is reasonable to assume that the changes of slope of the lattice constants reflect various superstructures or small changes of the lattice, appearing as a function of the nonstoichiometry. The corresponding changes of the total energy are very small and can easily be masked by traces of impurities. Thus, it is possible that the DO method leads to intermediate phases, whereas the CAR method destabilizes these phases. In any case, the nonstoichiometric ranges $x < 6.75$, $6.75 < x < 6.91$ and $x > 6.91$ must be discussed from the point of view of the phase separation. It is well established by now, that the $x \cong 6.91$ anomaly corresponds to $T_{\max}$ (Fig. 5.3) and is characteristic for the transition from the underdoped to the overdoped region. The existing lattice constant data do not allow to resolve a separate anomaly at $x \geq 6.95$, where the splitting of the derivative of the magnetization curve starts (section 5.). Assuming that the above oxygen concentrations are the results of a chemical phase separation, a rough three phase model could be envisaged (Fig. 3.5).

The existence of the $2a_0$ superstructure has been discussed several times in the literature as a separate thermodynamic phase at $x \cong 6.5$, responsible for the lower plateau of the $T_c$ vs. $(x)$ curve. On the other hand, candidates for the high oxygen content phase of Fig. 3.5 could be the tetragonal phase with 120K diamagnetic transition [11] or the recently found $x = 7.7$ tetragonal phase [12]. In the later sections we discuss possible evidence for this model.

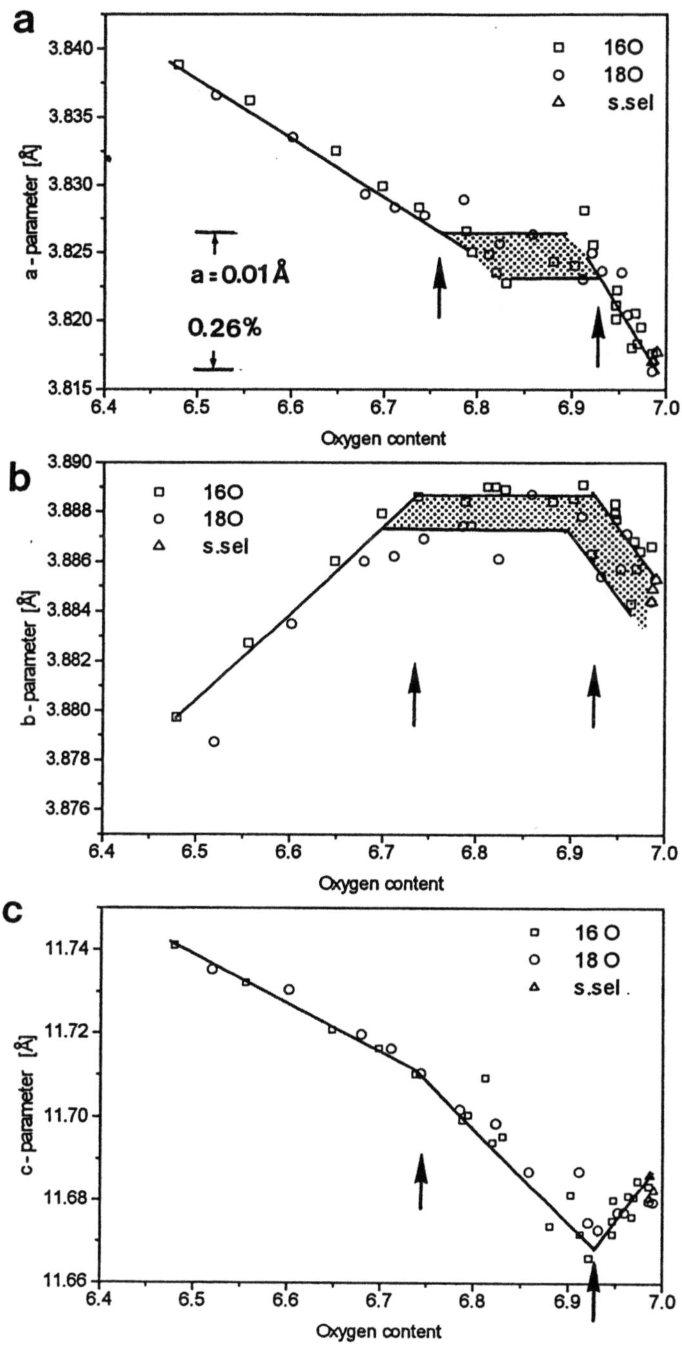


**Fig. 3.1.** (a) a-lattice parameter of 123 vs. oxygen content, as obtained from samples synthesized by direct oxidation of the metals (DO). Samples prepared with $^{18}O_2$, $^{16}O_2$ and site selectively exchanged (apex and chain sites $^{16}O$, planes $^{18}O$) are marked with different points. (b) b-lattice parameters, (c) c-lattice parameters.

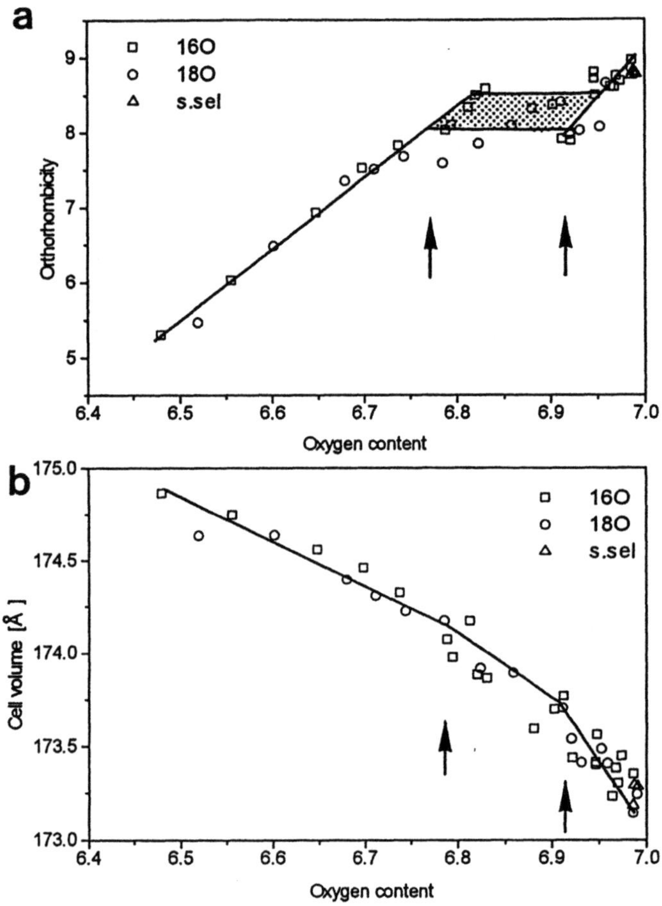


**Fig. 3.2.** (a) Orthorhombicity as in Fig. 3.1, (b) Cell volume.

## 4. Neutron diffraction

To find out if the first reported anomaly of the lattice constants at $x \cong 6.91$ [1] was the result of a structural distortion leading to changes of the bondlengths, we have investigated 12 samples [13] with oxygen stoichiometries $6.967 \geq x \geq 6.615$ determined volumetrically [6]. The samples have been synthesized according to the procedure described in [1] and summarized in section 2.

### 4.1 Experimental

For the neutron diffraction investigations the powder or pellet samples were enclosed under a He gas atmosphere in cylindrical vanadium containers of ⊘ 10mm. The neutron diffraction measurements were performed at several

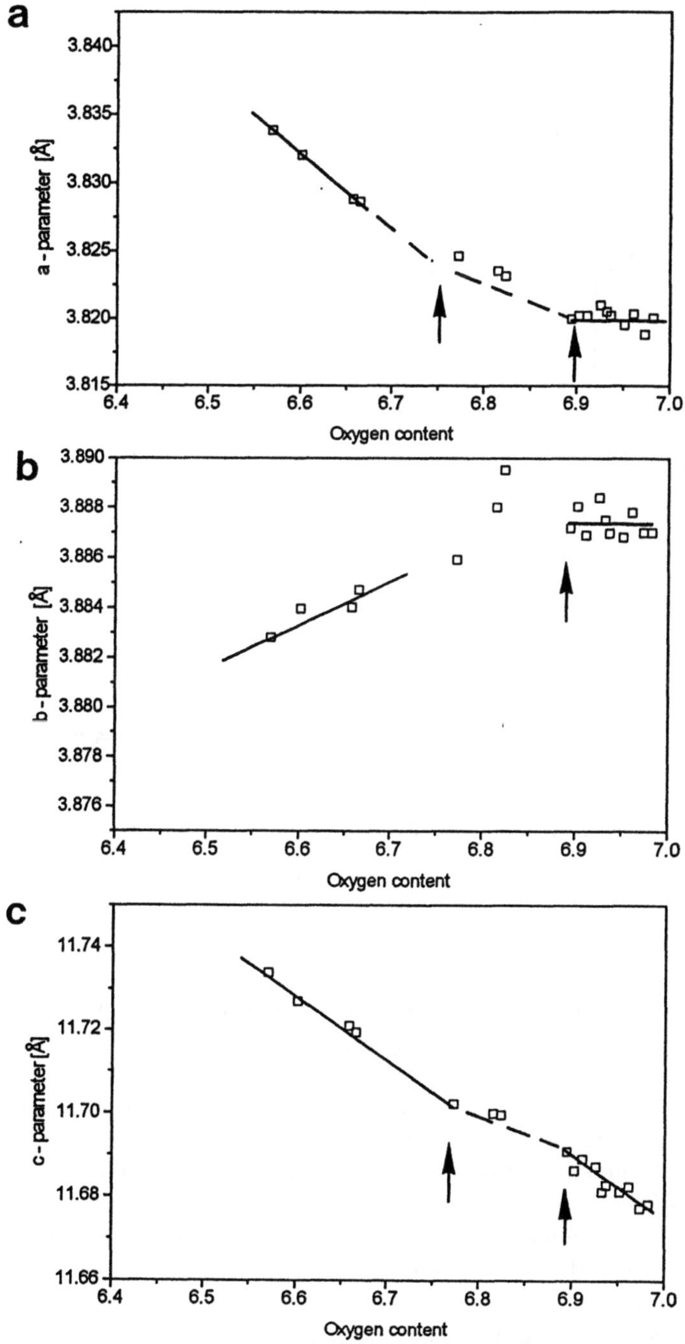


**Fig. 3.3.** (a) a-lattice parameter of 123 vs. oxygen content, as obtained from samples synthesized with $BaCO_3$ (CAR method), (b) b-lattice parameters, (c) c-lattice parameters.

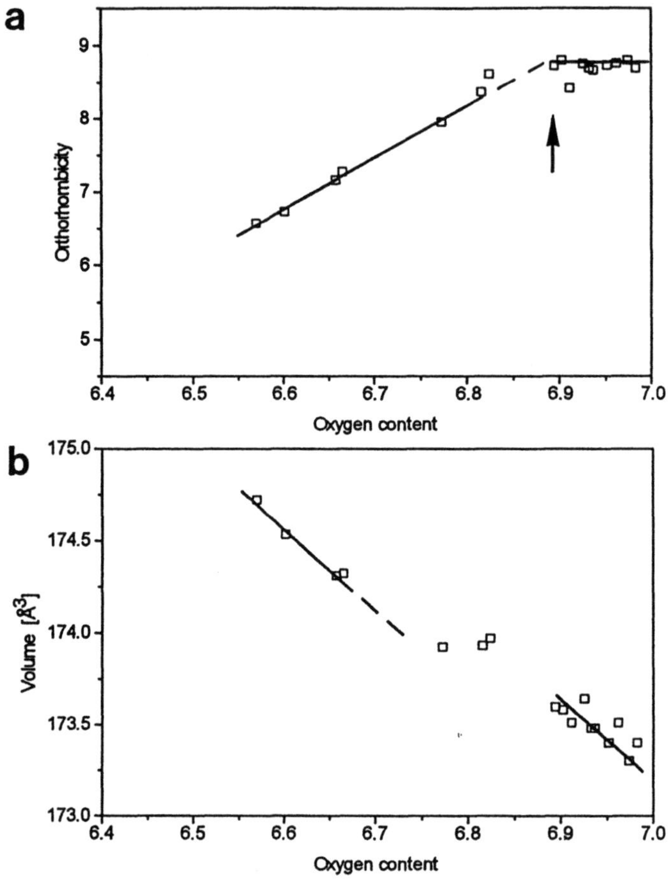


**Fig. 3.4.** (a) Orthorhombicity as in Fig. 3.3, (b) Cell volume.

temperatures in the range of 5 to 300K at ILL on the high-resolution neutron diffractometer D2B in the high intensity mode. Since the samples were large ( > 1cm$^3$) a short wavelength (lower intensity) neutron beam (1.0506Å instead of 1.5946Å) was used to obtain more precise positional and thermal motion parameters. Absorption corrections were performed according to the measured transmission (product of linear absorption coefficient and sample radius $\mu r = 0.115$ and 0.163 for powder and pellet samples respectively).

All samples were measured at 5K, and also at 300K. Some of the samples were also measured at 100K and at 200K, but since these results did not reveal anything new, they are not reported here.

The measured diffraction patterns were analyzed by means of the Wiles-Young program [14] including automatic background fit, using the scattering lengths published in [15] and the split-atom model for the chain oxygen O4 to model the strong anisotropic disorder of this atom [16].

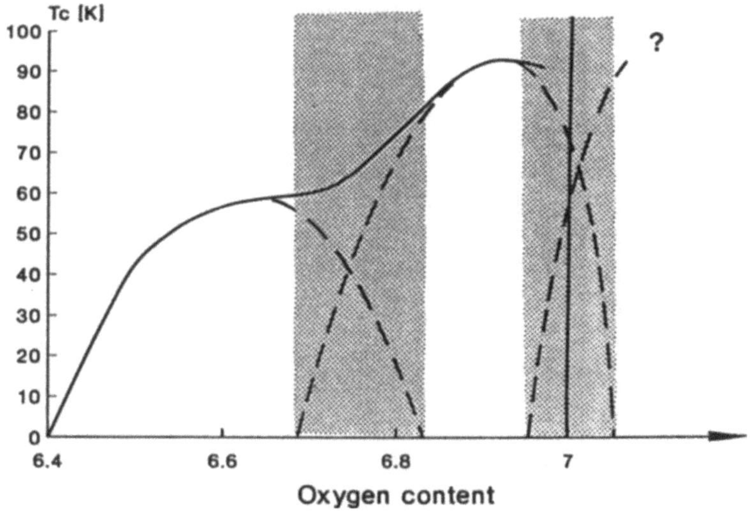


**Fig. 3.5.** Phase diagram ($T_c$) of $YBa_2Cu_3O_x$ according to an assumed three-phase model.

## 4.2 Anomalies in the Cu-O bond lengths

The structural parameters have been used to obtain the bond lengths in order to understand the origin of the $x \cong 6.9$ anomaly. Fig. 4.1 shows all the contributions to the c-axis drawn to the same scale. As expected, the largest changes occur in the Cu2-O1 apical bond distance. However, the apical bond continues to contract for$x > 6.9$, with only a slight inflection at this point, whereas the c-axis contracts. This is due to a sudden contraction of the distance between Cu2 and the O2,O3-plane. This contraction is the structural transformation corresponding to the $T_{c(\max)}$ and the underdoped-overdoped transition. The only other contributions to the c-axis are the Cu1-O1 distance and the distance of Y from the O2,O3-plane, both of which vary smoothly.

This, probably, first order structural transition is phenomenologically related to the increased buckling of the CuO-planes, measured by the displacement of Cu2 out of its own O2,O3-plane. As the bridging oxygen O1 and Cu2 move toward each other with increasing $x$, the O2,O3-plane moves away, but more slowly so that the net effect is an increase of the apparent valence charge on Cu2 [16]. However, for $x > 6.9$ the O2,O3-plane moves towards Cu2 more quickly, negating the contribution of the shortening Cu2-O1. This phenomenology implies that there is a critical distance between the bridging oxygen and the O2,O3-plane which can not be overcome. These results illustrate clearly that the lattice constant slope changes, correspond to bondlength changes (section 3.).

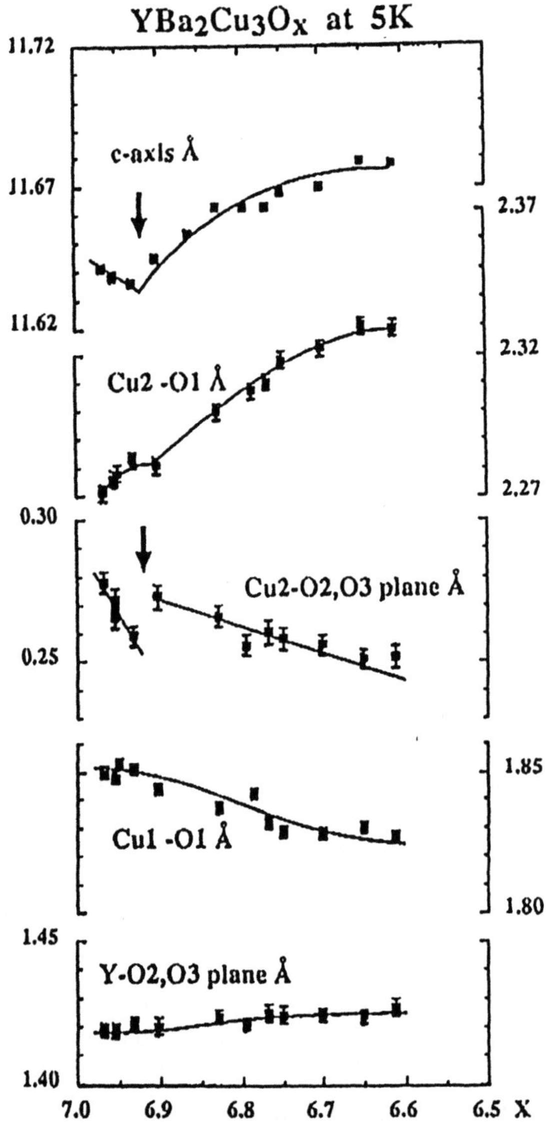


**Fig. 4.1.** C-lattice parameter and all the bondlengths contributing to it in the same scale vs. oxygen content.

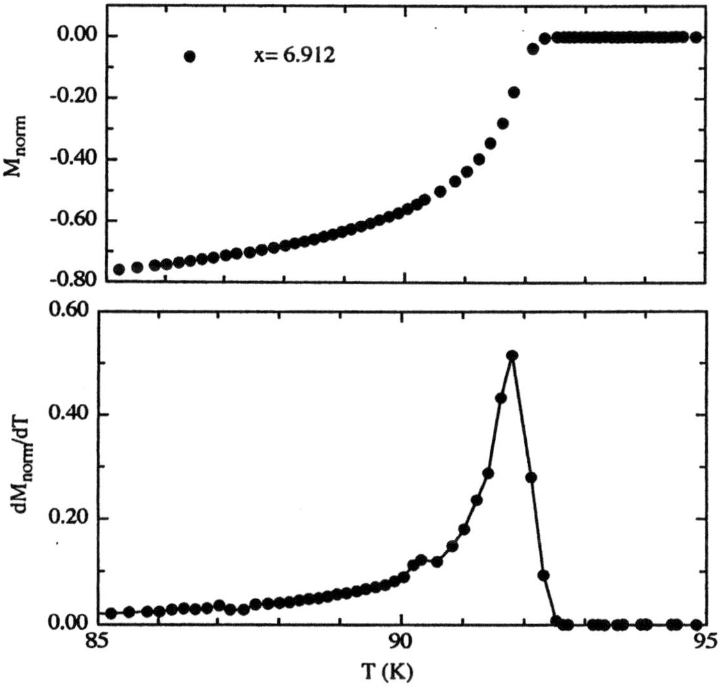


**Fig. 5.1.** Normalized magnetization $M_{norm}$ and its derivative $dM_{norm}/dT$ as a function of temperature for a sample with oxygen content $x = 6.912$.

## 5. Magnetization measurements

Magnetization measurements were performed on samples of $YBa_2Cu_3O_x$ with $x$ between 6.50 and 6.98 using a SQUID magnetometer. The samples were mounted in a closed sample holder made out of quartz in order to minimize oxygen losses and contamination with water both strongly affecting the superconducting properties of the material. All the measurements were taken under identical conditions in an external field of 1 mT by slow decreasing the temperature in small steps from 120 K to 10 K. The high temperature resolution (0.1 K) of the measurements around $T_c$ allowed a precise estimate of the onset of the diamagnetic signal ($T_{onset}$), which in general is not identical with the thermodynamic critical temperature $T_c$.

A sensitive quantity, describing the shape of the magnetization curve, is the derivative $dM/dT$, which can be determined numerically from measured magnetization curves $M(T)$. A typical example is shown in Fig. 5.1, where $M(T)$ and the corresponding $dM(T)/dT$ are plotted for a sample with $x = 6.912$. Note that $dM(T)/dT$ exhibits a sharp **asymmetric** peak at $T_p$ near $T_{onset}$ and decreases almost linearly when $T$ reaches $T_{onset}$.

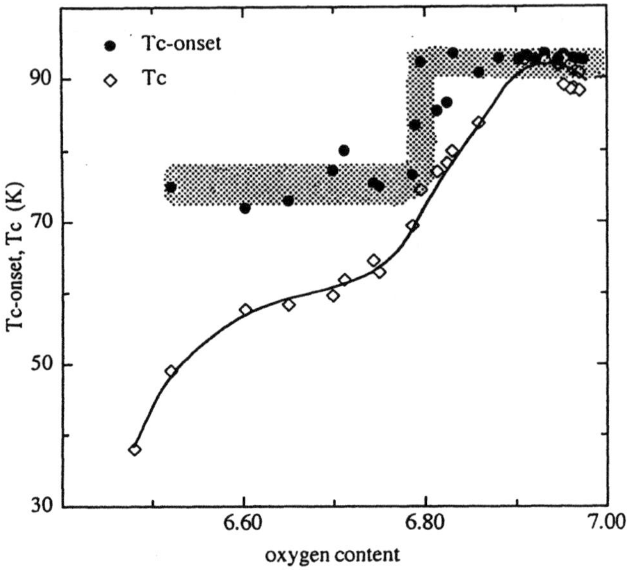


**Fig. 5.2.** Transition temperature $T_c$ (diamonds) and corresponding onset temperature $T_{onset}$ (circles). For $x > 6.96$ a splitting of the superconducting transition was observed in all samples investigated (see text).

A more detailed discussion of the shape of $M(T)$ and its derivative $dM/dT$ will be published elsewhere [17]. As shown in Ref. [17] the quantities $(T_{onset} - T_p)$ and $(T_{onset} - T_p)/2$ are reasonable measures of the transition width and the effective critical temperature $T_c$, respectively. Here we use this definition of $T_c$. In Fig. 5.2 the critical temperature $T_c$ is displayed as a function of the oxygen content $x$.

For $x = 6.92$ the critical temperature exhibits a broad maximum with a maximal value of $T_c = 92.5$K (see Fig. 5.3). With decreasing $x$ the critical temperature decreases and reaches a plateau ($T_c \sim 60$ K) in the region $x \cong 6.6 - 6.7$. For $x < 6.6$ the critical temperature $T_c$ is fast decreasing down to 38K at the lowest value $x = 6.520$ investigated. It is interesting to note that for all samples with high oxygen content ($x > 6.95$), the derivative $dM_{norm}/dT$ exhibits **a broad unresolved double-peak** structure with a non-linear decrease for $T > T_p$ (Fig. 5.4), in contrast to optimal doped samples with $x \cong 6.92$ (Figs. 5.1 and 5.4).

These results support certain aspects of the working hypothesis shown in Fig. 3.5. It is suggested that this broad double-peak may be associated with two different microscopic phases with a different effective critical temperatures $T_c$ (phase separation). This interpretation is supported by the work of Claus et al. [9]. In their x-ray diffraction investigation of single crystal $YBa_2Cu_3O_x$ with high oxygen content ($x > 6.92$) they found strong evidence that the crystals consist of two (or more) finely dispersed phases with slightly

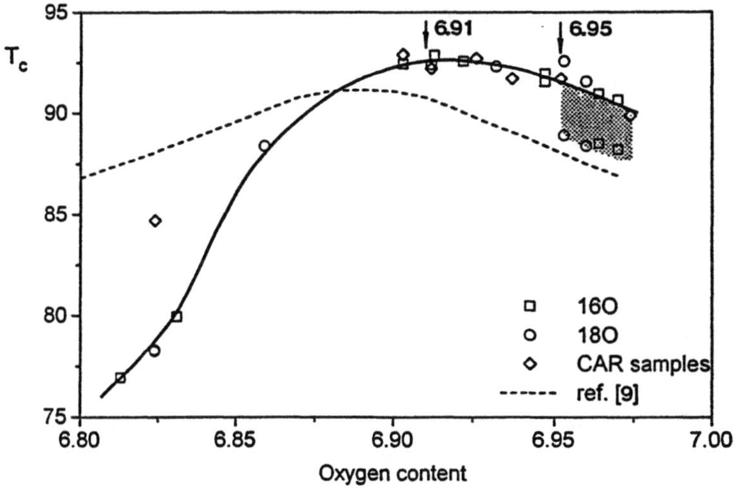


**Fig. 5.3.** Transition temperature $T_c$ in the oxygen concentration region $6.8 < 7 - x < 7$.

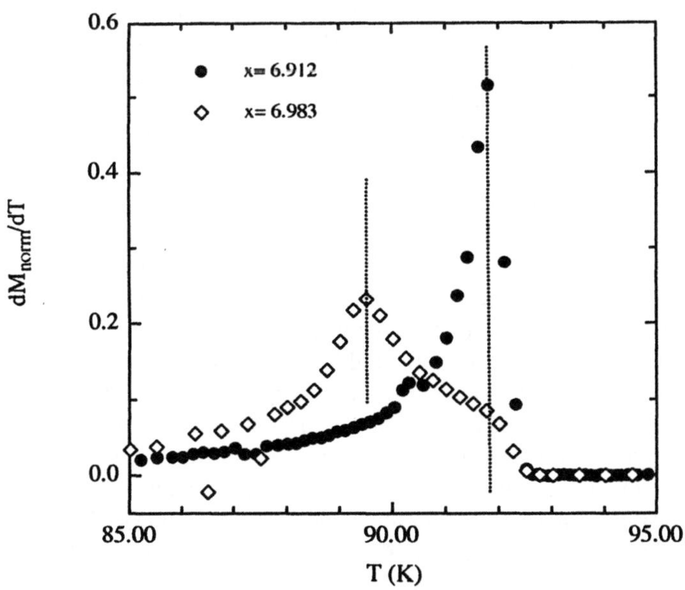


**Fig. 5.4.** Derivative $dM_{\mathrm{norm}}/dT$ for two samples with $x = 6.912$ (solid circles) and $x = 6.983$ (diamonds). The unresolved double-peak structure for $x = 6.983$ may be interpreted as arising from a phase separation with two different critical temperatures.

different c-axis parameters. Moreover, resistivity and magnetization measurements taken on such samples show a pronounced broadening of the superconducting transition. As shown in Figs. 3.1 and 3.2 the x-ray data of our powder samples also seem to indicate a scattering of the lattice parameters in the region $x > 6.92$. Furthermore, the onset temperatures $T_{c-onset}$ (Fig. 5.2) show an abrupt change at $x \cong 6.75$, supporting the existence of a phase boundary, in accordance with Fig. 3.5.

## 6. Oxygen diffusion and $T_c$ splitting

Contrary to the model of Fig. 3.5, Janot et al. [5] suggest that the observed splitting of the magnetization curve and specific heat anomaly are caused by the inhomogeneity of the sample (shell model). First, we refer to our oxygen isotope exchange experiments [ref. [10], Fig. 2]. If a diffusion barrier would exist on the surface of each grain, an isotope exchange process **would not be possible** or would be stopped after some initial time necessary for the formation of this barrier shell. However, already after 150 h at about 320°C nearly 100% of the chain and apex atoms have been exchanged (see also Fig. 8 ref. [10], for the confirmation of the exchange selectivity). Further, we decided to make susceptibility measurements on samples with different grain sizes [18]. A 123 sample has been annealed in oxygen at 950°C then cooled with 5°C/h down to 500°C and further with 4°C/h to room temperature. We divided this sample in three different grain fractions ($< 15\mu$m, 15–50$\mu$m and $> 50\mu$m), using ultra-sound sieving equipment. We calculated the oxygen content differences within the grains of different dimensions. A numerical solution of the Fick's second low equation was used [19], as during cooling the equilibrium oxygen concentration on the sample surface is increasing. Equilibrium values of the oxygen concentration on the surface were taken [20] and values of a diffusion coefficient from previous measurements [18, 21, 22, 23]. The calculations show that oxygen content differences do not exceed 0.001, 0.003 and 0.017 in $x$ for the 15, 30 and 75$\mu$m grains, respectively. Lattice parameters refinements by the Rietveld method for our samples with different grain size showed the samples to have no significant lattice constant variations. Lattice parameters were: $a = 3.819(1)$Å, $b = 3.884(1)$Å, $c = 11.680(2)$Å; orthorhombicity $8.46 \pm 0.05$ ($R_{wp} = 0.06 - 0.1$). No second phases have been detected. The magnetization measurements (SQUID) were performed with the sample before sieving and the fractions containing the smallest and the largest grains. Essentially the same results (see Fig. 6.1) showing two slopes have been obtained for all three samples.

These results clearly show that not an inhomogeneity of the oxygen content of 123 is responsible for the magnetization anomalies discussed in this work.

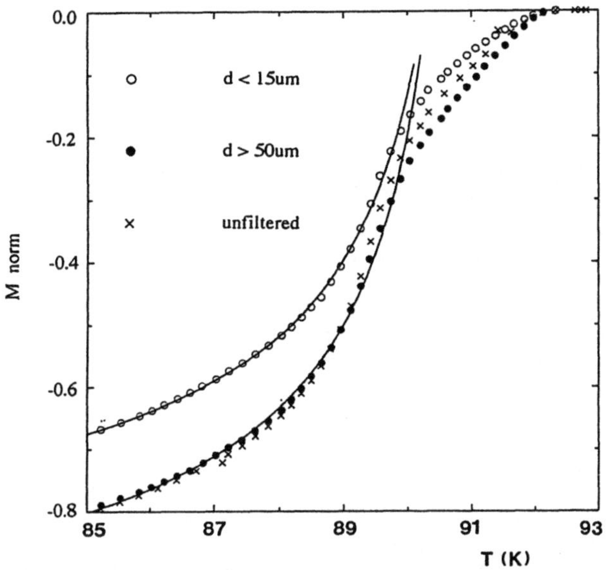


**Fig. 6.1.** Susceptibility curves for 123 samples with different grain size fractions. Note the two different slopes ($T < 90$K, $T > 90$K) leading to the double peak of the derivative. No changes appear for the different grain size fractions.

# 7. Conclusions

Carefully prepared samples, with accurate oxygen content determination, show changes of the slope of their lattice constants with $x$, indicating the existence of phase separation in 123:

- A phase boundary at $x \cong 6.75$ is supported by lattice constants (Fig. 3.1 and 3.2) and an abrupt $T_{c-\text{onset}}$ step (Fig. 5.2).
- The phase-transition underdoped-overdoped at $x \cong 6.91$ is clearly associated not only with the $T_{\max}$ (Fig. 5.3), but also with the structural transformation of the Cu2-O2/O3 distance (Fig. 3.5). Most lattice constants show this effect (Fig. 3.1–3.4).
- At $x > 6.95$, the double unresolved peak of the $\Delta M_{\text{norm}}/\Delta T$ curve is in agreement with the phase separation results by Claus et al. [9].

Concluding, it can be said that whereas there are strong indications for phase separation in 123, much more work is necessary to prove that finally. The extreme chemical complexity of the high temperature superconductors, makes, however, this goal very difficult to achieve.

*Acknowledgement.* The authors are indebted to Prof. K.A. Müller for interesting discussions. The support of the Swiss National Fonds NFP30" is greatfully acknowledged.

## References


1. S. Rusiecki, E. Kaldis, E. Jilek, C. Rossel, J. Less, *Common Metals* **164**, 31 (1990)
2. B. Pümpin, H. Keller, W. Kündig, I.M. Savic, J.W. Schneider, H. Simmler, P. Zimmermann, E. Kaldis, S. Rusiecki, C. Rossel and E.M. Forgan, J. Less, *Common Metals* **164&165**, 994 (1990)
3. J.W. Loram, J.R. Cooper and K.A. Mirza, *Supercond. Sci. Technol.* **4**, (1991) S391
4. H. Schwer, J. Karpinski, K. Conder, Ch. Krüger, E. Kaldis, Ch. Rossel, Proceedings Workshop NFP 30 *Hochtemperatur-Supraleitung* Baden 4.-5.2.1993
5. E. Janot, A. Junod, T. Graf, K.-Q. Wang, G. Triscone and J. Muller, *Physica C* **216**, 129 (1993)
6. K. Conder, S. Rusiecki and E. Kaldis, *Mater. Res. Bull.* **24**, 581 (1989)
7. H. Schwer, E. Kaldis, J. Karpinski and C. Rossel, *Physica C* **211**, 165 (1993)
8. H. Claus, M. Braun, A. Erb, K. Röhberg, B. Runtsch, H. Wühl, G. Bräuchle, P. Schweib, G. Müller-Vogt, H.v. Löhneysen,, *Physica C* **198**, 42 (1992)
9. H. Claus, U. Gebhard, G. Linker, K. Röhbrg, S. Riedling, J. Frannz, T. Ishida, A. Erb, G. Müller-Vogt, H. Wühl,, *Physica C* **200**, 271 (1992)
10. K. Conder, E. Kaldis, M. Maciejewski, K.A. Müller and E.F. Steigmeier, *Physica C* **210**, 282 (1993)
11. E. Kaldis, J. Karpinski, S. Rusiecki, B. Bucher, K. Conder and E. Jilek, *Physica C* **185-189**, 190 (1991)
12. B. Okai, *Jap. Journ. Appl. Phys.* **29**, 2193 (1990)
13. A.W. Hewat in *Materials and crystallographic Aspects of HTSC* E. Kaldis ed., Kluwer 1994
14. D.B. Wiles and R.A. Young, *J. Appl. Cryst.* **14**, 149 (1981)
15. V.F. Sears *Methods of Experimental Physics* 23A, 533 (1986), Eds. K. Sköld and D.L. Price, Academic Press, London 1986
16. R.J. Cava, A.W. Hewat, E.A. Hewat, B. Batlogg, M. Marezio, K.M. Rabe, J.J. Krajewski, W.F. Peck Jr., J.W. Rupp, *Physica C* **156**, 419 (1990)
17. D. Zech et al., to be published
18. K. Conder, Ch. Krüger, E. Kaldis, D. Zech and H. Keller, *Physica C* , (1994)
19. S.S.M. Wong, *Computational Methods in Physics and Engineering*, Prentice Hall, New Jersey 1992
20. K. Conder, J. Karpinski, E. Kaldis, S. Rusiecki and E. Jilek, *Physica C* **196**, 164 (1992)
21. S.J. Rothman, J.L. Routbort and J.E. Baker, *Phys. Rev. B* **40**, 8852 (1989)
22. S.J. Rothman, J.L. Routbort, U. Welp and J. E.Baker, *Phys. Rev. B* **44**, 2326 (1991)
23. J.L. Tallon and M.P. Staines, *J. Appl. Phys.* **68**, 3998 (1990)

# Superconductivity Versus Oxygen Concentration in 123 Compounds: Influence of RE Ionic Radii Studied by Cu NQR

H. Luetgemeier[1], I. Heinmaa[2], D. Wagener[3], and S.M. Hosseini[3]

[1] Institut f. Festkoerperforschung Juelich GmbH, 52428 Juelich
[2] Institute of Chemical Physics and Biophysics EE0100, Tallinn/Estonia
[3] Institut fuer Strahlen- und Kernphysik der Universitaet Bonn

**Summary.** The NQR spectra of Cu in 123 compounds $REBa_2Cu_3O_x$ containing either the large RE ion Nd or the smaller ones Gd, Y, or Tm are investigated at 4.2 and 1.2 K for different oxygen content $x$. In our powder samples of Tm123 we find a maximum of $T_c$ at $x = 6.92$ but the sharpest lines are at $x = 7.0$, indicating that no phase separation exists at 7.0. For the larger Nd ion the maximum of $T_c$is obtained for 7.0 and $T_c$ decreases fast at reduced $x$. Different schemes of oxygen ordering in the two types of 123 compounds are derived from the NQR spectra: whereas the oxygen defects in the Cu-O chains are distributed randomly in Nd123, the holes cluster in the compounds with the small RE ions and form extended pieces of empty chains surrounded by full chains, and near $x = 6.5$ the ortho-II structure. Correspondingly the mean length $n$ of the full chain fragments depends in a different way on $x$ in both types of 123 compounds. A unique dependence of $T_c$ on $n$, however, is found with $n = 5$ oxygen for the onset of superconductivity.

## 1. Introduction

In the high-$T_c$ superconductor $YBa_2Cu_3O_x$ (Y123) the orthorhombic structure ortho-I at oxygen concentration $x$ near 7.0 with $T_c$ of about 90 K is formed since only half of the oxygen sites in the Cu(1) layer is occupied. The occupied sites, designed as O(1), form the Cu-O-Cu chains along the b axis whereas the O(5) sites along the a axis remain empty. At $x = 6.4$ a transition to the tetragonal structure occurs with random occupation of both sites. The arrangement of the oxygen ions in the Cu(1) layer at $x < 7$ has an important influence on the superconducting transition in this compound and the related ones obtained when Y is replaced by rare earth ions. The plateau of $T_c$ at 60 K for Y123 is connected with the ordered ortho-II structure at $x$ near 6.5, characterized by doubling of the unit cell in a-direction due to alternating oxygen full and empty chains. Since the dependence of $T_c$ on $x$ is different for large and small ions on the Y site [1, 2], it must be expected that the oxygen arrangement changes with the ionic radii.

We have investigated previously this arrangement for Gd123 and Y123 by using the Cu nuclei as local probes in the nuclear quadrupole resonance (NQR) [3]. The NQR frequency is determined by the electric field gradient (EFG) of the charge distribution at the nuclear site. Since the intensity of the resonance line is proportional to the number of nuclei in corresponding sites,

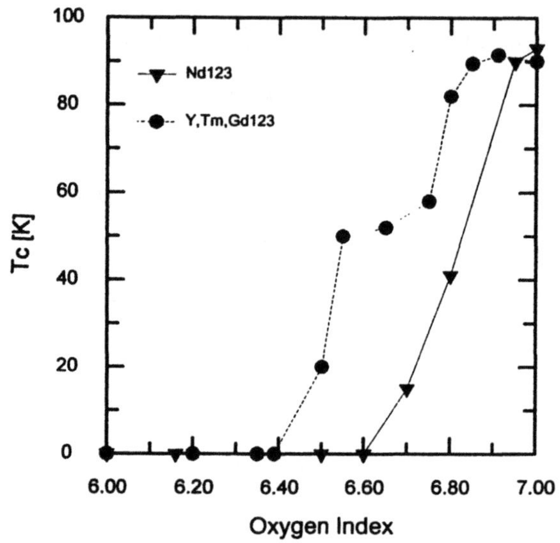


**Fig. 2.1.** Dependence of $T_c$ on the oxygen content in the 123 samples containing either Nd or Gd, Y, Tm.

one can study the distribution of different copper sites created by various oxygen arrangements. Here we have applied the copper NQR to Nd123 with the aim to compare the oxygen arrangement in the compound with RE of large ionic radius to that in the compounds with small RE (Y-, Gd-, and Tm123). A special effort was made in order to follow the structural changes in Tm123 in the region of the 90 K plateau.

## 2. $T_c$ versus $x$ for different RE in $REBa_2Cu_3O_x$

In the series of the the superconducting 123 compounds $REBa_2Cu_3O_x$ (RE123), the critical temperature $T_c$ obtains its largest value of 95 K for the largest ion $Nd^{3+}$ [4]. The two plateaus in the depence of $T_c$ on $x$ are best developed for the smaller ions like $Gd^{3+}$, $Y^{3+}$, $Tm^{3+}$ etc. [1, 2] (Fig. 2.1). For Nd the 60 K plateau is missing completely and $T_c$ decreases very fast at reduced $x$ and disappears already at $x < 6.7$. Electron diffraction has proved the missing of the 60 plateau in Nd123 to be connected with the absence of the ordered ortho-II structure [2]. In contrast to the fast decay of $T_c$ with $x$ for Nd123, the small RE ions reveal a weak maximum of $T_c$ at $x$ slightly reduced to about 6.9 [5, 6, 7], as shown for the case of Tm in Fig. 2.2. This maximum is generally taken as an indication for the transition from the underdoped region at low $x$ to the slightly overdoped at $x = 7$. Up to now it is not clear if the maximum of $T_c$ is connected also with a structural transition [6, 7].

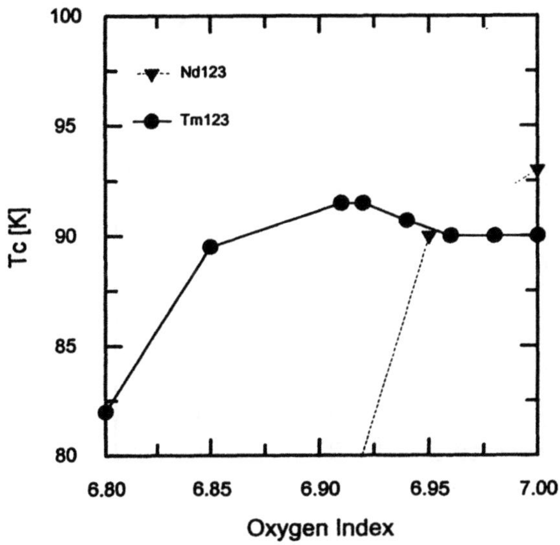


**Fig. 2.2.** $T_c$ versus oxygen content for Tm123 and Nd123 in the region of the maximum of $T_c$.

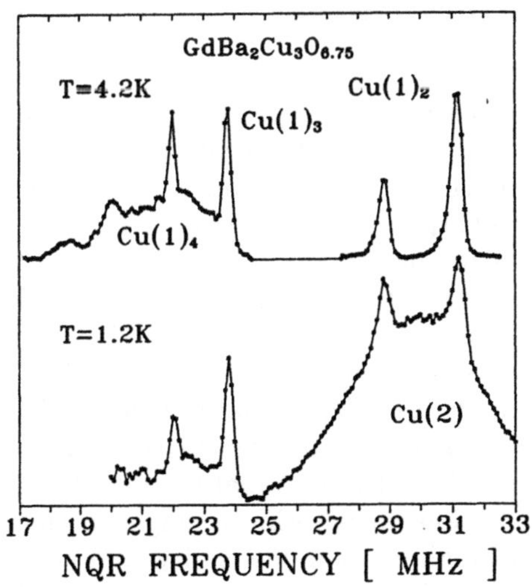


**Fig. 3.1.** NQR spectra of Cu in Gd123 with $x = 6.75$ with the doublets for the isotopes $^{63}$Cu (higher intensity and frequency) and $^{65}$Cu (lower). At 4.2 K the broad signal from Cu(2) sites seen at 1.2 K is suppressed by fast relaxation due to the fluctuations of the Gd 4f spins above the Neel temperature at 2.2 K.

## 3. Cu NQR lines in different configurations

The NQR spectra of Cu have been discussed in detail for RE = Y and Gd [3]. Since the EFG determining the NQR frequencies depends on the charge dis-

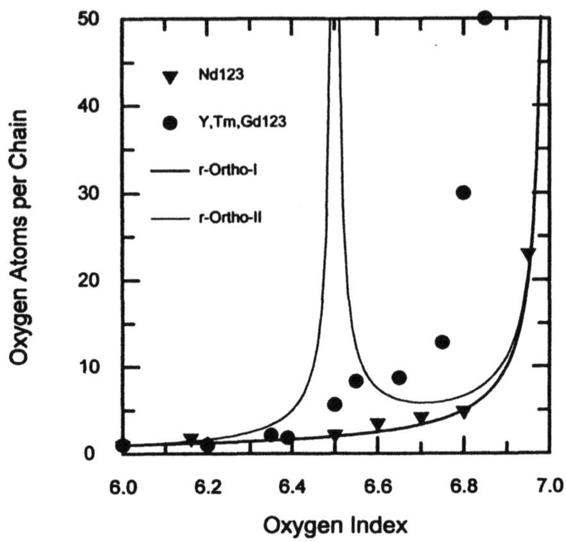


**Fig. 3.2.** Mean number $n$ of oxygen ions in the Cu-O-Cu chains versus oxygen content in the different 123 compounds. Model calculations are shown for random dilution of the ortho-I structure and for random dilution for $x < 6.5$ and random filling for $x > 6.5$ of the ortho-II structure.

tribution at the lattice site under inspection, one pair of NQR lines for the two isotopes is expected for each configuration of Cu. Three components in the spectra could be ascribed to Cu(1) sites with 2, 3, or 4 oxygen neighbors $Cu(1)_2$, $Cu(1)_3$ and $Cu(1)_4$, respectively (Fig. 3.1). $Cu(1)_4$ is the only NQR line of the Cu(1) sites in the ortho-I structure at $x = 7.0$ whereas $Cu(1)_2$ appears in the tetragonal antiferromagnetic state at $x = 6.0$. In the pure ortho-II structure the same amount of $Cu(1)_2$ and $Cu(1)_4$ is to be expected. $Cu(1)_2$ appears in the same frequency region as the Cu(2) NQR line but can be distinguished by the longer transversal relaxation time at 4.2 and 1.2 K when magnetic RE ions like Gd are used in the place of Y [3]. The NQR line of $Cu(1)_3$ corresponds to the terminal Cu ion of a Cu-O chain and can be especially the Cu(1) ion neighboring a single oxygen ion at low $x$ or a single oxygen defect at high $x$. This line is observed nearly in the full region $6.0 < x < 7.0$. From the relative intensities $I_2$, $I_3$, and $I_4$ of the three components we have concluded on the mean number $n$ of oxygen atoms in the Cu-O chains given by the relation

$$n = 1 + 2I_4/I_3 \tag{3.1}$$

From a large number of samples with RE = Y or Gd [3] the sytemetic variation of $n$ with $x$ has been found (Fig. 3.2): $n$ increases with $x$ at the onset of superconductivity at 6.3 and $n > 5$ holds in all superconducting samples. In the region of the 60 K plateau $n$ stays constant at about 8 and increases only at $x > 6.8$ where $I_3$ decreases quickly. The relative intensities can be discussed also in another way. Normally $Cu(1)_2$ is ascribed to the monovalent state, whereas $Cu(1)_3$ and $Cu(1)_4$ are assumed to be divalent. This means that $I_2$ corresponds to the monovalent component of the Cu(1) layer which has been determined by X ray absorption spectroscopy (XAS) [8]. The dependence on $x$ of the resonance of the Cu(2) ions in the double layers is quite different [3]. In the antiferromagnetic state for $x < 6.3$ the spins localized at the Cu(2) ions induce the hyperfine field at the nuclei which is observed by antiferromagnetic NMR at frequencies much higher than the NQR frequencies. With increasing $x$ this line disappears at about $x = 6.3$ [9]. The resonance of the Cu(2) sites appears again as NQR at $x > 6.6$. In the region $6.3 < x < 6.6$ no nuclear resonance of Cu(2), neither NMR nor NQR, is observed at 4.2 K, probably due to strong relaxation by the slow fluctuations of the electronic spins localized at the Cu(2) sites. For $x$ at about 6.6 the characteristic NQR spectrum of Cu(2) in the ortho-II structure is identified by two slightly different NQR frequencies corresponding to the Cu(2) sites attached to either a full or empty chain. At larger $x$ a broad line from a highly distorted ortho-I structure is superimposed on the spectrum. With $x$ approaching 7.0 this line becomes the single sharp line of the pure ortho-I structure.

## 4. Experimental details

In the following we describe experiments with a series of Tm123 samples prepared with small steps for $x > 6.9$ of only 0.02 and a series of Nd123. All samples were polycrystalline powders prepared by the technique as described previously [1, 2]. The NQR spectra have been recorded by standard spin-echo technique at 1.2 and 4.2 K in the region from 18 to 35 MHz. By measuring the spectra with short (30 $\mu$s) and long (150 $\mu$s) delay time between both rf. pulses, the fast and slowly relaxing contributions of Cu(2) and Cu(1), respectively, were separated. In the comparison of the intensities at different frequencies these were normalized by $f^2$. $T_c$ was determined from the DC susceptibility at 10 Oe measured by a SQUID magnetometer.

## 5. Results

### 5.1 NQR of Tm123

The spectra of Tm123 show the same evolution with $x$ as those observed previously with the Y and Gd compounds [3]. In the region of high $x$ the systematic variation of $T_c$ with $x$ (Fig. 2.2) and of the spectra (Fig. 5.1) confirms the high accuracy of determining differences in the oxygen concentration. The absolute values of $x$, however, may be affected with larger errors.

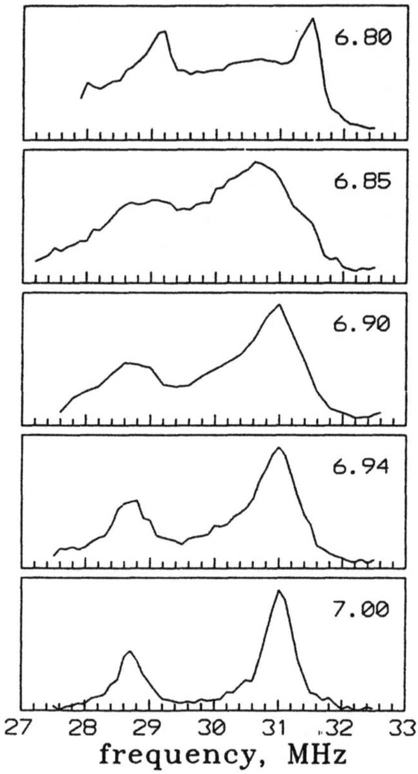


**Fig. 5.1.** NQR spectra of Cu in Tm 123 in the frequency range of the Cu(2) signal for large $x$. The narrow peaks for $x = 6.8$ correspond to $Cu(1)_2$ and can be detected up to 6.90.

A clear maximum of $T$ is found at $x = 6.92$ which is 1.5 K above the value for 7.0. In the NQR spectra of the Cu(2) sites we find the narrowest line for $x = 7.0$ and a systematic broadening when $x$ is reduced to 6.8, without any special feature in the region of the $T_c$ maximum at $x = 6.92$. In the spectra observed at short delay time (Fig. 5.1), the $Cu(1)_2$ signal is evident the first time at $x = 6.85$ as a narrow shoulder, but a weak $Cu(1)_2$ line could be detected even at $x = 6.90$. This line is the main line of $Cu(1)_2$ in the ortho-II structure and corresponds to a piece of an empty chain between filled chains [3]. The $Cu(1)_3$ line, in contrast, is absent already for $x > 6.85$. This means the oxygen defects are not isolated but clustered into empty fragments containing at least 4 vacant sites. It is difficult to estimate the mean length of the Cu-O chains in the limit of high $x$, but $n > 50$ holds for $x = 6.85$ (Fig. 3.2).

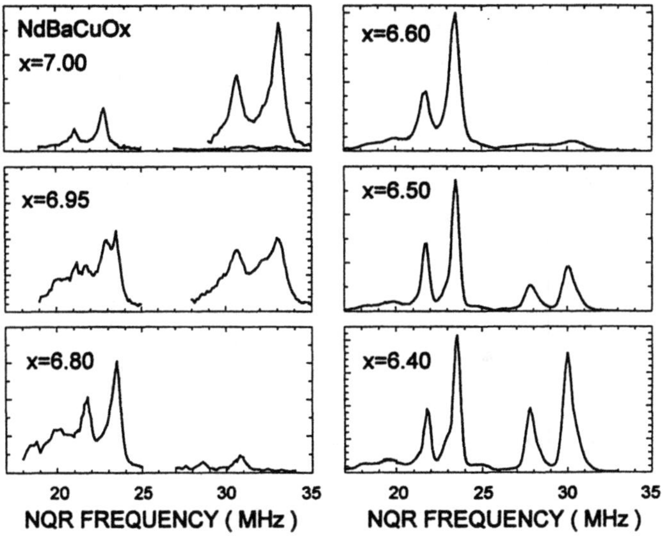


**Fig. 5.2.** NQR spectra of Nd123 at different $x$ at 1.2 K. From the relaxation rate it is evident that for $x < 6.8$ no signal of Cu(2) is detected and for $x > 6.8$ no signal from $Cu(1)_2$.

### 5.2 NQR of Nd123

In the Nd123 samples $T_c$ is reduced very fast at decreasing $x$ (Fig. 2.1) and superconductivity disappears already at $x = 6.6$. Due to the low antiferromagnetic ordering temperature of about 1 K for the Nd spins, the Cu(2) NQR signal reveals strong relaxation and can be observed only at a very short delay time (Fig. 5.2). This fast relaxing signal disappears already for $x = 6.8$ in a sample with $T_c = 40$ K, that is the same region of $T_c$ where for Gd123 and Y123 no Cu(2) signal can be observed [3]. For the Cu(1) signals the lines corresponding to the three different configurations can be distinguished. It is evident that the intensity of the line of $Cu(1)_3$ at 23.6 MHz is the largest in the region of $x$ between 6.8 and 6.4. Correspondingly the mean length of chains decreases very fast at decreasing $x$ (Fig. 3.2) with $n = 5$ at $x = 6.8$ for the sample with $T_c = 40$ K.

## 6. Discussion

The different $n$ vs. $x$ dependence for Nd and the smaller RE ions reflect different schemes of removing oxygen. If the oxygen are distributed randomly over all O(1) sites of the ortho-I structure, the dependence $n = 1 + 1/(7 - x)$ results. As can be seen in Fig. 3.2 this relation, indicated as r-ortho-I (random

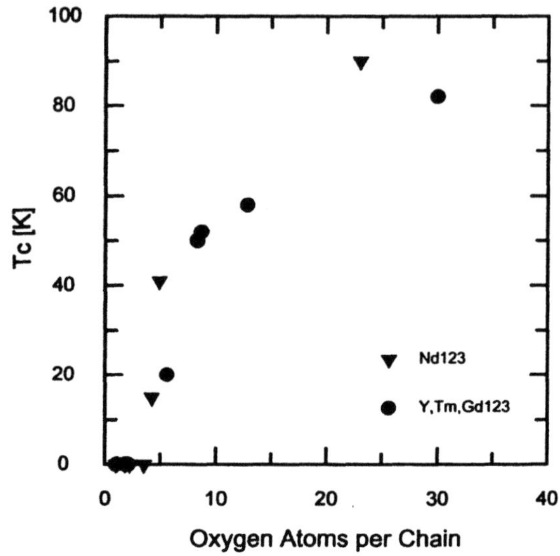


**Fig. 6.1.** Dependence of $T_c$ on the mean number $n$ of oxygen ions in the Cu-O-Cu chains in the different 123 compounds with the onset of superconductivity for $n = 5$.

dilution of ortho-I) holds for the case of Nd123. Since no correlation exists between the oxygen vacancies the ortho-II structure cannot develop.

The other simple oxygen ordering scheme is based on the ortho-II structure where at $x > 6.5$ the oxygen ions are removed only from every other chain leaving the remaining ones complete. Only at further reduction below 6.5, oxygen is removed randomly also from the remaining chains. This leads to a dependence of $n(x)$ with infinite chain length at 6.5 and a deep minimum at $x = 6.7$ as indicated by the curve r-ortho-II in Fig. 3.2. It is evident that $n(x)$ for Y, Gd, and Tm follows this dependence rather well for $x < 6.5$, but remains constant at about 10 for $6.5 < x < 6.7$ where it starts to increase again. As we pointed out before [3] this behavior must be ascribed to the coexistence of domains of ortho-I and ortho-II. A surprising result of the comparison of the two types of 123 compounds is that the onset of superconductivity appears in both cases of ortho-I (Nd) or ortho-II (Gd,Y,Tm) at the same $n$ of about 5 as indicated in the dependence of $T_c(n)$ in Fig. 6.1. This confirms the idea that a minimal length of the Cu-O chains is needed for doping holes into the Cu(2) layers.

The difference of both types of 123 compounds can also be seen from the relative numbers of the twofold coordinated Cu(1) ions, $I_2$, corresponding to the amount of monovalent Cu ions (Fig. 6.2). Three limiting cases can be considered. If all oxygen ions are arranged in chains of "infinite" length which means $I_3 \ll I_2$, a linear decay $I_2 = (7 - x)$ would result as is seen for the case of Gd, Y, Tm for $x > 6.5$. For $x < 6.5$ and for Nd the dependence of $I_2$ on

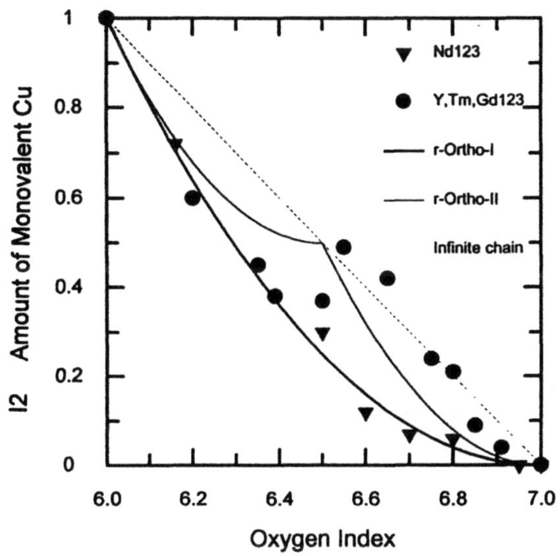


**Fig. 6.2.** Relative intensity $I_2$ of twofold coordinated Cu(1) corresponding to the amount of monovalent Cu versus oxygen content $x$. The curves indicate different models for oxygen order: random dilution of the ortho-I structure, random dilution and filling of the ortho-II structure, and ordering of the oxygen ions in chains of infinite length.

$x$ follows much better the dependence expected for the case of the random dilution of the ortho-I structure leading to the relation $I_2 = (7-x)^2$. For the random filling of ortho-II values between these limiting cases are expected.

Our results for the amount of monovalent Cu(1) can be compared with the data deduced from XAS for Y123 [8]. Good agreement is found for the region $x < 6.3$ where the oxygen ions remain separated, but a clear difference appears for $x > 6.6$. Here Tolentino et al. find only a small amount of monovalent Cu corresponding to a random distribution of the oxygen defects whereas our experiments for the small RE ions show the strong tendency of the oxygen defects to cluster in empty fragments of the chains and to form monovalent Cu(1).

There remains the open question which parameters are responsible for the different behavior of the large and small RE ions. On the one hand it is possible that by the increased lattice parameters for Nd123 the interaction energies between the oxygen ions in the Cu(1) plane are modified leading to different ordering schemes. On the other hand it might be possible that the larger Nd ions can replace some Ba sites. In any case it is evident that the formation of the ortho-II structure is suppressed in the case of Nd123 and that already at high $x$ the length of the filled chains becomes so short that the superconductivity is suppressed.

The investigation of polycrystalline samples of Y123 [6, 9] and of single crystals [7] has revealed anomalies in the lattice parameters and the pressure dependence of $T_c$ in the region of the $T_c$ maximum at $x$ near 6.9. From this and from two steps found in the temperature dependence of the magnetic shielding at high $x$ it has been concluded that the system consists of two phases for $x$ above the maximum of $T_c$. The NQR spectra of our Tm123 samples show the narrowest lines of the Cu(2) sites for the largest oxygen concentration and a continuous broadening at decreasing $x$ as expected when the number of lattice defects in an ordered structure increases. The NQR spectra of Cu(1) show that the maximum of $T_c$ appears at the oxygen concentration where the NQR line of Cu(1) in empty chain fragments appears for the first time, which means the formation of small domains of the ortho-II structure within the ortho-I structure. Thus the anomalies at the maximum of $T_c$ are probably connected with the coexistence of the ortho-I and ortho-II phases but not with another phase at higher oxygen concentration.

All results presented here are based on the fact that all the Cu(1) sites can be observed by NQR which means that the expectation value of the magnetic moment disappears for times longer than $10^{-12}$ s. This holds especially also for the $Cu(1)_3$ configuration at the chain ends. In many models the chain ends are assumed to be paramagnetic and to be the reason for the reduced $T_c$ by pair breaking. Our results show that at least most of these sites do not carry a stable magnetic moment.

## 7. Conclusion

The NQR spectra of Cu in the oxygen depleted 123 compounds allow a direct determination of the relative amounts of the Cu(1) sites with two, three or four oxygen neighbors. From these it is evident that the arrangement of the oxygen defects depends on the radii of the RE ions on the Y sites. For the large ion of Nd the oxygen vacancies show no tendency to cluster. Correspondingly only short pieces of Cu-O-Cu chains exist already at slightly reduced oxygen content x which leads to the fast decay of $T_c$. For the smaller ions like Gd, Y, Tm we have found a strong clustering of the defects in fragments of empty chains which keeps the mean length of the filled chains large and leads to the ortho-II structure with alternating full and empty chains near $x = 6.5$. In spite of this different arrangement of the defects we found a unique dependence of $T_c$ on the mean length of the filled chains with the onset of superconductivity at 5 oxygen ions per chain.

*Acknowledgement.* We are indebted to C. Osthoever, KFA Juelich, and E. Bruecher, MPI FKF Stuttgart for the measurement of the susceptibility. Financial support by the BMFT given through the DAAD for the stay of I.H. at the KFA Juelich is thankfully acknowledged. I.H. likes to thank Prof. Zinn for the hospitality he received in his institute.

## References


1. Buchgeister, M., Thesis Universitaet Bonn (1991).
2. Krekels, T. et al., Physica C196 (1992) 363.
3. Luetgemeier, H. and Heinmaa, I. in "Phase Separation in Cuprate Superconductors", Proc. of the Erice Workshop 1992, p. 243. K.A. Mueller and G. Benedek eds., (World Scientific 1993) Heinmaa I. et al., Appl. Mag. Res. 3 (1992) 689.
4. Buechner, B. et al., Sol. State Comm. 73 (1990) 357.
5. Graf, T., Triscone, G. and Mueller, J, J. Less-Common Met. 159 (1990) 349.
6. Rusiecki S. et al., J. Less-Common Met. 164-165 (1990) 31.
7. Claus, H. et al., Physica C200 (1992) 271.
8. Tolentino, H. et al., Physica C192 (1992) 115.
9. Mendels, P., Alloul, H., Marucco, J.F., Arabski, J. and G. Collin, Physica C171 (1990) 429.

# The Role of the Additional Oxygen Atoms on the Superconducting Properties of $La_2CuO_4$-Related Compounds

J.C. Grenier[1], F. Arrouy[1], J.P. Locquet[2], C. Monroux[1], M. Pouchard[1], A. Villesuzanne[1], and A. Wattiaux[1]

[1] Laboratoire de Chimie du Solide du C.N.R.S., Universite de Bordeaux I, 33 405 TALENCE-Cedex, FRANCE
[2] IBM Research Division, Zürich Research Laboratory, CH-8803 Rüschlikon, Switzerland

**Summary.** Thanks to the electrochemical oxidation in alkaline solution at room temperature, oxygen atoms were intercalated in ceramic materials, thin films and crystals of $La_2CuO_4$ (or La substituted). Physical properties of these materials show that at least two and probably three phases are formed with different compositions in oxygen (the $\delta$ value). The role of these additional oxygen atoms is discussed in terms of both steric and electronic effects. Electrochemical studies show that the electroactive species formed during the oxidation process, likely $O^-$, have a rather high diffusion coefficient. On the basis of the existence of a mixed valence of oxygen in these materials, we propose that the $O^-$ species are the mobile species thanks to a coupling with the conduction electrons via the electron transfer O $\rightarrow$ metal.

## 1. Introduction

The usual way for changing the stoichiometry of an oxide is to vary the oxygen pressure at constant temperature or/and the annealing temperature at a given oxygen pressure. This was largely used in the chemistry for cuprates for adjusting their oxygen content in the so-called "charge reservoir" layers and consequently their hole concentration in the $CuO_2$ planes. Otherwise the stabilization of high oxygen states of 3d cations such as, for instance $Cu^{III}$, depends on the considered structural type of the oxide and the counter cations present in the matrix, alkaline-earth cations as $Sr^{2+}$ or $Ba^{2+}$ being usually required for that purpose. Nevertheless in some cases, high oxygen pressure is necessary; the best examples are the superconducting materials with the simplest structures such as $La_2CuO_{4+\delta}$ [1] or the so-called "infinite layers" in the (Sr,Ca)-Cu-O system [2] or with more complex formulations and highest $T_c$ such those in the Ba-Ca-Hg-Cu-O system [3]. In the latter cases, the oxygen pressure allows to stabilize both the crystal structure and the mixed valence of Cu since in $La_2CuO_4$ it allows to insert oxygen into interstitial sites up to a value of $\delta \approx 0.05$.

More recently other ways of preparation belonging to the "Chimie Douce" were investigated and some of them developed in Bordeaux [4]–[9]. They use either a strong chemical agent at room or moderate ($T \leq 200°C$) temperature or the electrochemical potential. Therefore the range of oxygen stoichiometry

($\delta$ in $La_2CuO_{4+\delta}$) was considerably extended ($\delta$ up to $\approx 0.12$). We report here the most significant results that we obtained using the electrochemical process and discuss the role of the inserted oxygen atoms on the superconducting properties of this material.

## 2. Experimental and Results

Electrochemical experiments were described elsewhere in detail [4],[5],[10]; they are achieved at room temperature, in a solution of 1 M NaOH or KOH under air, in a three compartment cell. The working electrode is:

- either a rotating electrode (500–5000 rpm), which allows to perform electrochemical studies,
- or a fixed electrode, the surface area of which can be very large allowing to prepare a large quantity of material.

This working electrode can be a dense pellet, a crystal or even a thin film. Dense pellets were obtained from powders of the starting material $La_2CuO_4$ prepared by solid state reaction of a stoichiometric mixture of calcinated $La_2O_3$ and CuO heated in air at 1050°C for 18 h. The powder was then ground and pelleted as a disc and finally annealed at the same temperature for 24 h and quenched.

The electrochemical reaction is achieved either at steady potential ($E =$ 600–700 mV, *potentiostatic mode*) or at constant current intensity ($I < 1$ mA, *galvanostatic mode*). Typical $(I, E)$ curves exhibit a current plateau before the oxygen evolution reaction,

$$4\ OH^- \longrightarrow O_2^{\nearrow} + 2\ H_2O + 4\ e^-\ ,$$

which is assigned to the oxidation process of the material according to the overall reaction:

$$La_2CuO_4 + 2\delta OH^- \longrightarrow La_2CuO_{4+\delta} + \delta H_2O + 2\delta e^- \tag{2.1}$$

According to the Nernst equation, the oxygen activity can be calculated: for usual applied overpotentials ($\eta_{O_2} \approx 0.3$ V), it is equivalent to very high oxygen pressures ($p_{O_2} \approx 10^{15}$ Pa), which might explain why such a reaction is achieved.

### 2.1 Potentiostatic experiments on ceramics

The first experiments (previously reported [10],[11],[12]) were carried out on ceramics (typical densities of 70 to 95%) in the potentiostatic mode. The reaction was followed by the variation of the open circuit potential ($E_{OCV}$) as a function of the polarization time. In the electrochemical cell, only the working electrode is modified during the experiments and more especially

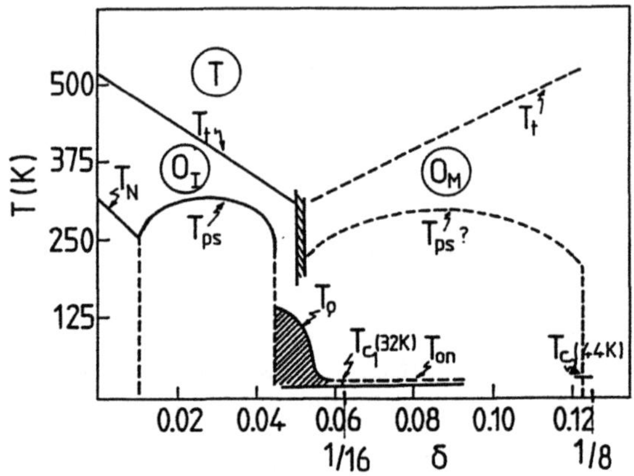


**Fig. 2.1.** Phase diagram of $La_2CuO_{4+\delta}$.

the chemical potential of the oxygen which can be correlated to the electrochemical potential of the oxygen or in other words to the Fermi level in the material. Therefore any variation of $E_{OCV}$ should result from the variation of the Fermi level and be related to the $\delta$ value or to the amount of $Cu^{III}$. It was indeed observed a direct correlation between $E_{OCV}$ and the amount of trivalent copper measured by chemical analysis [12].

Samples of various compositions ($0 \leq \delta \leq 0.10$) were thus prepared and studied using several techniques. On the basis of these results a phase diagram (Fig. 2.1) was proposed [11]. The following points can be pointed out:

– the existence of a critical concentration $\delta_c \approx 0.05$ separating the $O_I$ **insulating phase** ($\delta < 0.05$) from the $O_M$ **metallic phase** ($\delta > 0.05$). Samples of compositions $0 \leq \delta < 0.05$ ($O_I$ phase) exhibit a semiconducting-type behaviour of the electrical resistivity, at least near the room temperature. For increasing $\delta$, the values of the room temperature resistivity and of the positive Seebeck coefficient lessen, which well agrees with the occurrence of holes in the $CuO_2$ planes. However, for decreasing temperatures , a hump as well as a thermal hysteresis are observed in the range $200 < T < 250$ K. Such a behaviour has been intensively studied using various techniques (neutron diffraction, NMR, $\chi$(T), $\sigma$(T), EPR etc. [17],[18] and is now definitively assigned to a phase separation occuring at temperature $T_{ps}$ ranging between 200 and 300 K according to the composition $\delta$. This phase separation involves a migration of the interstitial oxygen atoms probably magnetically driven, which results in the formation of insulating antiferromagnetic $La_2CuO_4$-type domains and of oxygen rich metallic domains whose composition would be close to $La_2CuO_{4.05}$. This phase separation must be also distinguished from the dynamic charge fluctuation of electronic origin taking place at lower temperature, $T_\rho$, typically in the range 120–60 K. These metallic domains

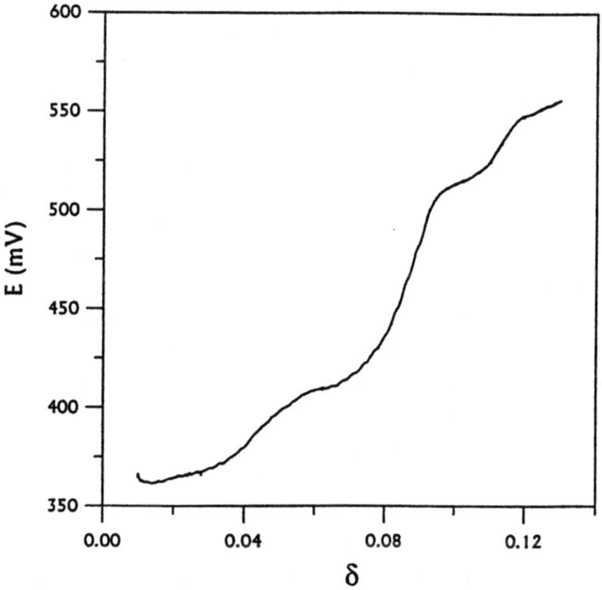


**Fig. 2.2.** Variation of the potential during a galvanostatic oxidation ($I = 5\ \mu A$) as a function of the oxygen composition $\delta$.

become superconducting at 32 K and start to percolate for an average composition (obtained by iodometric titration) close to $\delta \approx 0.04$.
- at higher $\delta$ values ($\delta \geq 0.06$, $O_M$ phase), the materials show a metallic-type behaviour in the normal state. Then, at about 50 K, the resistivity drops down and the superconducting state — i.e. a diamagnetic susceptibility — appears below 44 K; the $T_c(\rho = 0)$ being at 32 K. The superconducting transition always displays two steps that are also observed in the ac susceptibility measurements [11]. According to these results, we speculate the presence of at least two phases microscopically intergrown, the first one having a pseudo-tetragonal symmetry with $T_c = 32$ K, the second one having a strong orthorhombic distortion with $T_c = 44$ K. These results were recently corroborated by Chou et al. and Casan–Pastor et al. [19, 20]

These two superconducting phases seem to coexist, their compositions being close to the ideal compositions 1/16 and 1/8. respectively, for which one could expect the superconductivity to be suppressed as in the (La,Ba)$CuO_4$. Whether a phase separation occurs in the $O_M$ phase or not remains an open question.

## 2.2 Galvanostatic experiments on ceramics

In a second stage, galvanostatic experiments were performed in order to control the kinetics of the electrochemical oxidation process and therefore the value of $\delta$. Fig. 2.2 displays a typical curve obtained on a ceramic of $La_2CuO_4$ ($m \approx 0.4$ g, s.a. 0.5 $cm^2$, density 70%) with an anodic current intensity of 5 $\mu$A. In such conditions the kinetics of the electrochemical reaction is very slow and the process can be considered as very close to the thermodynamic equilibrium. The starting value of the potential is almost the same as the $E_{OCV}$ before polarization, which means that the overpotential $\eta$ is very small. Then one can notice on the $(E, t)$ curve the presence of three jumps of the potential leading to plateaus at about 410, 540 and 560 mV, respectively. In these conditions the upper limit of the potential seems to be reached after 45 days. By analogy with similar experiments of the electrochemical cationic intercalation (for instance $Li^+$), we conclude that three line phases are formed during the electrochemical oxygen intercalation in $La_2CuO_4$.

In addition the overall charge transfer $Q$ can be calculated; it is close to 0.26 $e^-$/f.u. and the iodometric titration of the washed sample leads to a concentration of holes to 0.22 provided that all species containing holes can be analyzed in such a way. The discrepancy between both values may be explained if we assume that not only reaction (2.1) takes place but that a part of the current intensity is used for the reaction of oxygen evolution, more especially at the end of the oxidation when the applied potential has raised.

Assuming that the hole concentration $[p] = 2\delta$, one can estimate the composition of the final oxidized materials to be close to $La_2CuO_{4.12\pm0.01}$. Therefore the compositions of the three line phases would be:

- phase I: $\delta \approx 0.05 - 0.06$
- phase II: $\delta \approx 0.10 - 0.11$
- phase III: $\delta \approx 0.12 - 0.13$

On the basis of the previous remark relative to the chemical analysis, one can suppose that the compositions of phases II and III are somewhat overestimated by coulometry.

## 2.3 Electrochemical oxidation of thin films

Thin insulating films of $La_2CuO_4$ were prepared using the molecular beam epitaxy (MBE) technique and provided by IBM (Zürich) [21]. They were grown in a layer-by-layer deposition mode on a $SrTiO_3$ single crystal substrate using a device equipped with two beam evaporators supplying the La and the Cu fluxes that are controlled by thanks to a quadrupole mass-spectrometer. The deposition was performed on a substrate heated at 700°C and under an oxygen background pressure of $10^{-10}$ Pa. The grown films of 100 nm were characterized *in situ* using the RHEED technique and *ex situ* with X-ray diffraction, atomic force microscope (TFM) and TEM. Four small gold contacts were also deposited on one side for performing electrical conductivity measurements and electrochemical experiments.

The as-prepared films have a reddish colour and are grown with the following orientation [001] $La_2CuO_4$ || [001] $SrTiO_3$, the in-plane orientation

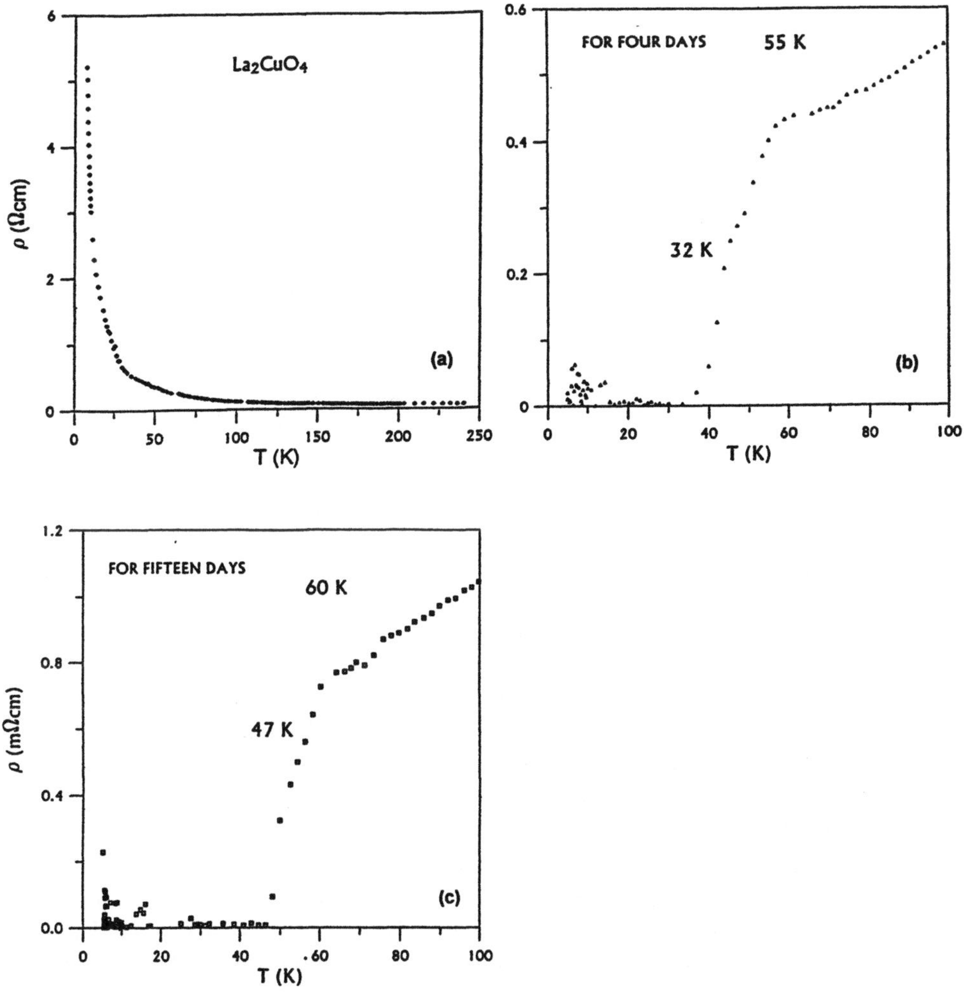


**Fig. 2.3.** Thermal dependence of the resistivity of a thin film of: (a) $La_2CuO_4$, (b) after electrochemical oxidation for 4 days ($E = 500$ mV), (c) after electrochemical oxidation for 15 days ($E = 700$ mV)

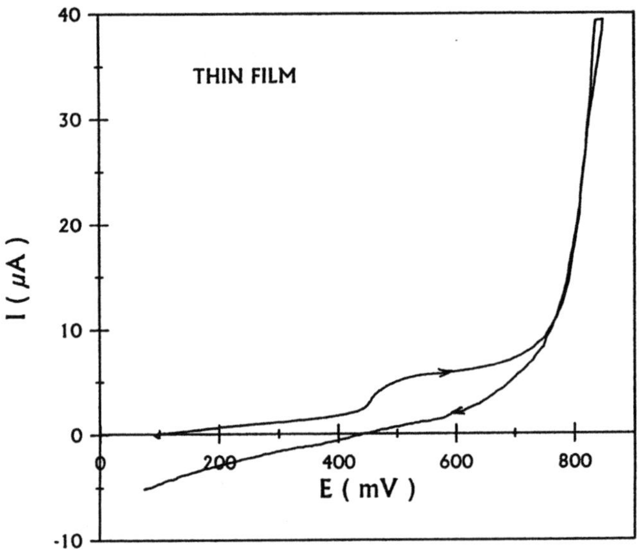


**Fig. 2.4.** Cyclic voltammogram of a thin file of $La_2CuO_4$ showing the oxidation plateau.

being (110) $La_2CuO_4$ || (010) $SrTiO_3$. The XRD analysis shows a single phase $c$ axis $La_2CuO_4$ with a value $c = 13.137 \pm 0.005$ Å. The resistivity curve vs. temperature (Fig. 2.3a) displays a semi-conductor-type behavior, which characterizes a stoichiometric $La_2CuO_4$ compound and indicates the almost no oxygen doping has occurred during the growth or cooling of the film.

Electrochemical experiments were carried out as previously described for the fixed electrodes. As for ceramic materials, $(I, E)$ cyclic voltammograms typically show an oxidation plateau between 500 mV and 700 mV (/Hg/HgO reference electrode) (Fig. 2.4). Therefore a steady potential of 700 mV was applied for various times (a few minutes to several days). The achievement of the reaction was followed by measuring from times to times the thermal variation of the electrical resistivity. The results after 4 and 15 days are reported in Figs. 2.3b and c. Nice superconducting transitions are observed, the zero resistivity occurring at 32 K and 47 K, respectively. However, steps are again observed in the transition and the onsets appear at higher temperatures than those of ceramic materials, which would indicate the possible existence of a third phase in the system transiting at about 55–60 K. This would also agree with the results of our galvanostatic experiments. In addition XRD data show an increase of the value of the $c$ parameter ($\Delta c \approx 0.06$ Å) in good agreement with the results obtained for ceramic materials.

One should also stress the slowness of the kinetics, which likely denotes that the intercalation of oxygen through the $(a,b)$ plane is much more difficult than perpendicularly to the $c$ axis as it can be easily seen in Fig. 2.5.

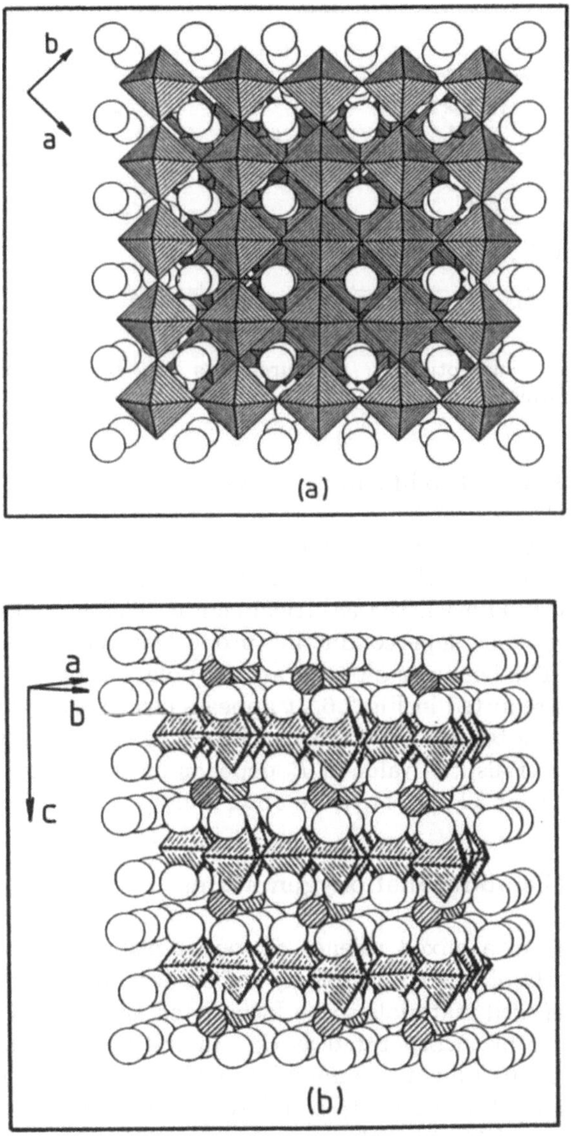


**Fig. 2.5.** Schematic representation of the $(a, b)$ and $(b, c)$ planes of $La_2CuO_4$: the large circles and hatched smaller circles represent the $La^{3+}$ cations and the additional oxygen atoms (here the $A_2MO_{4.25}$ composition) respectively.

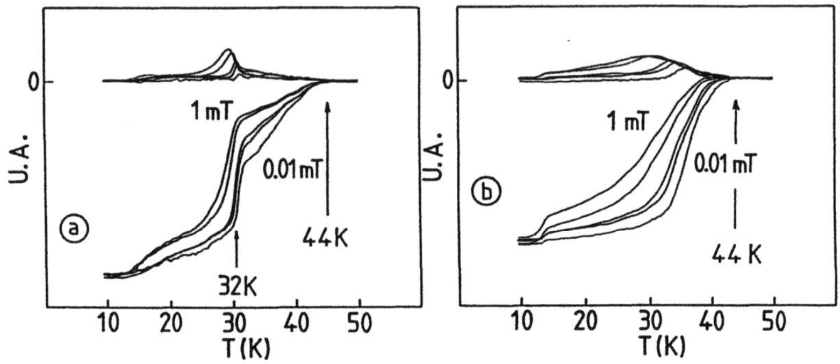


**Fig. 2.6.** A.c. susceptibility measurements of electrochemically oxidized $La_2CuO_{4+\delta}$ crystals: (a) for 15 days, (b) for 30 days.

## 2.4 Electrochemical oxidation of crystals

Preliminary experiments were carried out on a single crystal of small size ($\approx$ 30 mg), supplied by C. Oseroff and Z. Fisk. They were oxidized at constant potential (700 mV), the electrical contact was provided mechanically by golden tongs. The superconducting properties were measured using a.c. susceptibility. The behavior after an anodic polarization of 15 and 30 days, respectively, is reported in Fig 2.6. It appears clearly that the process has a very slow kinetics but that crystals can be oxidized. As previously observed for ceramic materials, the value of $T_c$ depends on the polarization time, but again transitions at 32 and 44 K are observed.

## 2.5 Role of the additional oxygen atoms

The occurrence of a mixed valence of copper using high oxygen pressures and subsequently of superconductivity led earlier authors [22],[23] to postulate the as-prepared materials to be formulated $La_2CuO_{4+\delta}$ and the excess oxygen atoms to be incorporated as $O^{2-}$ interstitial defects. However they did not advance the way of accommodating this non-stoichiometry. Later on Jorgensen et al. [13] using neutron diffraction showed that these additional atoms were located in unusual (even unexpected!) interstitial sites between two LaO layers since the La/Cu ratio remains equal to $2.00 \pm 0.01$. In addition they claimed for the first time the existence of the phase separation (mentioned above) occurring at or even below room temperature. This phase separation into an oxygen-poor ($La_2CuO_{4+\epsilon}$) and oxygen-rich ($La_2CuO_{4.08}$) phases implicitly involved the (surprising) migration of oxygen atoms. These results were confirmed by Chaillout et al. [24] who refined neutron diffraction data of a single crystal; they emphasized the existence of strong O-O bondings. However an alternative view was recently proposed by Galy [25] who argued that several kinds of non-stoichiometries can be imagined; they are based on both cationic and/or anionic non-stoichiometry and the existence of classic shear mechanisms leading to the formation of various kinds of crystallographic sites for La and Cu. The presence of oxygen vacancies would explain the easy variation of stoichiometry of these materials.

In a recent work concerned with electrochemically oxidized $La_2NiO_{4+\delta}$ ($\delta$ up to 0.25), we definitively showed that the oxygen atoms can be intercalated in such networks. For the upper limit of $\delta$ ($\delta = 0.25$), we even observed a perfect ordering of these additional atoms leading to the formation of a new definite compound $La_8Ni_4O_{17}$, the structure of which was refined using neutron diffraction and TEM [26].

Several questions can be then addressed:

- why such an oxygen intercalation is possible as one should remind that oxygen anions have an ionic radius of 1.40 Å, so large that they likely cannot diffuse at or below room temperature?
- what is the role of these additional atoms from a structural and/or an electronic point of view?
- a subsequent question would be the nature of the intercalated species.

**2.5.1 Structural viewpoint.** The main parameter governing the stability of the $K_2NiF_4$-type oxides is the well-known Goldschmidt tolerance factor

$$t = \frac{r_A + r_M}{\sqrt{2}(r_M + r_O)},$$

where $r_A$, $r_M$, $r_O$ are the ionic radii of $A^{n+}$, $M^{m+}$ and $O^{2-}$ ions for a given $A_2M0_4$ compound. In the ideal condition ($t = 1$), the symmetry is tetragonal ($I_{4/mmm}$ S.G.) [27]. However in most of the $A_2MO_4$ oxides, as in the perovskite compounds, the size of the A-cation is too small, which markedly lessens the $t$-value ($t \ll 1$) and a strong orthorhombic distortion appears ($B_{mab}$ S.G.). This is the case for $La_2CuO_{4.00}$ and $La_2NiO_{4.00}$ for which $t$ is equal to 0.868 and 0.885, respectively. Thus, instead of forming rather long weak bonds with respect to its size, the lanthanum cation prefers to form four strong and four weak bonds and in addition a ninth very strong bond (La-O $\approx$ 2.33 Å) with the apical oxygen atom of the perovskite layer, which provides the cohesion of the structure. This results also from the acidic character of $La^{3+}$ and the easy tendency to form the complex $LaO^+$ ion. Nevertheless such a linkage does not give rise to a particular stress in the structure. Reversibly, within the ($La_2O_2$) layers, the very long La-O bond lengths result from the mismatch with the too long M-O distances; this gives rise to a tensile stress in ($La_2O_2$) layers and as a consequence to an equatorial compressive stress within the $MO_2$ sheets, which leads to a buckling of the octahedra — i.e. to an alternating tilting of these octahedra — causing the orthorhombic distortion.

From a chemical viewpoint, these stresses can be relaxed:

- by the oxidation of $M^{2+}$ into $M^{3+}$ ions ($r_{M^{2+}} < r_{M^{3+}}$) so that the equatorial compressive stress gradually lessens as $t$ increases.

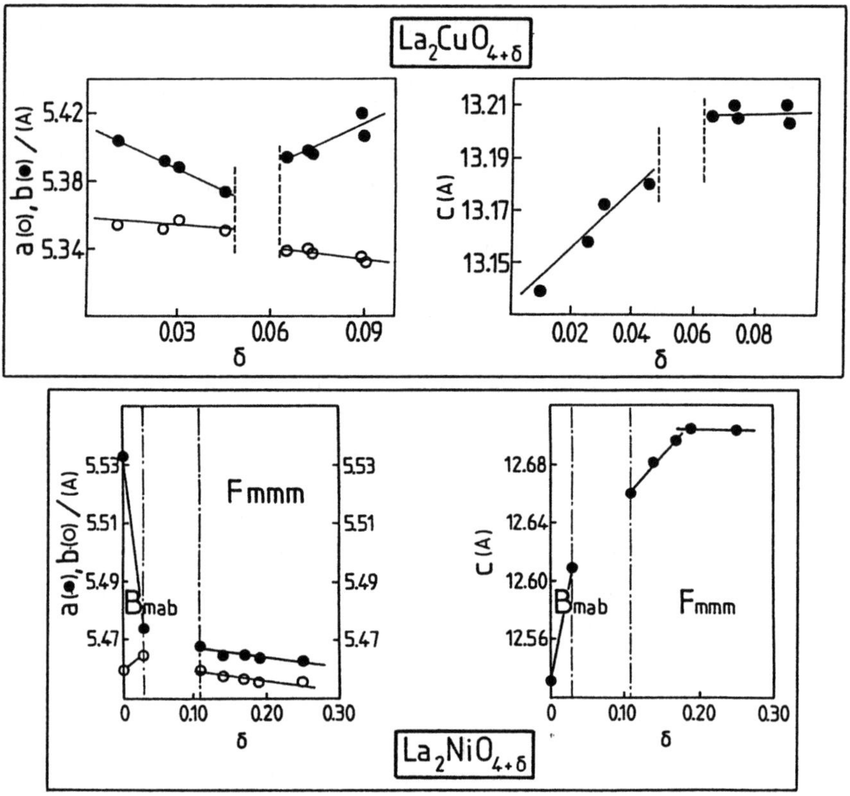


**Fig. 2.7.** Compositional dependence of the unit cell parameters of $La_2CuO_{4+\delta}$ ($0 \leq \delta \leq 0.09$) and $La_2NiO_{4+\delta}$ ($0 \leq \delta \leq 0.25$) [11],[26].

– or/and by the insertion of additional oxygen atoms which leads to an increase of the lanthanum coordination and consequently the mean La-O distance.

Both factors tend to increase $t$ and would explain that the intercalation of oxygen is easily achieved in air. This stabilizing effect is clearly illustrated by the variation of the cell parameters vs. $\delta$ (Fig. 2.7), the ideal pseudo-tetragonal symmetry ($a \approx b$) being reached for $\delta$ equal to $\approx 0.05$ and 0.03 for the Cu and Ni compounds respectively. Above these values the variation of $a$ and $b$ is quite different. In the same way the enhancement of the $c$ parameter is a consequence of the oxygen intercalation but for high $\delta$ values it seems to be much less, even null. All these features will be discussed further on.

**2.5.2 Nature of the diffusing species.** As quoted above $O^{2-}$ species are not likely the diffusing species with respect to their size; on the other hand the absence of protons or $OH^-$ in the oxidized materials was checked by

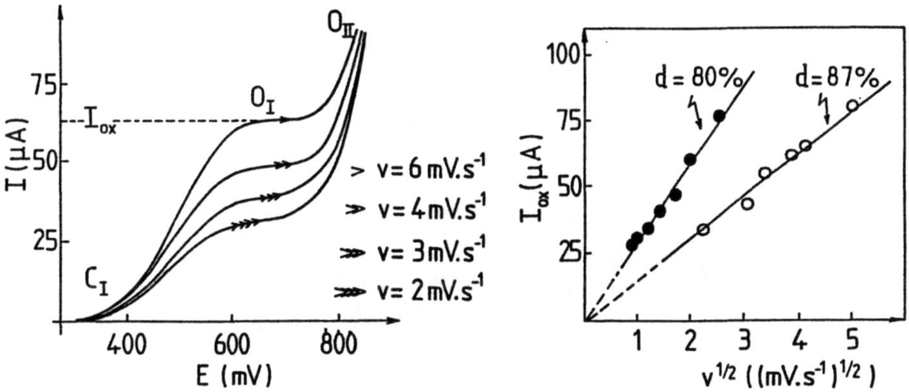


**Fig. 2.8.** Cyclic voltammetry study of $La_{1.70}Nd_{0.30}CuO_4$ for various sweep rates and variation of $I_{ox}$ vs. $v^{1/2}$ for ceramics of various densities.

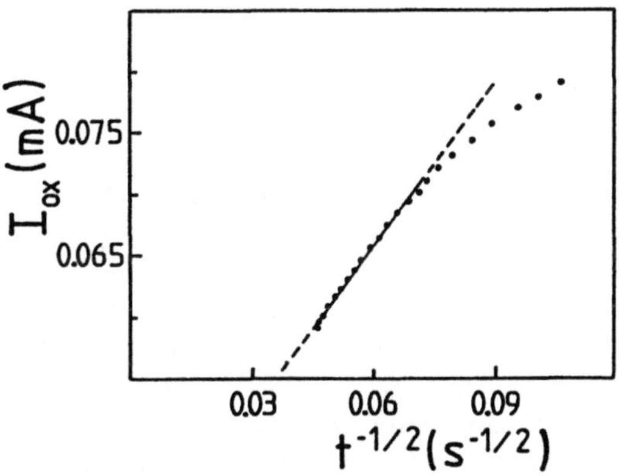


**Fig. 2.9.** Variation of the diffusion intensity current $I_{ox}$ vs. $t^{-1/2}$ at constant potential ($E = 700$ mV) for $La_2CuO_{4.09}$.

proton NMR and IR studies and seems to rule out a mechanism based on a proton migration as proposed earlier [28].

One should remind that the oxidation process occurs in the vicinity of the oxygen evolution reaction and that, at such potentials, the activity of oxygen at the surface of the electrode is very high (theoretically several orders of magnitude). Therefore one can suppose that the oxidation of the material probably involves:

- in a first step, the formation at the surface of electroactive species that are concerned with the mechanism of the oxygen evolution.
- in a second step, the diffusion of these species into the bulk of the electrode, which may also imply electron transfers.

The electrochemical behavior of $La_2CuO_{4+\delta}$ of Nd-substituted compounds ($0 \leq x \leq 0.60$) was more particularly examined [29]. Using a rotating

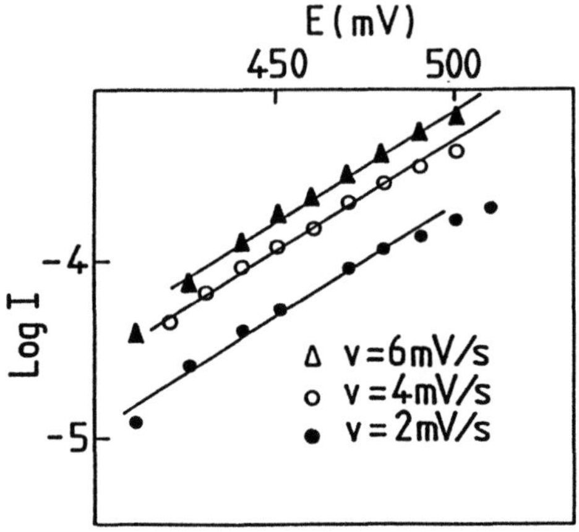


**Fig. 2.10.** Tafel behaviour ($\log(I) = f(E)$ of a ceramic of $La_{1.70}Nd_{0.30}CuO_4$ before the oxidation plateau for various sweep rates.

disc electrode (RDE) we showed that the value of the intensity current $I_{ox}$ remains constant in a large range of potentials whatever the rotation speed, which means that $I_{ox}$ characterizes processes relative to the electrode and not relative to the electrolyte. Moreover the linear variation of $I_{ox}$ as a function of $\sqrt{v}$ ($v$ being the sweep rate of the cyclic voltammogram) (Fig. 2.8) seems to corroborate a diffusion process. Then a potentiostatic study ($E = 650$ mV) was carried out and the diffusion intensity current plotted as a function of $t^{-1/2}$ (Fig. 2.9). Using the relation of Cottrell in the linear part of the curve, the value of the diffusion coefficient $D_O$ of the electroactive species was found to be close to $10^{-9}$ cm$^2$/s, which is somewhat greater than the value usually found for $O^{2-}$ at room temperature ($< 10^{-15}$ cm$^2$/s).

In addition the Tafel slope ($b$) was determined for various compositions of the $La_{2-x}Nd_xCuO_{4+\delta}$ from the ($\log i, E$) curves for potentials lower than the diffusion plateau (Fig. 2.10). Almost the same value of $b$ was found, $b \approx 70$ mV/dec. According to our previous studies of the oxygen evolution on oxide materials [30], we can propose for the cuprates the following mechanism of oxidation decomposed into five steps:

$$Cu^{2+}_{surf} + (OH^-)_{aq} \longleftrightarrow Cu^{2+}_{surf}\text{—}(OH^-)_{ads} \tag{2.2}$$
$$Cu^{2+}_{surf}\text{—}(OH^-)_{ads} \longleftrightarrow Cu^{2+}_{surf}\text{—}(OH)_{ads} + e^- \tag{2.3}$$
$$Cu^{2+}_{surf}\text{—}(OH^-)_{ads} + (OH^-)_{aq} \longleftrightarrow Cu^{2+}_{surf}\text{—}(O^-)_{ads} + H_2O \tag{2.4}$$
$$Cu^{2+}_{surf}\text{—}(O^-)_{ads} + 2\,Cu^{2+}_{net} \longleftrightarrow Cu^{2+}_{surf} + O^-_i\text{—}Cu^{2+}_{net} + Cu^{3+}_{net} + e^- \tag{2.5}$$
$$Cu^{2+}_{net}\text{—}O^-_i \longleftrightarrow Cu^{3+}_{net}\text{—}O^{2-}_i \tag{2.6}$$

The value of $b$ involves that the step 2.4 is the limiting step of the reaction [44]. It implies the formation at the surface of the electrode of $O^-$

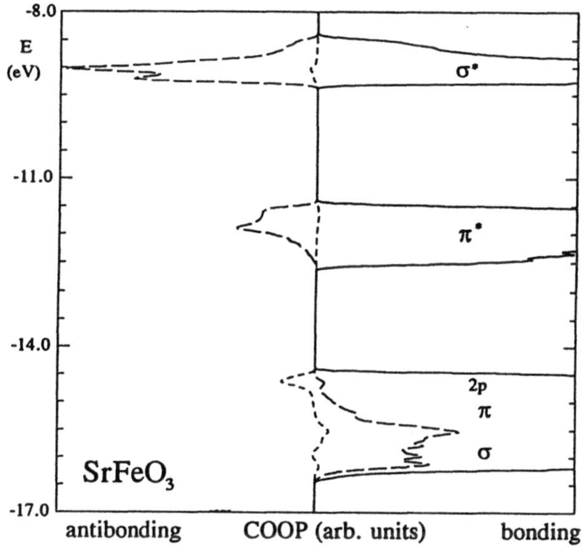


**Fig. 2.11.** $SrFeO_3$: Crystal Orbital Overlap Populations (COOP) as a function of energy, for Fe-O bonds (long dash) and O-O bonds (short dash). Electronic states at the top of "oxygen" bands, around −14.5 eV, contribute mainly to O-O bonds, with an antibonding character

species that are immediately consumed. Therefore are the $O^-$ species the mobile species?

**2.5.3 Mixed valence of oxygen.** These results led us to consider for these materials the existence of a mixed valence for oxygen ($O^{2-}/O^-$), or more generally of the following equilibrium:

$$M^{(n+1)+} + O^{2-} \longleftrightarrow M^{n+} + O^- , \quad (2.7)$$

characterized by an energy $\Delta$, namely the "charge transfer energy". Using the framework of Zaanen et al., J.B Torrance et al. [31], on the basis of a simple ionic model, showed that this $\Delta$ value can be sometimes very low ($\approx$ 0 eV), which implies that this equilibrium may be somewhat shifted toward the right. It seems now well-known that $La_2CuO_{4+\delta}$ and related compounds belong to this class of oxides in which the electron transfer from the oxygen to the metal is easy [32].

Do we have such a scheme on materials that we prepared using the electrochemical oxidation?

- $SrFeO_3$: For instance, we calculated the charge transfer energy $\Delta$ corresponding to the reaction (2.7) for $SrFeO_3$: it is close to zero, which means that we can expect a mixed valence of both iron ($Fe^{III}/Fe^{IV}$) and oxygen ($O^{2-}/O^-$). Some experimental results also support this mixed valence as, for example, the Mössbauer resonance study that showed the

isomer shift of iron, at room temperature, to be close to +0.08 mm/s; this value is inbetween the value obtained for $Fe^{4+}$ in a high spin state in $Sr_{0.5}La_{1.5}Li_{0.5}Fe_{0.5}O_4$ ($\delta = -0.20$ mm/s) and the usual value of $Fe^{3+}$ in octahedral sites (+0.40 mm/s) [33],[34]. Moreover neutron diffraction data of $SrFeO_3$ [35] have demonstrated the presence of a magnetic moment on each oxygen atom ($\approx 0.3\mu_B$); this was assigned to a charge transfer from the oxygen atoms towards iron. More recently XPS and UPS measurements [36] have corroborated the presence of holes on oxygen atoms, the ground state of iron in $SrFeO_3$ being close to $d^5\ \underline{L}$ states and reflecting the large covalence of the Fe-O bonds. Our band structure calculation, using the extended Hückel method, confirms these results and shows that the top of the oxygen 2p bands is of mainly (O-O) antibonding character with very low oxygen-iron mixing (Fig. 2.11). Due to the strong localization, these 2p states should be at higher energy when doubly occupied as expected from an $U_{pp}$ Hubbard parameter [37] and give rise to the equilibrium (2.7). The occurrence of holes in these antibonding 2p bands would moreover induce a stabilizing effect and some bonding character within the oxygen network.
Then we can conclude that $SrFeO_3$ seems to be a typical "mixed valence oxygen" compound.

- $La_2NiO_{4+\delta}$ and $La_2CuO_{4+\delta}$: As quoted above, the insertion of oxygen into the $La_2O_2$ layers of both $La_2CuO_{4+s}$ and $La_2NiO_{4+\delta}$ involves mainly "steric effects", at least for low values of $\delta$. But for higher amounts of excess oxygen ($\delta \geq 0.06$ and 0.18 respectively), the $c$ parameter remains almost constant and such a behavior can be explained only in terms of "electronic effects" as follows.
  A detailed study of the $\delta = 0.25$ composition of the $La_2NiO_{4+\delta}$ system, more especially a Rietveld structure refinement of the neutron diffraction data [26] and TEM observations [38], have clearly shown:
  - the existence of a long range ordering of the interstitial oxygen atoms both along the $[111]_T$ direction of the $I_{4/mmm}$ unit cell and in the $La_2O_2$ planes, which leads to a monoclinic supercell ($C_2$ S.G.) corresponding to a definite compound $La_8Ni_4O_{17}$.
  - the interstitial oxygen atom forms with two adjacent apical oxygen atoms a $O_3^{5-}$ polyoxide characterized by short O-O bond lengths ($\approx 2.2$ Å) and the presence of one hole over these three oxygen atoms. In addition calculations of the Madelung site potentials corroborate that the $O_{2p}$ electrons should be less stabilized in these interstitial sites [26].
  - the occurrence of such species is strongly correlated to the existence of an electron transfer between these clusters and the $NiO_2$ planes which can be written,

$$Ni^{2+} + 3O^{2-} \longleftrightarrow Ni^{+} + O_3^{5-} \, ,$$

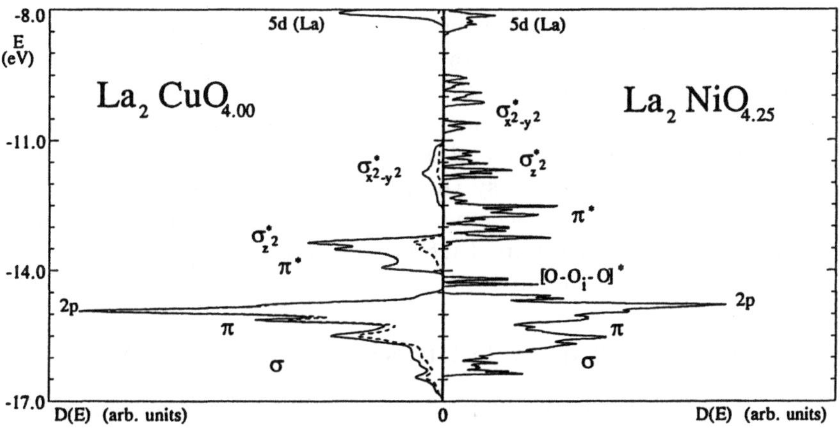


**Fig. 2.12.** Curves of density of states for $La_2CuO_{4.00}$ and $La_2NiO_{4.25}$, as a function of energy. Dashed line shows the global oxygen contribution to $La_2CuO_{4.00}$ total density of states. The top of "oxygen" bands corresponds to only oxygen states. Slightly antibonding interstitial oxygen-related states occur for $La_2NiO_{4.25}$, between -14.0 and -14.5 eV.

implying furthermore the disproportionation of nickel ($Ni^{(2+\delta)+} \Longrightarrow Ni^{+}, Ni^{2+}, Ni^{3+}$) (a kind of charge density wave in these planes). Such a coupling reinforces the oxygen ordering that occurs through the formation of $Ni^{+}$-$[O_3^{5-}]$-$Ni^{+}$-$[O_3^{5-}]$-$Ni^{+}$-... files along given $[111]_T$ directions.

In the same way, for the $La_2CuO_{4+\delta}$ cuprates, thermoelectric measurements show that the number of holes in the $CuO_2$ planes increases with $\delta$ and a decrease of the value of the thermoelectric power $\alpha$ is actually observed up to $\delta \approx 0.05$ [11]. But for higher $\delta$ values, $\alpha$ remains almost constant and this was assigned to the fact that additional oxygen intercalation does not generate new holes in the $CuO_2$ sheets but rather $O^-$ species — i.e. holes on the oxygen atoms in the $La_2O_2$ layer.

- Conclusion: Band structure calculations for both $La_2NiO_{4.25}$ and $La_2CuO_4$ (Fig. 2.12) show also that 3d states lie close to $O_{2p}$ states and that localized oxygen states occur at the top of 2p bands. Moreover intercalated oxygen in $La_2NiO_{4.25}$ gives rise to slightly antibonding $O_{ap}$-$O_i$-$O_{ap}$ states just above the 2p bands; the very low energy dispersion of these states indicates their strong localized character and, as in $SrFeO_3$, should lead to a small transfer energy $\Delta$ of equilibrium (2.7). One can expect similar features as oxygen is intercalated in $La_2CuO_{4+\delta}$.
  The formation of $O^-$ species, — i.e. of holes on oxygen — as $\delta$ increases results for both compounds in a constant value of the $c$ parameter and the relaxation of the stresses is obtained for nickelate thanks to an electron transfer in the $NiO_2$ planes and the disproportionation of nickel cations. Correlatively the structure remains undistorted, $a \approx b$. On the contrary

such a disproportionation seems unlikely to occur in the cuprates, due to the $3d^{10}$ electronic configuration of $Cu^{+}$, which results in a strong orthorhombic distortion (Fig. 2.7).

From these previous examples one can draw the conclusion that in such *charge transfer transition metal oxides* one may expect that the mixed valence of oxygen makes easy the formation of $O^{-}$ species that would be involved in the electrochemical oxidation process as well as in the phase separation. Last item that can be addressed is the mobility of these species within the lattice.

## 2.6 Mobility of "$O^{-}$" in oxide lattices

Assuming that $O^{-}$ is formed, at least temporarily, several evidences of the mobility of oxygen can be given. The first example was reported in 1988 by Jorgensen et al. [13] who claimed the occurrence of a phase separation below room temperature in $La_2CuO_{4+\delta}$ phases into a nearly stoichiometric phase $La_2CuO_{4.00}$ and an oxygen-rich one $La_2CuO_{4.08}$. This was confirmed later on by structural studies of C. Chaillout et al. on crystals [24], then by J. Ryder et al. [14] who observed an hysteresis of the electrical conductivity according to the cooling rate, which they assigned to a spinodal decomposition involving oxygen migration in the material. We repeated these experiments on $La_2CuO_{4.03}$ prepared by electrochemical oxidation and corroborated these previous results. Several other experiments have been reported [36] but most of them are concerned with slightly doped materials. For highly doped materials ($[p] > 0.10$, i.e. the metallic phase) this feature is less obvious.

Nevertheless we recently reported TEM observations of $La_2CuO_{4.09}$ at low temperature ($T = 100$ K) [40] revealing the presence of commensurate and incommensurate modulated superstructures that might be assigned to some oxygen ordering into microdomains. Such superstructures were observed only below 250 K in a reversible way. Large superstructures were also recently detected by P.G. Radaelli et al. using neutron diffraction [41].

Therefore one can conclude that the migration of oxygen atoms in these materials is rather easy since it may occur around room temperature. One should also point out that calculations of the migration energy of oxygen anions (assuming $O^{2-}$) between interstitial sites in $La_2CuO_4$, reported by N. L. Allan et al. [42] give a rather low value ($\approx 0.48$ eV), which confirms the observed facile oxygen diffusion in these oxides. However, this diffusion can be suppressed by substituting Nd for La in $La_2CuO_{4+\delta}$ phases. This was observed by Takeda et al. [43] for phases prepared under high oxygen pressures, which we also confirmed by studying the thermal behaviour of the electrical resistivity of $La_{2-x}Nd_xCuO_{4.04}$ ($x = 0$–$0.40$). As $x$ increases, the thermal hysteresis previously noticed for $La_2CuO_{4.03}$ disappears for $x \approx 0.40$, which can be assigned to the fact that Nd atoms would inhibit the migration of oxygen due to the smaller size of the channels as evidenced by the lessening of the $c$ parameter of these phases as $x$ increases [44].

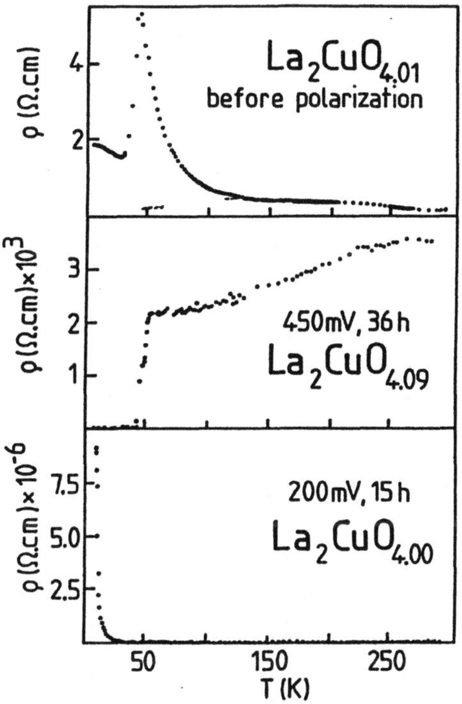


**Fig. 2.13.** Thermal dependence of the resistivity of $La_2CuO_{4+\delta}$ after anodic oxidation ($E = 450meV$) and cathodic reduction ($E = 200meV$) demonstrating the reversibility of the process.

Such a diffusion allows to perform not only the electrochemical oxidation but also the electrochemical reduction as reported in Fig. 2.13. Starting from the as-prepared material $La_2CuO_{4.01}$ (semiconducting material), the electrochemical oxidation (450 mV, 36 h) leads in a first stage to $La_2CuO_{4.09}$ (metallic), then by electrochemical reduction (200 mV, 15 h) to $La_2CuO_{4.00}$ (insulator). This clearly evidences the reversibility of the oxygen intercalation and consequently its mobility in oxide networks.

## 3. Conclusion

The insertion of oxygen in $La_2CuO_4$ and related materials has been achieved thanks to the electrochemical oxidation and new superconducting materials have been prepared. The overstoichiometry was extended up to 0.12, even in substituted compounds (Sr or Nd). For the first time, metallic compounds showing superconducting properties at a temperature as high as 44 K were obtained. Several groups successfully used this new method [19],[20],[28]. In addition, we also succeeded to insert oxygen in thin films as well as in $La_2CuO_4$

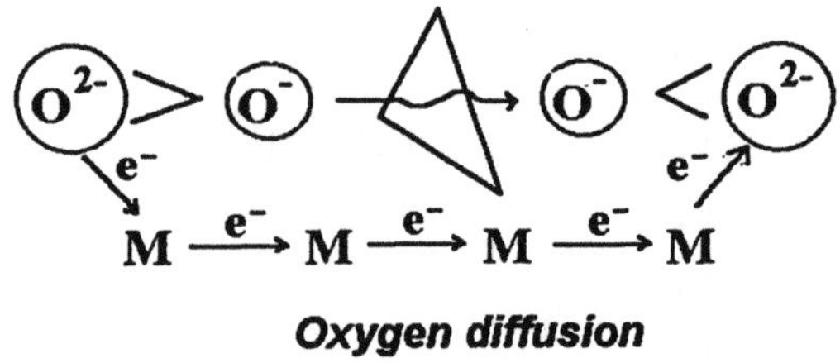


**Fig. 3.1.** Schematic representation of the coupling between the diffusion of $O^-$ species and the electronic conduction thanks to the electron transfer O → Metal.

crystals with again $T_c = 44$ K and even higher 47 K. This clearly shows that "Chimie Douce" processes can be relevant in chemistry of cuprates.

These materials belonging to the class of "*charge transfer oxides*" (small $\delta$), we propose that the $O^-$ species are the diffusing species via an electronic exchange with the metal cations according to the equilibrium: $Cu^{3+} + O^{2-} \longleftrightarrow Cu^{2+} + O^-$ as schematized in Fig. 3.1.

*Acknowledgement.* The authors are very grateful to B. Darriet, A. Demourgues, P. Dordor, J-P. Doumerc, J. Etourneau, L. Fournès, N. Lagueyte, E. Marquestaut, F. Weill and J.B. Goodenough (Univ. of Texas at Austin, USA) for their participation in this work. They also thank C. Oseroff(San Diego State Univ.,Ca., USA), Z. Fisk (Los Alamos National Laboratory, New Mexico, USA) and J-M. Bassat (CRPHT, Orléans, France) for providing crystals.

## References

1. J. Beille, R. Cabanel, C. Chaillout, B. Chevalier, G. Demazeau, F. Deslandes, J. Etourneau, P. Lejay, C. Michel, J. Provost, B. Raveau, A. Suplice, J-L. Tholence and R. Tournier, *C. R. Acad. Sc. Paris* **304(II)**, 1097(1987)
2. Z. Hiroi, M. Takano, M. Azuma and Y. Takeda, *Nature* **364**, 315(1993)
3. J-L. Tholence, B. Souletie, O. Laborde, J.J. Capponi, C. Chaillout and M. Marezio, *Physics Lett.A* , (in press)
4. A. Wattiaux, J.C. Park, J.C. Grenier and M. Pouchard, *C.R.Acad.Sci.Paris* **310(II)**, 1047(1990)
5. J.C. Grenier, A. Wattiaux, N. Lagueyte, J.C. Park, E. Marquestaut, J. Etourneau and M. Pouchard, *Physica C* **173(3-5)**, 139(1991)
6. B. Chevalier, A. Tressaud, B. Lépine, K. Amine, J.M. Dance, L. Lozano, E. Hickey and J.Etourneau, *Physica C* **167**, 97(1990)
7. B. Chevalier, A. Tressaud and B. Lépine, *J. Less-Common. Met.* **164/165**, 832(1990)
8. A. Tressaud, C. Robin, B. Chevalier, L. Lozano and E. Etourneau, *Physica C* **177**, 330(1991)
9. E. Takayama-Muromachi, T. Sasaki and Y. Matsui, *Physica C* **207**, 97(1993)
10. N: Lagueyte, A. Wattiaux, J.C.Park, J-C. Grenier, L. Fournès and M.Pouchard, *J. Phys. III* **1**, 1755(1991)

11. J.C. Grenier, N. Lagueyte, A. Wattiaux, J.P. Doumerc, P. Dordor, J. Etourneau, M. Pouchard, J.B. Goodenough and J.S. Zhou, *Physica C* **202**, 209 (1992)
12. J.C. Grenier, A. Wattiaux and M. Pouchard, Proc. Phase Separation in Cuprate Superconductors, Ed. K.A. Müller and G. Benedek, World Scientific Publ., 187 (1992)
13. J.D. Jorgensen, B. Dabrowski, S. Pei, D.G. Hinks, L. Soderholm, B. Morosin, J.E. Schirber, E.L. Venturini and D.S. Ginley, *Phys. Rev. B* **38(16)**, 11337 (1988)
14. J. Ryder, P.A. Midgley, R. Exley, R.J. Beynon, D.L. Yates, L. Afalfiz and J.A. Wilson, *Physica C* **173**, 9 (1991)
15. P.C. Hammel, E.T. Arhens, A.P. Reyes, J.D. Thompson, D.E. McLaughlin, Z. Fisk, P.C. Canfield and J.E. Schirber, Proc. Phase Separation in Cuprate Superconductors, Ed. K.A. Müller and G. Benedek, World Scientific Publ., 139 (1993)
16. G. Wübbeler and O.F. Schirmer, *Phys. Stat. Sol.(b)* **174**, K21 (1992)
17. R.K. Kremer, V. Hizhnyakov, E. Sigmund, A. Simon and K.A. Müller, *Z. Phys. B* **91**, 169 (1993)
18. M. Mehring, M. Baehr, P. Gergen, J. Grob and C. Kessler, *Physica Scripta (in press)* ,
19. F.C. Chou, J.H. Cho and D.C. Johnston, *Physica C* **197**,, 303 (1993)
20. N. Casan-Pastor, P. Gomez-Romero, A. Fuertes, J.M. Navarro, M.J. Sanchis and S. Ondono, *Physica C* **216**, 478 (1993)
21. J.P. Locquet, C. Gerber, A. Cretton, Y. Jaccard, E. Williams and E. Mächler, *Appl. Phys. A (accepted)* ,
22. P.M. Grant, S.S.P. Parkin, V.Y. Lee, E.M. Engler, R.L. Ramirez, J.E. Varquez, G. Lim, R.D. Jacowitz and R.L. Greene, *Phys. Rev. Lett.* **58**, 2482 (1987)
23. J.W. Rogers, N.D. Shinn, J.E. Schirber, E.L. Venturini, D.S. Ginley and B. Morosin, *Phys. Rev. B* **38**, 5021 (1988)
24. C. Chaillout, J. Chenavas, S.W. Cheong, Z. Fisk, M. Marezio, B. Morosin and J.E. Schirber, *Physica C* **170**, 87 (1990)
25. J. Galy, *Acta Cryst* **B48**, 777 (1992)
26. A. Demourgues, F. Weill, B. Darriet, A. Wattiaux, J.C. Grenier and M. Pouchard, *J. Sol. State Chem.* **106**, 317 (1993)
27. K.K. Singh, P. Ganguly and J.B. Goodenough, *J. Sol. State Chem.* **52**, 254 (1984)
28. P. Rudolf, W. Paulus and R. Schöllhorn, *Adv. Mater.* **3(9)**, 438 (1991)
29. F. Arrouy, A. Wattiaux, C. Cros, G. Demazeau, J.C. Grenier, M. Pouchard and I. Etourneau, *Physica C* **175**, 342 (1991)
30. A. Wattiaux, J.C. Grenier, M. Pouchard and P. Hagenmuller, *J. Sol. State Chem.* **134**, 1714 (1987)
31. J.B. Torrance, P. Lacorre, C. Asavaroengchai and R.M. Metzger, *J. Sol. State Chem.* **90**, 168 (1991)
32. J.B. Goodenough, J.S. Zhou, K. Allan, *J. Mater. Chem.* **1(5)**, 715 (1991)
33. L. Fournes, Y. Potin, J.C. Grenier, G. Demazeau and M. Pouchard, *Sol. State Commun.* **62**, 239 (1987)
34. A. Wattiaux, L. Fournis, A. Demourgues, N. Bernaben, J.C. Grenier and M. Pouchard, *Sol. State Comm.* **77(7)**, 489 (1991)
35. H. Oda, Y. Yamaguchi, H. Takei and H. Watanabe, *J. Physique C1* **38**, 121 (1977)
36. A.E. Bocquet, A. Fujimori, T. Misokawa, T. Saitoh, H. Namatame, S. Suga, N. Kimizuka, Y. Takeda and M. Takano, *Phys. Rev. B* **45**, 1561 (1992)
37. Kuiper et al., *Phys. Rev. Letters* **2**, 342 (1989)

38. A. Demourgues, F. Weill, J.C. Grenier, A. Wattiaux and M. Pouchard, *Physica C* **192**, 425 (1992)
39. Proc. Phase Separation in Cuprate Superconductors. Ed. K.A. Müller and G. Benedek, World Scientific Publ., (1993)
40. N. Lagueyte, F. Weill, A. Wattiaux and J.C. Grenier, *Eur. J. Sol. State Inorg. Chem* **30**, 859 (1993)
41. P.G. Radaelli, J.D. Jorgensen, A.J. Schultz, B.A. Hunter, J.L. Wagner, F.C. Chou and D.C. Johnston, *Phys. Rev. B* **48(1)**, 499 (1993)
42. N.L. Allan and W.C. Mackrodt, *Phil. Mag.* **64(5)**, 1129 (1991)
43. Y. Takeda, K. Yoshikawa, O. Yamamoto and M. Takano, *J. Sol. State Chem.* **92**, 241 (1991)
44. F. Arrouy, Doctoral Thesis, Univ. of Bordeaux I,779 (1992) and F. Arrouy, A. Wattiaux, E. Marquestaut, C. Cros, G. Demazeau, J.C. Grenier and M. Pouchard,*J. Sol. State Chem.*(to be published).

# Structural Disorder and Phase Separation in TmBaCuO as Seen from Tm NMR

O.N. Bakharev[1], A.V. Dooglav[1], A.V. Egorov[1], O.B. Marvin[1], V.V. Naletov[1], M.A. Teplov[1], A.G. Volodin[1], and D. Wagener[2]

[1] Kazan State University, 420008 Kazan, Russia
[2] Universität Bonn, Institut für Strahlen- und Kernphysik, W-5300 Bonn, Germany

**Summary.** The 1-2-3 compounds are known to have a disordered crystal structure. The "chain" oxygen atoms have a tendency to aggregate in clusters and to create microdomains with different oxygen content, three main types of these domains being of OI ($x = 1.0$), OII ($x = 0.5$) and $T$ ($x = 0$) structure. In the present paper the "enhanced NMR" of $^{169}$Tm nuclei (spin I=1/2, 100% natural abundance) is used to study the structure and magnetic properties of $TmBa_2Cu_3O_{6+x}$ compounds (TmBCO) at temperatures from 4.2 K down to 40 mK. The Tm NMR results in combination with the data obtained by other methods support the idea that the structure and properties of underdoped compounds are governed by a random alternation of CuO basal planes (or small areas of these layers) of OI and OII structure along the c-direction. The superconductivity seems to be provided only by "sandwiches" consisting of identical (OI-OI or OII-OII) planes that are symmetrical with respect to the rare-earth (RE) planes. The boundaries between superconducting (SC) and non-superconducting (NSC) microdomains are shown to be enriched with $Cu^{2+}(2)$ paramagnetic centers (PC). The Tm nuclear spin-lattice relaxation (NSLR) data make it possible to estimate the PC's concentration and the mean length of CuO chain segments in the series of TmBCO samples ($x$ =0–1).

## 1. Experimental

The ceramic TmBCO samples were prepared as described in [1], crushed to powder and aligned in the magnetic field of 94 kOe using a rotating container with melt paraffin [2],[3]. The ac diamagnetic susceptibility measurements at the frequency of 1 kHz (the field $H_1 \approx 1$ Oe perpendicular to the c-axis) had shown the critical temperatures for our samples to coincide with those for $YBa_2Cu_3O_{6+x}$ ceramic samples (Figs. 1.1 and 1.2). A conventional home-built pulsed spectrometer was used for Tm NMR measurements. The experiments at temperatures below 1.5 K were performed using a $^3$He-$^4$He dilution refrigerator.

The ground energy level of the $Tm^{3+}$ 4f-electron shell in the crystal electric field (CEF) is a singlet [2]. The NMR spectra at $T \leq 4.2$ K (see Figs. 1.3 and 1.4) are described by the Hamiltonian $H = -\hbar \sum_{(i=\mathrm{a,b,c})} \gamma_i H_i I_i$ whose parameters $\gamma_\mathrm{b}$ and $\gamma_\mathrm{c}$ are very sensitive to CEF local symmetry changes ($|\gamma_\mathrm{b}/2\pi| = 6.8(1)$, 6.1(1) and 5.3(1) kHz/Oe, $|\gamma_\mathrm{c}/2\pi| = 2.20(5)$, 2.56(5) and 3.05(5) kHz/Oe for OI, OII and T structures, respectively [4],[5], whereas $|\gamma_\mathrm{a}/2\pi| = 5.3(1)$ kHz/Oe remains constant in all three main phases). In fact, five different combinations of "filled" and "empty" nearest-neighbor CuO

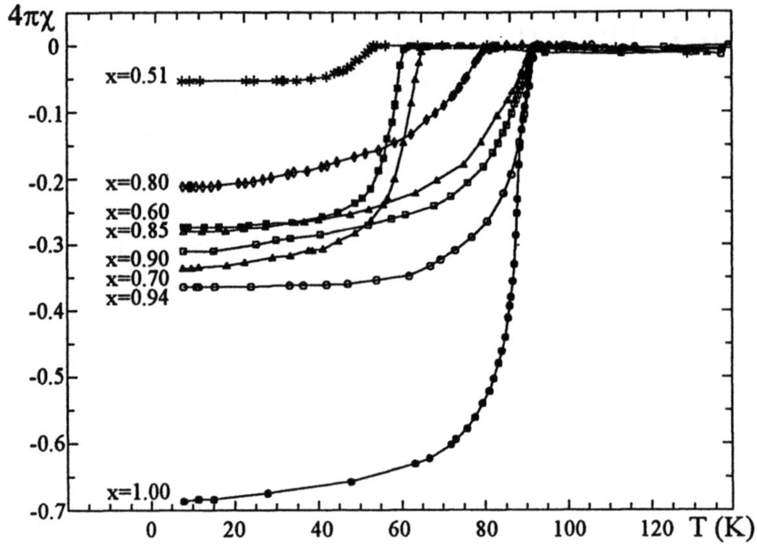


**Fig. 1.1.** Diamagnetic susceptibility of $TmBa_2Cu_3O_{6+x}$ as a function of temperature.

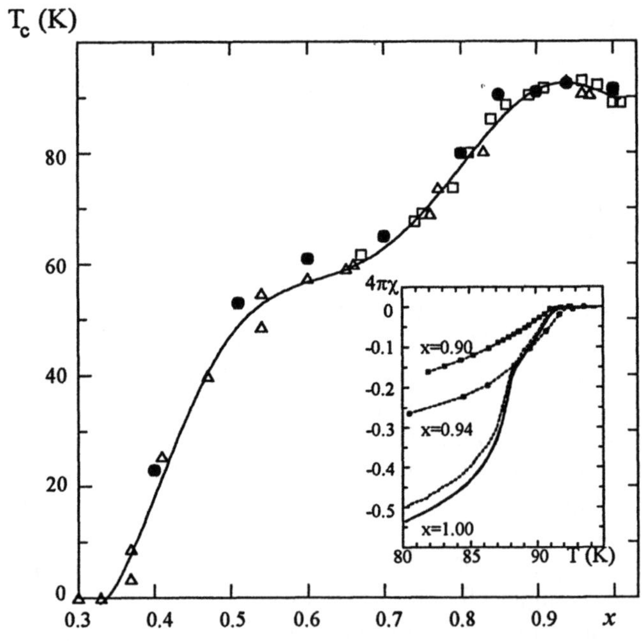


**Fig. 1.2.** Critical temperatures of $YBa_2Cu_3O_{6+x}$ ($\Delta,\square$ - Ref. [3]) and $TmBa_2Cu_3O_{6+x}$ ($\bullet$ - our data) as a function of oxygen content. The temperature dependences of the diamagnetic susceptibility of $TmBa_2Cu_3O_{6+x}$ are shown in the insert: dotted line - the same as solid line but measured five months later (see text).

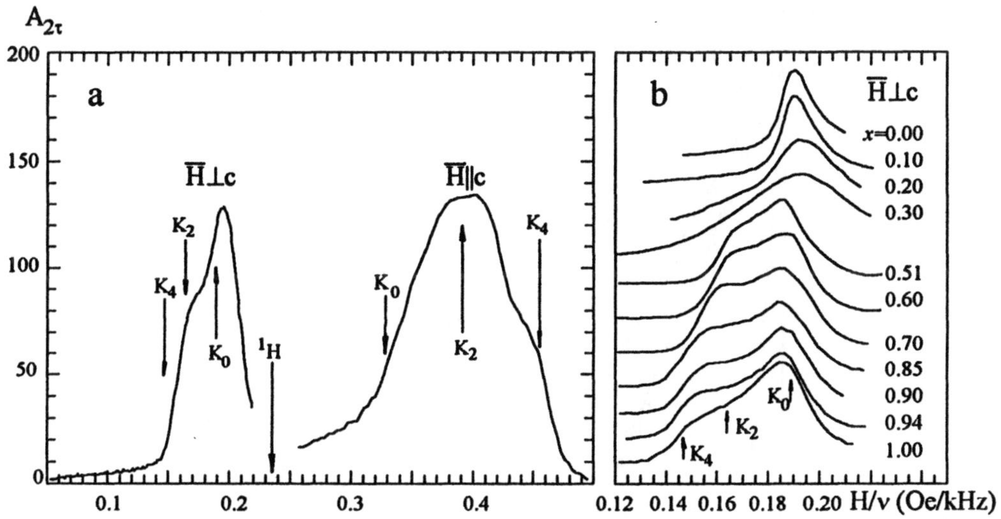


**Fig. 1.3.** [169]Tm NMR spectra of $TmBa_2Cu_3O_{6.8}$ (a - T=4.2 K) and $TmBa_2Cu_3O_{6+x}$ (b - T=1.5 K) aligned powders.

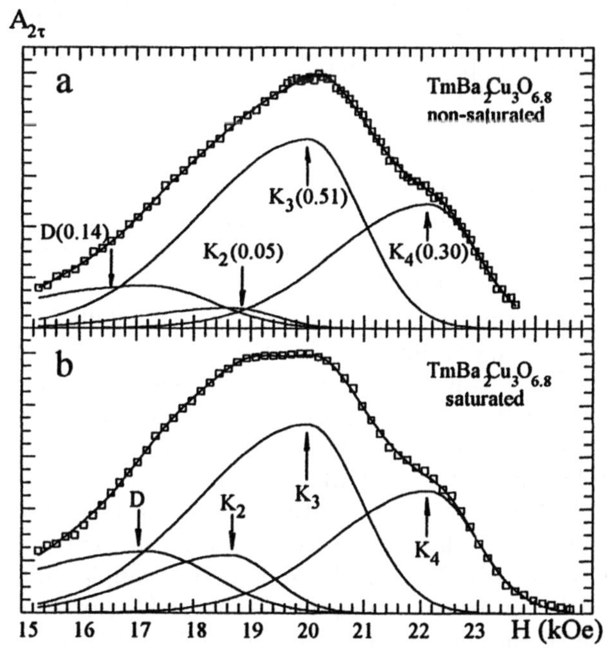


**Fig. 1.4.** [169]Tm NMR spectra of $TmBa_2Cu_3O_{6.8}$ in the field $H \| c$ at $T = 4.2$ K; $\nu$=49.7 MHz, the pulse sequence 1.5-20-3.0 $\mu$s, the pulse repetition frequency 1 Hz (a), 330 Hz (b). The parameters of fitting curves are: (a) $K_4$ - $\gamma_c/2\pi$ = 2.163 kHz/Oe, $d$ = 826 Oe, ; $K_3$ - $\gamma_c/2\pi$ = 2.373 kHz/Oe, $d$ = 930 Oe; $K_2$ - $\gamma_c/2\pi$ = 2.560 kHz/Oe, d=753 Oe; D - $\gamma_c/2\pi$ = 2.708 kHz/Oe, $d$ = 1134 Oe (the line seems to belong to the disordered structure).

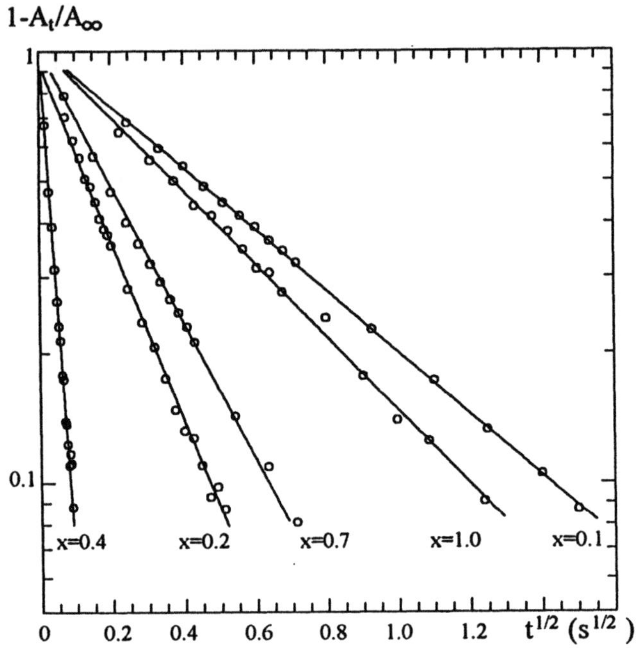


**Fig. 1.5.** $^{169}$Tm nuclear spin-lattice relaxation in $TmBa_2Cu_3O_{6+x}$ samples in the field $H \| a$ at T=1.5 K (solid line - best fit by Eq. (1.1)).

chain segments form five types of $RE^{3+}$ centers $K_i$ ($i$ = 0,1,2,3,4 denote the number of "filled" chains, $i$ = 0,2,4 corresponding to T, OII and OI structures, respectively) in relative amounts well consistent with a model of 2D-ordering of the chains in CuO basal planes [6].

A strong inhomogeneous broadening is typical for thulium NMR spectra. As for other Van Vleck paramagnets [7], it is caused partly by the scattering of $\gamma_i$-parameters due to local CEF distortions, and this contribution from the crystal lattice defects is most pronounced at high NMR frequencies [4]. However, besides this there is a large frequency-independent contribution to the Tm NMR linewidths in TmBCO caused by nondipolar local fields produced by $Cu^{2+}(2)$ ions. Even for the most perfect crystal structures of $x = 0$ compounds, the halfwidth of the Tm NMR line extrapolated to zero NMR frequency exceeds 300 Oe [4], and in the samples with $x > 0$ the local fields can be an order of magnitude larger [8] (for example, see the weak-field wing of the Tm spectrum in Fig. 1.3a. The existence of such a strong inhomogeneous Tm NMR line broadening due to $Cu^{2+}(2)$ local fields makes it unnecessary to take into account the inhomogeneity of internal fields ($\approx$ 20 Oe) caused by the vortex lattice in the SC samples [9],[10].

In different samples $T_1$ and $T_2$ relaxation times were measured at frequencies $\nu$ from 42 to 50 MHz, but when measuring in the field $H \perp$ c the $H$ value always was chosen according to $H/\nu = 2\pi/\gamma_a = 0.189$ Oe/kHz condition, in

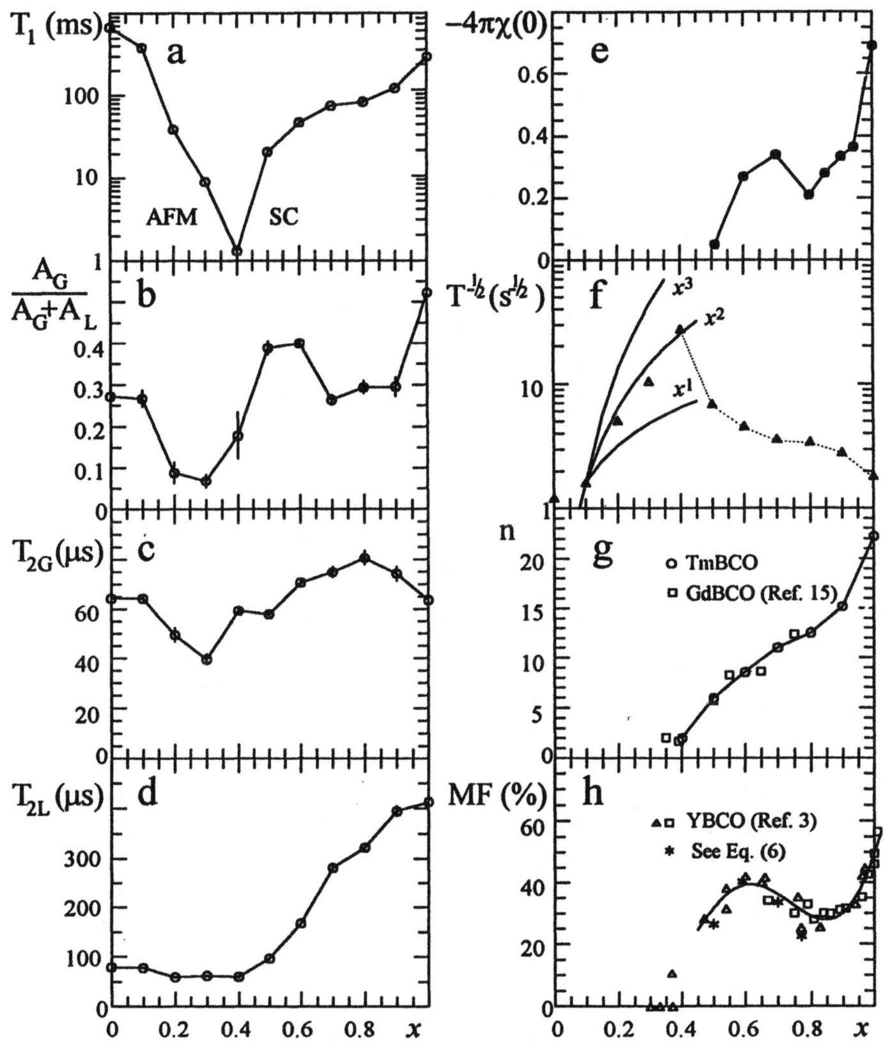


**Fig. 1.6.** Parameters of $^{169}$Tm relaxation (a-d,f), diamagnetic susceptibility (e), and CuO-chain mean length (g) and Meissner fraction (h) of $TmBa_2Cu_3O_{6+x}$ as a function of oxygen content; h - the Meissner fraction of $YBa_2Cu_3O_{6+x}$ [3].

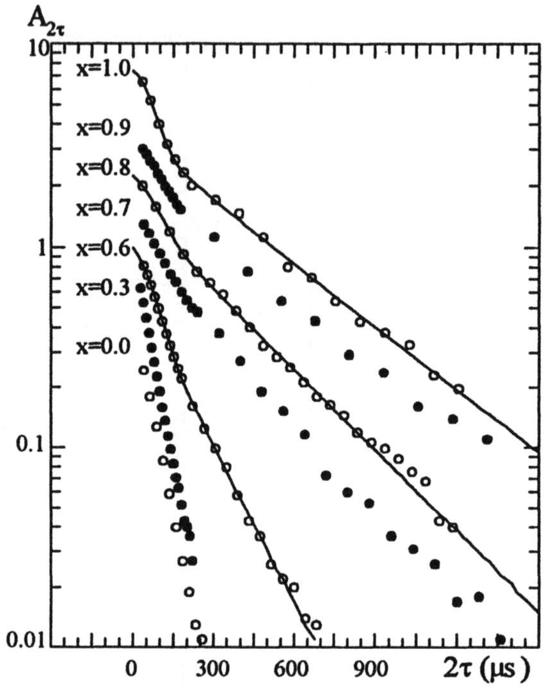


**Fig. 1.7.** $^{169}$Tm transverse relaxation in $TmBa_2Cu_3O_{6+x}$ in the field $H \| a$ at T=1.5 K (solid line - best fit by Eq. (1.2)).

this case only powder particles with the a axes lying close to $H$ direction were influenced by the resonant $H_1 \approx 100$ Oe radio-frequency (RF) magnetic field (see Fig. 1.3b. The thulium NLSR times $T_1$ were measured using the $\pi/2 - \pi - t - \pi/2 - \pi$ pulse sequence and observing the recovery of the spin echo intensity $A_t$ when the time $t$ between RF pulse pairs was increased. It appeared that for all samples at $T \leq 4.2$ K the recovery curve is fairly well described by the law (see Fig. 1.5):

$$1 - \frac{A_t}{A_\infty} = a \exp(-\sqrt{\frac{t}{T_1}}) . \qquad (1.1)$$

Due to a big inhomogeneous Tm NMR linewidth the longitudinal magnetization component could not be completely rotated by 90$^o$ by the RF pulses, thus the $a$ value was always less than 1 (0.65 - 0.85). For comparing the results in Fig. 1.5, all the $a$ values are taken to be 1. The dependence of $T_1$ on the oxygen content in the field $H \| a$ at $T$=1.5 K (Fig. 1.6a) is very like to the phase diagram for 1-2-3 compounds.

The relaxation times $T_2$ of the Tm transverse magnetization were measured by the Hahn method ($\pi/2 - \tau - \pi -$ echo$A_{2\tau}$). The results of the measurements at $T$=1.5 K in the field $H \| a$ (Fig. 1.7) are fitted well by the following expression:

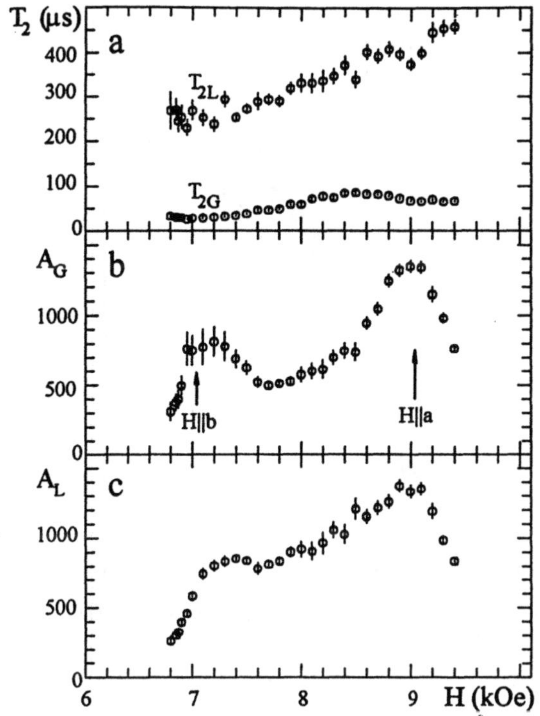


**Fig. 1.8.** Decomposition of Tm NMR spectrum in the overdoped (x=1.0) sample into fast relaxing (Gaussian) and slow relaxing (Lorentzian) components: T=1.5 K, $\mathbf{H} \perp c$, $\nu = 48$ MHz, the RF pulses are 1.5 and 3 μs

$$A_{2\tau} = A_{\mathrm{G}} \exp\left[-\frac{1}{2}\left(\frac{2\tau}{T_{2\mathrm{G}}}\right)^2\right] + A_{\mathrm{L}} \exp\left[-\frac{2\tau}{T_{2\mathrm{L}}}\right]. \tag{1.2}$$

The dependencies of $A_{\mathrm{G}}/(A_{\mathrm{G}}+A_{\mathrm{L}})$, $T_{2\mathrm{G}}$ and $T_{2\mathrm{L}}$ on the oxygen content are presented in Figs. 1.6b, c, d. The decomposition (1.2) of the $TmBa_2Cu_3O_7$ spectrum in the field $H \perp c$ is shown as an example in Fig. 1.8. Here the external field values of 7.05 kOe and 9.05 kOe correspond to positions of NMR lines which could be observed in a single crystal in the fields $H\|b$ and $H\|a$, respectively. Using the lattice sums from Ref. [5] for calculations of the second moments $M_2$ of Tm NMR lines [7], one can estimate the relaxation times $T_2 \approx 1/\sqrt{M_2}$ for the ideal crystal lattice as follows: $T_2 \approx 17$ μs and 43 μs for $H\|[100]$ and $H\|[110]$, respectively. It can be seen in Figs. 1.6c, d, and Fig. 1.8 that the measured $T_2$ values are much larger than the calculated ones. This especially coincides with $T_{2\mathrm{L}}$ times in SC samples that exceed the calculated $T_2$'s by an order of magnitude. It is known, that the spin-echo decay is influenced not only by the "flip-flop" process, but also by "spectral diffusion" and "instantaneous diffusion" processes [11]. If the inhomogeneous

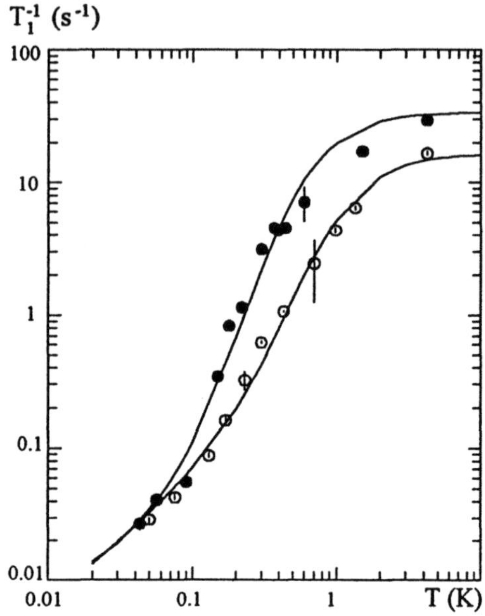


**Fig. 2.1.** $^{169}$Tm NSLR rate in $TmBa_2Cu_3O_{6.8}$ in the field $H \| c$: • - $\nu = 24$ MHz, $H_0 = 9.5$ kOe, $\Delta = 2.9$ kOe; ○ - $\nu = 39$ MHz, $H_0 = 16$ kOe, $\Delta = 5.9$ kOe.

broadening of the NMR line is so strong that only a small part of a total amount of spins can be excited by RF pulses, all these processes become as slow as in a magnetically diluted system. The experimental results shown in Figs. 1.6–1.8 allow to conclude that thulium $T_2$ times in TmBCO are strongly influenced by the inhomogeneous broadening which increases in higher external fields. However, the large difference in $T_{2G}$- and $T_{2L}$-values as well as the clearly pronounced maximum in the $T_{2G}$(H) curve (Fig. 1.8) corresponding to $\gamma$-values between $\gamma_a$ and $\gamma_b$ (a feature expected for a "powder spectrum" of a good crystalline material; see above) allows to distinguish two types of $Tm^{3+}$ centers. They can be naturally ascribed to two different fractions of the samples, i.e., slightly disordered ($A_G$) and heavily disordered ($A_L$) ones. Furthermore, the obvious correlation between $A_G$ and the SC volume fraction in TmBCO samples, as can clearly be seen in Figs. 1.6b and 1.6e at $x > 0.5$, allows to attribute the slightly disordered fraction to the SC volume. The common behavior of the nuclear spin-lattice relaxation for all $^{169}Tm^{3+}$ centers gives an evidence that both fractions are closely mixed together in a finely dispersed form.

## 2. Mean length of CuO chain segments

It is known [12] that the type (1.1) kinetics is typical for disordered systems for which a strong inhomogeneous NMR line broadening takes place and a nuclear spin diffusion is hampered due to the Larmor frequencies' difference of the nuclear spins located in the neighboring crystal lattice sites. In such systems the nuclear spins at low temperatures relax directly via PC's, well coupled with phonons. When PCs with $S = 1/2$ are randomly distributed in a crystal lattice, the NSLR rate is determined as follows [12]:

$$\frac{1}{T_1} = \frac{32}{15}\pi^3 \mu_B^2 N_0^2 \gamma_\perp^2 g_\parallel^2 \frac{\tau_c}{1+\omega^2\tau_c^2}\left(1 - \tanh^2 \frac{g_\parallel \mu_B H}{2kT}\right), \qquad (2.1)$$

where $\mu_B$ is the Bohr magneton, $N_0$ is the PC's concentration, $\omega = \gamma H$ is the NMR frequency, $\tau_c$ is a correlation time for PC's local field fluctuations created at nuclei's sites (usually the shortest one of PC's relaxation times $\tau_1$ or $\tau_2$). The Eq.(2.1) is valid in the case of purely dipolar I-S interaction, the $g_\parallel$- and $\gamma_\perp$-values should be interpreted as $g$- and $\gamma$-tensor components, corresponding to the directions along and transverse to the spins quantizing axis, i.e., the external field $H$.

Figs. 1.6a and f show the results of NSLR rate measurements done for $H \| a$ at $T = 1.5$ K. As one can see in Fig. 1.6f the nearly quadratic dependence of $\sqrt{1/T_1}$ on $x$ is valid at $x = 0.1$–$0.4$. If $\tau_c$ =const, for $x = 0.1$–$0.4$ one obtains $N_0 \sim x^2$. It is natural to assume that the role of acceptors in thulium NSLR is played by PCs connected with $Cu^{2+}(1)$ ions in CuO chains, because the concentration of the latter increases with $x$ increasing. However, if PC-acceptors are connected with all $Cu^{2+}(1)$ ions, including those located inside Cu-O-...-O-Cu chains, one should have $N_0 \sim x$ [13]. The $N_0 \sim x^2$-type dependence is typical for randomly distributed atoms forming pairs. In our case one can assume that paramagnetic centers responsible for thulium NSLR are localized near $Cu^{2+}(1)$ ion pairs at the chain ends. This assumption allows us to explain the most interesting peculiarity of Fig. 1.6f, namely, the rate peak taking place at the tetragonal and orthorhombic phase boundary ($x = 0.4$): at $x < 0.4$ only short chain segments are formed, and at $x > 0.4$ they are joined in long segments, so that the "terminal" $Cu^{2+}(1)$ ions and their pairs' concentration drops sharply (the compound becoming superconducting).

In order to confirm the validity of $\tau_c$ =const condition, let us first obtain the $\tau_c/(1 + \omega^2\tau_c^2)$ value using the $T_1$ time measured for the $x = 0.4$ sample. For the further analysis it is important to estimate the CuO mean chain length $n_{0.4}$ for our sample with $T_c = 23$ K [4] correctly. Superconductivity can take place only when the $CuO_2$ planes are doped with holes. This only can be provided by CuO chain segments not shorter than $n = 4$ (here n is the oxygen atoms number) [14]. However the superconducting fraction in samples with $x = 0.4$ is small (see Fig. 1.6h). Therefore it seems reasonable to assume $n_{0.4} = 2$, also because this value (see Fig. 1.6g) was obtained from Cu(1) NQR measurements made for a series of ceramic GdBaCuO samples [15], which were prepared analogous to the TmBCO samples. For $n = 2$ two of three Cu(1) atoms are "terminal", the relative amount of them in $x = 0.4$ sample is equal to $(2/3) * 0.4 = 0.27$, and consequently the concentration of pairs of such atoms has to be equal to $(0.27)^2$ of the total Cu(1) concentration, i.e., $N_0 = 4.1 \cdot 10^{20}\mathrm{cm}^{-3}$. Substituting this value as well as $\gamma_\perp^2 = (\gamma_b^2 + \gamma_a^2)/2$, $g_\| = 2$ and the measured $T_1 = 1.3$ ms into Eq.(2.1), we obtain the equation $\tau_c/(1+\omega^2\tau_c^2) = 1.1 \cdot 10^{-9}$ s. For the frequency $\omega/2\pi = 48$ MHz the roots are $\tau_c = 8.7 \cdot 10^{-9}$ s and $1.3 \cdot 10^{-9}$ s. Another estimation of the $\tau_c$ value is provided by the low-temperature $T_1$ measurements made for the sample with $x = 0.8$ in the field $H \| c$ (Fig. 2.1). Comparing Eq.(2.1) with the measured ratio $[T_1(39\ \mathrm{MHz})/T_1(24\ \mathrm{MHz})]_{T=4.2\ \mathrm{K}} = 2$ we obtain $\tau_c = 8.9 \cdot 10^{-9}$ s. This coincide with one of the roots given above.

Since $\tau_c$ is constant, one can use the NSLR rates (Fig. 1.6f) to estimate the mean chain length for the SC samples. Assuming that the measured $(1/T_1)^{1/4}$ values are proportional to the amount $p$ of $Cu^{2+}(1)$ terminal ions. If $q$ is the amount of internal Cu(1) atoms in the CuO chains, then the length of the segments is equal to $n = 1 + 2q/p$ [15]. Assuming again the mean chain length for the $x = 0.4$ sample is equal to $n_{0.4} = 2$, in this case $(p+q)/p = 3/2$; then for samples with $x > 0.4$ the chain length can be calculated as follows:

$$n_x = 1 + 2\left[\frac{3}{2}\frac{c(x)}{c(0.4)}\left(\frac{T_1(x)}{T_1(0.4)}\right)^{\frac{1}{4}} - 1\right], \tag{2.2}$$

where $c(x)$ is the relative amount of divalent copper ions in $CuO_2$ planes. Fig. 1.6g shows the mean chain mean length for superconducting TmBCO samples calculated by using Eq.(2.2) with the help of $c(x)$-values measured for YBaCuO [13]. A good agreement of $n_x$ values for TmBCO and GdBaCuO in the $x = 0.4$–$0.75$ region can be considered as an indirect confirmation of the correctness of the assumptions underlying our analysis. Let us notice that if $\tau_c =$const one can estimate also the concentration of the Cu paramagnetic centers for the samples studied using the simple relation: $N_0(x) \approx N_0(0.4)[T_1(0.4)/T_1(x)]^{1/2}$. Taking again $N_0(0.4) = 4.1 \cdot 10^{20}\ \mathrm{cm}^{-3}$ we find, in particular for the overdoped sample ($x = 1.0$), $N_0 = 2.4 \cdot 10^{19}\ \mathrm{cm}^{-3}$. This value agrees well with the data obtained for the most pure oxygenrich $YBa_2Cu_3O_7$ samples by other authors [16],[17].

## 3. Localized $Cu^{2+}$ centers

Let us return to the discussion of the correlation time $\tau_c$. As estimations show it does not depend on $x$ and $N_0$ and is approximately equal to $10^{-8}$ s. Its origin is not clear for us at present. It is possibly determined by the interaction energy $E_p$ of terminal $Cu^{2+}(1)$ ions coupled along the c-axis. Assuming that Cu(1) moments are coupled by a purely dipolar interaction we can use the relation $\tau \sim \hbar/E_p = \hbar r^3/(S(S+1)g^2\mu_B^2)$ in order to estimate the distance between $Cu^{2+}(1)$ spins in pairs: if $\tau = 8.9 \cdot 10^{-9}$ s, $S = 1/2$ and $g = 2$, then $r \sim 13$ Å, i.e., the distance obtained is very close to the c lattice parameter. It should be emphasized here that the results obtained at ultra-low temperatures can be fitted by Eq.(2.1) only if we assume that PC acceptors are influenced by very strong (of the order of few kOe) inhomogeneous internal magnetic fields. Assuming that the local field distribution is Gaussian with a dispersion $\Delta$ we fit the NSLR rate temperature dependence (Fig. 2.1) by the following formula:

$$\frac{1}{T_1} = \frac{B}{\Delta} \int_{-25\ \mathrm{kOe}}^{25\ \mathrm{kOe}} \exp\left(-\frac{H^2}{2\Delta^2}\right) \left[1 - \tanh^2\left(0.0806\frac{H_0 + H}{T}\right)\right] dH\,, \quad (3.1)$$

where $H_0$ is the external field. As one can see in Fig. 2.1, the $\Delta$ values are taken in kOe. It is difficult to expect that so strong fields (definitely non-dipolar in origin) can be attributed to the chain $Cu^{2+}(1)$ ions, because for them dipole fields of $\sim$ 400 Oe are typical which explain fairly well the copper EPR linewidth observed at liquid helium temperatures [18]. On the contrary, strong local fields are typical for $CuO_2$ planes and are seen by RE probes [8],[19]. So all the facts and estimations indicate that in superconducting samples paramagnetic copper centers, responsible for thulium NSLR at low temperatures, are located on the boundaries between superconducting and non-superconducting (oxygen-deficient) microdomains. It is known that $Cu^{2+}(1)$ moments are ferromagnetically coupled with $Cu^{2+}(2)$ moments in the adjacent $CuO_2$ planes, and in oxygen-deficient samples ($0.2 < x < 0.4$) this leads to a formation of type-II magnetic structure with doubled lattice parameter 2c [13]. In a regular type-II structure the directions of $Cu^{2+}(1)$ spins, as of $Cu^{2+}(2)$ spins become frozen at liquid helium temperatures, the mean Cu(1) moment being much less than that for Cu(2) ($0.65\mu_B$). However in our case the motion of the terminal $Cu^{2+}(1)$ spins coupled in pairs along the c axis is probably not frozen even at ultra-low temperatures, similar to flip-flop processes in paramagnetic systems. If it is so, then this motion can entail the reorientation of Cu(2) spins, coupled with Cu(1), and thus produce the low-frequency ($\sim 10^8$ s$^{-1}$) fluctuations of strong local fields of Cu(2) at the RE sites.

One can realize that the reorientation of Cu(2) spins due to the ferromagnetic coupling with Cu(1) terminating spins results in the appearance of five-spins ferromagnetic clusters (ferrons) [20] in $CuO_2$ planes. It was suggested earlier [21] that antiferromagnetically coupled pairs of ferrons localized in adjacent $CuO_2$ planes near RE impurity ions in the $YBa_2Cu_3O_{6.85}$ compound can be responsible for the local field $\mathbf{h} \| [001]$ at RE sites ($\sim$ 160 Oe) observed at low temperatures in the NSC fraction of the sample. A RE impurity ion located between $CuO_2$ planes leads to the appearance of defects (deformations) symmetrical in both planes. Therefore it was reasonable to expect that in the NSC material weakly doped with oxygen holes the localization of ferrons

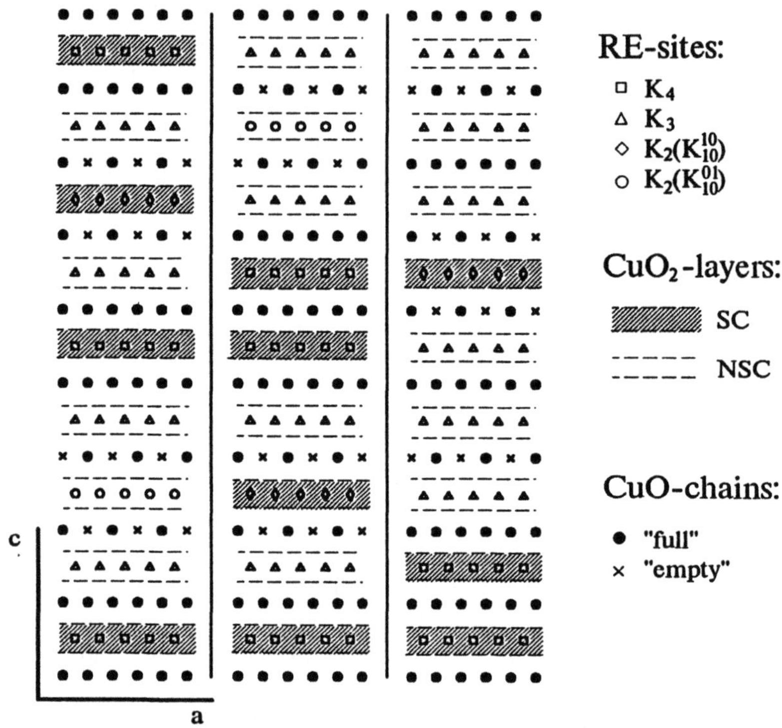


**Fig. 3.1.** The plausible structural fragments of $RBa_2Cu_3O_{6.8}$.

has to take place at such symmetrical defects. As for the TmBCO underdoped compounds which do not contain artificial defects like RE impurity ions, the ferrons are reasonably expected to localize near natural defects, i.e., oxygen vacancies in CuO chains at the boundaries between SC and NSC microdomains.

In any case $Cu^{2+}(2)$ centers localized at the SC domain boundaries should be considered as true acceptors of the energy absorbed by Tm nuclei at resonance. It is worth to note that $T_1$ times for Tm nuclei in TmBCO measured at ultra-low temperatures are very sensitive to the sample's state. At the end of a five-months experiments' period the $T_1$ value for $x = 1.0$ sample at $T = 1.5$ K appeared to be almost two times less than at the beginning, and this corresponds to the CuO chain shortening from 26b to 22b during five months. At the same time the diamagnetic susceptibility measured at liquid helium temperatures decreased only by $\sim 2\%$, i.e., from 0.69 to 0.67 (see also insert in Fig. 1.2). These changes definitely give evidence that the overdoped 1-2-3-7 state is unstable.

## 4. Phase separation in underdoped $TmBa_2Cu_3O_{6+x}$

Let us collect together some facts that can be deduced from the Tm NMR experiments and the data obtained earlier by other methods in investigating underdoped 1-2-3 ceramics.

- The dependence of the Meissner fraction (MF) of YBaCuO samples on the oxygen content has a local minimum at $x \approx 0.8$ (Fig. 1.6h).
- The "slightly disordered" fraction of TmBCO samples (Fig. 1.6b) reflecting the SC volume fraction of the same samples (Fig. 1.6e) has a minimum at $x$ around 0.7–0.8.
- The relative intensities of the $K_3$-lines in the $Gd^{3+}$ EPR spectrum of (Y,Gd)BaCuO [6] and the $^{169}Tm^{3+}$ NMR spectrum of TmBCO (Fig. 1.4, see number in brackets) have a maximum at $x \approx 0.75$.
- The intensity of the $Yb^{3+}$ EPR spectrum observed at $T < T_c$ due to the NSC fraction of (Y,Yb)BaCuO samples has a maximum at $x \approx 0.7$ [21].
- The value of 34 $\mu$s for $T_{2G}(K_3)$ time of Tm nuclei in $TmBa_2Cu_3O_{6.8}$ measured at $T = 1.5$ K in the field $H \perp c$ corresponding to the $K_3$-line (intermediate between $2\pi\nu/\gamma_b(K4)$ and $2\pi\nu/\gamma_b(K2)$, see Fig. 1.3a) is only slightly larger than the time $T_{2G}(K4) = 30\mu s$ measured for K4-centers in $TmBa_2Cu_3O_{7.0}$in the field $H = 2\pi\nu/\gamma_b(K4)$ (see Fig. 1.8a).

Taken together, these facts can be summarized in the following way: (i) the $K_3$-centers belong to the regular crystal structure that is, however, disordered to a larger extent as compared to the OI structure; (ii) the volume fraction of the SC sample occupied by the $K_3$-centers is not superconducting. Having in mind that the mean length of CuO chain segments changes monotonically in the region of $x \sim 0.8$ (Fig. 1.6g), one can assume the NSC volume fraction occupied by the $K_3$-centers to be formed by pairs of $CuO_2$ basal planes of OI and OII structures, as is shown in Fig. 3.1. This assumption is supported by the fact that there is only a partial correlation between oxygen distributions in adjacent $CuO_2$ basal planes [22].
A reasonable question arises then: why is it necessary for the superconductivity that a pair of $CuO_2$ layers lie in between identical $CuO_2$ layers (of OI- or OII-type), but not in between different ones? The answer could be simple: in the samples with an oxygen content corresponding to the step between 90 K- and 60 K-plateaus the ordered regions in "filled" and "half-filled" $CuO_2$ planes are too small, i.e., the "structural coherence length" [23] is smaller than SC coherence length ($\xi_{ab} \approx 15$ Å ). It is known, for example, from ultra-high resolution electron microscopy experiments [22] that in the YBaCuO sample with $x = 0.4$ corresponding to the edge of the 60K-plateau the oxygen tends to become ordered in clusters of very small size (from 8 to 30 Å). However, another explanation is possible too: even if the oxygen ordering regions are big enough, a full identity of doping $CuO_2$ layers, i.e., their mirror symmetry with respect to RE ion plane, is necessary for high-$T_c$ superconductivity to

appear. If this idea is valid, a half of OII-OII "sandwiches" containing $K_2$ centers of $K_{10}^{10}-$ type [6] has to be superconducting, and another half with $K_2$ centers of $K_{10}^{01}$-type-non-superconducting, as is shown in Fig. 3.1. The SC volume fraction can be determined then by $K_4$- and $K_2$- resonance lines with relative intensities as follows:

$$\text{Meissner fraction} = I(K_4) + \frac{1}{2}I(K_2)\,. \tag{4.1}$$

The open symbols in Fig. 1.6h show the results of the most detailed (as one can judge this according to available literature) measurements of Meissner effect in YBaCuO ceramics [3], and the asterisks correspond to the sum Eq.(4.1) of impurity $Gd^{3+}$ EPR lines intensities for the same compounds [6]. As one can see, in the range of the 60 K-plateau ($0.5 < x < 0.8$) the agreement of the data, obtained by different methods, is excellent. This allows us to conclude that the hypothesis about the necessity of the translation symmetry along the c-axis for doping $CuO_2$ layers is not meaningless and is worth to be checked seriously.

Discussing about ordered and "slightly disordered" crystal structures one can actually speak only about "sandwiches" made up of small areas of ordered $CuO_2$ layers of different types: "filled" (OI) and "half-filled" (OII). The Tm NSLR data make it possible to roughly estimate the mean size of SC "sandwiches" in the b-direction: in our TmBCO samples it varies from 6b (or 23 Å) to 13b (50 Å) at $x$ values ranging from 0.5 to 0.8. Thus, the results of our experiments support the idea [24],[25], that the superconductivity of underdoped 1-2-3 compounds seems to be provided by a percolative network of mesoscopic platelike OI and OII domains. The walls of this percolative cluster are enriched by $Cu^{2+}(2)$ paramagnetic centers, which are characterized by a very slow ($\approx 10^8$ s$^{-1}$) spin dynamics. Since the size of SC domains in underdoped 1-2-3 compounds is rather small, the $Cu^{2+}(2)$ paramagnetic centers may have a strong influence on the magnetic relaxation of all kinds of nuclear probes situated inside a SC percolative cluster.

*Acknowledgement.* The authors gratefully acknowledge H. Lütgemeier, K.A. Müller, A. Furrer, J. Mesot, M. Mali, D. Brinkmann, J. Roos, V.F. Gantmakher, H. Brom, M.V. Eremin, I.F. Schegolev and H. Alloul for the discussion of experimental results and for fruitful critical remarks. This work was done in the framework of Russian State Program "High-temperature superconductivity" (the Project No 91001). Five of us (O.N.B., A.V.D., A.V.E., M.A.T., A.G.V.) are grateful to the American Physical Society for a financial support of the work.

## References

1. M. Buchgeister, P. Herzog, S.M. Hosseini, K. Kopitzki, and D. Wagener, Physica C 178 (1990) 105.

2. O.N. Bakharev, A.V. Dooglav, A.V. Egorov, V.V. Naletov, M.P. Rodionova, M.S. Tagirov, and M.A. Teplov, Appl.Magn.Reson. 2 (1991) 559.
3. T. Graf, G. Triscone, and J. Muller, J.Less-Common Met. 159 (1990) 349.
4. A.V. Egorov, H. Lütgemeier, D. Wagener, A.V. Dooglav, and M.A. Teplov, Sol.St.Comm. 83 (1992) 111.
5. O.N. Bakharev, A.V. Dooglav, A.V. Egorov, H. Lütgemeier, M.P. Rodionova, M.A. Teplov, A.G. Volodin, and D. Wagener, Appl.Magn.Reson. 3 (1992) 613.
6. S. Pekker, A. Janossy, and A. Rockenbauer, Physica C 181 (1991) 11.
7. L.K. Aminov and M.A. Teplov, Sov.Phys.Usp. 28 (1985) 762.
8. J.A. Hodges, P. Bonville, P. Imbert, G. Jehanno, and P. Debray, Physica C 184 (1991) 270.
9. H. Brom and H. Alloul, Physica C 185-189 (1992) 1789.
10. P. Carretta and M. Corti, Phys.Rev.Letters 68 (1992) 1236.
11. J.R. Klauder and P.W. Anderson, Phys.Rev. 125 (1962) 912.
12. M.R. McHenry, B.G. Silbernagel and J.H. Wernick, Phys.Rev.B 5 (1972) 2958.
13. J.M. Tranquada, in: Earlier and Recent Aspects of Superconductivity, eds. J.G. Bednorz and K.A. Müller (Springer-Verlag Berlin, Heidelberg, 1990), pp.422-440.
14. G.V. Uimin, V.F. Gantmakher, A.M. Neminsky, L.A. Novomlinsky, D.V. Shovkin, and P.Brül, Physica C 192 (1992) 481.
15. I. Heinmaa, H. Lütgemeier, S. Pekker, G. Krabbes, and M. Buchgeister, Appl.Magn.Resonance 3 (1992) 689.
16. W.C. Lee, K. Sun, L.L. Miller, D.C. Johnston, R.A. Klemm, S. Kim, R.A. Fisher, and N.E. Phillips, Phys.Rev.B 43 (1991) 463.
17. M. Lee, N. Tea, Y.-Q. Song, W.P. Halperin, J. Thiel, M.A. Kennard, K.R. Poeppelmeier, M.M. Fang, and U. Welp, Physica C 201 (1992) 95.
18. J. Genossar, D. Shaltiel, V. Zevin, A. Grayevsky, and B. Fisher, J.Phys.: Condensed Matter 1 (1989) 9471.
19. M. Guillaume, U. Staub, F. Fauth, J. Mesot, and A. Furrer, $Ho^{3+}$ dimer excitations and $Cu^{2+}$ spin fluctuations in $Ho_{0.1}Y_{0.9}Ba_2Cu_3O_x (6.6 < x < 7.0)$. Preprint, 1993.
20. V. Hizhnyakov and E. Sigmund, Physica C 156 (1988) 655.
21. I.N. Kurkin, I.Kh. Salikhov, L.L. Sedov, M.A. Teplov, and R.Sh. Zhdanov, JETP 76 (1993) 657.
22. S. Horiuchi, Jpn.J.Appl.Phys. 31 (1992) L1335; S. Horiuchi, Y. Matsui, and B. Okai, Jpn.J.Appl.Phys. 31 (1992) L59.
23. J.D. Jorgensen, D.G. Hinks, P.G. Radaelli, S. Pei, P. Lightfoot, B. Dabrowski, C.U. Segre, and B.A. Hunter, Physica C 185-189 (1991) 184.
24. J. Mesot, P. Allenspach, U. Staub, A. Furrer and H. Mutka, Phys.Rev.Letters 70 (1993) 865.
25. M. Iliev, C. Thomsen, V. Nadjiev, and M. Cardona. Resonant Raman scattering of oxygen-deficient $YBa_2Cu_3O_{7-\delta}$: Evidence for the co-existence of OI, OII and T microstructures, Preprint, High-$T_c$ Update 7, No.6 (1993) 8.

# The Coupling of a Wigner Polaronic Charge Density Wave with a Fermi Liquid Arising from the Instability of a Wigner Polaron Crystal: A Possible Pairing Mechanism in High $T_c$ Superconductors

A. Bianconi and M. Missori

University of Roma, Department of Physics, P. A. Moro 2, 00185 Roma, Italy

**Summary.** We report evidence that the metallic phase in high $T_c$ cuprate superconductors results from an instability of the quasi 2D-electron gas at densities $\delta_{tot}$ lower or larger than the critical density $\delta_c$ for the formation of a Wigner polaronic crystal. At $\delta_{tot} \neq \delta_c$ the electron gas undergoes a microscopic phase separation between 1) a polaron gas with density $\delta_l$ at the *Wigner localization limit* and 2) a *Fermi liquid confined in a superlattice of quantum stripes* with density $\delta_i$.

The 2D-polaron gas condensed into a paired unidimensional Wigner charge density wave (CDW) at $T < T^* \sim 1.4T_c$. Below the superconducting critical temperature the two types of carriers are coupled, and the pairing mechanism involves virtual polaron pairs. This mechanism drives the condensate to the maximum critical temperature for a Fermi liquid. In addition the pairing model proposed here correctly predicts the dependence of $T_c$ onto the charge carrier concentration (Uemura plot), and the variation of the isotope effect from the weak coupling limit $\alpha \sim 0.5$ where at low doping $\delta_i < \delta_l$ to $\alpha \sim 0.1$ where $\delta_i > \delta_l$.

## 1. Introduction

Starting from the recent experimental determination [1] of the polaron area $S_p$ we present a new model a) for the quantum state of the electron gas in the cuprate superconductors and b) for the microscopic pairing mechanism. The quantum state of the electron gas in the metallic phase is described by a microscopic phase separation between a Fermi liquid and a polaron gas close to the Wigner localization limit. This phase separation results from the instability of the Wigner polaron gas for densities close but different from the critical density $\delta_c$ for a generalized Wigner polaronic crystal. The pairing mechanism results from the coupling between the two types of charge carriers.

We want to show here that the present model allows us to understand some key phenomenological points of the physics of cuprate superconductors that up to now remained obscure:

1. the insulator to metal transition at finite doping [2];
2. the electron lattice instability at the critical doping 1/8 [3];
3. the anomalous normal metallic phase;
4. the opening of the spin gap at the temperature $T^* > T_c$ [4];

5. the variation of the exponent $\alpha$ of the isotope effect from the weak coupling value $\alpha \sim 0.5$ at low doping and $\alpha < 0.1$ at optimum doping [5];
6. the relation of $T_c$ versus $n_s/m^\star$, known as Uemura plot [6], where $n_s$ is the density of superconducting condensate, and $m^\star$ the effective electron mass.

As shown in some recent works [1],[7]–[9] the metallic phase of the electron gas in the cuprate superconductors results from a microscopic phase separation between a polaron gas and a Fermi liquid. By doping the antiferromagnetic insulator with a total number of holes per Cu site $\delta_{tot}$ two types of holes are formed: a polaron gas with average hole doping $\langle\delta_l\rangle$ and a doped spin fluid forming a Fermi liquid with an average hole doping $\langle\delta_i\rangle$.

The phases of the quantum state of the electron gas in cuprate high $T_c$ superconductors as function of doping are described by the polaron pair filling factor $\nu = 2\rho S_p$, where $\rho$ is the total surface hole density in the $CuO_2$ plane. The homogeneous phases of the polaron gas are formed at integer values of the polaron pair filling factor:

1. (a) $n = 1$ indicates the formation of a generalized Wigner [10] charge density wave (CDW) where the polarons cover half of the surface of the $CuO_2$ plane. (b) The 1/8 doping instability [3] is shown to be due to the formation of a generalized Wigner [10] polaronic crystal, where the polarons cover the full surface of the $CuO_2$ plane, i.e. $n = 2$.
2. In the insulator phase at very low doping, defined by the polaron filling factor in the range $\nu < 1$, the polaron gas is in the Wigner low density insulating phase. In fact we demonstrate that its density parameter $r_s' > 37$ [10] and the insulator to metal transition takes place at $\nu = 1$.
3. The metallic phase is assigned to the instability of the electron gas at non integer values of $\nu$. The anomalous normal metallic state is due to the coexistence of a polaron gas condensed into a polaronic generalized Wigner charge density wave, where the local doping is $\delta_l = 1/8$ and a Fermi liquid with local doping $\delta_i < 1/8$ or $\delta_i > 1/8$ at the metallic low or optimum doping regime respectively.

   *The nature of the microscopic pairing mechanism at $T_c$ is revealed by the following experimental results:*
4. The opening of the spin gap [5] at the temperature $T^\star > T_c$ corresponds to the critical temperature $T^\star$ for the formation of the self trapped polarons [1] and therefore it is assigned to the spin pairing of polarons.
5. The superconductivity shows two doping regimes defined by the relation $\delta_i < \delta_l$ and $\delta_i > \delta_l$ where the exponent of the isotope effect is $\alpha \sim 0.5$ and $\alpha \sim 0.1$ respectively.
6. The superconducting condensate with density $n_s$ is formed by *a portion of the Fermi liquid coupled with the polaron pairs*, where the polaron pairs play the role of virtual excitations. On the other hand the condensate can be also described as *polaron pairs which decay into superconducting*

*electron pairs of the Fermi liquid.* Therefore the portion of the electron gas which condenses will satisfy the condition that the coherence length is the same as the interatomic distance $\xi_0 \sim 2r_s a_B$ that is valid for an electron gas close to the Wigner crystallization [11],[12], and where $a_B$ is the Bohr radius and $r_s$ is a measure of the electron density. This condition is close to the condition giving the maximum critical temperature for a Fermi liquid $k_F\xi_0 \sim 2\pi$ [13],[14]. We show that this pairing mechanism correctly predicts the experimental dependence of $T_c$ versus $n_s/m^\star$.

## 2. The Polaron Size and its Temperature Dependence

There are many experimental data on cuprate superconductors showing that the holes (or electrons) in the $CuO_2$ plane induced by doping can form 1) a polaronic charge [15]–[19] and 2) itinerant charge carriers forming a Fermi liquid both by doping the half filled band that gives the antiferromagnetic phase at half filling [20]–[23].

While the presence of polaronic charges is well established in the insulating phase at very low doping regime, the presence of itinerant charges forming a correlated Fermi liquid with a large Fermi surface is well established in the optimum doping regime.

To solve the puzzle of the quantum state of high $T_c$ superconductivity we have performed two key experiments to determine 1) the size of the polaron and 2) the ordering of the polaron gas in the metallic phase. In these experiments we have measured 1) the polaron size in a $Bi_2Sr_2CaCu_2O_{8+y}$ (Bi 2212) crystal at optimum doping (the total hole doping was about 0.18 holes per Cu site in our crystal) by EXAFS and 2) the ordering of the polaronic impurity states by diffraction.

We have focused our interest to $Bi_2Sr_2CaCu_2O_{8+y}$, because it is the ideal case to study the non homogeneous $CuO_2$ plane. In fact in this system the acceptors out of plane (additional oxygens in the BiO plane) are ordered along lines in the a-axis direction in the metallic phase. On the contrary, the ordering of the acceptors takes place in small domains in most of the families of cuprate superconductors and it can not be observed with most of the available experimental methods.

The polaronic domain that we have found is shown in Fig. 2.1. It can be described by a configurational coordinate, the Cu-O(apical) distance which becomes shorter, mainly due to the displacement of the Cu ion out of plane, and the tilting of the $CuO_4$ square planes. The $CuO_2$ plane has been shown to be decorated by stripes of polarons associated to holes trapped by the acceptors in linear arrays [7]–[9] at optimum doping. Two Cu-O (apical) distances, 2.40±0.02 and 2.57±0.02 Å below 120 K, have been found by polarized EXAFS, in agreement with diffraction works [24] confirming the presence of domains with different Cu site configurations in the $CuO_2$ plane. We have measured the width $W$ of the polaronic stripes by measuring the

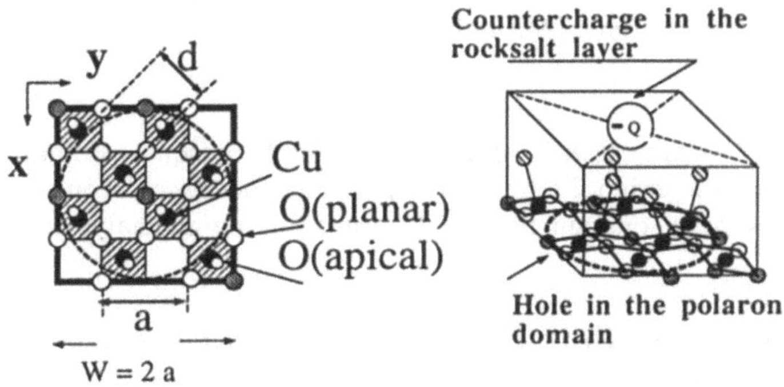


**Fig. 2.1.** Pictorial view of a single polaronic state where the hole L(a) in the $CuO_2$ plane with mixed $b_1$, $a_1$ and e symmetry trapped in polaronic domain of width $W = 2a$ including 8 Cu sites. The hole is trapped in a polaronic impurity state by the negative counter-ions on the BiO layer.

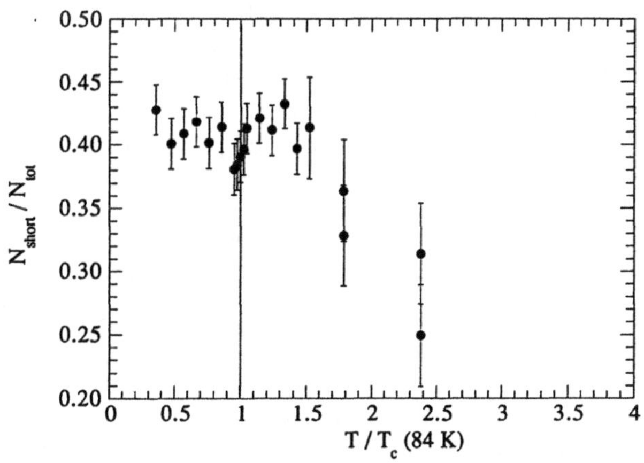


**Fig. 2.2.** The measured values of the effective number of short Cu-O (apical) bonds $N_{short}/N_{tot}$ as function of temperature in a monodomain crystal Bi2212 and the values of the width of the polaron stripe $W/a = (W/\lambda_p)\ \lambda_p/a = (N_{short}/N_{tot})\lambda_p/a$. Where $\lambda_p/a$ is 4.65 Åas measured by electron diffraction. The width $W$ at $T < T^* =$ 110 K$\sim 1.3T_c$ is $W = 10.54 \pm 0.6$ Å.

ratio between the number $N_{short}$ of short Cu-O (apical) bonds and the total number $N_{tot}$ of the Cu-O (apical) bonds in a monodomain oriented crystal by polarized EXAFS.

In Fig. 2.2 we have obtained $W$ by the measured values of $N_{short}/N_{tot}$ and by measuring the superstructure period $\lambda_p$ by electron diffraction

$$W/a = (W/\lambda_p)\lambda_p/a = (N_{short}/N_{tot})\lambda_p/a\,.$$

The width $W$ is about $a = 5.4$ Å at room temperature in agreement with neutron diffraction data [24] and it becomes larger at $T < T^{\star} = 120$ K $\sim 1.4T_c$.

$$W = 10.54 \pm 0.6\mathring{A}$$

Therefore $W \sim 2a = 10.8\mathring{A}$, which is the side of the square domain shown in Fig. 2.1, and we assign $T^{\star}$ to the temperature for the polaron self trapping; in fact $W \sim 2a$ is constant below $T^{\star}$ except for the anomaly around $T_c$.

## 3. The Polaron Pair Filling Factor and the Insulator to Metal Transition

The structure of the polaronic domain shown in Fig. 2.1 is characterized by $CuO_4$ square planes with a LTT-type distortion i.e. similar to the low temperature tetragonal phase (LTT) observed in $La_{1.85}Ba_{0.125}CuO_4$ [3] and a displacement of the Cu ion out of plane, giving a short Cu-O (apical) bond. These two sets of distortions of the $CuO_5$ pyramid introduce a mixing of the molecular orbitals of $b_1$ symmetry ($3d_{x^2-y^2}$ O (planar) $2p_{x,y}$), with orbitals of $a_1$ (Cu $3d_{3z^2-r^2}$ O (planar) $2p_{x,y}$, O (apical) $2p_z$) and orbitals with e (Cu $3d_{xz}$ $3d_{yz}$ O (planar) $2p_z$) symmetry forming new molecular orbitals with larger components of the charge density out of plane giving the polaronic hole that we denote as $\underline{L}$(a). For these reasons this polaron was called a $3d_{3z^2-r^2}$ polaron [7].

The polaron domain, shown in Fig. 2.1 and Fig. 3.1a, is extended over an even number of Cu sites, which confirms that it is associated with a hole of symmetry different from the holes of $b_1$ symmetry forming the antiferromagnetic lattice in the parent compound (see Fig. 3.1a). A polaron associated with a ligand hole $\underline{L}$(b1) of same symmetry as the holes of $b_1$ symmetry forming the antiferromagnetic lattice in the parent compound is expected to be centered on a Cu ion and extended over an odd number of Cu sites. See Fig. 3.1b.

The polaron mass $\pm$ can be estimated from the polaron radius to be about $16m_0$. The polaron area is found to be $S_p = 4a^2$ (where $a = 5.4$ Å). Usually the doping in the cuprates is measured as the number of holes per Cu site $\delta_{tot}$. Therefore the surface density is given by $\rho = \delta_{tot}/d^2$, where $d$ is the Cu-Cu distance, about 3.8 Å, and therefore $\rho = 2\delta/a^2$ with $a = d\sqrt{2}$.

Having determined the polaron area we can now measure the hole density by the polaron pair filling factor which is given by

$$\nu = 2\rho S_p = 16\delta_{tot} \, .$$

We have reported in Fig. 3.2 the critical temperature as a function of the polaron pair filling factor [25]–[27] of the $Bi_2Sr_2Ca_{1-x}Y_xCu_2O_{8+y}$ system, and it is compared with the similar curves for other superconducting families.

We clearly show that 1) the system is an insulator for $\nu < 1$, the metal insulator transition takes place at the critical integer number $\nu = 1$ which

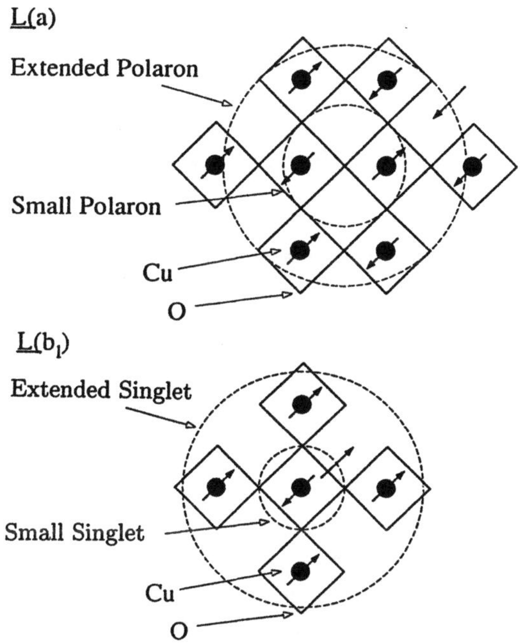


**Fig. 3.1.** Pictorial view (a) of extended and small Zhang Rice singlets $3d^9L(b_1)$ on an odd number of Cu sites and (b) of the polaron found in this work extended over an even number of Cu sites associated with L(a) holes in orbitals with mixed symmetry.

corresponds to the critical density for percolation in two dimensions, suppressing the long range antiferromagnetic order in the undistorted lattice outside the polaron domain walls, and forming a spin liquid.

At the critical value of $\nu = 1$, where the polaron gas covers half of the plane surface, the ordering of the polarons driven by the short range attraction and long range Coulomb repulsion is expected. If the Coulomb repulsion overcomes the kinetic energy at this critical density a generalized Wigner charge density wave could appear as will be shown below. The plots of $T_c$ versus the filling factor for all superconducting families show that the insulator to metal transition occur always at $\nu = 1$. This result clearly supports our hypothesis that the size of the polaron is a characteristic property of the $CuO_2$ plane and therefore it is the same in all families of cuprate superconductors.

## 4. Ordering of Polarons as Function of Doping

The ordering of the polarons along linear arrays in the metallic phase of Bi2212 has been already discussed in the first workshop on phase separation [9]. The polarons form parallel stripes resulting in a unidimensional

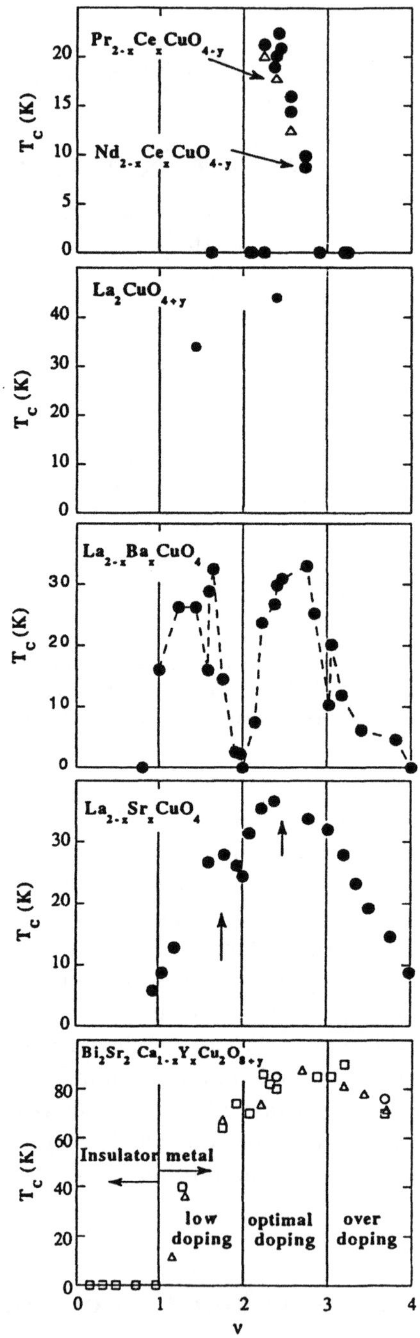


**Fig. 3.2.** $T_c$ versus the polaron pair filling factor for $Bi_2Sr_2Ca_{1-x}Y_xCu_2O_{8+y}$ system as measured by Quitmann et al. (triangles), Groen et al. (squares) and by Pettiti et al. (circles). $T_c$ versus $\nu$ for hole doped LaSrCuO, LaBaCuO, $La_2CuO_{4+y}$ and electron doped systems.

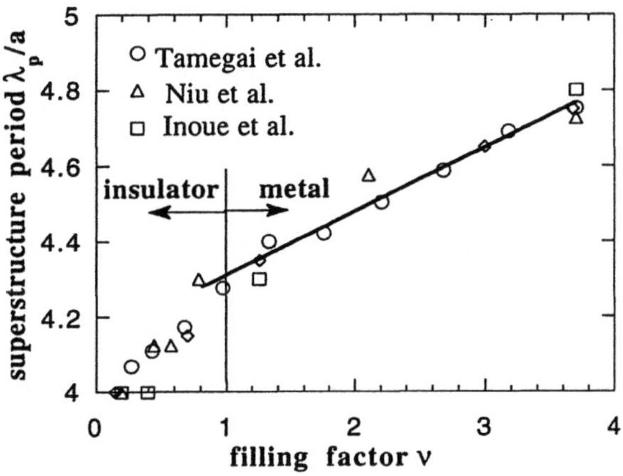


**Fig. 4.1.** The superstructure period $\lambda_p$ of $Bi_2Sr_2Ca_{1-x}Y_xCu_2O_{8+y}$ as function of the doping measured by the polaron pair filling factor $\nu$. The diamonds are data collected by our group and the solid line is the fitting in the range of the metallic phase for $\nu > 1$.

charge density wave in the $CuO_2$ plane and its period is the same as the superstructure period of the full crystal. The incommensurate period $\lambda_p$ as function of doping has been measured by x-ray diffraction and electron diffraction in samples where the hole doping is controlled by Y doping $Bi_2Sr_2Ca_{1-x}Y_xCu_2O_{8+y}$.

The superstructure period is $4a$ in the insulating case due to the misfit of the BiO and the $CuO_2$ planes but in the metallic phase it shows a linear behaviour from $\lambda_p = 4.3a$ at the insulator to metal transition to about $\lambda_p = 4.8a$ in the over doped regime as shown in Fig. 4.1.

The variation of the incommensurate superstructure period $\lambda_p$ as function of Y content reported in Fig. 4.1 in the *metallic phase* has been measured by several groups [28]. A correlation between $T_c$ and the superstructure period has been remarked by Pham et al. [29] in different heat treatments of the Bi2212 system. To obtain $\lambda_p$ as function of $\delta_{tot}$ shown in Fig. 4.1 we have fitted the data of the Y content $x$ versus hole doping $\delta_{tot}$ reported by Groen et.al. [25], that is in good agreement with our experience on the characterization of this class of materials.

Let us assume that the polaron size that we have measured is not doping dependent. This can be justified in two ways. First, the polaron size is controlled by an intrinsic instability of the $CuO_2$ lattice for a structural transition from the LTO phase to the LTT phase where the Fermi energy is tuned at the van Howe singularity [30], which fixes the local density at 1/8 and therefore the polaron area. Second, the condition that the radius of the impurity state trapped by the acceptor should be the same as the polaron radius gives a fixed value for the polaron radius.

The experimental result that the ratio $N_{\text{short}}/N_{\text{tot}}$ is $\neq 0.5$ is a strong indication for the anharmonic electron lattice interaction giving an anharmonic CDW. The anharmonicity can be associated with the electron lattice interaction at the critical local doping 1/8. In the presence of an anharmonic electron lattice interaction of pseudo Jahn Teller type [31] the anharmonic CDW drives the system toward a microscopic phase separation between polaronic holes $\underline{\mathrm{L}}(\mathrm{a})$ condensed in linear arrays separated by a Fermi liquid formed by the holes $\underline{\mathrm{L}}(\mathrm{b}_1)$.

## 5. Partial Charge Density for the Polaron Gas and the Fermi Liquid

Here we will show that starting from these two experimental data 1) $S_{\mathrm{p}}$ and 2) the relation between $\lambda_{\mathrm{p}}$ and doping, we can solve the quantum state of the cuprate high $T_{\mathrm{c}}$ superconductors.

In fact by starting from the experimental data of $\lambda_{\mathrm{p}}(\nu)$ in Fig. 4.1 and by assuming that the polarons form stripes of width $W = 2a$ constant with doping, we can calculate the charge densities of the polaron gas and of the Fermi liquid as function of doping at $T < T^*$ shown in Fig. 5.1.

The average number of holes per Cu site $\langle\delta_{\mathrm{l}}\rangle$ forming the polaron gas and the average number of holes per Cu site $\langle\delta_{\mathrm{i}}\rangle$ forming the Fermi liquid are given by,

$$\langle\delta_{\mathrm{l}}\rangle = a^2/(2W\lambda_{\mathrm{p}}) = 1/4(\lambda_{\mathrm{p}}/a) \qquad (5.1)$$
$$\langle\delta_{\mathrm{i}}\rangle = \delta_{\mathrm{tot}} - \langle\delta_{\mathrm{l}}\rangle$$

and the local hole density $\delta_{\mathrm{i}}$ in the stripes where the Fermi liquid is confined and the local density $\delta_{\mathrm{l}}$ for the polaron gas are given by:

$$\delta_{\mathrm{l}} = \langle\delta_{\mathrm{l}}\rangle(\lambda_{\mathrm{p}}/W) = 1/8 \qquad (5.2)$$
$$\delta_{\mathrm{i}} = \langle\delta_{\mathrm{i}}\rangle((\lambda_{\mathrm{p}}/(\lambda_{\mathrm{p}} - W))$$

where $W$ and $\lambda_{\mathrm{p}}$ are given in units of $a$.

In the metallic phase the polaron gas with average density $\delta_{\mathrm{l}}$ is expected to give a small Fermi surface. The presence of a small Fermi surface of area $A = \pi k_{\mathrm{F}}'^2 = 0.048 \pm 0.008\ \text{Å}^2$, giving $k'_{\mathrm{F}} \sim 0.123\ \text{Å}^{-1} = 0.106\ (2\pi/a)$ has been measured by de Haas-van Alphen (dHvA) experiments [32] in $\mathrm{YBa_2Cu_3O_7}$ (Y123) and Tl2201 systems. The incommensurate superstructure in Bi2212 with period $\lambda_{\mathrm{p}}/a \sim 4.7$ and wavevector $q = 2\pi/\lambda_{\mathrm{p}} = 0.21\ (2\pi/a)$ at doping of about 18% is assigned here to a CDW driven by the nesting at the small Fermi surface for the average electronic structure in fact $q \sim 2k'_{\mathrm{F}}$. The small pocket is expected to be at the S point if it arises only from an intrinsic electron-lattice instability of the $\mathrm{CuO_2}$ plane. Because the molecular orbitals, contributing to these states, point out of the plane, the contribution of the atomic structure of the rocksalt layer can move this pocket at the M point where it appears in Bi2212.

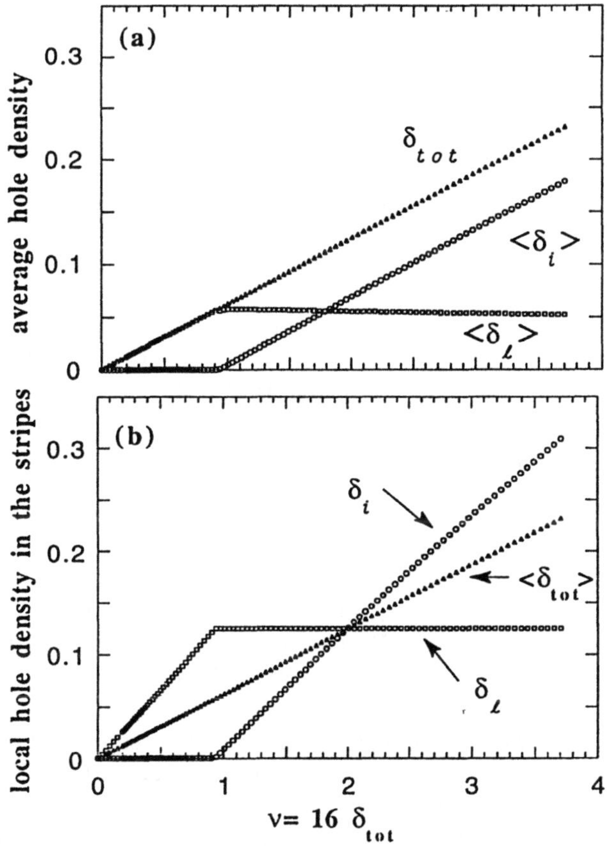


**Fig. 5.1.** Panel (a) The distribution of the total hole density $\delta_{\rm tot}$ into the average hole density $\langle\delta_{\rm i}\rangle$ of the itinerant electrons and the average hole density $\langle\delta_{\rm l}\rangle$ of the polaronic charges. Panel (b) The local hole density $\delta_{\rm i}$ of the itinerant electrons in the stripes of width $L$ and the local hole density $\delta_{\rm l}$ of polaronic charges.

## 6. The Polaron Gas Condenses in a Generalized Wigner CDW at $\nu = 1$ and in a Generalized Wigner Crystal at $\nu = 2$

The experimental research on the instabilities of the electron gas close to condense into a Wigner crystal [10] at a critical density of charge carriers $\rho < \rho_{\rm c}$ has been at the basis of several unexpected phenomena in solid state physics in these last years. The formation of a Wigner crystal has been reported in the case of extreme low carrier concentration of the two dimensional (2D) electron gas on the surface of liquid helium [33]. The Wigner localization limit for a 2D-electron gas has been predicted at values of the electron density parameter $r_{\rm s} > 37$ where $\rho = 1/\pi(r_{\rm s}a_{\rm B})^2$ [10]. If an external field localizes the electrons, a 2D-electron gas can reach the Wigner localization

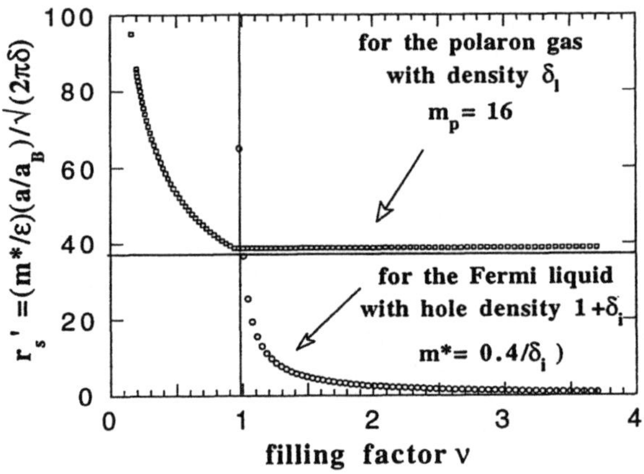


**Fig. 6.1.** The effective electron density parameter $r'_s$ for the polaron gas and for the Fermi liquid as function of doping.

limit at an electron density much higher than the expected $\rho_c$ for a free electron gas. In the quantum Hall effect [34],[35] the external perpendicular magnetic field confines the electrons to the magnetic length $l_c = (\hbar/eB)^{1/2}$ i.e. within a surface $S = p\hbar/eB$. The quantized anomalies in the Hall coefficient are observed at critical integer or rational numbers of filling factors $\nu = 2S/(\pi(r_s a_B)^2) = 2\rho S = \rho h/eB$.

Here we show that in the high $T_c$ cuprate superconductors the polaronic impurity state, trapping the holes within the polaron area $S_p$, acts as the external force to drive this 2D-polaron gas toward the Wigner localization limit. The polaron gas will reach the Wigner localization limit at the critical density $\rho_c = 1/S_p$. In these conditions the Coulomb repulsion between polarons overcomes the kinetic term and the condition $r'_s = r_s \pm /\varepsilon_\infty > 37$ for Wigner localization of a 2D-electron gas is satisfied, where $r_s$ is the density parameter of the polaron gas of density $\rho_l$, $\pm$ is the polaron mass, and $\varepsilon_\infty$ the dielectric constant. The phase diagram of the polaron gas will be described as in the quantum Hall effect by the polaron pair filling factor $\nu = 2\rho S_p$ which predicts the formation of homogeneous phases of the polaron gas at integer values of $\nu$: a Wigner charge density wave (CDW) at $\nu = 1$, a Wigner crystal at $\nu = 2$, and a bipolaron crystal is expected at $\nu = 4$, which however, will be shown to be forbidden by the Coulomb repulsion.

In order to see if the polaron gas is in the Wigner localization limit we have calculated the effective polaron gas density parameter $r'_s = r_s \pm /\varepsilon_\infty$, where $\pm$ is found to be $16m_0$ and $1/\pi(r_s a_B)^2 = \rho_l = 2\delta_l/a^2$ is the polaron density shown in Fig. 5.1. The dielectric constant is $\varepsilon_\infty = 4.75$ [36]. The variation of $r'_s$ for the polaron gas drawn in Fig. 5.1 clearly shows that it is always in the Wigner localization limit $r'_s > 37$ [10].

For the Fermi liquid we assume that the effective mass is $m^* = 0.4m_0$ in absence of electron-electron correlations. Because the Fermi liquid results from doping a spin fluid with large electron-electron correlation giving an insulator at half filling, we assume that the renormalized effective mass is doping dependent following the standard expected law $m^{*\prime}/m_0 = 0.4/\delta_i$.

The resulting effective density parameter for the Fermi liquid, drawn in Fig. 5.1, shows that this liquid clearly is in the regime of an high density electron gas, except in a very narrow region close to the insulator-metal transition.

## 7. On the Pairing Mechanism

*We show here that the pairing mechanism arises from the coupling of a portion of the Fermi liquid with the Wigner polaronic CDW.* From Fig. 3.2 and Fig. 5.1 it is clear that the cuprates exhibit high $T_c$ superconductivity with a large Meissner fraction at two main critical hole (or electron) concentrations i.e. in the low doping region $1 < \nu < 2$ at $\nu \sim 1.8$ where the maximum critical temperature is $T_{c1max}$ and in the optimum doping region $2 < \nu < 3$ at $\nu \sim 2.5$–3 where the maximum critical temperature is $T_{c2max}$. These two phases can be described as resulting from the instability of the electron gas close to critical hole concentration for the polaron Wigner crystal at $\nu = 2$.

At $T_{c1max}$ the local density of the Fermi liquid $\delta_i$ in the Fermi liquid domain is smaller than the local density of the polaron gas $\delta_l$ in the polaronic domain, i.e. $\delta_l > \delta_i$; on the contrary at $T_{c2max}$ the opposite situation appears, where $\delta_l < \delta_i$.

The experimental evidence for the coupling between the Fermi liquid and the polaron gas giving superconducting pairs is shown by the fact that the optimum conditions for high $T_c$ superconductivity are realized at $T_{c1max}$ and $T_{c2max}$ where the ratio between the average density of two gas $\langle\delta_i\rangle/\langle\delta_l\rangle$ takes the integer values 1 and 2 respectively.

Let us consider the simplest case where $\langle\delta_i\rangle = \langle\delta_l\rangle$. This situation is realized for example in a single layer material ($La_2CuO_{4+y}$ with oxygen doping $y$ of about 6% giving a superconducting material at $T_{c1max}$). For each additional interstitial oxygen in the rocksalt layer about two holes are injected in the $CuO_2$ plane, one hole is trapped in the polaronic impurity state and the second hole contributes to the Fermi liquid of free carriers. It is interesting to remark that the charge neutrality of the metallic phase for the two different charge carriers can be imagined to be provided by the same oxygen atom out of the plane.

The pairing mechanism can be described as in Fig. 7.1. Let us consider a pair of electrons of the Fermi liquid moving in the $(0\pi)$ direction, i.e. they are in GM direction of the Fermi surface of Bi2212 [21], with momentum $-k$ and $+k$ as shown in Fig. 7.1.

The wavefunction of the electron moving in the $(0\pi)$ direction has not a pure $b_1$ character (at the Cu site $3d_{x^2-y^2}$) as for the electrons moving in the $(\pi\pi)$ direction. The wavefunction of the electron moving in this direction

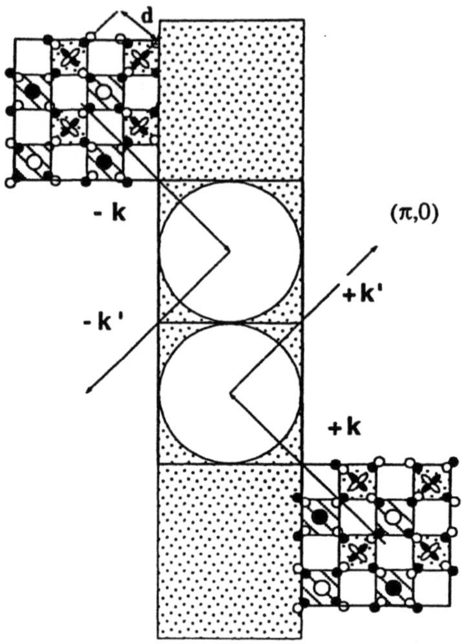


**Fig. 7.1.** Pictorial view of the pairing mechanism involving pairing of electrons in the Fermi surface moving in the $(0\pi)$ direction mediated by a pair of $3d_{3z^2-r^2}$ polarons. The polarons have condensed at $T^* > T_c$ into a Wigner unidimensional charge density wave, forming stripes indicated by the shaded region. The symmetry of the wavefunction of the electrons moving in the $(0\pi)$ direction is shown (from ref. [23]) in order to stress the point that the electrons moving in this direction has a mixed character of $b_1$ and $a_1$ local symmetry.

have a mixed $b_1$ and $a_1$ character ($3d_{3z^2-r^2}$ and $3d_{x^2-y^2}$) as calculated by Mahan [23] and it is shown in Fig. 7.1. The localized polarons have a large effective mass resulting from the lattice distortion which forms a molecular orbital with a similar mixture of $b_1$ and $a_1$ character. The mobile electron on the large Fermi surface in the GM direction is coupled below $T_c$ with the localized polaron in the small Fermi surface centered at the M point in hole doped superconductors.

The pairing between electrons in the Fermi liquid is mediated by the virtual polaron pairs condensed in a Wigner paired CDW. Therefore they will act as a negative Hubbard center for the Fermi liquid. Because the polaron gas and the Fermi liquid are coupled below $T_c$ a more exotic description of the pairing mechanism that we propose here can be given.

*A 2D-electron gas condensed into a generalized Wigner CDW formed in the region $T_c < T < T^*$ can form a superconducting condensate at $T_c$ if it results from an instability of a 2D generalized Wigner crystal for a microscopic phase separation between a Wigner CDW and a Fermi liquid.*

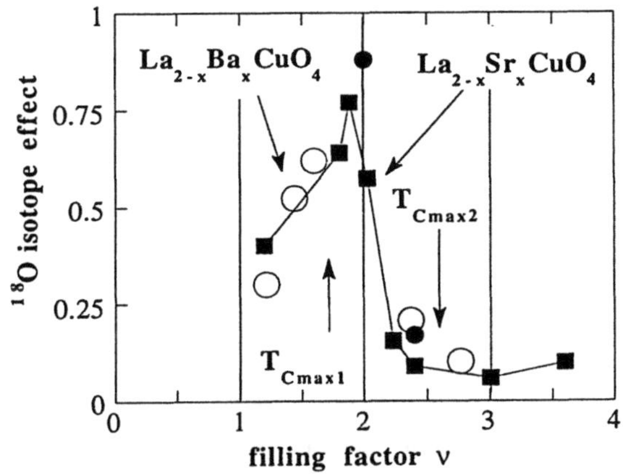


**Fig. 7.2.** The isotope exponent for LaBaCuO (open circles) and LaSrCuO (filled squares) systems from ref. [5]. In low doping regime, $1 < \nu < 2$, the two electrons of the pair remain in two different stripes (see Fig. 7.1) therefore the Coulomb repulsion is close to zero and the isotope effect is close to 0.5.

Now we want to show that this pairing model for the cuprate superconductors is supported by the experiments on the isotope effect. The isotope effect is described by the oxygen isotope exponent $\alpha$ measured by the shift in transition temperature where $T_c = \text{const } \mathrm{M}^{-\alpha}$. The experimental investigations now converge towards the fact, that the isotope exponent in most of the families of high $T_c$ superconductors is in the range 0.5–1 in low doping regime while it is very small in the high doping regime.

The common interpretation for this variation is that the isotope exponent decreases with increasing the critical temperature. The data, however, do not support this interpretation because this predicts a smooth variation of $\alpha$ with $T_c$ that is not observed. We have plotted in Fig. 7.2 the isotope exponent for LaBaCuO and LaSrCuO systems as function of the polaron pair filling factor. We show that the isotope exponent is 0.5–1 where $\delta_{\mathrm{i}} < \delta_{\mathrm{l}}$ and it is constant at about 0.1 where $\delta_{\mathrm{i}} > \delta_{\mathrm{l}}$. These data strongly support the pairing mechanism proposed here. In the low doping regime $1 < \nu < 2$, and $\delta_{\mathrm{i}} < \delta_{\mathrm{l}}$, where the superconducting quantum state with $T_{\mathrm{c1max}}$ is realized for $\langle\delta_{\mathrm{i}}\rangle/\langle\delta_{\mathrm{l}}\rangle = 1$, the effective Coulomb repulsion $m^\star$ between the pairs is negligible. Therefore in spite of the short coherence length the isotope exponent is close to 0.5 like in the weak coupling regime for standard BCS superconductors, where $m^\star$ is negligible because of the very large coherence length. In fact the Coulomb repulsion for the charges condensed into the Wigner polaronic CDW is compensated by the crystallization in the pairing mechanism described in Fig. 7.3. The mobile pairs of the Fermi liquid are located in two different stripes and therefore the Coulomb repulsion is negligible.

The sharp drop of the isotope exponent to about 0.1 takes place where $\delta_{\mathrm{i}} > \delta_{\mathrm{l}}$. In this regime $2 < \nu < 4$, where the superconducting quantum

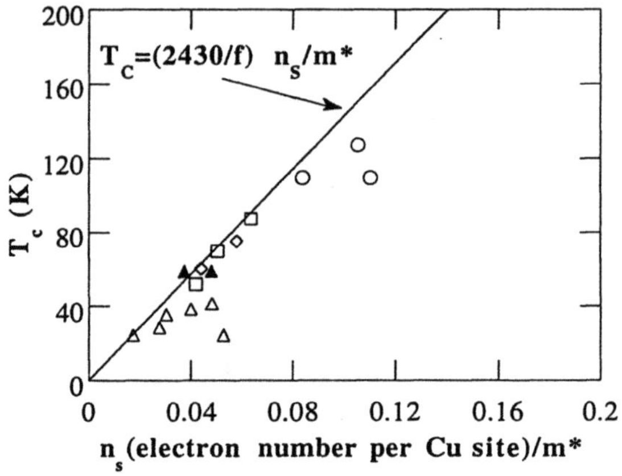


**Fig. 7.3.** The solid line is the calculated curve of $T_c$ versus $n_s/m^*$ for a condensate formed via the pairing mechanism described here $T_c(K) = 2430/f\ (n_s/m^*)$ by using $f = 1.7$, valid for Bi2212. Theoretical prediction is compared with the data of $T_c$ versus $n_s/m^*$ as measured by muon spin relaxation experiments by Uemura et al. (empty symbols) and Keller et al. (filled symbols) for different materials (triangles La124; diamonds Y123 (ortho II phase); squares Bi2212; circles Bi2223).

state with $T_{c2max}$ is realized for $\langle\delta_i\rangle/\langle\delta_l\rangle = 2$, the density of the Fermi liquid becomes high and the probability for the mobile pairs of the Fermi liquid to be located in the same stripe becomes relevant.

Now we want to show that the proposed pairing mechanism correctly predicts the Uemura plot [6].

A two dimensional (2D) Fermi liquid of fixed density $\rho$ can be brought to the maximum $T_c$ by driving the electron gas close to the Wigner localization limit, where each electron is confined inside a sphere of diameter $d = 2r_s a_B$. The Pippard coherence length of an eventual superconducting phase will be $\xi_0 = 2r_s a_B$. For a 2D Fermi liquid, $k_F = \sqrt{2\pi\rho}$, in the quantum state with $\xi_0 = 2r_s a_B$ the critical temperature will depend only on $\rho/m^*$

$$k_B T_c = \hbar^2/2m_0(\sqrt{2}e^\gamma/\pi)\rho/(f\,m^*)$$

where $f$ is a phenomenological measure of the deviation from the weak coupling limit $2\Delta_0//k_B T_c = 3.52f$, $\gamma$ is the Euler's constant 0.577 and $\sqrt{2}e^\gamma/\pi = 0.8$. If $\rho$ is measured in Å$^{-2}$, then $T_c(\mathrm{K}) = 35440.8\rho/f\,m^*$. In our model of high $T_c$ superconductivity *only the portion of the Fermi liquid coupled with the paired CDW contributes to the condensate. Therefore $T_c$ follows the formula given above where we have to insert as condensate density* $\rho_s = 2n_s/a^2$, where $n_s$ is the number of electrons per Cu site. We obtain an expression for the cuprates that can be compared with the experiments:

$$T_c(\mathrm{K}) = 2430 n_s/f\,m^*$$

$m^*$ is the effective mass for the Fermi liquid. In order to calculate the predicted $T_c$ for a particular family we have to measure the ratio $2\Delta_{0/}/k_B T_c$ to get $f$. In the Bi2212 family $2\Delta_{0/}/k_B T_c$ has been found to be around 6 giving $f = 1.7$. By using this constant we can quantitatively predict the so called "Uemura plot" of $T_c$ versus $n_s/m^*$ as measured via the muon-spin relaxation rate [6]. The predicted relation will be different for different families because it depends on the phenomenological parameter $f$, which measures the deviation from the weak coupling limit and which can also vary with doping. The comparison between the calculated curve and the data for Bi2212 is very good. The deviation from the calculated curve at high doping and for different families can be assigned to the variation of $f$.

In conclusion we have shown that the measure of the polaron size sheds a new light on many physical properties of cuprate superconductors and indicates a possible solution for the microscopic pairing mechanism in high $T_c$ superconductors.

## References


1. A. Bianconi, M. Missori, H. Oyanagi, H. Yamaguchi, D.H. Ha, S. Della Longa, to be published.
2. H. Takagi, R.J. Cava, M. Marezio, B. Batlogg, J.J. Krajewski, W.F. Peck, Jr., P. Bordet, and D.E. Cox in Proceedings of the workshop Phase Separation in Cuprate Superconductors edited by K.A. Müller, World Scientific Pub., Singapore, 1992
3. Y. Maeno, N. Kakehi, M. Kato, and T. Fijita Phys. Rev. B 44, 7753 (1991); J.D. Axe, A.H. Moudden, D.H. Hohlwein, D.E. Cox, K.M. Mohanty, A.R. Moodenbaugh, and Y. Xu Phys. Rev. Lett. 62, 2751 (1989)
4. R.E. Walstedt, W.W. Warren, Jr., R.F. Bell, R.J. cava, G.P.Espinosa, L.F. Schnneemeyer, and J.V. Waezczak, Phys. Rev. B 41, 9574 (1990); M. Takigawa et al. Phys. Rev. B 43, 247 (1991); R.E. Walstedt, W.W. Warren, Jr., R.F. Bell, G.P.Espinosa, Phys. Rev. B 40, 2572 (1989)
5. M.K. Crawford, M.N. Kunchur Phys. Rev. B 41, 282 (1990); M.K. Crawford, W.E. Farneth, E.M. McCarron, III, R.L. Harlow and A.H. Moudden Science 250, 1390 (1990); J.P. Frank, S. Harker, and J.H. Brewer Phys. Rev. Lett. 71, 283 (1993); J.P. Franck, S. Gygax, G. Sorensen, E. Altshuler, A. Hnatiw, J. Jung, M.A.-K. Mohamed, M.K. Yu, G.I. Sproule, J. Chrzanowski, and J.C. Irwin Physica C 185-189, 1379 (1991)
6. Y.J. Uemura et al. Phys. Rev. Letters 62, 2317, (1989), Phys. Rev. B 38, 909 (1988); Phys. Rev. Lett. 66, 2665 (1991); H. Keller, et al. Physica (Amsterdam) 185-189C, 1089 (1991); T. Schneider and H. Keller Phys. Rev. Letters 69, 3374 (1992)
7. A. Bianconi, A.M. Flank, P.Lagarde, C. Li, I. Pettiti, M. Pompa, and D. Udron in High Temperature Superconductivity, edited by J. Ashkenazi, S.E. Barnes, F. Zuo, G.C. Vezzoli and B.M. Klein (Plenum Press, New York, 1991) pag. 363
8. A. Bianconi, S. Della Longa, M. Missori, I. Pettiti, and M. Pompa in Lattice Effects in High-Tc Superconductors, (proc. meeting in Santa Fe, New Mexico, January 13-15 1992) edited by Y. Bar-Yam, T. Egami, J. Mustre de Leon and A.R. Bishop. World Scientific Pub., Singapore, (1992) pag. 95.

9. A. Bianconi in Phase Separation in Cuprate Superconductors, (proc. of the meeting held in Erice, Italy, May 6-12, 1992) edited by K.A. Müller and G. Benedek World Scientific Pub., Singapore, 1992.
10. E.P. Wigner Phys. Rev. 46, 1002 (1934); Trans. Faraday Soc. 34, 678 (1938); B. Tanar, and D.M. Ceperly Phys. Rev. B 39, 5005 (1989)
11. K. Moulopoulos and N. W. Ashcroft Phys. Rev. Lett. 69, 2555 (1992)
12. Y. Takada J. Phys. Soc. Japan 61, 3849 (1992).
13. G. Strinati, F. Pistolesi, Proc. of Forum Euroconfernce on Superconductivity in Fullerenes Oxides and Organic Materials, Pisa, Italy, January 25–29 1993; F. Pistolesi PhD thesis, Scuola Normale di Pisa 1993; and to be published
14. C. Di Castro private communication
15. C. Taliani, R. Zamboni, G. Ruani, F.C. Matacotta, and K.I. Pokhodnya Sol. Stat Commun. 66 (1988) 487.
16. R. Micnas, J. Ranninger, and S. Robasziewicz Rev. Mod. Phys. 62, 113 (1990); N.F. Mott Physica C 205, 191 (1993)
17. G. Yu, A.J. Heeger, G. Stuky, N. Herron, and E.M. McCarron Phys. Rev. Lett. 67, 2581 (1991) and Phys. Rev. B 45, 4964 (1992)
18. X.-X. Bi, and P.C. Eklund Phys. Rev. Lett. 70, 2625 (1993)
19. C.Y. Chen, R.J. Birgeneau, M.A. Kastner, N.W. Preyer, and T. Thio Phys. Rev. B 43, 392 (1991); J.P. Falk, A. Levy, M.A. Kastner, and R.J. Birgeneau Phys. Rev. Lett. 69, 1109 (1992); J.P. Falk, A. Levy, M.A. Kastner, and R.J. Birgenau Phys. Rev. B 48, 4043 (1993)
20. T. Takahashi, H. Matsuyama, H. Katayama-Yoshida, Y. Okabe, S. Hosoya, K. Seki, H. Fujimoto, M. Sato, and H. Inokuchi Nature 334 (1988) 691
21. B.O. Wells, Z.-X.Shen, D.S. Dessau, W.E. Spicer, C.G. Olson, D.B. Mitzi, A. Kapitulnik, R.S. List, and A. Arko, Phys. Rev. Lett. 65 (1990) 3056.
Z.X. Shen J. Phys. Chem. Solids 53, 1583 (1992); B.O. Wells, Z.X. Shen, D.S. Dessau, W.E. Spicer, D.B. Mitzi, L. Lombardo, A. Kapitulnik, A.J. Arko, Phys. Rev. B 46, 11830 (1992); Z.-X. Shen et al., Phys. Rev. Lett. 70, 1553 (1993); D.S. Dessau et al., Phys. Rev. Lett. 71, 2781 (1993).
22. R. Liu, B.W. Veal, A.P. Paulikas, P.J. Kostic, S. Fleshler, U. Welp, C.G. Olson, X. Wu, A.J. Arko, and J.J. Joyce Phys. Rev. B 46, 11056 (1992); J.C. Campusano, L.C. Smedskjaer, R. Benedek, G. Jennings, A. Basil Phys. Rev. B 43, 2788 (1991); J.C. Campusano, K. Gofron, R. Liu, H. Ding, B.W. Veal and B. Dabrowski J. Phys. Chem. Solids 53, 1577 (1992); A.A. Abrikosov, J.C. Campusano, and K. Gofron Physica C 214 , 73 (1993)
23. C. Di Castro, L.F. Finer and M. Grilli Phys. Rev. Lett. 66, 3209 (1991) ; and J.B. Grant and A.K. McMahan Phys. Rev. B 46,8440 (1992)
24. A.I. Beskrovnyi, M. Dlouhà, Z. Jiràk and S. Vratislav, Physica C 171 (1990) 19; and A.I. Beskrovnyi, M. Dlouhà, Z. Jiràk, S. Vratislav and E. Pollert, Physica C 166 (1990) 79.
25. W.A.Groen, D.M. de Leeuw and L.F. Finer Physica C 165, 55 (1990)
26. C. Quitmann, D. Andrich, C. Jachow, M. Fleuster, B. Beschoten, G. Güntherodt, V.V. Moshchalkov, G. Mante, and R. Manzke Phys. Rev. B 46, 11813 (1992)
27. I. Pettiti, Laurea Thesis, Department of Chemistry, Universiy of Rome, (1990).
28. T. Tamegai, K. Koga, K. Suzuki, M. Ichihara, F. Sakai, Y. Iye Jpn. J. Appl. Phys. 28, L112 (1989); H. Niu, N. Fukushima, S. Takeno, S.-I. Nakamura and K. Ando Jpn. J. Appl. Phys. 28, L784–L786 (1989); Y. Inoue, Y. Shichi, F. Munakata and M. Yamanaka, Phys. Rev. B (1989) 40, 7307.
29. A.Q. Pham, M. Hervieu, A. Mignan, J. Provost, and B. Raveau Physica C 194, 243 (1992)

30. W.E. Pickett, R.E. Cohen, and K. Krakauer Phys. Rev. Lett. 67, 228 (1991); R.S. Markiewicz, Physica C 200, 65 (1992)
31. K.I. Kugel and D.I. Khomskii Sov. Phys. Usp. 25( 1982) 321 and Sol. State Commun. 13 (1973)763; M.D. Kaplan and D.I. Khomskii Pis'ma Zh Eksp. Teor. Fiz. 47 (1988) 631
32. G. Kido, H. Katayama-Yoshida and T. Takaashi J. Phys. Chem. Solids 53, 1555 (1992); F.M. Mueller J. Phys. Chem. Solids 53, 1549 (1992); C.M. Fowler, B.L. Freeman, W.L. Hults, J.C. King, F.M. Mueller, and J.L. Smith Phys. Rev. Lett. 68, 534 (1992)
33. C.C. Grimes and G. Adams Phys. Rev. Lett. 42, (1979) 795
34. K. von Kilzing Rev. Mod. Physics 58, (1986) 519;
35. F.I.B. Williams, G. Deville, D.C. Glatti, P.A. Wright, E.Y. Andrei, R.G. Clark, B. Etienne, C.T. Foxon, and J.J. Harris Suface Science 263, 23 (1992)
36. R.T. Collins, Z. Schlesinger, G.V. Chandrashekhar and M.W. Shafer Phys. Rev. B 39, 2251 (1989)

# Pairing Interaction and Phonon Renormalization Favored by Phase Separation

A. Bill[1,2], V. Hizhnyakov[2,3,4], and E. Sigmund[1,2]

[1] Institut für Theoretische Physik, Universität Stuttgart, Pfaffenwaldring 57, 70550 Stuttgart
[2] Institut für Physik, Technische Universität Cottbus, PBox 101344, 03013 Cottbus
[3] Insitute of Physics, Estonian Academy of Sciences, Riia 142, Tartu, Estonia
[4] Physics Department, Tartu University, Tähe 4, Tartu, Estonia

**Summary.** A general expression for the effective electron-electron interaction mediated by electron-lattice coupling is derived in the framework of the BCS theory. It is shown that for the Cooper channel this interaction can be divided into two parts: a retarded and a non-retarded one. Depending on the strength and range of the electron-phonon coupling one or the other part is adequate for the description of the superconducting state. Special attention is focused on the non-totally screened long-range interaction which is important in the high-$T_c$ materials. The same derivation technique is used in a second step for the determination of the renormalization of the phonon energies in the superconducting phase. The simple expression obtained for $\mathbf{q} = 0$ phonons allows for the study of the influence of the gap anisotropy on this renormalization. Further, a modified version of the Eliashberg equations with explicit $\mathbf{k}$-dependence of the gap comes out from the calculation in a natural way.

## 1. Introduction

According to the BCS theory [1] in conventional superconductors the effective interaction between charge carriers leading to the superconducting state is mediated by phonons. This interaction was derived by Fröhlich [2] who applied a special unitary transformation to the hamiltonian that allowed to exclude the linear electron-phonon coupling in first order and to replace it by an effective second order electron-electron interaction and higher order electron-phonon interaction terms. The electron-electron interaction obtained depends in an essential way on the energies of the charge carriers. Indeed, this interaction is attractive when the energy difference of the two carriers forming a Cooper pair is smaller than the energy of the phonons $\hbar\omega$ and repulsive in the opposite case. This fact is related to phonon retarding effects. Since the conventional superconductors have high Fermi energy $E_F$ and because of the Pauli principle only charge-carriers with energies inside the thin shell $[E_F - \hbar\omega, E_F + \hbar\omega]$ will attract each other. The result of Fröhlich is valid in the weak coupling limit only, when third and higher order terms can be neglected.

There exist also other transformations which like the Fröhlich one allows to remove the linear electron-phonon coupling ([3]–[7]). The obtained particle-particle interactions have in general however different forms. Since in the BCS theory this interaction is considered as a part of an effective hamiltonian for *quasiparticles* the physical difference between the various forms of it should be understood as corresponding to different quasiparticles associated with different unitary transformations. In [6] and [7] it was noticed that among the transformations mentioned, there is one which is of special interest for superconductivity because it leads to a purely attractive coupling of the charge carriers in the Cooper channel ($\mathbf{k}\uparrow, -\mathbf{k}\downarrow$). This non-retarded interaction can be associated with a static distortion of the lattice around the Cooper pair rather than with the exchange of phonons.

In general the unitary transformation is rapidly converging only in the weak coupling limit. However, in the case of a non-totally screened long-range electron-phonon coupling this transformation can be used also in the intermediate and even strong coupling case as will be shown above. Such non-totally screened interactions are extremely weak in usual metals but they are essential in high-$T_C$ superconductors ([8] and references therein). Experimentally they reveal themselves in these materials e.g. in the superconductivity induced shifts and widths of phonons with wave vector $q = 0$ observed in Raman and infrared spectroscopy [9, 10]. The reason for the enhancement of the long-range electron-phonon coupling in high-$T_C$ superconductors is usually connected with the fact that these materials have a strong ionic character, a low symmetry, a low charge carrier concentration and that they are strongly anisotropic [8]. Here we want to pay attention to one more reason for the existence of non-totally screened long-range interactions in the high-$T_C$ materials, namely the appearance of the electronic phase separation caused by strong electronic correlations. According to theory [11], [3] and experiments [12], [13] the doped holes locally destroy the antiferromagnetic order in these materials and build up a conducting subphase of fractal-percolative origin (at least in the low doping case). As a result the system gets a strongly inhomogeneous structure with characteristic size $L_0 \sim 10a$ (determined by the size of the magnetic hole-polaron; a is the lattice constant). The structures are dynamical on a long time scale but quasistatic on the short time scale ($\leq 10^{-12}$ sec) which is the important one for the formation of Cooper pairs. While the conducting and non-conducting subphases interpenetrate each other on the characteristic scale $L_0$ the Madelung interactions cannot be screened on this distance. As a result the electron-phonon interaction function should have a maximum at $q \sim \frac{\pi}{L_0} \sim 0.1 q_{max}$ where $q_{max} = \frac{\pi}{a}$ is the maximal q value.

In the following we consider in more detail the class of unitary transformations that remove the linear electron-phonon coupling and which allow to obtain a purely attractive interaction between the charge carriers in the Cooper channel. We also give a new description of the $q = 0$ phonon renormalization caused by the superconducting phase transition (we remind that in the Fröhlich calculation this renormalization vanishes in the $q = 0$ limit). In our description we use the unitary transformation approach in combination with a Hartree-Fock type mean-field approximation. The obtained formula for this renormalization takes explicitly into account the effects of the superconducting gap anisotropy and is therefore more general than the corresponding expression used in [14] for the calculation of the superconductivity induced shifts and widths of the q = 0 phonons.

## 2. Unitary transformations

We consider a metal (superconductor) with a single conducting band but include all phonon branches. Account of several electronic bands is given in [3]. The system is described by the hamiltonian

$$H = H_e + H_L + H_{eL} \tag{2.1}$$

where

$$H_e = \sum_{\mathbf{k},\sigma} \varepsilon_{\mathbf{k}} c^{\dagger}_{\mathbf{k},\sigma} c_{\mathbf{k},\sigma} \tag{2.2}$$

is the hamiltonian of the charge-carriers (electrons, holes); $c^{\dagger}_{\mathbf{k},\sigma}(c_{\mathbf{k},\sigma})$ is the creation(destruction) operator of the particle with wave vector $\mathbf{k}$ and spin index $\sigma$ (which has the values $\uparrow$ and $\downarrow$),

$$H_L = \sum_{\mathbf{q},\nu} \hbar\omega_{\nu\mathbf{q}} \left( b^{\dagger}_{\nu\mathbf{q}} b_{\nu\mathbf{q}} + \frac{1}{2} \right) \tag{2.3}$$

is the hamiltonian of the lattice vibrations; $b^{\dagger}_{\nu\mathbf{q}}(b_{\nu\mathbf{q}})$ is the creation(destruction) operator of the phonon of branch $\nu$, wave vector $\mathbf{q}$ and frequency $\omega_{\nu\mathbf{q}}$ and

$$H_{eL} = \sum_{\mathbf{k},\mathbf{q}} \sum_{\sigma,\nu} g_{\nu}(\mathbf{q}) c^{\dagger}_{\mathbf{k},\sigma} c_{\mathbf{k}+\mathbf{q},\sigma} b^{\dagger}_{\nu\mathbf{q}} + h.c. \tag{2.4}$$

is the electron-lattice interaction which is linear in the phonon operators. $g_{\nu}(\mathbf{q})$ is the interaction function assumed here to be only dependent on $\mathbf{q}$.

As mentioned before there is a whole class of unitary transformations which allow to exclude the linear electron-phonon coupling $H_{eL}$ replacing it by a series of electron-electron as well as higher order electron-lattice interaction terms. In the case of the Fröhlich transformation

$$H_F = e^{S_F} H e^{-S_F} \tag{2.5}$$

where the operator $S_F$ is defined by the condition

$$[S_F, (H_e + H_L)] = -H_{eL} \quad . \tag{2.6}$$

This condition however, is obviously not the most general one for removing the linear coupling in the hamiltonian. It is always possible to add a higher order term to the right hand side of (2.6). If for example one adds a second order term describing a particle-particle interaction

$$[S,(H_e+H_L)] = -H_{eL} + H^{(2)} \tag{2.7}$$

it will still lead to the exclusion of $H_{eL}$.

Another possibility to introduce a more general transformation is to divide the operator $S_F$ into two parts [4]

$$S_F = S_1 + S_2 \tag{2.8}$$

and to perform two subsequent transformations

$$\tilde{H} = e^{S_1} H e^{-S_1}, \qquad \approx\!\!\!\!H = e^{S_2}\tilde{H}e^{-S_2} = e^{S_2}e^{S_1}He^{-S_1}e^{-S_2}. \tag{2.9}$$

This two step transformation also removes the linear interaction but leads to another particle-particle interaction and to higher order electron-phonon interaction terms. This can be seen immediately if e.g. $H_F$ and $\overset{\approx}{H}$ are calculated up to second order terms. In this approximation one has

$$e^{-\bar{S}} = e^{S_2}e^{S_1} \approx e^{S_F - \frac{1}{2}[S_1,S_2]} \tag{2.10}$$

i.e.

$$\bar{S} \approx S_F - [S_1,S_2]/2 \tag{2.11}$$

and

$$\overset{\approx}{H} = H_F + \frac{1}{2}[H_e + H_L,[S_1,S_2]].$$

The operator $\bar{S}$ satisfies the condition (2.7) where $H^{(2)}$ is given by the second term of the r.h.s. of (2.11). Therefore, the two ways presented to obtain a generalized form of the transformation are equivalent. Below we use a two step transformation with $S_1$ and $S_2$ being linear operators. The first transformation $S_1$ is purely distortive (i.e. leading to ionic shifts only) and removes the linear electron-phonon interaction term only partly. The residual term will be removed by a second transformation.

a) First unitary transformation

The first unitary transformation is given by

$$S_1 = \sum_{\mathbf{k},\mathbf{q}}\sum_{\nu} \gamma_\nu(\mathbf{q}) c^\dagger_{\mathbf{k},\sigma} c_{\mathbf{k}+\mathbf{q},\sigma} b^\dagger_{\nu\mathbf{q}} - h.c.. \tag{2.12}$$

For simplicity we omit here the spin indices. $\gamma_\nu(\mathbf{q})$ is an arbitrary function with values being of the order $|g_\nu(\mathbf{q})/\hbar\omega_{\nu\mathbf{q}}|$. The operator $S_1$ can also be presented in the form

$$S_1 = \sum_{\mathbf{n}} c^\dagger_{\mathbf{n}} c_{\mathbf{n}} \nabla_{\mathbf{n}} \tag{2.13}$$

where

$$c_{\mathbf{n}}^{\dagger} = \frac{1}{\sqrt{\mathrm{N}}} \sum_{\mathbf{n}} e^{i\mathbf{k}\mathbf{n}} c_{\mathbf{k}}^{\dagger}$$

is the particle creation operator in site representation, N is the number of unit cells of the lattice.

$$\nabla_{\mathbf{n}} = \sum_{r,\delta} a(r,\delta) \frac{\partial}{\partial x\binom{\mathbf{n}+\delta}{r}} \tag{2.14}$$

is the differential operator in the space of vibrational coordinates and

$$a(r,\delta) = \sum_{\mathbf{q},\nu} e^{-i\mathbf{q}\delta} \left(\frac{M_r \hbar\omega_{\nu\mathbf{q}}}{2\hbar}\right)^{-\frac{1}{2}} \gamma_\nu(\mathbf{q}) \xi_r(\nu,\mathbf{q}) \quad . \tag{2.15}$$

$x\binom{\mathbf{n}+\delta}{r}$ is the coordinate of ion $r$ in the unit cell $\mathbf{n}$ of the crystal, $M_r$ is the mass of the ion and $\xi_r(\nu,\mathbf{q})$ is the phonon polarization vector at the ion $r$ (we omit here the index for the cartesian coordinate of $x$ and $\xi$). In the case of a localized particle ($c_{\mathbf{n}}^{\dagger} c_{\mathbf{n}} = \delta_{\mathbf{n},\mathbf{n}_0}$) this transformation describes the shifts of the ions, the amplitude of the shift being given by $a(r,\delta)$.

We perform the transformation up to second order terms ($\sim \gamma^2$).

$$\tilde{H} \simeq H + [S_1, H] + \frac{1}{2}[S_1,[S_1, H_{\mathrm{e}} + H_{\mathrm{L}}]] \,. \tag{2.16}$$

The commutators reads

$$\begin{aligned}
[S_1, H_{\mathrm{L}}] &= -\sum_{\mathbf{k},\mathbf{q}\nu} \gamma_\nu(\mathbf{q}) \hbar\omega_{\nu\mathbf{q}} c_{\mathbf{k}}^{\dagger} c_{\mathbf{k}+\mathbf{q}} b_{\nu\mathbf{q}}^{\dagger} + h.c. \quad , & (2.17)\\
[S_1, H_{\mathrm{e}}] &= \sum_{\mathbf{k},\mathbf{q}\nu} \gamma_\nu(\mathbf{q}) \left(\varepsilon_{\mathbf{k}+\mathbf{q}} - \varepsilon_{\mathbf{k}}\right) c_{\mathbf{k}}^{\dagger} c_{\mathbf{k}+\mathbf{q}} b_{\nu\mathbf{q}}^{\dagger} + h.c. \quad , & (2.18)\\
[S_1, H_{\mathrm{eL}}] &= -\sum_{\mathbf{k}\mathbf{k}',\mathbf{q}\nu} \gamma_\nu(\mathbf{q}) g_\nu^*(\mathbf{q}) c_{\mathbf{k}}^{\dagger} c_{\mathbf{k}+\mathbf{q}} c_{\mathbf{k}'+\mathbf{q}}^{\dagger} c_{\mathbf{k}'} + h.c. \quad , & (2.19)\\
\frac{1}{2}[S_1,[S_1, H_{\mathrm{L}}]] &= \frac{1}{2} \sum_{\mathbf{k}\mathbf{k}',\mathbf{q}\nu} |\gamma_\nu(\mathbf{q})|^2 \hbar\omega_{\nu\mathbf{q}} c_{\mathbf{k}}^{\dagger} c_{\mathbf{k}+\mathbf{q}} c_{\mathbf{k}'+\mathbf{q}}^{\dagger} c_{\mathbf{k}'} + h.c. & (2.20)\\
\frac{1}{2}[S_1,[S_1, H_{\mathrm{e}}]] &\equiv \tilde{H}_{\mathrm{eL}}^{(2)} \\
&\simeq \frac{1}{2} \sum_{\mathbf{k},\mathbf{q}\nu} |\gamma_\nu(\mathbf{q})|^2 \left(\varepsilon_{\mathbf{k}+\mathbf{q}} - \varepsilon_{\mathbf{k}}\right) c_{\mathbf{k}}^{\dagger} c_{\mathbf{k}} b_{\nu\mathbf{q}}^{\dagger} b_{\nu\mathbf{q}} \quad . & (2.21)
\end{aligned}$$

In the last expression terms $\sim c_{\mathbf{k}}^{\dagger} c_{\mathbf{k}'}$ for $\mathbf{k} \neq \mathbf{k}'$ as well as terms $\sim b_{\mathbf{q}}^{\dagger} b_{\mathbf{q}'}$ for $\mathbf{q} \neq \mathbf{q}'$ and $\nu \neq \nu'$ are not taken into account since they give a contribution to the particle-particle interaction in higher orders only. The transformed hamiltonian up to second order terms can be written as

$$\tilde{H} = H_{\mathrm{e}} + H_{\mathrm{L}} + \tilde{H}_{\mathrm{ee}} + \tilde{H}_{\mathrm{eL}}^{(1)} + \tilde{H}_{\mathrm{eL}}^{(2)} \quad . \tag{2.22}$$

Here

$$\tilde{H}_{\mathrm{ee}} = \frac{1}{2}\sum_{\mathbf{kk}',\mathbf{q}\nu} W(\mathbf{q}) c^{\dagger}_{\mathbf{k}} c_{\mathbf{k}+\mathbf{q}} c^{\dagger}_{\mathbf{k}'+\mathbf{q}} c_{\mathbf{k}'} + h.c. \quad , \tag{2.23}$$

$$\tilde{H}^{(1)}_{\mathrm{eL}} = \sum_{\mathbf{k},\mathbf{q}}\sum_{\nu} \tilde{g}_{\nu}(\mathbf{k},\mathbf{q}) c^{\dagger}_{\mathbf{k}} c_{\mathbf{k}+\mathbf{q}} b^{\dagger}_{\nu\mathbf{q}} + h.c. \tag{2.24}$$

and

$$W(\mathbf{q}) = \sum_{\nu} \gamma_{\nu}(\mathbf{q}) \left(\hbar\omega_{\nu\mathbf{q}}\gamma^{*}_{\nu}(\mathbf{q}) - 2g^{*}_{\nu}(\mathbf{q})\right) \quad , \tag{2.25}$$

$$\tilde{g}_{\nu}(\mathbf{k},\mathbf{q}) = g_{\nu}(\mathbf{q}) + \gamma_{\nu}(\mathbf{q}) \left(\varepsilon_{\mathbf{k}+\mathbf{q}} - \varepsilon_{\mathbf{k}} - \hbar\omega_{\nu\mathbf{q}}\right) \quad . \tag{2.26}$$

b) Second unitary transformation

In a second step we remove the residual linear electron-lattice coupling $\tilde{H}^{(1)}_{\mathrm{eL}}$ with the transformation

$$\approx\!\!\!\!H = e^{S_2} \tilde{H} e^{-S_2} \tag{2.27}$$

where $S_2$ satisfies the condition

$$[S_2, (H_{\mathrm{e}} + H_{\mathrm{L}})] = -\left(H_{\mathrm{eL}} + [S_1, (H_{\mathrm{e}} + H_{\mathrm{L}})]\right) = -\tilde{H}^{(1)}_{\mathrm{eL}} \quad . \tag{2.28}$$

This operator has the form

$$S_2 = -\sum_{\mathbf{k},\mathbf{q}}\sum_{\nu} \frac{\tilde{g}_{\nu}(\mathbf{k},\mathbf{q})}{(\varepsilon_{\mathbf{k}+\mathbf{q}} - \varepsilon_{\mathbf{k}} - \hbar\omega_{\nu\mathbf{q}})} c^{\dagger}_{\mathbf{k}} c_{\mathbf{k}+\mathbf{q}} b^{\dagger}_{\nu\mathbf{q}} - h.c. \quad . \tag{2.29}$$

Performing the transformation up to second order terms we obtain

$$\approx\!\!\!\!H = H_{\mathrm{e}} + H_{\mathrm{L}} + H^{(2)}_{\mathrm{eL}} + H_{\mathrm{ee}} \quad . \tag{2.30}$$

Here

$$\tilde{H}^{(2)}_{\mathrm{eL}} = \sum_{\mathbf{q}\nu} \left\{ \sum_{\mathbf{k}} 2\hbar\omega_{\nu\mathbf{q}} \left(\varepsilon_{\mathbf{k}+\mathbf{q}} - \varepsilon_{\mathbf{k}}\right) \times \left[ |\gamma_{\nu}(\mathbf{q})|^2 - \frac{|\tilde{g}_{\nu}(\mathbf{k},\mathbf{q})|^2}{\left(\varepsilon_{\mathbf{k}+\mathbf{q}} - \varepsilon_{\mathbf{k}}\right)^2 - \left(\hbar\omega_{\nu\mathbf{q}}\right)^2} \right] c^{\dagger}_{\mathbf{k}} c_{\mathbf{k}} \right\} b^{\dagger}_{\nu\mathbf{q}} b_{\nu\mathbf{q}} \tag{2.31}$$

describes the quadratic electron-phonon coupling (terms in $c^{\dagger}_{\mathbf{k}} c_{\mathbf{k}'}$ for $\mathbf{k} \neq \mathbf{k}'$ and $\nu \neq \nu'$, $\mathbf{q} \neq \mathbf{q}'$ as well as terms $\sim b^{\dagger}_{\nu\mathbf{q}} b^{\dagger}_{\nu\mathbf{q}'}, b_{\nu\mathbf{q}} b_{\nu\mathbf{q}'}$ are neglected) and

$$H_{\mathrm{ee}} = \frac{1}{2}\sum_{\mathbf{kk}',\mathbf{q}} V(\mathbf{k},\mathbf{k}',\mathbf{q}) c^{\dagger}_{\mathbf{k}} c_{\mathbf{k}+\mathbf{q}} c^{\dagger}_{\mathbf{k}'+\mathbf{q}} c_{\mathbf{k}'} + h.c. \tag{2.32}$$

describes the electron-electron (hole-hole) interaction with

$$V(\mathbf{k},\mathbf{k}',\mathbf{q}) = W(\mathbf{q}) + \frac{1}{2}\sum_{\nu} \frac{\tilde{g}_\nu(\mathbf{k},\mathbf{q})\tilde{g}^*_\nu(\mathbf{k}',\mathbf{q})}{\varepsilon_{\mathbf{k}+\mathbf{q}} - \varepsilon_{\mathbf{k}} - \hbar\omega_{\nu\mathbf{q}}} \tag{2.33}$$

As one can see, depending on the choice of $\gamma_\nu(\mathbf{q})$ this interaction function (as well as higher order electron-phonon couplings) may have different forms. This is associated with different choices of the basic quasiparticles corresponding to different unitary transformations.

## 3. The Cooper pair interaction

According to the BCS theory superconductivity results from the formation of Cooper pairs caused by an interaction of the form

$$H_{\mathrm{SC}} = -\sum_{\mathbf{k}\mathbf{k}'} V_{\mathrm{C}}(\mathbf{k},\mathbf{k}')c^\dagger_{\mathbf{k}\uparrow}c^\dagger_{-\mathbf{k}\downarrow}c_{-\mathbf{k}'\downarrow}c_{\mathbf{k}'\uparrow} \tag{3.1}$$

(it corresponds to the condition $\mathbf{k}' + \mathbf{q} = -\mathbf{k}$ in (2.23)). In our case the interaction function $V_{\mathrm{C}}(\mathbf{k},\mathbf{k}')$ equals

$$\begin{aligned} V_{\mathrm{C}}(\mathbf{k},\mathbf{k}') &= V(\mathbf{k},\mathbf{k}',\mathbf{k}'-\mathbf{k}) \\ &= \mathrm{Re}\sum_{\nu} g^*_\nu(\mathbf{q})\left[2\gamma_\nu(\mathbf{q}) + \hbar\omega_{\nu\mathbf{q}}\frac{2\gamma_\nu(\mathbf{q})\hbar\omega_{\nu\mathbf{q}} - g_\nu(\mathbf{q})}{(\varepsilon_{\mathbf{k}} - \varepsilon_{\mathbf{k}'})^2 - (\hbar\omega_{\nu\mathbf{q}})^2}\right] \end{aligned} \tag{3.2}$$

($\mathbf{q} = \mathbf{k} - \mathbf{k}'$). If one chooses $\gamma_\nu(\mathbf{q}) = \alpha_\nu(\mathbf{q})g_\nu(\mathbf{q})/2\hbar\omega_{\nu\mathbf{q}}$ with real and positive $\alpha_\nu(\mathbf{q})$ then

$$V_{\mathrm{C}}(\mathbf{k},\mathbf{k}') = \sum_{\nu} \frac{|g_\nu(\mathbf{q})|^2}{\hbar\omega_{\nu\mathbf{q}}}\left[\alpha_\nu(\mathbf{q}) - \frac{(1-\alpha_\nu(\mathbf{q}))\,(\hbar\omega_{\nu\mathbf{q}})^2}{(\varepsilon_{\mathbf{k}} - \varepsilon_{\mathbf{k}'})^2 - (\hbar\omega_{\nu\mathbf{q}})^2}\right] \tag{3.3}$$

The first term in brackets is a purely attractive (i.e. non-retarded) interaction while the second is a sign-alternating (retarded) interaction. We note that the Fröhlich interaction corresponds to the choice $\alpha_\nu(\mathbf{q}) = 0$, while a purely attractive interaction is obtained if $\alpha_\nu(\mathbf{q}) = 1$ is chosen. Consequently, the procedure we present here allows to divide the particle-particle interaction in the Cooper channel into two parts : a retarded and a non-retarded one.

These two parts of the interaction result from two subsequent transformations performed only up to second order of the coupling constant that have, in general, different convergence conditions. The first transformation can be used if $\sum_{\nu\mathbf{q}} |\alpha_\nu(\mathbf{q})g_\nu(\mathbf{q})/\hbar\omega_{\nu\mathbf{q}}|^2 \ll 1$ while the second one if $\sum_{\nu\mathbf{q}} |g_\nu(\mathbf{q})|^2/\hbar\omega_{\nu\mathbf{q}} \ll \mathrm{E_F}$ (Migdal theorem). In the weak coupling limit these conditions can be fulfilled for any $\alpha_\nu(\mathbf{q})$ in the interval $[0,1]$. Therefore, the division of the interaction into two parts is arbitrary and is mainly determined by our convenience. The simplest form of the interaction function $V_C$ corresponds to the case $\alpha_\nu(\mathbf{q}) = 1$ i.e. to the non-retarded interaction.

In the general case the condition for the validity of the non-retarded interaction depends on the type of electron-phonon coupling. If this coupling is mainly short-range (i.e. the maximum of $|g_\nu(\mathbf{q})|^2$ is near $q \sim \frac{\pi}{a}$) the non-retarded interaction can be used only if the condition

$$\sum_{\nu} \sum_{q=0}^{k_F} |g_\nu(\mathbf{q})/2\hbar\omega_{\nu\mathbf{q}}|^2 \ll 1 \tag{3.4}$$

($k_F$ is here the Fermi vector) is fulfilled (i.e. $\alpha_\nu(\mathbf{q}) = 1$). In the opposite case, when non-screened long-range interactions are dominating $k_F$ instead of $k_0$ appears (the reversed characteristic length of the screening) as the upper cutoff value of $q$. If $k_0$ is small the non-retarded interaction may be used even in the case of a rather strong electron-phonon coupling.

If the electron-phonon coupling is of intermediate character and short-range as well as long-range interactions are present then by choosing $\alpha_\nu(q) \simeq 1$ for $q \leq k_0$ and $\alpha_\nu(q) \ll 1$ for $q \gg k_0$ one can account for the long-range part via the non-retarded particle-particle interaction and for the short-range part via the retarded interaction.

Considerations about the superconducting state with account of both mentioned interactions is given in [3], where the following expression for the superconducting gap is derived

$$\Delta(T = 0) = 2(\mathrm{BE_F})^{\frac{1}{2}} \exp\left(-\frac{1}{\mathrm{w}}\right) \tag{3.5}$$

B is the bandwidth, $\mathrm{E_F}$ the Fermi energy, and w the effective interaction. w as well as the corresponding expression for the critical temperature are given in [3].

## 4. The phonon renormalization

We have seen that the transformed hamiltonian contains the term $H_{\mathrm{eL}}^{(2)}$ which in the mean-field approximation can be considered as a renormalization of the electron and phonon energies in second order of the electron-lattice coupling. For example, the renormalization of the frequency of the phonon of branch $\nu$ and wave vector $\mathbf{q}$ is given by the factor in figured brackets of expression (2.31) with $\langle c_{\mathbf{k}}^\dagger c_{\mathbf{k}} \rangle = f_{\mathbf{k}}$ instead of the operator $c_{\mathbf{k}}^\dagger c_{\mathbf{k}}$ ($f_{\mathbf{k}}$ is the thermal occupation). This renormalization does not take into account effects caused by the superconducting transition since only parameters of the normal state appear in (2.31). Moreover, it vanishes for IR and Raman active phonons with $q = 0$ [14]. To describe the experimentally observed influence of superconductivity onto these phonons in the calculations of the renormalization $\delta\omega_{\nu\mathbf{q}}$ the modification of the electronic spectrum (appearance of the gap) in the superconducting state must be included. For that we modify the hamiltonian in the zeroth approximation. The introduced parameters have to be determined selfconsistently. More specifically, the idea of our approach is the following : we add and subtract to the hamiltonian the three terms $H_\phi$, $H_\chi$ and $\delta H_\mathrm{L}$ where

$$H_\phi = \sum_{\mathbf{k}} \phi_{\mathbf{k}} c^\dagger_{\mathbf{k}\uparrow} c^\dagger_{-\mathbf{k}\downarrow} + h.c. \tag{4.1}$$

$$H_\chi = \sum_{\mathbf{k},\sigma} \chi_{\mathbf{k}} c^\dagger_{\mathbf{k},\sigma} c_{\mathbf{k}\sigma} \tag{4.2}$$

$$\delta H_\mathrm{L} = \sum_{\mathbf{q},\nu} \delta\omega_{\nu\mathbf{q}} b^\dagger_{\nu\mathbf{q}} b_{\nu\mathbf{q}} \tag{4.3}$$

$\phi_{\mathbf{k}}$ and $\chi_{\mathbf{k}}$ describe the superconducting (gap) and Hartree-Fock mean-fields, $\delta\omega_{\nu\mathbf{q}}$ is the renormalization of the phonon energy (which is a second order effect and is thus considered to be small). We finally introduce a renormalization of the electron-phonon interaction $\delta g_\nu(\mathbf{q}) = \bar{g}_\nu(\mathbf{q}) - g_\nu(\mathbf{q})$. $\bar{g}_\nu(\mathbf{q})$ is the renormalized and $g_\nu(\mathbf{q})$ the non-renormalized part of the interaction. The residual coupling $\delta g_\nu(\mathbf{q})$ is considered to be a small second order effect too.

Our zeroth-order hamiltonian is given by

$$H_0 = H_\mathrm{SC} + \bar{H}_L \tag{4.4}$$

where

$$\begin{aligned} H_\mathrm{SC} &= H_\mathrm{e} + H_\phi + H_\chi - \mu N \\ &= \sum_{\mathbf{k}} E_{\mathbf{k}} \left( a^\dagger_{1\mathbf{k}} a_{1\mathbf{k}} + a^\dagger_{2\mathbf{k}} a_{2\mathbf{k}} \right) \quad . \end{aligned} \tag{4.5}$$

We have written the hamiltonian in Bogoliubov operators $a_{1\mathbf{k}}, a_{2\mathbf{k}}$ in which it is diagonal. They are defined by

$$\begin{aligned} a_{1\mathbf{k}} &= u_{\mathbf{k}} c_{\mathbf{k}\uparrow} + v_{\mathbf{k}} c^\dagger_{\mathbf{k}\downarrow} \\ a_{2\mathbf{k}} &= v_{\mathbf{k}} c^\dagger_{\mathbf{k}\uparrow} - u_{\mathbf{k}} c_{\mathbf{k}\downarrow} \end{aligned}$$

with

$$u_{\mathbf{k}} = \sqrt{\frac{1}{2}\left(1 + \frac{\bar{\varepsilon}_{\mathbf{k}}}{E_{\mathbf{k}}}\right)}, \qquad v_{\mathbf{k}} = \sqrt{\frac{1}{2}\left(1 - \frac{\bar{\varepsilon}_{\mathbf{k}}}{E_{\mathbf{k}}}\right)} \quad .$$

$u_{\mathbf{k}}$ and $v_{\mathbf{k}}$ are chosen real to simplify the notation. In (4.5) $\mu$ is the chemical potential which can be found by standard methods (see e.g. [15]),

$$N = \sum_{\mathbf{k},\sigma} c^\dagger_{\mathbf{k},\sigma} c_{\mathbf{k},\sigma} = \sum_{\mathbf{k}} \left( a^\dagger_{1\mathbf{k}} a_{1\mathbf{k}} + a^\dagger_{2\mathbf{k}} a_{2\mathbf{k}} \right)$$

is the particle number operator,

$$E_{\mathbf{k}} = \sqrt{\bar{\varepsilon}_{\mathbf{k}}^2 + \Delta_{\mathbf{k}}^2} \tag{4.6}$$

is the energy of the one particle excitation in the superconducting state, $\bar{\varepsilon}_{\mathbf{k}} = \varepsilon_{\mathbf{k}} - \chi_{\mathbf{k}} - \mu$, $\Delta_{\mathbf{k}} = |\phi_{\mathbf{k}}|$ is the gap parameter and $\bar{H}_L = H_{\mathrm{L}} + \delta H_{\mathrm{L}}$ is the renormalized phonon hamiltonian. The total hamiltonian then has the form

$$\begin{aligned} H &= H_0 + H' \quad , & (4.7)\\ H' &= \bar{H}_{\mathrm{eL}} - H_\phi - H_\chi - \delta H_{\mathrm{L}} + \delta H_{\mathrm{eL}} \quad . & (4.8) \end{aligned}$$

To obtain the quantities $\phi_{\mathbf{k}}, \chi_{\mathbf{k}}, \delta\omega_{\nu\mathbf{k}}$ and $\delta g_\nu(\mathbf{q})$ associated with the superconducting state, we perform a unitary transformation to exclude $\bar{H}_{\mathrm{eL}}$ (the renormalized electron-phonon coupling) and require that all the terms of the transformed hamiltonian except $H_0$ give no contribution to the electron and the phonon self energies in lowest order. In the simplest case of a one-step unitary transformation performed up to second order, the unknown $\phi_{\mathbf{k}}$ and $\chi_{\mathbf{k}}$ can be obtained from the following condition : we demand that the contribution of the electron-electron coupling part to the electron self-energy $\Sigma$ that arise from the renormalized electron-phonon interaction in the superconducting state, i.e. the term coming from $[S, \bar{H}_{\mathrm{eL}}]/2$ cancels with the contribution of $H_\phi$ and $H_\chi$

$$\Sigma_{\frac{1}{2}[S,\bar{H}_{\mathrm{eL}}]_{\mathrm{ee}}} = \Sigma_{H_\phi + H_\chi} \quad .$$

$\delta\omega_{\nu\mathbf{q}}$ is obtained from the condition that the quadratic electron-phonon coupling part coming from $[S, \bar{H}_{\mathrm{eL}}]/2$ gives the same contribution to the phonon self-energy $\Pi$ as $\delta H_{\mathrm{L}}$. Finally, the renormalization $\delta g_\nu(\mathbf{q})$ of the interaction function can be found from the condition that the terms

$$\delta H_{\mathrm{eL}} + [S, H_\phi + H_\chi] + \frac{1}{2}[S,[S, H_\phi + H_\chi]] \tag{4.9}$$

altogether give no contribution to the phonon self-energy. The procedure to find all these parameters ($\phi_{\mathbf{k}}, \chi_{\mathbf{k}}, \delta\omega_{\nu\mathbf{q}}$ and $\delta g_\nu(\mathbf{q})$) is practically the same for the two-step transformation.

In the first non-vanishing approximation the value of $\delta\omega_{\nu\mathbf{q}}$ does not depend on whether a one or a two-step transformation is used. One obtains (some details are given in the appendix)

$$\begin{aligned} \delta\omega_{\nu\mathbf{q}} &= |\bar{g}_\nu(\mathbf{q})|^2 \sum_{\mathbf{k}} \\ & \left[ \frac{\bar{f}_{\mathbf{k}} (E_{\mathbf{k}+\mathbf{q}} - E_{\mathbf{k}}) (E_{\mathbf{k}} E_{\mathbf{k}+\mathbf{q}} + \bar{\varepsilon}_{\mathbf{k}} \bar{\varepsilon}_{\mathbf{k}+\mathbf{q}} - \Delta_{\mathbf{k}} \Delta_{\mathbf{k}+\mathbf{q}})}{E_{\mathbf{k}} E_{\mathbf{k}+\mathbf{q}} \left[ (\hbar\omega_{\nu\mathbf{q}})^2 - (E_{\mathbf{k}} - E_{\mathbf{k}+\mathbf{q}})^2 \right]} \right. \qquad (4.10) \\ &+ \left. \frac{(1 - 2\bar{f}_{\mathbf{k}}) (E_{\mathbf{k}+\mathbf{q}} + E_{\mathbf{k}}) (E_{\mathbf{k}} E_{\mathbf{k}+\mathbf{q}} - \bar{\varepsilon}_{\mathbf{k}} \bar{\varepsilon}_{\mathbf{k}+\mathbf{q}} + \Delta_{\mathbf{k}} \Delta_{\mathbf{k}+\mathbf{q}})}{E_{\mathbf{k}} E_{\mathbf{k}+\mathbf{q}} \left[ (\hbar\omega_{\nu\mathbf{q}})^2 - (E_{\mathbf{k}} + E_{\mathbf{k}+\mathbf{q}})^2 \right]} \right] \end{aligned}$$

where $\bar{f}_{\mathbf{k}} = (exp(E_{\mathbf{k}}/k_B T) + 1)^{-1}$. For phonons with $q = 0$ it reduces to the following simple formula

$$\delta\omega_{\nu 0} = -4|\bar{g}_\nu(0)|^2 \sum_{\mathbf{k}} \frac{\Delta_{\mathbf{k}}^2(1 - 2\bar{f}_{\mathbf{k}})}{E_{\mathbf{k}}\left[(2E_{\mathbf{k}})^2 - (\hbar\omega_{\nu 0})^2\right]} \tag{4.11}$$

This result is more general than the analogous one obtained in [14] for numerical calculations since it takes the dependence of the gap $\Delta_{\mathbf{k}}$ on the wave vector $\mathbf{k}$ into account and thus the gap anisotropy. Numerical results of (4.11) will be presented elsewhere.

## 5. Appendix

We consider the unitary transformation applied to (4.7)

$$\tilde{H} = e^S H e^{-S} = H_0 + \tilde{H}' + \tilde{H}'' \tag{5.1}$$

where in the Bogoliubov operators

$$\begin{aligned} S &= \sum_{\mathbf{k},\mathbf{q}} \sum_{\nu} \bar{g}_\nu(\mathbf{q}) \left(\alpha_1 a_{1\mathbf{k}}^\dagger a_{1\mathbf{k}+\mathbf{q}} + \alpha_2 a_{2\mathbf{k}+\mathbf{q}}^\dagger a_{2\mathbf{k}} \right. \\ &\qquad \left. + \alpha_3 a_{1\mathbf{k}}^\dagger a_{2\mathbf{k}+\mathbf{q}}^\dagger + \alpha_4 a_{2\mathbf{k}} a_{1\mathbf{k}+\mathbf{q}}\right) b_{\nu\mathbf{q}}^\dagger - h.c. \quad , \qquad (5.2) \\ \tilde{H}' &= \frac{1}{2}\left[S, \bar{H}_{\mathrm{eL}}\right] - H_\Delta - H_\chi - \delta H_{\mathrm{L}} \quad , \qquad (5.3) \\ \tilde{H}'' &= -\delta H_{\mathrm{eL}} - [S, H_\Delta + H_\chi] - \frac{1}{2}[S,[S, H_\Delta + H_\chi]] \quad . \qquad (5.4) \end{aligned}$$

Here and in the following one has $\alpha_i \equiv \alpha_i(\mathbf{k}, \mathbf{q}\nu)$, $\beta^{(\prime)} \equiv \beta^{(\prime)}(\mathbf{k}, \mathbf{q})$ and

$$\begin{aligned} \alpha_1 &= \frac{\beta}{E_{\mathbf{k}} - E_{\mathbf{k}+\mathbf{q}} + \hbar\omega_{\nu\mathbf{q}}} \quad , \quad \alpha_2 = -\frac{\beta}{E_{\mathbf{k}} - E_{\mathbf{k}+\mathbf{q}} - \hbar\omega_{\nu\mathbf{q}}} \quad , \\ \alpha_3 &= \frac{\beta'}{E_{\mathbf{k}} + E_{\mathbf{k}+\mathbf{q}} + \hbar\omega_{\nu\mathbf{q}}} \quad , \quad \alpha_4 = -\frac{\beta'}{E_{\mathbf{k}} + E_{\mathbf{k}+\mathbf{q}} - \hbar\omega_{\nu\mathbf{q}}} \quad , \qquad (5.5) \\ \beta &= u_{\mathbf{k}} u_{\mathbf{k}+\mathbf{q}} - v_{\mathbf{k}} v_{\mathbf{k}+\mathbf{q}} \quad , \quad \beta' = u_{\mathbf{k}} v_{\mathbf{k}+\mathbf{q}} + u_{\mathbf{k}+\mathbf{q}} v_{\mathbf{k}} \quad . \end{aligned}$$

Taking into account that in the Bogoliubov operator representation the electron-phonon interaction has the form

$$\begin{aligned} \bar{H}_{\mathrm{eL}} &= \sum_{\mathbf{k}} \sum_{\mathbf{q}\nu} \bar{g}_\nu(\mathbf{q}) \left[\beta \left(a_{1\mathbf{k}}^\dagger a_{1\mathbf{k}+\mathbf{q}} + a_{2\mathbf{k}+\mathbf{q}}^\dagger a_{2\mathbf{k}}\right)\right. \\ &\quad + \left. \beta' \left(a_{1\mathbf{k}}^\dagger a_{2\mathbf{k}+\mathbf{q}}^\dagger + a_{2\mathbf{k}} a_{1\mathbf{k}+\mathbf{q}}\right)\right] b_{\nu\mathbf{q}}^\dagger + h.c. \qquad (5.6) \end{aligned}$$

one finds after simple calculations

$$\frac{1}{2}\left[S,\bar{H}_{\mathrm{eL}}\right]_{\mathrm{ee}} = \sum_{\mathbf{k}}\sum_{\mathbf{q}\nu} |\bar{g}_\nu(\mathbf{q})|^2 \times \tag{5.7}$$
$$\left[\alpha\left(a^\dagger_{1\mathbf{k}}a_{1\mathbf{k}+\mathbf{q}} + a^\dagger_{2\mathbf{k}+\mathbf{q}}a_{2\mathbf{k}}\right) + \alpha'\left(a^\dagger_{1\mathbf{k}}a^\dagger_{2\mathbf{k}+\mathbf{q}} + a_{2\mathbf{k}}a_{1\mathbf{k}+\mathbf{q}}\right)\right] \times$$
$$\left[\beta\left(a^\dagger_{1\mathbf{k}+\mathbf{q}}a_{1\mathbf{k}} + a^\dagger_{2\mathbf{k}}a_{2\mathbf{k}+\mathbf{q}}\right) + \beta'\left(a^\dagger_{1\mathbf{k}+\mathbf{q}}a^\dagger_{2\mathbf{k}} + a_{2\mathbf{k}+\mathbf{q}}a_{1\mathbf{k}}\right)\right],$$

(5.8)

$$\frac{1}{2}\left[S,\bar{H}_{\mathrm{eL}}\right]_{\mathrm{eL}} = \sum_{\mathbf{q}\nu}\left\{\sum_{\mathbf{k}}\left[|\bar{g}_\nu(\mathbf{q})|^2\bar{\alpha}\beta\left(a^\dagger_{1\mathbf{k}}a_{1\mathbf{k}} + a^\dagger_{2\mathbf{k}}a_{2\mathbf{k}}\right)\right.\right.$$
$$+\bar{\alpha}'\beta'\left(a^\dagger_{1\mathbf{k}}a_{1\mathbf{k}} + a^\dagger_{2\mathbf{k}}a_{2\mathbf{k}} - 1\right)$$
$$\left.\left. + \left(\bar{\alpha}\beta' + \bar{\alpha}'\beta\right)\left(a^\dagger_{1\mathbf{k}}a^\dagger_{2\mathbf{k}} + a_{2\mathbf{k}}a_{1\mathbf{k}}\right)\right]\right\}b^\dagger_{\nu\mathbf{q}}b_{\nu\mathbf{q}} \tag{5.9}$$

where

$$\alpha = \frac{1}{2}(\alpha_1 + \alpha_2) \qquad \bar{\alpha} = \frac{1}{2}(\alpha_1 - \alpha_2)$$
$$\alpha' = \frac{1}{2}(\alpha_3 + \alpha_4) \qquad \bar{\alpha}' = \frac{1}{2}(\alpha_3 - \alpha_4) \tag{5.10}$$

Terms with $\mathbf{k} \neq \mathbf{k}'$ and $\mathbf{q} \neq \mathbf{q}'$ as well as terms $\sim b^\dagger_{\nu\mathbf{q}}b^\dagger_{\nu\mathbf{q}}$ and $b_{\nu\mathbf{q}}b_{\nu\mathbf{q}}$ are neglected. Expression (5.9) is an electron-electron interaction whereas (5.7) is a second order electron-phonon coupling. The quantities $\Delta_\mathbf{k}, \chi_\mathbf{k}, \delta\omega_{\nu\mathbf{q}}$ and $\delta g_\nu(\mathbf{q})$ are determined in such a way that the contributions of $\tilde{H}' + \tilde{H}''$ to the electron and phonon self-energies vanishes in the mean-field approximation.

Taking into account that the contributions to the electron-electron interaction arising from $\tilde{H}''$ are of higher order than those of $\tilde{H}'$, they can be neglected in the calculations of $\Delta_\mathbf{k}$ and $\chi_\mathbf{k}$. With

$$H_\Delta + H_\chi = \sum_{\mathbf{k}} E^{-1}_{\mathbf{k}}\left[\left(\chi_\mathbf{k}\bar{\varepsilon}_\mathbf{k} + \Delta^2_\mathbf{k}\right)\left(a^\dagger_{1\mathbf{k}}a_{1\mathbf{k}} + a^\dagger_{2\mathbf{k}}a_{2\mathbf{k}}\right)\right.$$
$$\left. + \left(\chi_\mathbf{k} - \bar{\varepsilon}_\mathbf{k}\right)\Delta_\mathbf{k}\left(a^\dagger_{1\mathbf{k}}a^\dagger_{2\mathbf{k}} + a_{2\mathbf{k}}a_{1\mathbf{k}}\right)\right] \tag{5.11}$$

and with mean-field approximation we find the following equations

$$\chi_\mathbf{k}\bar{\varepsilon}_\mathbf{k} + \Delta^2_\mathbf{k} = E_\mathbf{k}\sum_{\mathbf{q}\nu}|\bar{g}_\nu(\mathbf{q})|^2\left(1 - 2f_{\mathbf{k}+\mathbf{q}}\right)\left(\alpha\beta - \alpha'\beta'\right) \tag{5.12}$$

$$\Delta_\mathbf{k}\left(\chi_\mathbf{k} - \bar{\varepsilon}_\mathbf{k}\right) = E_\mathbf{k}\sum_{\mathbf{q}\nu}|\bar{g}_\nu(\mathbf{q})|^2\left(1 - 2f_{\mathbf{k}+\mathbf{q}}\right)\left(\alpha'\beta + \alpha\beta'\right) \ . \tag{5.13}$$

These equations can be considered as a version of the Eliashberg equations with an explicit account of the wave vector dependence of the superconducting parameters. In the general case these equations are rather complicated.

But in the case of small $\bar{\varepsilon}_{\mathbf{k}}$ one obtains a much simpler form. E.g. for $\bar{\varepsilon}_{\mathbf{k}} = 0$ one gets from (5.12) the usual BCS gap equation

$$\Delta_0 = -\sum_{\mathbf{q}} W_{\mathbf{q}} \frac{\Delta_{\mathbf{q}}}{2E_{\mathbf{q}}} \left(1 - 2f_{\mathbf{k}+\mathbf{q}}\right) \tag{5.14}$$

where

$$W_{\mathbf{q}} \approx \sum_{\nu} 2|\bar{g}_{\nu}(\mathbf{q})|^2 \frac{\hbar\omega_{\nu\mathbf{q}}}{\left[\bar{\varepsilon}_{\mathbf{q}}^2 - (\hbar\omega_{\nu\mathbf{q}})^2\right]} \tag{5.15}$$

Here we took into account that for $\bar{\varepsilon}_{\mathbf{k}} = 0$

$$\alpha\beta - \alpha'\beta' = \frac{\hbar\omega_{\nu\mathbf{q}}}{\left[(\hbar\omega_{\nu\mathbf{q}})^2 - E_{\mathbf{q}}^2\right]}$$

and that the main contribution in (5.14) is given by terms with $E_{\mathbf{q}} \simeq \bar{\varepsilon}_{\mathbf{q}} \gg \Delta_0, \Delta_{\mathbf{q}}$.

Finally let us turn to the calculation of $\delta\omega_{\nu\mathbf{q}}$ and $\delta g_{\nu}(\mathbf{q})$. The last quantity is fixed by the condition that the contribution of $\tilde{H}''$ to the phonon self-energy vanishes in mean-field approximation. Then $\delta\omega_{\nu\mathbf{q}}$ is given by the mean-field value of the factor in figured brackets of expression (5.9) . The corresponding formula is given by (4.10). The equation for $\delta g_{\nu}(\mathbf{q})$ is rather lengthy and will be presented elsewhere.

## References


1. J. Bardeen, L.N. Cooper, J.R. Schrieffer, *Phys.Rev.* **108**(1957)1175
2. H. Fröhlich, *Proc.Roy.Soc.*(London) **A215**(1952)291
3. V. Hizhnyakov, E. Sigmund, A. Bill, "Cooper Pair Formation by Distortive Electron-Lattice Coupling" in Proc. of the Workshop on "*Phase Separation in Cuprate Superconductors*" in Erice, (K.A. Müller, G. Benedek eds.) World Scientific 1992
4. M. Wagner, this volume
5. M. Wagner, "*Unitary Transformations in Solid State Physics*", (V.M. Agranovich, A.A. Maradudin eds.) North-Holland 1986
6. V. Hizhnyakov, E. Sigmund, *Proc. of the Estonian Acad. Sci., Physics, Mathematics* **40**(1991)337
7. V. Hizhnyakov, E. Sigmund, *Phil.Mag.B* **65**(1992)775
8. R. Zeyher, *Z.Phys.B Cond.Matter* **80**(1990)187
9. C. Thomsen, M. Cardona, B. Gegenheimer, R. liu, A. Simon, *Phys.Rev.***B37**(1988)9860
10. L. Genzel, A. Wittlin, M. Bauer, M. Cardona, E. Schönherr, A. Simon, *Phys.Rev.***B40**(1989)2170
11. V. Hizhnyakov, E. Sigmund, *Physica C***156**(1988)655
12. R.K. Kremer, E. Sigmund, V. Hizhnyakov, F. Hentsch, A. Simon, K.A. Müller, M. Mehring, *Z.Phys.B Cond.Matter* **86**(1992)319

13. R.K. Kremer, V. Hizhnyakov, E. Sigmund, A. Simon, K.A. Müller, *Z.Phys.B* **91**(1993)169
14. R. Zeyher, G. Zwicknagel, *Z.Phys.B Cond.Matter* **78**(1990)175
15. J.R. Schrieffer, *Theory of Superconductivity*, W.A. Benjamin(ed.) 1964

# Electronic Excitations in Undoped and Lightly Oxygen Doped $La_2CuO_{4+y}$

J.P. Falck, J.D. Perkins, A. Levy, M.A. Kastner, J.M. Graybeal*, and R.J. Birgeneau

Department of Physics MIT, Cambridge MA 02139

**Summary.** The electronic excitations from 0.1 to 2.5 eV in pure and lightly oxygen doped single-crystal $La_2CuO_{4+y}$ are studied by means of polarization and temperature dependent reflectivity and transmission measurements. We discuss the current understanding of these excitations. In particular, we demonstrate that both extended and localized holes interact with optical phonon modes and form large polarons. Moreover, in undoped $La_2CuO_4$ we show the existence of weakly electric dipole allowed excitons with multimagnon sidebands.

## 1. Introduction

$La_2CuO_4$ is one of the simplest antiferromagnetic semiconductors into which introduction of charge-carriers gives rise to high-temperature superconductivity. Accordingly, the elementary electronic excitations of doped and undoped $La_2CuO_4$ have received a great deal of attention [1],[2]. Working with high-quality single crystals we have, in a series of measurements, studied the electronic excitations of undoped and lightly oxygen doped $La_2CuO_{4+y}$ in the energy range 0.1 to 2.5 eV. In this paper we review our current understanding of these excitations.

In order to understand the lowest-energy electronic excitations of undoped $La_2CuO_4$ it is sufficient to consider the $CuO_2$ planes, including only the Cu $3d$ and O $2p$ orbitals. In the ground state the hole is localized in the Cu $3d_{x^2-y^2}$ orbital with excited states consisting of both Cu $d$ and O $p$ states. The crystal-field transitions to the excited states of the $3d^9$ $Cu^{+2}$ ion are electric dipole forbidden because of inversion symmetry at the Cu site. The lowest energy electric-dipole-allowed excitation is a charge-transfer excitation of the hole on the $Cu^{+2}$ ion to the excited O $2p_\sigma$ states.

In Fig. 1.1(a) we display the in-plane reflectivity for undoped ($y = 0$) and lightly oxygen doped ($y = 0.014$) $La_2CuO_{4+y}$ crystals at $T = 10\,K$. The corresponding $\epsilon_2(\omega)$ and $\sigma(\omega)$ spectra are shown in panels (b) and (c) respectively. The charge-transfer transitions give rise to the characteristic peak at 2.2 eV and a corresponding exciton band at 1.75 eV. In section 2 we review our detailed study of the charge-transfer excitation spectrum [3]. In particular, we provide clear evidence for polaron formation of photoexcited electrons and holes. Recently we have also identified crystal-field excitons

* Present address: University of Florida, Department of Physics, 215 Williamson, Gainesville, FL 32611

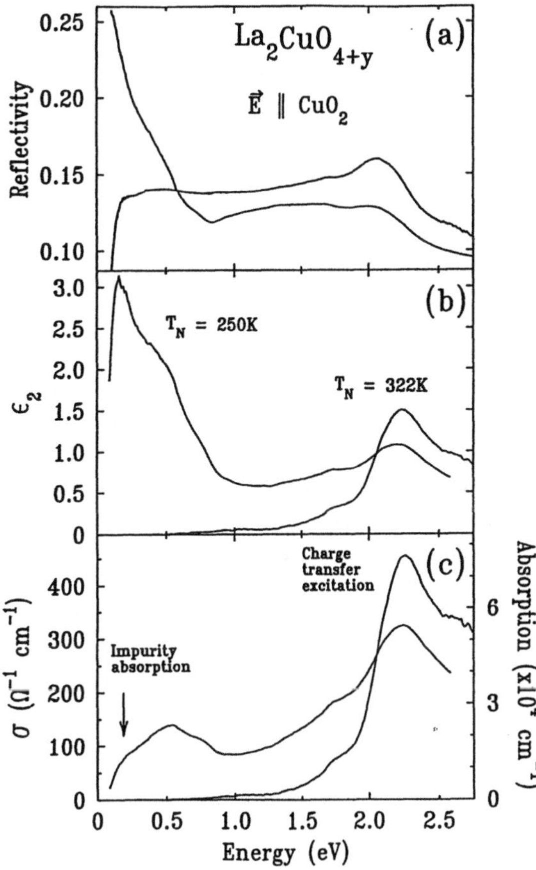


**Fig. 1.1.** **(a)** In-plane reflectivity spectra of $La_2CuO_{4+y}$ with $y = 0$ and $y = 0.014$ at 10 K. **(b)** The in-plane $\epsilon_2(\omega)$ spectra extracted from the reflectivity spectra shown in (a). **(c)** The same spectra as in (b) but displayed as both conductivity and absorption coefficient.

near 0.4 and 1.5 eV [4],[5]. They are orders of magnitude weaker than the charge-transfer excitations in the undoped material and are only detected in transmission spectra. The latter measurements are reviewed in section 3. We observe strong interaction between these crystal-field excitons and the magnetic excitations of the Cu spin system.

The introduction of holes into $CuO_2$ layers causes insulating $La_2CuO_4$ to become metallic at doping levels above $\sim$ 2% of the Cu-ion density [6],[7]. For lower doping levels, $La_2CuO_{4+y}$ with excess oxygen behaves like a doped semiconductor with an impurity binding energy of 0.035 eV as measured in transport measurements [8]. As seen in Fig. 1.1, two separable bands emerge with doping within the charge-transfer gap, at 0.13 and 0.5 eV, respectively. In addition, one sees a broad continuum spanning the energy range 1.0 – 2.0 eV.

In section 4 we review a study [9] in which we demonstrate transfer of spectral weight between free carrier conductivity and the 0.13 eV band. We show that the line shape as well as the temperature dependence is consistent with optical ionization of a polaronic impurity state. We close in section 5 with conclusions and some remarks concerning the origin of the remaining 0.5 to 2.0 eV sub-gap oscillator strength.

## 2. Charge-Transfer Excitation

In Fig. 2.1(a) we show spectra of the optical reflectivity at 122 K and 447 K for polarization parallel and perpendicular to the $CuO_2$ sheets. The prominent charge-transfer feature near 2.2 eV is observed only for the in-plane spectrum. The out-of-plane spectrum shows no structure in this energy region demonstrating the 2D nature of the excitations. The charge-transfer peak is seen to sharpen and move to higher energies upon cooling. In addition to the charge-transfer peak, a kink in the reflectivity near 1.75 eV is observed in the in-plane spectrum at low temperatures.

The imaginary part of the dielectric function, $\epsilon_2$, at 122 K for the in-plane spectrum is shown in Fig. 2.1(b), together with the photoconductivity at 132 K for an identically prepared crystal [10]. The photoconductivity has a threshold coinciding with that of the $\epsilon_2$ peak, indicating that the absorption is associated with the creation of free electrons and holes. The plateau on the low energy side of the 2 eV absorption edge results from the kink in the reflectivity (Fig. 2.1a). This second excitation, centered at 1.75 eV, has no associated photoconductivity indicating that the absorption results in the creation of excitons.

Even though the absorption above 2 eV leads to a separated electron and hole, the narrowness of the peak at 2.25 eV indicates that the interaction between the electron and hole is important in the interband transitions as well as in those of the bound exciton. It is well known that the electron-hole interaction enhances matrix elements for interband transitions just above threshold as discussed in Ref. [3]. We find that only a short range interaction between electrons and holes provides a consistent description of the peak in $\epsilon_2$. Specifically, we assume that the interaction potential $V(\boldsymbol{r}) = -g\,\delta(\boldsymbol{r})$, where $\boldsymbol{r}$ is the separation of the electron and hole, as is appropriate for Frenkel excitons.

The charge-transfer excitation peak in Fig. 2.1(a) broadens and shifts to lower energies with increasing temperature. A detailed measurement of this shift is displayed in Fig. 2.2 where we have plotted the energy of the in-plane reflectivity peak as a function of temperature. Cooling from 450 K, the peak position increases linearly with decreasing temperature at a rate of $6.8 \times 10^{-4}$ eV/K before it saturates at $T \sim 120$ K.

The shift of the reflectivity peak results from a shift of the absorption edge. A reduction of the effective band gap with increasing temperature is observed

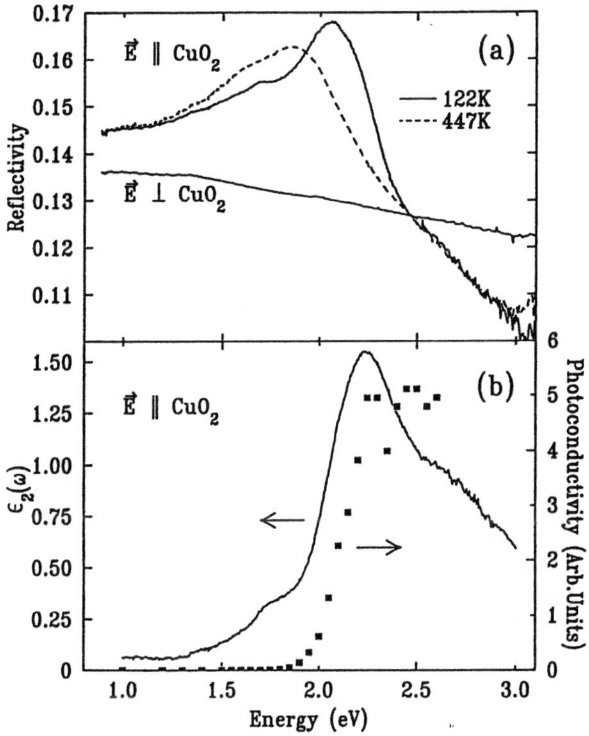


**Fig. 2.1.** (a) Reflectivity spectra of $La_2CuO_4$ at $T$ = 122 K and 447 K. (b) In-plane $\epsilon_2(\omega)$ at $T = 122$ K together with the photoconductivity at $T = 132$ K on an identically prepared sample (from Ref. [10]).

in covalent as well as ionic semiconductors; the dominant contribution in both systems comes from the electron-phonon coupling. In Fig. 2.2 we display the best fit assuming coupling to either acoustical or optical phonons. Only a coupling to optical phonons is consistent with the data; we extract an average phonon energy of $\hbar\omega_0 = 0.043 \pm 0.004$ eV. The absence of temperature dependence below $\sim 100$ K results from the freeze-out of optical phonons at low temperatures.

The temperature dependence of the absorption edge strongly suggests that photoexcited electrons and holes form polarons. This result is consistent with measurements of Chen et al. [7] which show that the static dielectric constant in $La_2CuO_4$ is large enough that electrons and holes form Fröhlich polarons. For such polarons, the $T$-dependence of the polaron self energy causes a shift of the band edge [11] given by

$$E_g(T) = E_g^0 - 2\,\hbar\omega_0\,\alpha_p\,(n(\hbar\omega_0/k_B T) + 1), \tag{2.1}$$

where $n$ is the Bose-Einstein occupation number. The assumption of coupling to a single LO phonon of frequency $\omega_0$, implicit in Eq. (2.1), will be discussed

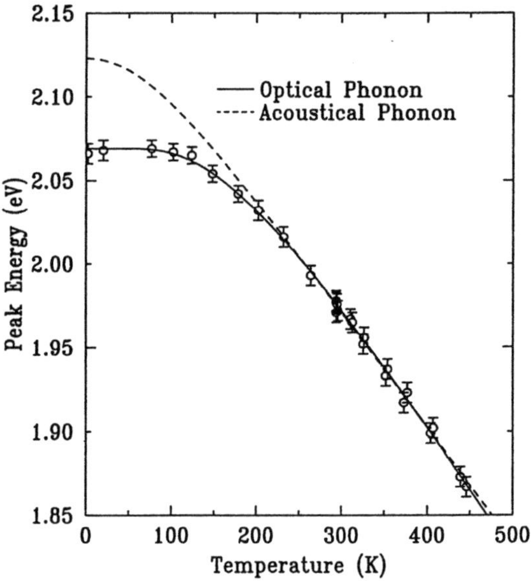


**Fig. 2.2.** The temperature dependence of the reflectivity peak. The solid and broken lines represent the best fits to an edge shift induced by optical and acoustic phonons, respectively.

below. The polaron coupling constant is defined as

$$\alpha p = \frac{1}{2}\frac{e^2}{\hbar\omega_0}\left(\frac{1}{\epsilon_\infty} - \frac{1}{\epsilon_s}\right)\sqrt{\frac{2\,m\,\omega_0}{\hbar}}. \tag{2.2}$$

where $\epsilon_s$ is the static dielectric constant and $\epsilon_\infty$ the electronic contribution. For simplicity, we assume that the electron and the hole have the same mass and hence the same $\alpha_p$.

Equation (2.1) shows that the electron-phonon scattering lowers the energy of the quasiparticle. The phonon scattering also induces a decay rate of the quasiparticles. We assume that the decay rate for the correlated state of electron and hole is the sum of the individual polaron relaxation rates [12]. This gives

$$\gamma(T) = 2^{3/2}\,\omega_0\sqrt{\hbar\omega_0/E}\,\alpha p\,(2\,n(\hbar\omega_0/k_B T) + 1). \tag{2.3}$$

Here $E$ denotes the total kinetic energy of the electron and hole, that is, $\hbar\omega - E_g(T)$. We emulate the effect of the electron-phonon interaction phenomenologically by introducing the lifetime broadening into the dielectric function for the Frenkel excitonic band edge where the band edge is shifted according to equation (2.1). A detailed account on this model and how it is applied to $La_2CuO_4$ can be found in Ref. [3].

Fig. 2.3 displays the experimental in-plane dielectric function for four temperatures together with the predictions of our model. With $\hbar\omega_0$ fixed at

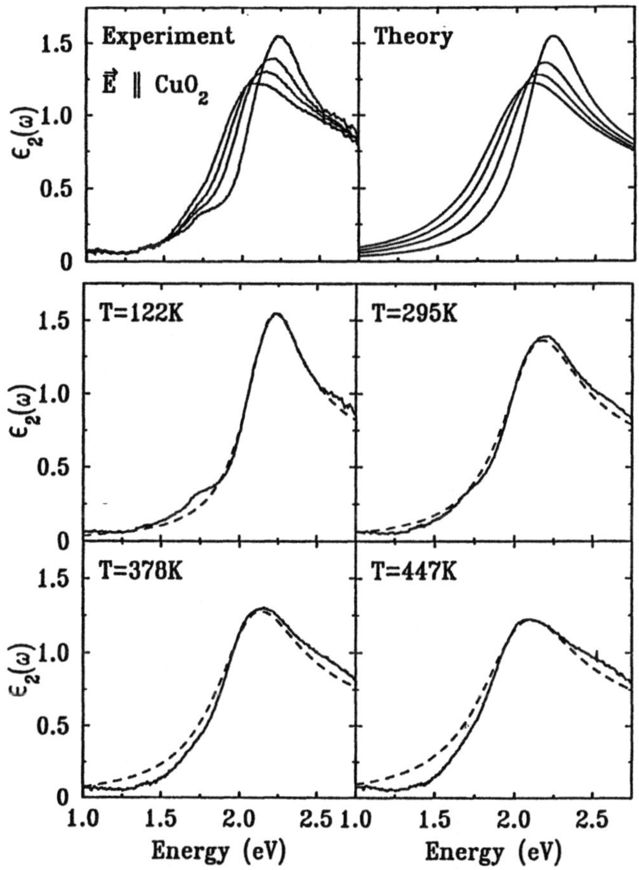


**Fig. 2.3.** The in-plane $\epsilon_2(\omega)$ for four different temperatures together with the best fit to the data using the model and the parameters discussed in the text. All four temperatures are shown in the upper panel; theory and experiment are compared for individual temperatures in the lower panel.

0.043 eV, the model has four adjustable parameters: $g$, $\alpha$p, $E_g^0$ and the overall scale factor. Two of these parameters, $g$ and $\alpha$p, control the shape of $\epsilon_2(\omega)$ including its temperature dependence. To test the model, we first fit $\epsilon_2(\omega)$ at the lowest temperature, $T = 122$ K, with all four parameters adjustable. As seen in Fig. 2.2, the model describes the measured $\epsilon_2(\omega)$ at $T = 122$ K quite well. The dashed lines for $T = 295$ K, 378 K and 447 K in Fig. 2.3 are calculated from our model with the parameters fixed at their $T = 122$ K values. With the polaron coupling constant determined by the broadening at $T = 122$ K the theory predicts the change of the peak shape and its position as well as its overall strength all the way up to $T = 447$ K. Moreover, the optimal values of the fitting parameters are found to be consistent with independent measures as well as theoretical estimates [3]. Clearly, the theory works very

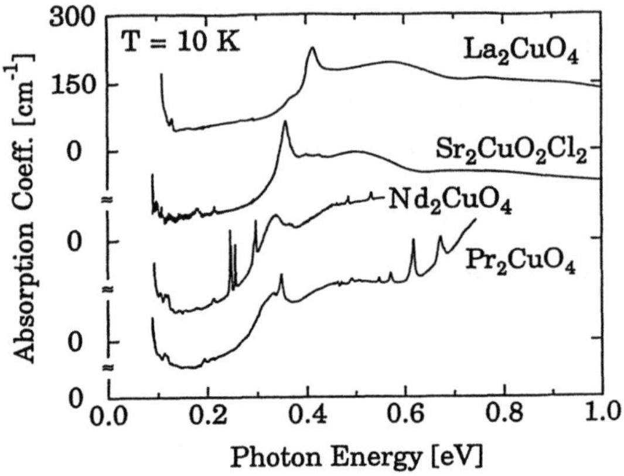


**Fig. 2.4.** Absorption coefficient vs. photon energy in $\alpha$-polarization for $La_2CuO_4$, $Sr_2CuO_2Cl_2$, $Nd_2CuO_4$, and $Pr_2CuO_4$. Common absorption features are seen in the range 0.2 to 0.6 eV. The additional narrow lines which complicate the Nd- and $Pr_2CuO_4$ spectra corresponds to intra-f-shell transitions.

well, indeed much better than one might reasonably expect given the simple form and the small number of adjustable parameters.

To justify the use of a single phonon frequency $\omega_o$, consider the optical phonons in $La_2CuO_4$. Only longitudinal optical modes of $E_u$ symmetry couple to the charge carriers in the $CuO_2$ plane. $La_2CuO_4$ has four such modes. Inelastic neutron scattering measurements [13] find that of these four, two Cu-O bond-bending modes display large LO-TO splittings at the zone center, indicating large associated dipole moments. The zone-center energies of the respective LO phonons are 0.034 and 0.057 eV, respectively. Thus, our measured $\hbar\omega_0 = 43$ meV is consistent with a weighted average of these two frequencies, and since the two frequencies are not too different, a description assuming coupling to one effective phonon mode provides an excellent fit to the data as shown above.

## 3. Crystal-Field Excitations

In Fig. 2.4 we show the absorption coefficients in the energy range 0.1–1.0 eV at 10 K for four undoped materials [4]. The absorption spectra are obtained from transmission measurements through $\sim$ 100$\mu$m thick single crystals. Common absorption features are seen in the range 0.2 to 0.6 eV. The additional narrow lines which complicate the Nd- and $Pr_2CuO_4$ spectra result from intra-$f$-shell transitions, which are electric-dipole allowed because the rare-earth sites do not possess inversion symmetry. Absorption structure is also observed near 1.5 eV as shown in Fig. 3.1. The absorption coefficient for

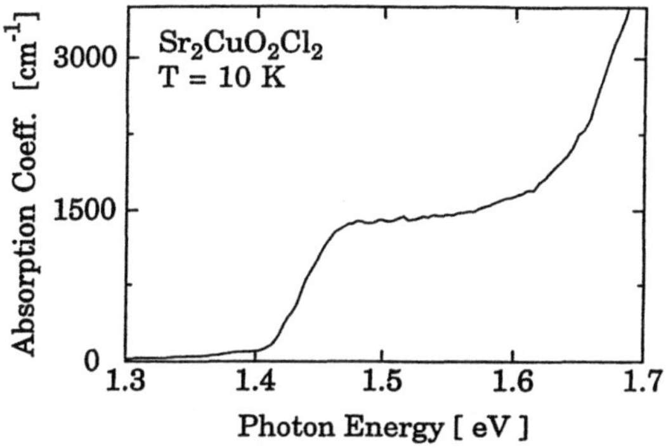


**Fig. 3.1.** Strong $\sim 1.5$ eV absorption band observed in $Sr_2CuO_2Cl_2$ at 10 K, identified as arising from a $d_{x^2-y^2} \to d_{xy}$ exciton. The charge-transfer absorption edge begins at 1.65 eV.

this band is so large that we can only observe it in $Sr_2CuO_2Cl_2$, which can be cleaved to $\sim 10\,\mu$m thickness.

In Fig. 3.2 we display the polarization dependent low-energy absorption spectra for undoped $La_2CuO_4$. A linear background which results from surface scattering or absorption is subtracted. The dominant feature is a sharp line near 0.4 eV with a set of broader bands at higher energies. Identical spectra are obtained in both $\alpha$ ($\boldsymbol{E} \parallel CuO_2$, $\boldsymbol{k} \perp CuO_2$) and $\sigma$ ($\boldsymbol{E} \parallel CuO_2$, $\boldsymbol{k} \parallel CuO_2$) polarization while all features are absent in the $\pi$ ($\boldsymbol{E} \perp CuO_2$, $\boldsymbol{k} \parallel CuO_2$) polarization. This tells us that the absorption bands are electric-dipole allowed when the electric field vector lies in the $CuO_2$ layer. Identical spectra are obtained for $La_2CuO_4$ grown in a Pt crucible (solid lines) and by the floating-zone method using no crucible (dashed line) demonstrating that the absorption features are independent of the impurity concentration. Thus we conclude that the weakly dipole-allowed excitations are intrinsic.

The sharp feature near 0.4 eV clearly resembles an exciton. One strong candidate is the lowest energy transition between the crystal-field split Cu $3d$ bands. Since all the four materials are $2D$, $S = \frac{1}{2}$ Heisenberg antiferromagnets, it is natural to ask whether the broad high energy sidebands could result from simultaneous exciton-magnon absorption. With a zone-boundary energy of $\sim 2J \simeq 0.3$ eV for $La_2CuO_4$, such a mechanism is certainly compatible with the broad structure.

For the crystal-field excitations both the initial and final state have even parity, making the transition electric dipole forbidden. Creation of a magnon, on the other hand, requires a spin-flip which is also electric dipole forbidden. However, the joint excitation of an exciton and a magnon is weakly electric-dipole-allowed [14]. With the exchange integral for the excited Cu $3d^9$ states much lower than $J$, the excess energy cost of creating one magnon in addition to the exciton is less than $2J$. A simple Ising cluster calculation yields $\sim 3\,J/2 \simeq 0.19$ eV. We show in Fig. 3.3 that the measured offset to the first

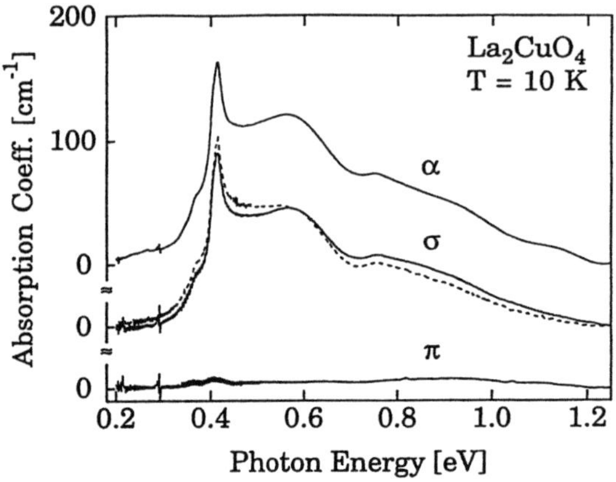


**Fig. 3.2.** $La_2CuO_4$ absorption coefficient vs. photon energy for $\alpha$, $\sigma$ and $\pi$ polarizations, after subtracting the linear background. Additional $\sigma$ trace (dashed line) is for a floating-zone grown crystal.

peak is 0.15 eV (0.14 eV) in $La_2CuO_4$ ($Sr_2CuO_2Cl_2$). The predicted positions for the 2-, 3-, and 4-magnon peaks relative to the one-magnon peak assuming a constant energy difference of $3J/2 \simeq 0.19$ eV are also shown. Clearly the predicted 2-magnon position agrees well with the observed second peak in both $La_2CuO_4$ and $Sr_2CuO_2Cl_2$. In the $La_2CuO_4$ absorption there are also pronounced shoulders at the predicted 3- and 4-magnon positions.

A full theory for the exciton-multimagnon sidebands does not exist. We show in Fig. 3.3 the prediction for the exciton-1-magnon absorption spectrum for classical spins [15]. This classical theory predicts the peak position well, but underestimates the peak width. Currently no theory exists for the higher order sidebands.

Weakly electric dipole allowed exciton-magnon excitations appear in, for example, $RbMnF_3$ as discussed in Ref. [4]. By comparing the oscillator strength for the 1-magnon sideband in this material to the strength of the 1-magnon sideband in the copper oxides we find that not only the energies, but also the intensities of the magnon sidebands agrees with expectations. Thus the magnon sidebands, and by implication the exciton itself, are intrinsic excitations of the $CuO_2$ layer.

While identification of the sidebands seems unambiguous, the electronic origin of the exciton is somewhat more difficult to ascertain. Both the $d_{xy}$ and the $d_{zx,yz}$ states are coupled to the $d_{x^2-y^2}$ through the angular momentum $\boldsymbol{L}$. If any of these states were just 0.4 eV above the ground state the spin-orbit coupling would give rise to extremely large anisotropies in the $g$ value, and thereby the spin susceptibility and the anisotropic exchange, as well as a large anisotropic Van Vleck susceptibility [4]. None of these are observed thus ruling out identification of the 0.4 eV peak as involving transitions to the excited $d_{xy}$ and $d_{zx,yx}$ states. The only Cu-ion crystal-field level not coupled

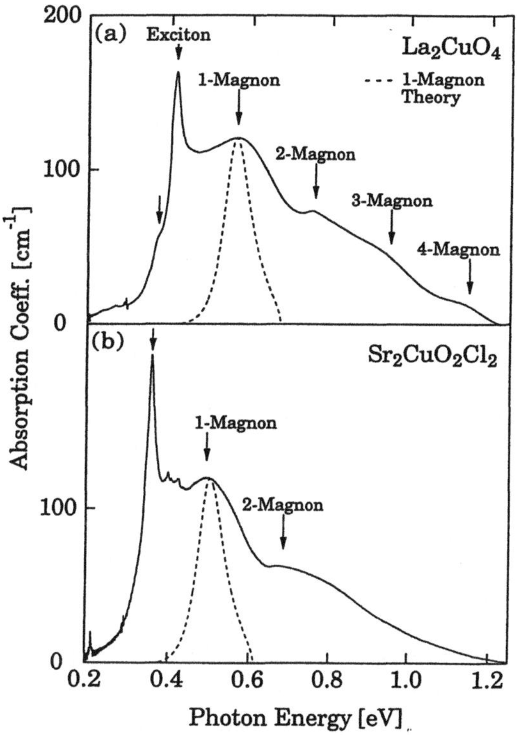


**Fig. 3.3.** $La_2CuO_4$ and $Sr_2CuO_2Cl_2$ absorption coefficients in $\alpha$ polarization. The characteristic energies $\sim 3\,n\,J/2$ for the magnon sidebands are noted. The predicted functional form for a 1-magnon sideband (dashed line) is also shown. The sharp peak near 0.4 eV most likely results from the $d_{x^2-y^2} \rightarrow d_{3z^2-r^2}$ transition.

to $d_{x^2-y^2}$ through $\boldsymbol{L}$ is the $d_{3z^2-r^2}$ state. Thus the exciton near $\sim 0.4$ eV most likely results from $d_{x^2-y^2} \rightarrow d_{3z^2-r^2}$ transitions. This conclusion is consistent with recent electroreflectance measurements [5] where we positively identify the $d_{x^2-y^2} \rightarrow d_{xy}$ and $d_{x^2-y^2} \rightarrow d_{zx,yz}$ transitions in $La_2CuO_4$ near 1.4 and 1.6 eV respectively. The weak absorption band observed near 1.5 eV in Fig. 3.1 is then most likely due to transitions to the $d_{xy}$ excited state.

## 4. Impurity Absorption

In Fig. 4.1(a) we display the reflectivity spectra for undoped and lightly doped $La_2CuO_{4+y}$. The out-of-plane spectrum shows no structure except for a sharp reduction in the reflectivity for $\hbar\omega \leq 0.19$ eV. This decrease results from optical phonons near 0.08 eV [2]. The in-plane reflectivity spectrum for the undoped crystal is almost identical to the out-of-plane spectrum. For the doped crystal, however, the in-plane reflectivity spectrum is enhanced throughout the region from 0.1 to 0.75 eV with a maximum at 0.1 eV. The strength of the 0.1 eV reflectivity peak decreases rapidly as the temperature increases.

In Fig. 4.1(b) we display the imaginary part of the dielectric constant, $\epsilon_2$. Two separate absorption peaks near 0.13 and 0.5 eV are clearly distinguished. Of these, only the 0.13 eV absorption band displays any temperature dependence below $\sim$ 300 K. The decrease of the oscillator strength in the 0.13 eV band between 100 K and 300 K coincides with an increase of the free carrier density as carriers are thermally excited from the impurity band to the extended states. Specifically, we find that the decrease in spectral weight of the 0.13 eV band is, within error bars, identical to the increase in spectral weight of the dc conductivity as evinced by the temperature dependence of the Hall coefficient [8]. Moreover, we will show that the detailed temperature dependence of the 0.13 eV band is predicted by that of the Hall density. Together this provides strong evidence that the 0.13 eV band results from photo-ionization of the impurities.

The optical excitation energy $\sim$ 0.13 eV is almost a factor 4 larger than the thermal activation energy $E_b = 0.035$ eV measured from the Hall effect. This situation is quite common in ionic insulators in which the charge carrier, bound to the impurity, is strongly coupled to the optical phonons, causing a lattice distortion around the impurity site. During optical ionization of the impurity, the lattice is effectively frozen, obeying the Frank-Condon principle. Thus, the charge carrier must overcome its polaronic binding energy in addition to the impurity binding energy in order to delocalize.

The localization length of the impurity state in lightly oxygen doped $La_2CuO_4$ is measured to be 8 Å [7], implying that the localized polarons are still large. We therefore assume the carriers are strongly enough coupled to the optical modes for the lattice to be frozen during an optical excitation. A detail discussion of photoionization of polaronic impurities as applied to $La_2CuO_4$ is published elsewhere [9]. The model contains four parameters; two of these, the impurity binding energy $E_b = 0.035$ eV and the average phonon frequency $\hbar\omega_0 = 0.043$ eV are determined from other experiments. The two fitting parameters are the number of phonons involved in the excitation, $S$, and the ratio of donors to acceptors.

The model can best be compared to the data by studying the difference $\epsilon_2(\omega, T) - \epsilon_2(\omega, 238\text{ K})$ displayed in Fig. 4.2. The reflectivity data in Fig. 4.1(a) show a slight decrease in the strength of the 0.5 eV band between 238 K and 295 K. Thus we use the 238 K scan to subtract off the 0.5 eV absorption. The resulting peak at $\sim$ 0.13 eV has a half-width of $\sim$ 0.07 eV. The long high energy tails may result from a weak temperature dependence of the 0.5 eV band. The dominant response to increasing temperature is a reduction in the overall oscillator strength in the 0.13 eV band with little change in shape as the localized carriers are thermally ionized.

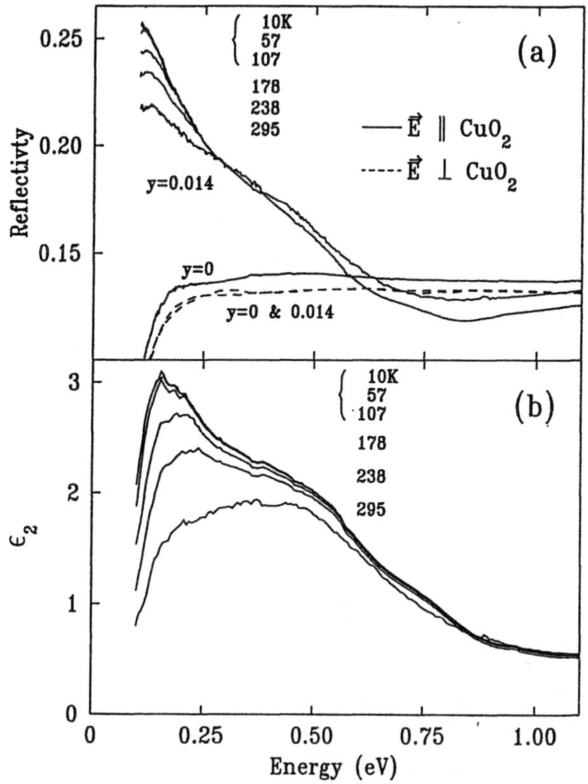


**Fig. 4.1.** (a) Reflectivity spectra of $La_2CuO_{4+y}$ with $y = 0$ and $y = 0.014$. The in-plane spectra for the doped sample are indistinguishable below $\sim$ 100 K. Above 0.3 eV they are only displayed at 10 K and 295 K with the latter having the higher reflectivity. (b) In-plane $\epsilon_2(\omega)$ for the oxygen doped sample.

The peak position, the line width and the line shape are all properties determined solely by S. The best fit is displayed in Fig. 4.2 which clearly recreates the position of the center frequency and the overall width as well as the general temperature dependence [9].

To study the temperature dependence in more detail we compare in Fig. 4.3 the in-plane reflectivity at 0.11 eV as a function of temperature with that calculated from our model. The reflectivity is temperature independent below $\sim$ 100 K reflecting the negligible fraction of carriers thermally excited from the localized states. At higher temperatures this fraction increases, decreasing the strength of the 0.13 eV band. The real part of the dielectric response, $\epsilon_1(\omega)$, is extracted from the calculated $\epsilon_2(\omega)$ using the Kramer-Kronig relations. The reflectivity is then calculated from the total complex dielectric response, $\epsilon(\omega) = \tilde{\epsilon_o} + \epsilon_1(\omega) + i\epsilon_2(\omega)$ where $\tilde{\epsilon_o}$ is the effective dielectric constant. The best fit, displayed as a solid line in Fig. 4.3, recreates the

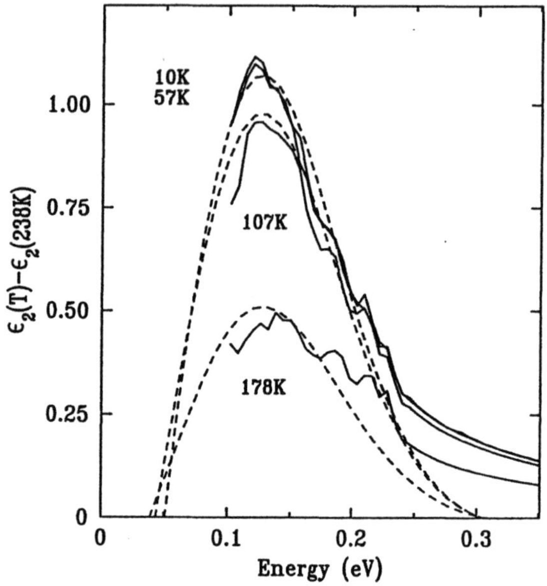


**Fig. 4.2.** The in-plane difference spectrum $\epsilon_2(w,T) - \epsilon_2(\omega, 238\text{ K})$ for different temperatures together with the best fit to the data for a model of photo-ionization of polaronic impurities (broken lines). The fit clearly recreates the peak position and the overall width as well as the general temperature dependence.

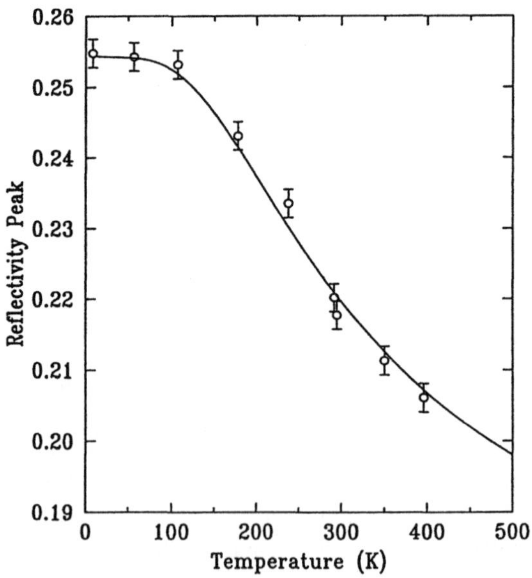


**Fig. 4.3.** The temperature dependence of the in-plane reflectivity at 0.11 eV. The solid line represent the best fit to the data using the model and parameters discussed in the text.

temperature dependence very well with fitting parameters consistent with independent measurements. For a complete discussion see Ref. [9].

The mid-infrared absorption is a common feature of the cuprates. In $La_2CuO_{4+y}$, $Nd_2CuO_{4-y}$ as well as $YBa_2Cu_3O_{6+y}$ two infrared absorption bands become stronger with doping [16], one in the energy range 0.5–0.75 eV and the other in the range 0.13–0.16 eV. It would be quite remarkable if the origin of these absorption processes were fundamentally different for the different materials. The lower energy band in $La_2CuO_{4+y}$ clearly comes from photo-ionization of polarons bound to impurities.

## 5. Conclusion

We have by means of polarization and temperature dependent reflectivity and transmission measurements developed new insight into the origin and nature of the infrared and optical electronic excitations in pure and lightly doped $La_2CuO_{4+y}$. In particular, we have demonstrated that both extended and localized holes interact with optical phonon modes to form large polarons. A remaining question concerns the origin of the doping enhanced 0.5 to 2 eV sub-gap absorption in the lightly doped materials. Perkins *et al.* [4] point out that introduction of carriers breaks the inversion symmetry causing the primary crystal-field excitons to grow with doping. Furthermore, the addition of holes is likely to introduce new matrix elements for dipole-allowed exciton-magnon excitations, possibly explaining the doping enhanced 0.5 eV mid-infrared band. Alternative mechanisms have been suggested by several theoretical workers [17]. Further research is required to establish unambiguously the origin of the 0.5 eV band in the doped samples.

*Acknowledgement.* The authors thank A. Cassanho and B. Keimer for supplying some of the crystals used in the measurements. We would like to thank D. Emin, P.A. Lee, A.K. McMahan and G. Sawatzki for helpful comments. This work was supported by the National Science Foundation Grants No. DMR 90-07825, DMR 90-22933 and DMR 90-14839. JMG also acknowledges partial support from the NTT Foundation.

## References

1. For a recent review see D.B. Tanner and T. Timusk in *Physical Properties of High Temperature Superconductors III*, edited by D. Ginsberg (World Scientific, Singapore 1992)
2. S. Uchida, T. Ido, H. Takagi, T. Arima, Y. Tokura, S. Tajima, Phys. Rev. B **43**, 7942 (1991)
3. J.P. Falck, A. Levy, M.A. Kastner, R.J. Birgeneau, Phys. Rev. Lett. **69**, 1109 (1992)

4. J.D. Perkins, J.M. Graybeal, M.A. Kastner, R.J. Birgeneau, J.P. Falck, M. Greven, Phys. Rev. Lett. **71**, 1621 (1993)
5. J.P. Falck, J.D. Perkins, A. Levy, M.A. Kastner, J.M. Graybeal, R.J. Birgeneau, unpublished
6. N.W. Preyer, M.A. Kastner, C.Y. Chen, R.J. Birgeneau, Y. Hidaka, Phys. Rev. B **44**, 407 (1991)
7. C.Y. Chen, R.J. Birgeneau, M.A. Kastner, N.W. Preyer, T. Thio, Phys. Rev. B **43**, 392 (1991)
8. N.W. Preyer, R.J. Birgeneau, C.Y. Chen, D.R. Gabbe, = H.P. Jenssen, M.A. Kastner, P.J. Picone, T. Thio, Phys. Rev. B **39**, 11563 (1989)
9. J.P. Falck, A. Levy, M.A. Kastner, R.J. Birgeneau, Phys. Rev. B **48**, 4043 (1993)
10. T. Thio, R.J. Birgeneau, A. Cassanho, M.A. Kastner, Phys. Rev. B **42**, 10800 (1990)
11. H.Y. Fan, Phys. Rev. **82**, 900 (1951)
12. H. Fröhlich, H. Pelzer, S. Zienau, Phil. Mag. **41**, 221 (1950); J. Appel, "Polarons", Solid State Physics **21**, 193–391 (Academic, New York, 1968)
13. L. Pintschovius, N. Pyka, W. Reichardt, A.Yu. Rumiantsev, N.L. Mitrofanov, A.S. Ivanov, G. Collin, P. Bourges, Physica B **174**, 323 (1991)
14. Y. Tanabe, T. Moriya, S. Sugano, Phys. Rev. Lett. **15**, 1023 (1965)
15. J.B. Parkinson, J. Phys. C **2**, 2012 (1969); J.B. Parkinson and R Loudon, J. Phys. C **1**, 1568 (1968)
16. G.A. Thomas, D.H. Rapkine, S.L. Cooper, S-W. Cheong, A.S. Cooper, L.F. Schneemeyer, J.V. Waszczak, Phys. Rev. B **45**, 2474 (1992)
17. See J. Lorenzana, L.Yu, Phys. Rev. Lett. **70**, 861 (1993) and references therein.

# Electronic Structure and High $T_c$ Superconductivity in the Cuprates

J. Bok and L. Force

Laboratoire de Physique du Solide, CNRS–ESPCI, 10, rue Vauquelin 75231 PARIS cedex 05

**Summary.** The origin of superconductivity in the cuprates is attributed to free carriers in the $CuO_2$ planes. These planes are weakly coupled in the c direction so that their electronic properties are nearly 2D. Electrons in a 2D periodic potential show a logarithmic singularity in the density of states called the van Hove singularity (v.H.s.). We assume that the Fermi level lies in the v.H.s. for optimum doping (highest $T_c$). We develop this van Hove scenario and compare its theoretical predictions to experiments.

We show that it explains, high $T_c$'s, variation of $T_c$ with doping, anomalous isotope effect, marginal Fermi liquid properties, and very small coherence length. We also take into account the Coulomb repulsion between electrons and show that it does not kill superconductivity if the v.H. peak is above a wide band. Finally we show that in this model we have two stable phases for half filling and optimum doping and perhaps phase separation for intermediate doping.

## 1. Introduction

It is now well-accepted that the origin of cuprates superconductivity is to be found in the $CuO_2$ planes which are weakly coupled together in the c direction, so that their electronic properties are nearly 2D. It is well-known since van Hove [1] (1953) that electrons in a 2D periodic potential show a logarithmic singularity (v.H.s.) in the density of states. The van Hove scenario is based on the assumption that, in the high $T_c$ cuprates, the Fermi level lies near such a singularity. Labbé and Bok [2] (1987) have shown that this assumption, in the frame work of the B.C.S. model, gives high $T_c$ and anomalous isotope effect.

Recently, several angular resolved photoemission spectroscopy (ARPES) experiments have revealed the presence of v.H.s. at the Fermi level in various cuprates the Bi 2122 compound and YBaCuO 1237 and 1248 [3],[4],[5].

This simple model already explained a certain number of experimental observations: variation of $T_c$ with doping, anomalous isotope effect, marginal Fermi liquid behaviour.

Another important point is the Coulomb repulsion which is neglected in the B.C.S. model. Anderson and Morel have shown [6] that the Coulomb potential is weakened by renormalization in broad bands, thus allowing the apparition of superconductivity. This suppresses the possibility to achieve a high $T_c$ by narrowing the band width. Nevertheless, we have shown that a peak in the density of states in the middle of a broad band, such as predicted by the v.H.s. can lead to a high $T_c$ [7].

An astonishing feature of high $T_c$ superconductors is the very low value of the coherence length $\xi$ from the B.C.S. formula and experimental values of Fermi velocity we find a higher value. We can solve this paradox by taking into account the anisotropy of the Fermi surface. The small value of $\xi$ corresponds to an averaging of $v_F$ over the whole Fermi surface and the singular points where $v_F = 0$ play an important role in reducing the coherence length.

## 2. Band structure of the cuprates

The simplest band structure, we can take for a square lattice is

$$\varepsilon_k = -2t(\cos k_x a + \cos k_y a) \tag{2.1}$$

$t$ is an interaction with nearest neighbours, this gives a square Fermi surface, and the v.H.s. corresponds to half filling (Fig. 2.1). We know that this is not a good representation of the high $T_c$ cuprates because for half filling (one electron per copper site) they are antiferromagnetic insulators. We think that the Fermi level is at the v.H.s. for the doping corresponding to maximum $T_c$, i.e. 16 % of holes per Cu in each $CuO_2$ plane or 0.42 filling of the first Brillouin zone (B.Z.). This can be achieved by taking into account the repulsive interaction between second nearest neighbours and the effect of the rhomboedric distortion [8].

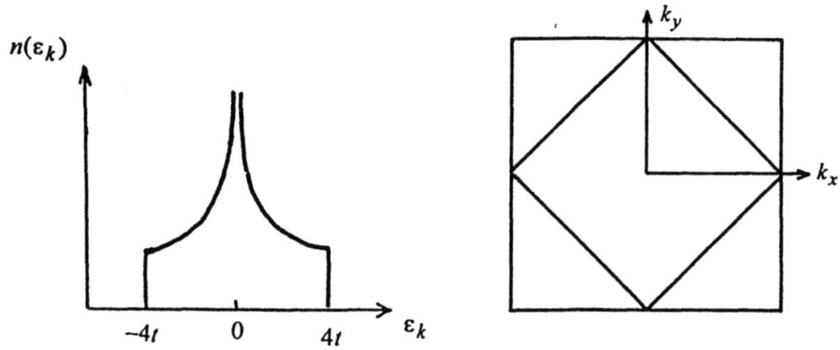


**Fig. 2.1.** Density of states $n(\varepsilon)$ and Fermi surface for a band given by eq. 2.1.

- repulsive interaction with second nearest neighbours (s.n.n.); the band structure becomes:

$$\varepsilon_k = -2t(\cos k_x a + \cos k_y a) + 4\alpha t \cos k_x a \cos k_y a \tag{2.2}$$

where $\alpha t$ is an integral representing the interaction with s.n.n. The singularity occurs for $\varepsilon = -4\alpha t$, there is a shift towards lower energy. The Fermi surface at the v.H.s. is no longer a square but is rather diamond-shaped (Fig. 2.2). For $a = 0.1$, this corresponds to a 46 % filling of the B.Z. .

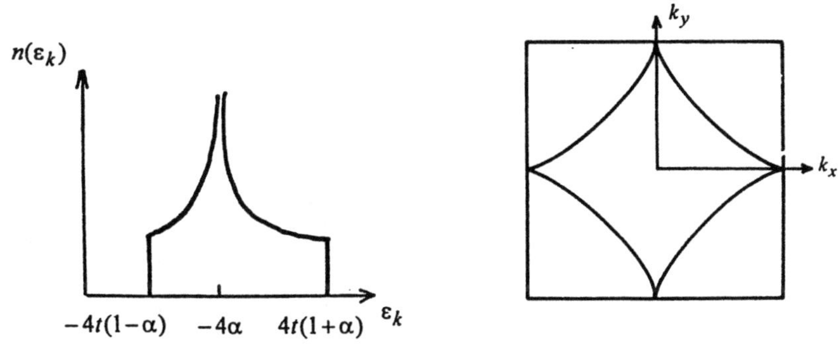


**Fig. 2.2.** Density of states $n(\varepsilon)$ and Fermi surface for a band given by eq. 2.2.

– rhomboedric distortion, the band structure is then:

$$\varepsilon_k = -2t(1+\beta)\cos k_x a - 2t\cos k_y a \tag{2.3}$$

where $\beta t$ represents the difference in the interaction with first neighbours in the x and y directions. The effects is now that the singularity is splitted in two (Fig. 2.3).

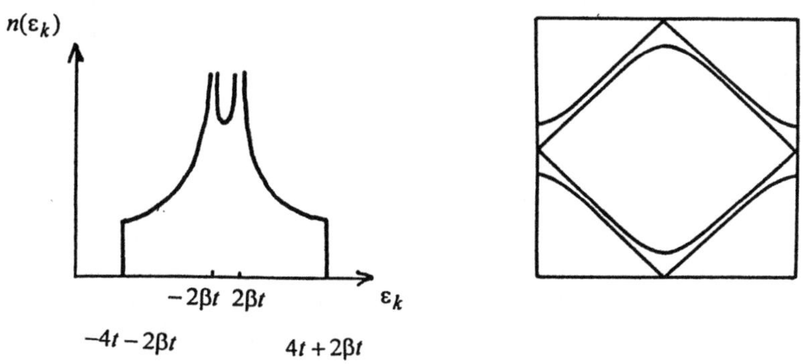


**Fig. 2.3.** Density of states $n(\varepsilon)$ and Fermi surface for a band given by eq. 2.3.

We may combine both effects and with $a = 0.1$ and $b = 0.1$, the first singularity is at 41 % of filling and the second at 51 % (Fig. 2.4). The second one has no physical meaning because near half filling electron–electron interaction opens a gap. Superconductivity is observed only when the Fermi level lies in the vicinity of the first v.H.s. .

Many more elaborate band structure calculations have been done for the cuprates [8, 9]. Most of them confirm the two dimensional character of the electronic structure of these compounds in the $CuO_2$ planes. They all find saddle points with singularities in the D.O.S. more or less near the Fermi

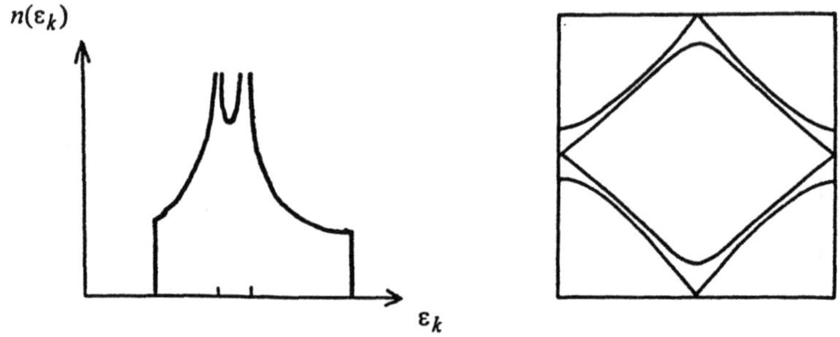


**Fig. 2.4.** Density of states $n(\varepsilon)$ for a band given by eq. 2.2 and 2.3 with $a = 0.1$ and $b = 0.1$.

level. A recent calculation of the band structure of the mercury cuprates [10] finds the Fermi level exactly pinned at the v.H.s. for the doping corresponding to the maximum $T_c$, thus confirming our simple model.

Recently, several experiments have been reported that measure the properties of the Fermi surface in the high $T_c$ cuprates, especially for Bi 2122 [3] and YBaCuO 1237 and 1248 [4]. The technique used is angle resolved photoemission spectroscopy (ARPES), which needs a careful preparation of the surface. All results obtained confirm the existence of a v.H.s. (or flat band, or saddle point) at the Fermi level for optimum doping. We present in Fig. 2.5 the results obtained by Dessau et al [3] for Bi 2122. Fig. 3.2 represents the energy of the electron measured from the Fermi level versus $\boldsymbol{k}$ three directions in the Brillouin zone $\Gamma$Y, Y$\overline{\text{M}}$ and $\overline{\text{M}}\Gamma$. A square B.Z. is used Y is the center of the square, $\Gamma$ the corner and M the middle of a side. Measurements are made only near regions A and B. Region A corresponds to the maximum of Fermi velocity. From the experimental results we find $v_{\mathrm{F,max}} = 3{\cdot}10^7$ cm/s. Region B is the v.H.s. it corresponds to a zero Fermi velocity. For comparison we put the curve representative of formula [1] on the same figure (in small dots). We see that the experimental band is flatter than that given by formula [1], the singularity is narrower. By comparison with experiment, we find $t = 0.25$ eV and for the band width $w = 8t = 4$ eV.

## Density of states (D.O.S.)

We know that near the v.H.s. the D.O.S. is logarithmic

$$n(\varepsilon) = n_0 \ln \left| \frac{D^*}{\varepsilon} \right| , \tag{2.4}$$

the values of $D^*$ and $n_0$ are determined from the band structure. From the simple model given by eq. 2.2, we find [11]

$$n_0 = 8/(\pi^2 D^*)$$

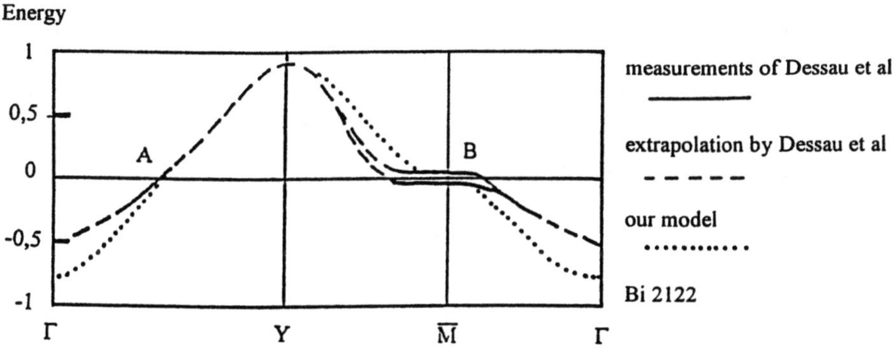


**Fig. 2.5.** Energy of the electrons measured from the Fermi level versus $k$ three directions in the Brillouin zone.

per spin, per unit cell and

$$D^* = 16t\sqrt{1-\alpha^2/4}.$$

We remark that $D^*$ is larger than the band width $8t$. We know that whatever the exact band structure in 2D, the D.O.S. is constant near the band edges ($\pm W/2$) and logarithmic near the v.H.s. . We shall thus take the following approximation:

- a constant D.O.S. $n_0$ between $-W/2$ and $+W/2$.
- a logarithmic peak $n = n_1 \ln|D/\varepsilon|$ between $-D$ and $+D$.

Near the v.H.s. this is equivalent to formula (2.4) with $D^* = D\exp(n_0/n_1)$. The constants $D$, $n_0$ and $n_1$ are not independent. The total number of states in the band is fixed. We have one orbital state (i.e. two spin states) per copper atom so that

$$n_0 W + 2n_1 D = 1$$

$n_0$ is also given by the effective mass at the band edge. For example, for $W = 2$ eV ($t = 0.25$ eV) we obtain $n_0 = 0.3$ states/(eV Cu atom) and $n_1 = 0.2$ which gives $D = 0.9$ eV. We shall also make calculations with $D = 0.3$ eV, corresponding to a much narrower v.H.s., as observed experimentally.

## 3. The Labbé Bok formula

This formula was obtained [2] using the following assumptions:

1. The Fermi level lies at the van Hove singularity.
2. The B.C.S. approximations:
   - The electron–phonon interaction is isotropic and so is the superconducting gap $\Delta$.

- The attractive interaction $V_p$ between electrons is nonzero only in an interval of energy $\pm\hbar\omega_0$ around the Fermi level where it is constant. When this attraction is mediated by emission and absorption of phonons, $\omega_0$ is a typical phonon frequency.

In that case, the critical temperature is given by

$$K_B T_c = 1.13 D \exp\left[-\sqrt{\frac{1}{\lambda} + \ln^2\left(\frac{\hbar\omega_0}{D}\right) - 1.3}\right] \tag{3.1}$$

where $\lambda = n_1 V_p/2$.

A simplified version of formula (3.1), when $\hbar\omega_0$ is not too small compared to $D$ is:

$$K_B T_c = 1.13 D \exp\left(-\frac{1}{\sqrt{\lambda}}\right)$$

The two main effects enhancing $T_c$ are

1. The prefactor in formula (3.1) is an electronic energy much larger than a typical phonon energy $\omega_0$.
2. $\lambda$ is replaced by $\sqrt{\lambda}$ in formula (4.1). In the weak coupling limit when $\lambda < 1$, the critical temperature given by formula (4.1) is very high. In fact it gives too high values of $T_c$, we shall see later that this is due to the fact that we have neglected Coulomb repulsion between electrons. Taking this repulsion into account we shall obtain values for $T_c$ which are very close to the observed one.

As it is however, this approach already explains many of the properties of the high $T_c$ cuprates near optimum doping.

- The variation of $T_c$ with doping:
  The highest $T_c$ is obtained when the Fermi level is exactly at the v.H.s. For lower or higher doping the critical temperature decreases. That is what is observed experimentally [12].
- The isotope effect:
  Labbé and Bok [2] showed using formula (3.1), that the isotope effect is strongly reduced for high $T_c$ cuprates. C.C. Tsui et al [13] have calculated the variation of the isotope effect with doping and shown that it explains the experimental observations (Fig. 3.1).
- Marginal Fermi liquid behaviour:
  In a classical Fermi liquid, the lifetime broadening $1/\tau$ of an excited quasi particle goes as $\varepsilon^2$. The marginal Fermi liquid situation is the case where $1/\tau$ goes as $\varepsilon$. Theoretically marginal behaviour has been established in two situations (a) the half filled nearest–neighbour coupled Hubbard model on a square lattice and (b) the Fermi level lies at a v.H. singularity [13]. Experimental evidence of marginal Fermi liquid behaviour has been seen in angle resolved photoemission [14], infrared data [15] and temperature dependence of electrical resistivity [16]. Marginal Fermi liquid

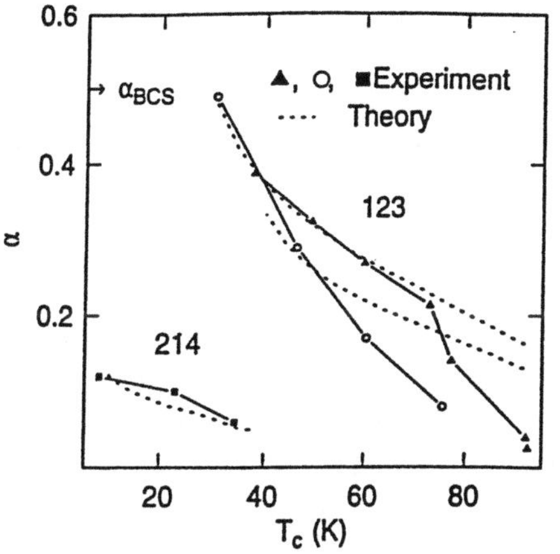


**Fig. 3.1.** Isotopic effect in $La_2CuO_4$ and YBaCuO (123).

theory, in the framework of v.H.s. predicts a resistivity linear with temperature $T$. This was observed by Kubo et al [16]. They also observe that the dependence of resistivity goes from $T$ for high $T_c$ material to $T^2$ for as the system is doped away from the $T_c$ maximum, which is consistent with our picture; in lower $T_c$ material the Fermi level is pushed away from the singularity (Fig. 3.2).

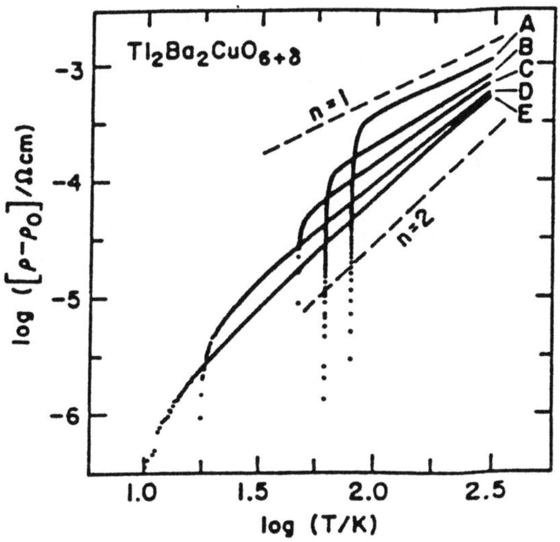


**Fig. 3.2.** Resistivity $\rho$ versus temperature. For the highest $T_c$ compounds $\rho$ varies linearly with $T$, for low $T_c$ compounds the variation is quadratic, from ref. [17].

## 4. The Coulomb repulsion between electrons

As soon as 1962 Anderson and Morel [6] have shown that the electron–electron repulsion plays a central role in superconductivity. Assuming a constant repulsive potential $V(\boldsymbol{k}, \boldsymbol{k}') = V_c$ from 0 to $E_F$ they find that $T_c$ is given by:

$$T_c \cong T_0 \exp\left[-\frac{1}{\lambda - \mu^*}\right] \tag{4.1}$$

with

$$\mu = N_0 V_c \quad \text{and} \quad \mu^* = \frac{\mu}{1 + \mu \ln E_F/\omega_0} .$$

M.L. Cohen and P.W. Anderson [17] assumed that for stability reasons $\mu$ is always greater than $\lambda$. V. Ginzburg [18] gave arguments that in some special circumstances $\mu$ can be smaller than $\lambda$. Nevertheless if we take $\mu \geq \lambda$, superconductivity only exists because $\mu^*$ is of the order of $\mu/3$ to $\mu/5$ for a Fermi energy of the order of 100 $\omega_0$. It is useless to reduce the width of the band $W$ ($E_F$= W/2 for a half–filled band) because $\lambda$ and $\mu$ vary simultaneously and $\mu^*$ becomes greater if $E_F$ is reduced, thus giving a lower $T_c$. Superconductivity can even disappear in a very narrow band if $\lambda - \mu^*$ becomes negative.

We have shown [17] that nevertheless high $T_c$ can be achieved in a metal containing almost free electrons (Fermi liquid) in a broad band, with a peak in the D.O.S. near the middle of the band.

Taking a D.O.S. which is a constant $n_0$ between energies $-W/2$ and $+W/2$, (the zero of energy is at the Fermi level) and is $n(\varepsilon) = n_1 \ln|D/\varepsilon| + n_0$ between $-D$ and $+D$ we find for $T_c$, the following formula:

$$K_B T_c = \frac{D}{2}\left[0.819 + \frac{n_0}{n_1} - \sqrt{F}\right] \tag{4.2}$$

where

$$F = \left(0.819 + \frac{n_0}{n_1}\right)^2 + \left(\ln\frac{\hbar\omega_0}{D}\right)^2 - 2 - \frac{2}{n_1}\left(n_0 \ln\frac{2.28D}{\hbar\omega_0} - \frac{1}{V_p - V_c^*}\right)$$

$$V_c^* = \frac{V_c}{1 + V_c\left[\frac{n_1}{2}\left(\ln\frac{D}{\hbar\omega_0}\right)^2 + n_0 \ln\frac{W}{2\hbar\omega_0}\right]}$$

We can have a few limiting cases for this formula:

- $n_1 = 0$: no singularity. We find the Anderson–Morel formula.
- $V_c = 0$ and $n_0 = 0$: this gives the Labbé–Bok formula.

There are many effects enhancing $T_c$:

$\lambda - \mu^*$ is reduced by the square root, down to $\sqrt{\lambda_1 - \mu_1^*}$ when $n_1$ is large enough. As $\lambda - \mu^* < 1$ the critical temperature is strongly increased because this factor appears in an exponential.

The prefactor before the exponential is $D$, the singularity width instead of $\hbar\omega_0$. We expect $D > \hbar\omega_0$. For instance $D$ may be of the order of 0.5 eV and $\omega_0$ about a few 10 meV ($D/\hbar\omega_0$ of the order of 5 to 10).

We have made some numerical calculations using formula (4.2) to illustrate the effect of Coulomb repulsion. We used two values of $D$: $D = 0.9$ eV corresponding to $t = 0.25$ eV and a much more smaller value $D = 0.3$ eV. This calculations are illustrated by figures 4.1 to 4.4.

These calculations show that the Coulomb repulsion does not kill superconductivity in the framework of the L.B. model. The general rule for high $T_c$ in this model is to have a peak in the density of states near the middle of a broad band to renormalize the effective repulsion $\mu$.

## 5. Coherence length and anisotropy effects

The relation between the Fermi velocity $v_F$ and the intrinsic coherence length $\xi_0$ is, in the framework of the B.C.S. theory, $\xi_0 = \hbar v_F/\pi\Delta$. This is valid for a spherical Fermi surface where $v_F$ is constant.

If we take for $v_F$, the maximum value as measured by Dessau et al, $v_F = 3{\cdot}10^7$ cm/s, we find with the B.C.S. formula $\xi = 30$ Å while the experimental value is between 10 and 15 Å. Actually, these calculations do not take into account the fact that the singular points corresponding to $v_F = 0$ have an important statistical weight. The density of states is given by:

$$n(\varepsilon)d\varepsilon = \int_\Gamma \frac{dk_t dk_\perp}{2\pi^2}$$

where $\Gamma$ is a constant energy surface and $dk_t$ and $dk_\perp$ are the tangential and orthogonal components of $\boldsymbol{dk}$. If, following B.C.S., we construct a wave packet of width $2\Delta$, the average Fermi velocity becomes:

$$\hbar\overline{v}_F = \frac{2\Delta \int_\Gamma dk_t}{2\pi^2 \int_{-\Delta}^{\Delta} n(\varepsilon)d\varepsilon}$$

with the energy band given by eq. (2.1) and the Fermi level at the singularity, we find:

$$\hbar\overline{v}_F = \frac{2\sqrt{2}\pi a t}{1 + \ln(16t/\Delta)}$$

or

$$\overline{v}_F = \frac{\pi v_{F,max}}{1 + \ln(16t/\Delta)} \tag{5.1}$$

In formula (5.1) all parameters $v_{F,max}$, $t$ and $\Delta$ have been measured experimentally with the following numerical values, $v_{F,max} = 3{\cdot}10^7$ cm/s, $8t = 1.5$ eV, $\Delta = 20$ meV, we find $v_F = 1.5{\cdot}10^7$ cm/s and $\xi_0 = 15$ Å. We insist that this observed value is obtained without any adjustable parameter.

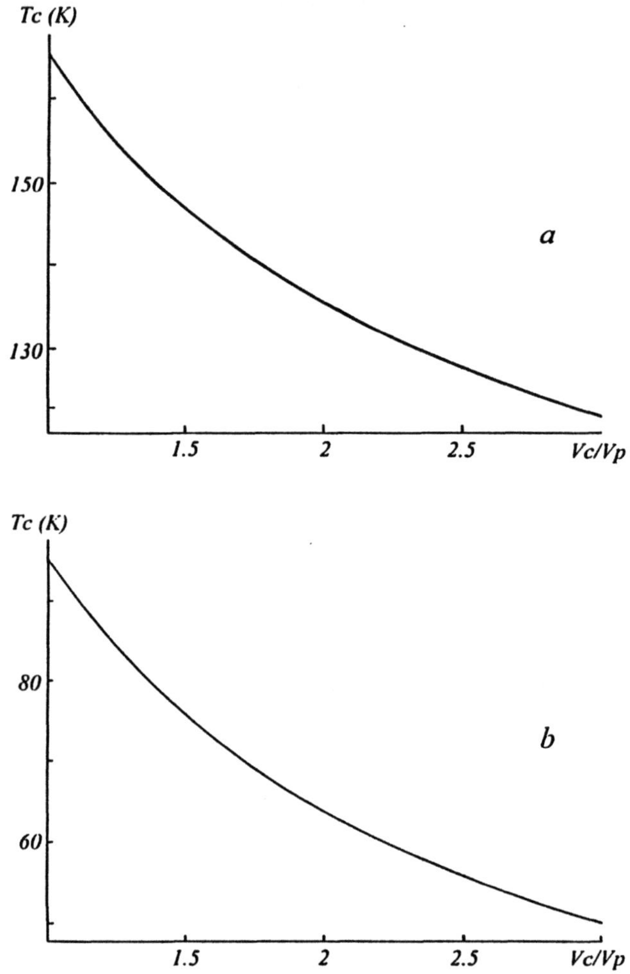


**Fig. 4.1.** Effect of Coulomb repulsion on $T_c$, the following numerical values have been used:
(a) $D = 0.9$ eV, $n_0 = 0.3$ states/(eV Cu atom), $W = 2$ eV, $n_1 = 0.2$.
(b) $D = 0.3$ eV, $n_0 = 0.3$ states/(eV Cu atom), $W = 2$ eV, $n_1 = 0.16$.
In both cases $\hbar\omega_0 = 0.05$ eV and $\lambda = 0.5$.

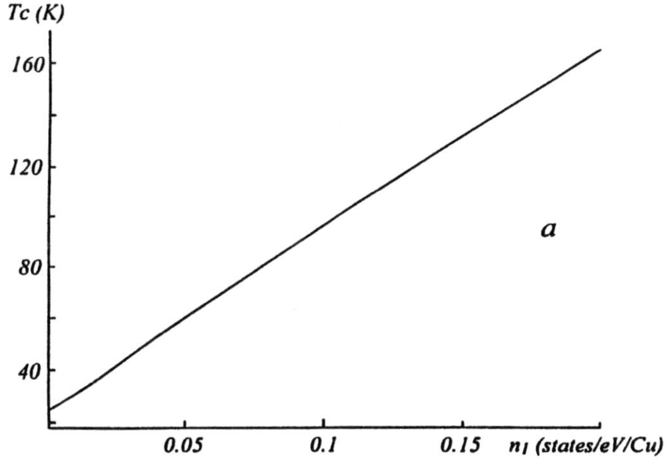


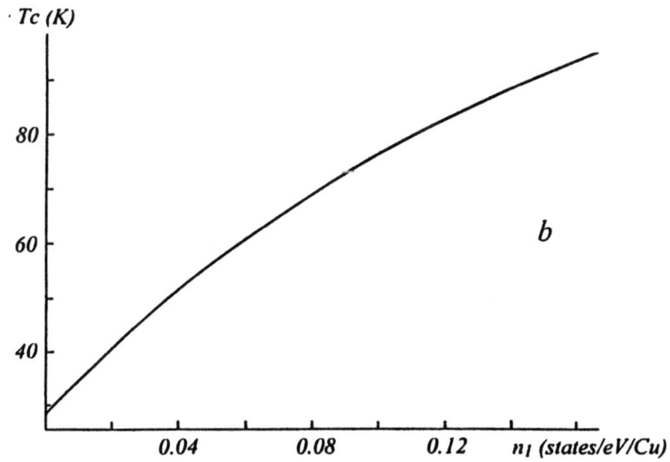


**Fig. 4.2.** Influence of the number of electrons in the singularity $n_1$, on the critical temperature $T_c$. Fig. a and b correspond to the same numerical values as in Fig. 4.1a and b respectively.

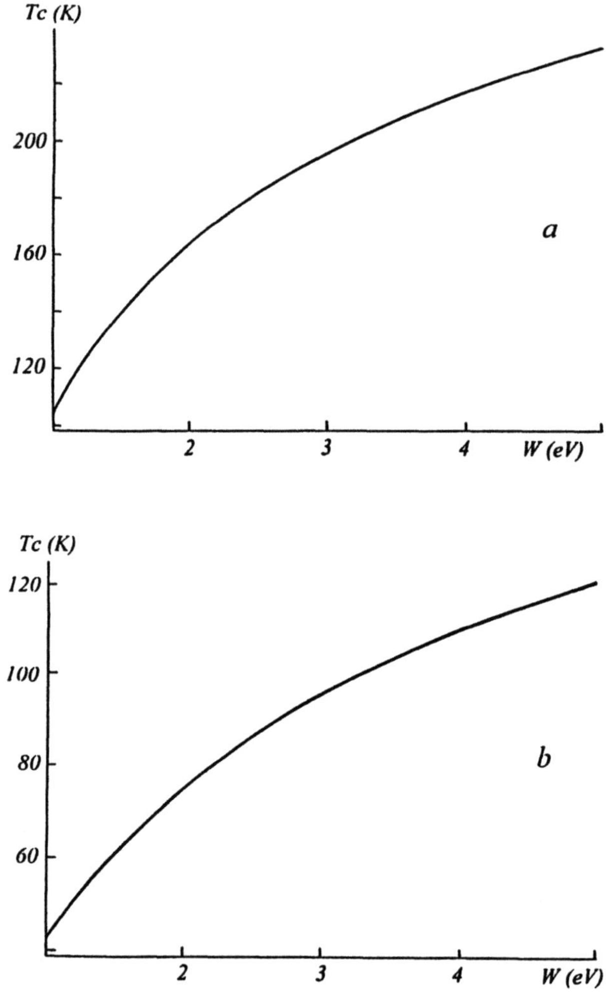


**Fig. 4.3.** Influence of the band width $W$ on $T_c$. The numerical values are the same as in Fig. 4.1.

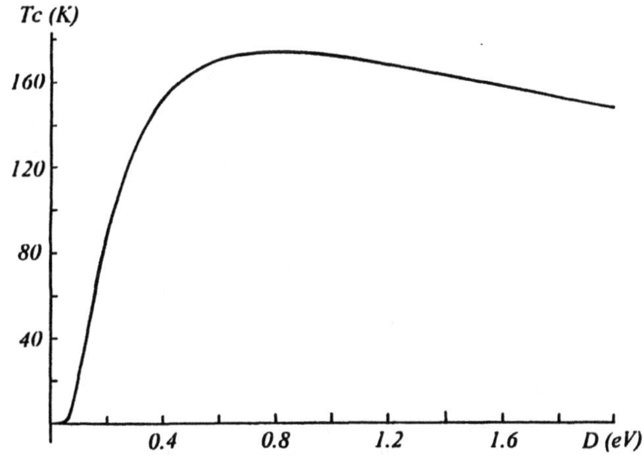


**Fig. 4.4.** Effect of the width of the singularity $D$ on $T_c$. $n_0$ and the total number of electrons per unit cell are maintained constant with this set of parameters.

## 6. Phase separation and conclusion

At low temperature, the free energy of the system is minimum in two situations:

1. At half filling, where a gap opens and the stable phase is insulating and antiferromagnetic.
2. When the Fermi level lies at the singularity where the condensation energy is of the order of $E_c \cong n_1 \Delta^2$, this is a large energy for HTCS where $\Delta$ is of the order of 20 meV. For intermediate doping, there is a natural tendency towards phase separation but this can be prevented if the dopants (oxygen for instance) are not mobile. In any case these phase transitions should be of first order and should be easily observable experimentally, by the existence of latent heat and by discontinuities in many properties such as elastic constants, sound velocity, etc.

In conclusion, we have shown that the van Hove scenario can explain many properties of the high $T_c$ cuprates. Its principal weakness is that it does not include strong correlations between electrons. These can perhaps be included as perturbations in a more complete theory.

## References

1. L. van Hove, *Phys. Rev.* **89**, 1189 (1953).
2. J. Labbé, and J. Bok, *Europhysics Letters* **3**, 1225 (1987).
3. D.S. Dessau et al, preprint.
4. K. Gofron, and J.C. Campuzano, *J. Phys. Chem. Sol.* **53**, 12, 1577 (1992).

5. K. Gofron et al, preprint.
6. P.W. Anderson, and P. Morel, *Phys. Rev.* **125**, 4 (1962).
7. L. Force, and J. Bok, *Solid State Comm.* **85**, 975 (1993).
8. D.L. Novikov, V.A. Gubanov, and A.J. Freeman, *Physica C* **210**, 301 (1993).
9. M. Hirao, T. Uda, and Y. Marayama, *Physica C* **195**, 230 (1992).
10. D.L. Novikov, and A.J. Freeman, *Physica C* **3**, 4 (1993).
11. D.Y. Xing, M. Liu, and C.D. Dong, *Phys. Rev. Lett.*, (comments), to be published.
12. Y. Koike et al, *Physica C* **159**, 105 (1989).
13. C.C. Tsuei, D.M. Newns, C.C. Chi, and P.C. Pattnaik, *Phys. Rev. Lett.* **65**, 2724 (1990); D.M. Newns, C.C. Tsuei, P.C. Pattnaik, and C.L. Kane, *Comments Cond. Mat. Physics* **15**, 273 (1992).
14. G.G. Olson et al, *Phys. Rev. B* **42**, 381 (1990).
15. Z. Schlesinger et al, *Phys. Rev. Lett.* **67**, 1657 (1991).
16. Y. Kubo et al, *Phys. Rev. B* **43**, 7875 (1991).
17. M.L. Cohen, and P.W. Anderson in: *Superconductivity in d and f band Metals*, ed.: D.H. Douglass A.I.P., New York (1972).
18. V. Ginzburg, *Comtemporary Physics* **33**, 1, 15–23 (1992).

# Apex Oxygen Anharmonicity and Low Energy Electronic Transitions Observed in 1-2-3 by IR-Excited Raman Spectroscopy

G. Ruani

Istituto di Spettroscopia Molecolare, C.N.R., Bologna(I)

**Summary.** The near infrared exited Raman scattering provides, because of very different resonant condition, new informations on high $T_c$ cuprate systems with respect to the usual visible excited one. In this paper the Raman scattering of $MBa_2Cu_3O_{6+x}$ (M = Y, Er) is reported. Their results obtained at different oxygen content (O≤x≤1), with different oxygen isotopes ($^{16}O$, $^{18}O$) and different annealing procedures are compared with those obtained by visible excitation. The peculiar behaviour of the apex O(4) oxygen atom in this system is emphasized by these results. In particular, the evidences for the existence of an anharmonic potential associated to the apex oxygen, the observation of electronic levels around the Fermi energy associated with this atom are presented.

## 1. Introduction

The Raman scattering study of high critical temperature superconductors, shortly after their discovery, has played a significant role in characterizing materials and in investigating such low energy excitations as phonons, electronic or magnetic transitions and their interactions. In spite of some limitation of Raman scattering (the first order transitions are limited to the $\Gamma$ point of the Brillouin zone) compared with other spectroscopic techniques, like neutron scattering, the high accuracy and resolution that can be achieved allow detailed study of the lineshape of the phonon bands, their interaction with low energy electronic excitations, and also the detection of some minor changes that can be determined by doping or temperature changes. The strong intensity dependence of the Raman scattering on laser excitation frequency can provide information on the nature and the interband transitions and/or, because of the resonantly enhanced scattering, on the microstructure of the sample. In this sense it is very important to study the Raman scattering in as wide range of excitation energy as possible. In this paper we report the Raman scattering in $MBa_2Cu_3O_{6+x}$ exciting at 1.16 eV and compare the results, obtained at different oxygen content, with different oxygen isotopes and different annealing procedures with those obtained by visible excitation. In particular we will discuss the peculiar behaviour of the apex O(4) oxygen atom in this system, showing the existence of an anharmonic potential associated to this atom, the evidence for the existence of electronic levels, related to the O(4), at the Fermi energy and the relevance of these observations on the formation of the superconducting state.

## 2. IR excited FT-Raman spectroscopy

Infrared excited Raman spectra were performed in a back-scattering configuration by means of an IFS 88 Bruker FT spectrometer equipped with the FT-Raman module at 4 $cm^{-1}$ resolution. A diode pumped c.w. Nd-YAG laser is used for excitation (1.16 eV). A cooled Ge detector allowed us to collect the scattered radiation in a wide spectral range around the excitation energy (from 100 to 3300 $cm^{-1}$ in Stokes side and from 100 to 2100 $cm^{-1}$ in the anti-Stokes one). In order to avoid heating and possible damaging effects due to absorption of the incident laser radiation, we kept the laser power as low as possible (less than 75 mW within a spot of about 0.3 mm in diameter). The temperature is estimated from the Stokes/anti-Stokes ratio of the observed Raman bands (duly corrected for the response of the detector at both frequencies).

## 3. $YBa_2Cu_3O_{6+x}$ ($x \leq 2$): IR-Raman scattering at low oxygen concentration

The study of the resonant Raman scattering at different excitation frequencies allows clarification of important feature of the electronic band structure and electron-phonon coupling of solids. The Raman excitation profiles in $YBa_2Cu_3O_{6+x}$ were measured in the region of 1.8–2.7 eV by E.T. Heyen et al. [1]. Good agreement was found with Raman intensities calculated in the local density approximation [2] for $x = 1$, indicating that electronic states of $YBa_2Cu_3O_7$ can be treated, in first approximation, within the mean-field approach as for ordinary metals. However, electronic correlations are strong in the tetragonal semiconducting phase [3]. So far there is no reliable quantitative description of the metal-insulator transition in the high-$T_c$ oxides. Moreover the origin of the insulating gap and the nature of the electronic states, filling the gap upon doping, are not clear. In this context, measurements of Raman spectra of the semiconducting phase with laser excitation at photon energy lower than the charge transfer gap (1.7 eV) can provide useful information. The first Raman spectra of $YBa_2Cu_3O_{6+x}$ ceramics ($x = 0.15$; 0.3 and 0.85) exciting at 1.16 eV were reported in Ref. [4] and [5]. The main spectral feature was a strong band at 505 $cm^{-1}$ with overtones at 1010 $cm^{-1}$ and 1515 $cm^{-1}$ for $x = 0.15$. This observation suggested the resonant nature of the Raman scattering at 1.16 eV as well as a strong electron-phonon coupling.

By site selective oxygen isotope substitution in the semiconducting $YBa_2Cu_3O_{6+x}$ we have unambiguously assigned the strong resonant Raman band at about 505 $cm^{-1}$ to the $A_g$ mode of the apex oxygen in the proximity of short chain segment. The relevant energy shift of this mode with respect to the unperturbed $YBa_2Cu_3O_6$ structure is a clear indication of the strong interaction of the apex oxygen atoms with electronic excitation.

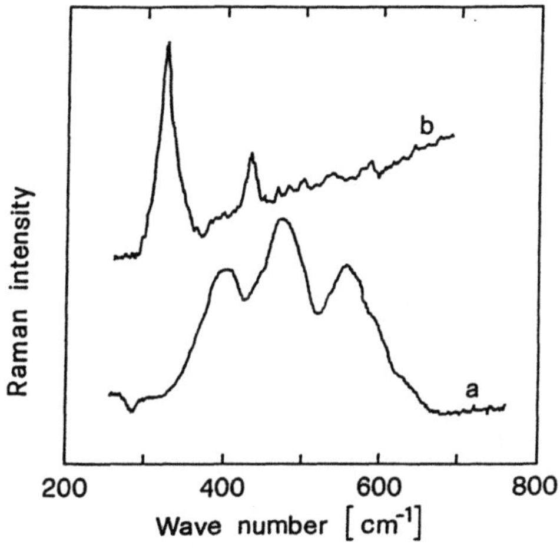


**Fig. 3.1.** Raman spectra of $YBa_2Cu_3O_{6.0}$ obtained (a) by exciting at 1.16 eV and (b) by exciting at 2.41 eV.

The behaviour of the Raman scattering of semiconducting $YBa_2Cu_3O_{6+x}$ samples is strongly dependent on laser excitation energy. As an example IR (1.16 eV) and visible (2.41 eV) excited Raman spectra of $YBa_2Cu_3O_6$ are reported in Fig. 3.1. The IR excited Raman spectrum (Fig. 3.1a) shows three strong bands at 387, 459 and 552 $cm^{-1}$ which are absent in the visible excited spectrum (Fig. 3.1b). At the same time for exciting energy 1.16 eV no signs are detected of the 320 $cm^{-1}$ band (340 $cm^{-1}$ in $YBa_2Cu_3O_6$, with $^{16}O$), which on the contrary dominates the Raman spectrum obtained with visible excitation. The latter band is usually present in Raman spectra of ceramics and crystals in the wide range of excitation energies from 1.8 eV to 2.7 eV. Such behaviour shows that the IR excited Raman scattering in these systems is characterized by peculiar resonant conditions.

The oxygen doping dependence of the visible excited Raman scattering of $YBa_2Cu_3O_{6+x}$ is well known [6]; there is only a minor shift of the bands (less then 2 $cm^{-1}$ at room temperature) and a small redistribution of their intensities by increasing the oxygen content from $x = 0$ to $x = 0.3$ (the same behaviour has been confirmed by us in these samples). On the contrary, the IR excited Raman scattering exhibits a strong dependence of the spectral features on oxygen content (see Fig. 3.2a,b). Also the strong bands clearly observed for excitation energy 1.16 eV, despite the fourth power dependence on frequency of the Raman cross-section, indicate their resonance enhanced character suggesting the presence of an electronic dipole transition deep inside the semiconducting charge transfer gap.

Upon doping a new resonant band appears at 477 $cm^{-1}$ (in $^{18}O$ substituted samples). For $x = 0.2$ the peak at 477 $cm^{-1}$ is the strongest (Fig. 3.2b), and its overtones are clearly seen. Inset of Fig. 3.2 represents the spectrum

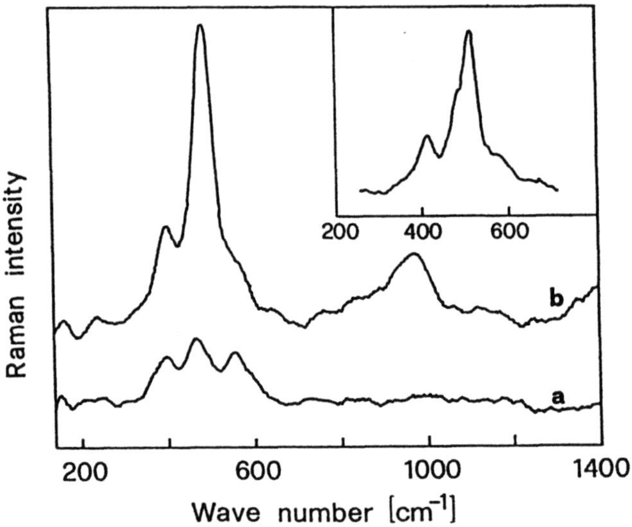


**Fig. 3.2.** FT-Raman spectra obtained by exciting at 1.16 eV of: (a) $YBa_2Cu_3O_{6.0}$ and (b) $YBa_2Cu_3O_{6.2}$. The spectrum of $YBa_2Cu_3^{16}O_{6.05}$ is shown in the inset.

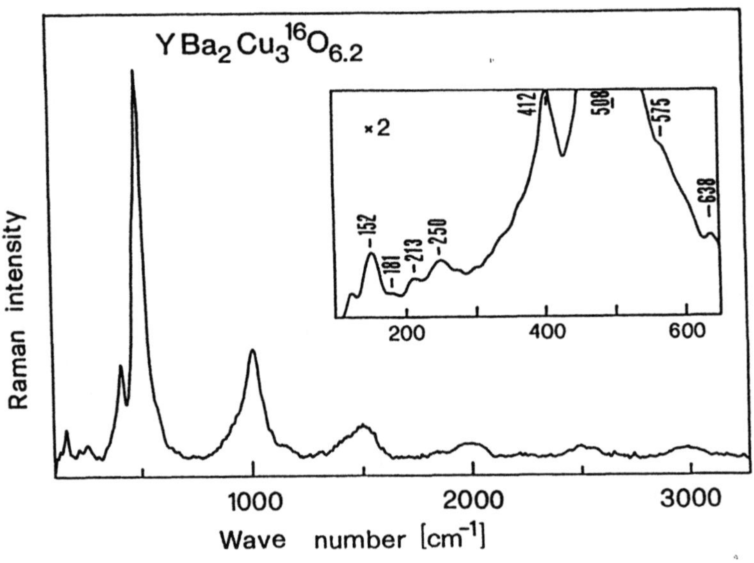


**Fig. 3.3.** FT-Raman spectrum exciting at 1.16 eV of $YBa_2Cu_3O_{6.2}$ (the spectrum is not corrected for the detector response in order to emphasize the presence of overtones up to the 6th order). In the inset is shown a blow-up of the 1st order phonons region.

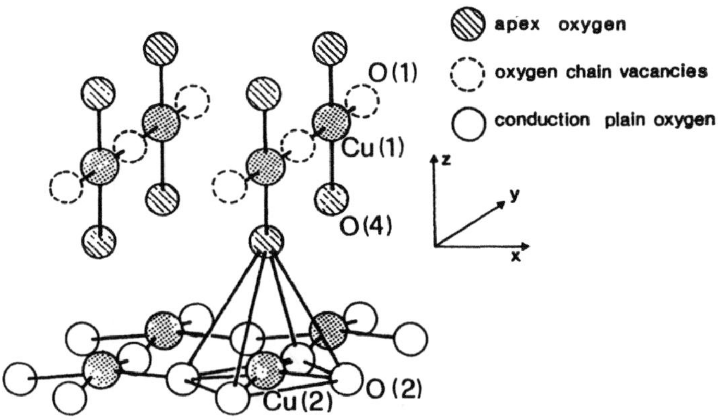


**Fig. 3.4.** Schematic representation of the crystal structure of conduction and basal planes in $YBa_2Cu_3O_{6+x}$.

of $YBa_2Cu_3^{16}O_{6.05}$. All frequencies are shifted with respect to the upper spectrum due to the isotope substitution of oxygen. The peak at 507 $cm^{-1}$ (477 $cm^{-1}$ for $^{18}O$) and its overtones are weaker. We can conclude that the new doping induced band at 507 $cm^{-1}$ (477 $cm^{-1}$ in $^{18}O$ substituted samples) is induced by the oxygen occupation of the O(1) chain sites in the basal plane (see Fig. 3.3). In some samples it's possible to observe overtones associated to this band up to the limit of our detector working range as is shown in Fig. 3.4 for $YBa_2Cu_3O_{6.2}$.

The frequency of the peak at 507 $cm^{-1}$ is very close to the frequency of the $A_g$ phonon mode of the O(4) atoms in the fully doped ($x = 1$) system, and it can be attributed to vibration of the O(4) atoms in microscopic domains of the orthorhombic phase. In order to confirm this assignment we have performed site selective isotope substitution of oxygen $^{16}O$ in the $YBa_2Cu_3^{18}O_6$ sample by means of low temperature annealing in air. Pristine $YBa_2Cu_3O_6$ IR excited Raman spectrum is shown in Fig. 3.5a. Fig. 3.5b shows the spectrum of the same sample after 30 min annealing at 200°C in air. As expected, the spectrum changes with the oxygen doping and a new peak at 477 $cm^{-1}$ ($\omega^{2*}$) accompanied by a weak overtone ($2\omega^{2*}$) appears. At this annealing temperature, oxygen $^{16}O$ (from the atmosphere) diffuses within the sample only in the basal plane but cannot jump in other oxygen sites occupied by $^{18}O$ atoms [7]. We notice that this new $\omega^{2*}$ band at 477 $cm^{-1}$ has the same frequency as the main band of the spectrum of the fully $^{18}O$ substituted sample $YBa_2Cu_3^{18}O_{6+x}$ (Fig. 3.2b). This is indeed a confirmation of our assignment. In fact, chain sites are occupied after annealing by $^{16}O$ and therefore the doping induced band cannot be assigned to vibrations of oxygen atoms on chain sites because in this case we should have observed the band characteristic of $^{16}O$ at 507 $cm^{-1}$. A second heat treatment of the sample is performed at 300°C. At this temperature O atoms in the O(1) sites can exchange with the

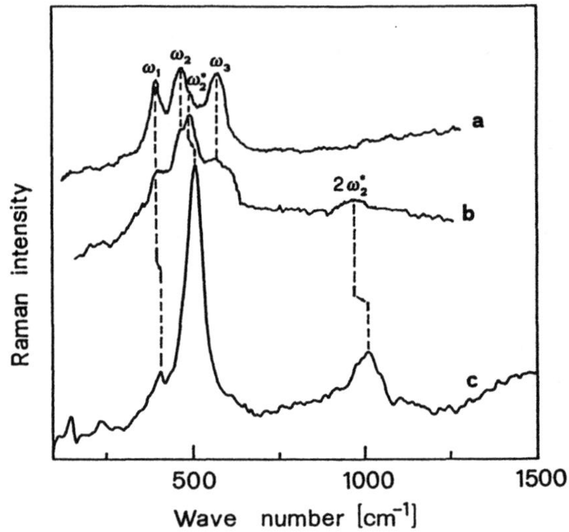


**Fig. 3.5.** FT-Raman spectra exciting at 1.16 eV of: (a) $YBa_2Cu_3O_{6.0}$, (b) the same sample after 30 minutes of annealing at 200°C in air and (c) the same sample after a second annealing of 60 minutes at 300°C in air.

nearest neighbour apex O(4) sites. At the same time, as reported in Ref. [7] and [8], the temperature is not so high as to allow oxygen in $CuO_2$ planes to diffuse out of the plane. After 60 minute of annealing at this temperature a dramatic change in the Raman spectrum is observed (Fig. 3.5c). A shift towards higher frequencies of the bands at 385cm$^{-1}$ ($\omega^1$), 477cm$^{-1}$ ($\omega^{2*}$) and the overtone of the latter band ($2\omega^{2*}$) is clearly seen together with the increase in intensity of $\omega^{2*}$ and $2\omega^{2*}$. The similarities between the spectra of the annealed sample in Fig. 3.5c with the one of $YBa_2Cu_3^{16}O_{6.2}$ in the inset of Fig. 3.2, is quite evident; the frequencies of the modes in the two spectra do not correspond exactly because of a not complete isotope substitution of $^{18}O$ on apical positions.

The spectrum in Fig. 3.5c looks like that of Fig. 3.2b approximately shifted by 20 cm$^{-1}$ towards higher energy. However, the band $\omega^3$ in Fig. 3.5a that is clearly seen as a shoulder at 552 cm$^{-1}$ in Fig. 3.2b is not present in Fig. 3.5c. If this band is related to vibrations of oxygen atoms on apex sites one should expect that the band at 575 cm$^{-1}$, which is seen in the spectrum of $YBa_2Cu_3^{16}O_{6.05}$ (see inset of Fig. 3.2 should be present also in Fig. 3.5c, at least as a shoulder, similar to Fig. 3.3. The absence of this feature indicates its relevance to vibrations of oxygen atoms in $CuO_2$ layers which were not isotopically substituted during the annealing of the sample. Thus we conclude that all oxygen vibrational bands that appear in the IR excited Raman spectra of $YBa_2Cu_3O_{6+x}$, with the exception of the highest frequency 575 cm$^{-1}$ peak (in $^{18}O$ samples), are associated with phonons whose eigenvectors are mainly associated to the apex oxygen atoms displacement.

The resonance nature of the scattering suggests that the resonance electronic transition near 1 eV must involve 2p-3d electronic orbitals of the O(4)-Cu(1)-O(4) complex which, for $x = 0$, seems to be weakly interacting with the $CuO_2$ planes due to the increased distance between planes and apex oxygen atoms.

Site selective substitution of oxygen isotope in the $YBa_2Cu_3O_{6.2}$ system therefore allows assignment of the bands observed in the IR excited Raman spectra. It is now widely recognized that the $\omega^2$ band in the semiconducting samples ($x = 0$) is assigned to a totally symmetric ($A_g$) axial vibration of the apex oxygen1. We assign the $\omega^{2*}$ peak appearing upon slight oxygen doping to an $A_g$ vibration of apex oxygen atoms adjacent to short chain segments.

The assignment of the $\omega^1$ and $\omega^3$ bands is not so straightforward. These bands were not present in the visible excited Raman spectra. This means that these bands are very weak or not at all Raman active when out of resonance.

The best candidate for $\omega^1$ (at 412 $cm^{-1}$ in $^{16}O$ samples) is the Raman forbidden IR Eu(LO) phonon calculated [9], at 414 $cm^{-1}$ and found in reflectance measurements [10],[11] near 420 $cm^{-1}$. According to calculations [9] this mode involves vibrations of the apex oxygen atoms parallel to the $CuO_2$ planes. It is reasonable to assume that the LO component of the Eu mode becomes Raman active near resonance due to the Frölich electron-phonon mechanism [12] through the quadrupole term. Other Eu(LO) modes, rather weak but reproducible in all samples, are also seen in the IR excited Raman spectra. In the spectrum shown in Fig. 3.3 (inset) at least three other peaks at 181 $cm^{-1}$, 250 $cm^{-1}$ and 638 $cm^{-1}$ can be assigned to Eu(LO) modes which were observed in infrared reflectance measurements [10] at 199 $cm^{-1}$, 266 $cm^{-1}$, 637 $cm^{-1}$ respectively. Confirmation of this assignment is given by the observation of the same four modes in $YBa_2Cu_3O_6$ when the red (676.4 nm) excitation light is used [13].

We envisage at least two possible assignments for the $\omega^3$ band at 575 $cm^{-1}$. Isotope substitution, as was discussed above, shows that this band is related to the vibrations of oxygen atoms O(2) in the $CuO_2$ planes. Calculations [9] predict only one Raman active mode close to 575 $cm^{-1}$ in $YBa_2Cu_3O_6$ which is assigned to the $E_g$ vibration at 577 $cm^{-1}$ of the O(2) oxygen atoms. Another possibility for the assignment of the 575 $cm^{-1}$ band is a defect induced local vibration found in reflectance measurements (see Ref. [10] and [11]), respectively at 595 $cm^{-1}$ and 588 $cm^{-1}$. Comparing calculated eigenvector of this mode (502 $cm^{-1}$) and the one of the Eu(LO) (414 $cm^{-1}$) vibration [6] one can see that both have a similar mixed character. Therefore, its increased Raman intensity for 1.16 eV excitation may be caused by the admixture of the resonantly enhanced Eu(LO) mode near 412 $cm^{-1}$. Raman measurements on single crystals in zx, zy geometries may allow a choice to be made between these two possible assignments.

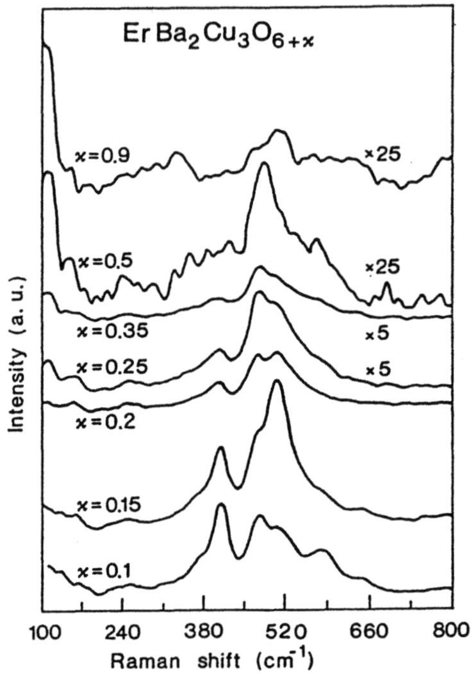


**Fig. 4.1.** Room temperature Raman spectra of ceramic $ErBa_2Cu_3O_{6+x}$ with $0.1 \leq x \leq 0.9$.

## 4. $ErBa_2Cu_3O_{6+x}$: $0.1 \leq x \leq 0.9$

The $ErBa_2Cu_3O_{6+x}$ ($x = 0.9$) were prepared by the standard solid state technique. The vacuum annealing of the oxygen reach material allows the oxygen deficient ($x = 0.1$) one to be obtained. The intermediate concentrations, $0.1 \leq x \leq 0.9$, were obtained by annealing the oxygen rich and the oxygen deficient materials mixed in appropriate proportions for two weeks at 400°C in sealed ampullas which were then slowly cooled over a period of two months [14].

As in the case of previous samples, at low oxygen concentration the IR excited Raman of $ErBa_2Cu_3O_{6+x}$ spectra are characterized by the structure at 510 $cm^{-1}$ (see Fig. 4.1 and 4.2). In this case the resonant behaviour with a sharp intragap band is more evident. In the sample with lower oxygen content ($x = 0.1$) the $\omega^{2*}$ 510 $cm^{-1}$ band appears as a shoulder of the $\omega^2$ 475 $cm^{-1}$ band, while the most intense band is the $\omega^1$ LO phonon at 408 $cm^{-1}$. As it is obtained by the fitting with a summation of Lorentzians of the spectra (see Fig. 4.2), at $x = 0.15$ the intensity of the $\omega^2$ remains practically constant, the $\omega^1$ slightly decreases and $\omega^{2*}$ becomes the most pronounced one. At $x = 0.2$, the $\omega^2$ band is still of the same intensity while both the $\omega^1$ and $\omega^{2*}$ decrease in intensity.

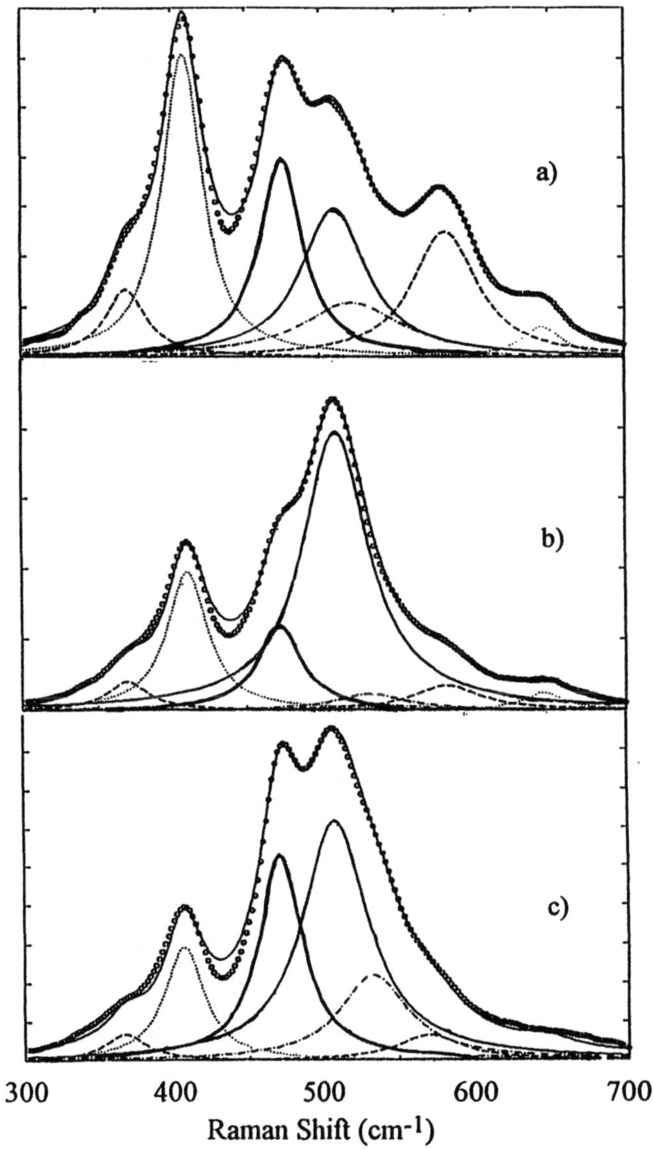


**Fig. 4.2.** Room temperature Raman spectra of ceramic $ErBa_2Cu_3O_{6+x}$ with $x =$ 0.1 (a), 0.15 (b) and 0.2 (c). The curve fitting has been executed using the same set of Lorentzians and only the intensities as free parameters.

At higher oxygen content the appearance of macroscopic orthorhombic domains is observed [14]; at $x = 0.25$ the total scattering intensity is strongly reduced (Fig. 4.1) because of the appearance of metallic islands associated with the orthorhombic domains and $\omega^1$ and $\omega^{2*}$ intensities are further reduced relatively to $\omega^2$. The profile of the Raman spectra does not change till $x = 0.4$. The Raman spectra of samples with $0.42 \leq x \leq 0.7$ show only a single band associated with the apex oxygen O(4) at about 480 $cm^{-1}$, above a broad "electronic" background, (see Fig. 4.1) that shifts towards higher energy with the increase of oxygen content. At $x = 0.75$ a band at about 340 $cm^{-1}$ (associated to the antisymmetric plane-dimpling mode of oxygen O(2)/O(3)) start to be visible in the spectra and, at $x = 0.9$, its intensity becomes comparable with the apex oxygen mode (see Fig. 4.2) whose energy at this oxygen concentration is comparable with the one indicated in the low oxygen concentration samples as $\omega^{2*}$. The relative intensity behaviour of these two $A_g$ modes observed in IR excited Raman is the opposite of that observed by visible excitation [1],[3]. As mentioned before, the resonant Raman conditions of these two modes are very different. Also LMTO calculations [1] show that the Raman efficiency of the $A_g$ O(2)/O(3) mode is reduced only by a factor of 4 while the Raman efficiency of the $A_g$ O(4) is reduced more than two orders of magnitude when the excitation energy is changed from 2.5 eV to 1 eV.

These results are in agreement with the recent NQR measurements by R. De Renzi et al. [15] on the same samples. They have observed that only a small number of well defined frequency lines, whose intensities are $x$ dependent, can be detected in all the samples. Such a result is not consistent with a macroscopic picture obtained by X-ray or neutron diffraction techniques that observe a continuous change of the lattice parameter with oxygen content, but indicates that the crystal structure of 1-2-3 systems obtained using diffraction techniques only an average of four coexistent phases: pure tetragonal (no oxygen in the chain sites), pure orthorhombic (ORTHO I), isolated chains and alternate empty and full chains (ORTHO II). The Raman scattering with excitation at 1.16 eV, because of its peculiar resonant condition in these compounds, allows the presence of these phases to be distinguished.

## 5. $^{18}O$ Metallic $YBa_2Cu_3O_{6+x}$ Substituted Samples

Metallic $YBa_2Cu_3O_{6+x}$ samples with different oxygen content ($0.48 \leq x \leq 0.974$) both with $^{16}O$ and with 96% of substituted $^{18}O$ have been prepared by using high purity Y, Ba and Cu metals and $^{16}O$ or $^{18}O$ oxygen as starting materials. The oxidation of Y, Ba and Cu and the following sintering of the oxides have been described elsewhere [7].

As mentioned before, the IR excited Raman spectrum of $MBa_2Cu_3O_{6+x}$ in the metallic phase (see Fig. 4.1 for $x \sim 0.8$) is characterized mainly by the presence of two phonon bands above a background between 250–600 $cm^{-1}$.

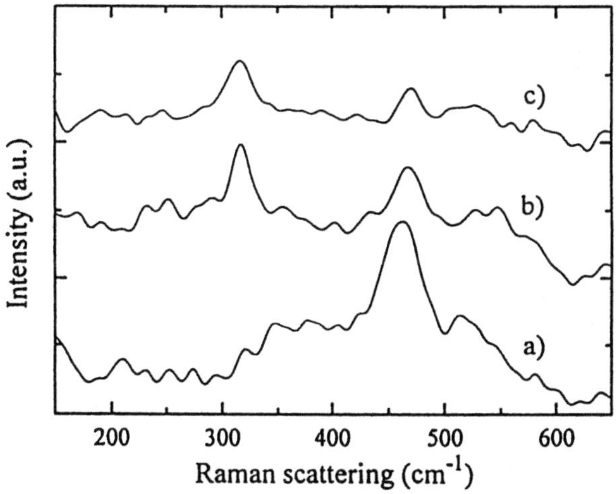


**Fig. 5.1.** RT Raman spectra of ceramic $YBa_2Cu_3^{18}O_{6+x}$ with $x = 0.52$ (a), 0.85 (b) and 0.953 (c).

The band at lower energy, around 340 cm$^{-1}$ in $^{16}$O samples, is associated with the $A_g$ out of phase displacement along c of the O(2)/O(3) atoms; the second band, around 490 cm$^{-1}$ in $^{16}$O samples, is the $A_g$ O(4) phonon mode. The relative intensities of these two bands change with oxygen content: at $x \leq 0.6$ only the $A_g$ O(4) phonon mode is strong enough to be detected while at $x = 1$ the dominant feature of the spectrum is the $A_g$ O(2)/O(3) phonon mode (see Fig. 5.1); this behaviour, opposite to what is observed by visible excited Raman [1],[3], is related to the very different resonant condition between 1.16 eV and 2.5 eV photon energy and to the change of low energy electronic states induced by doping.

The oxygen isotope substitution is observed in Raman by the shift of the phonon modes whose eigenvectors are represented by displacements of oxygen atoms. As is shown in Fig. 5.2 for the $YBa_2Cu_3O_{6+x}$ samples with $x \sim 0.8$, the almost complete (96%) homogeneous oxygen isotope substitution of $^{16}$O with $^{18}$O determines a systematic shift of all the phonon features of the spectrum [16].

The peak position of the $A_g$ O(2)/O(3) Raman band for different samples is reported in Fig. 5.3 for oxygen $^{16}$O (full squares) and for $^{18}$O (open dots) as a function of oxygen content. The full triangles represent the "calculated" frequency of the same band for a sample with 96% of $^{18}$O. The isotope shift of the energy relative to this phonon, and also of the $A_g$ O(4) one, has been calculated by several methods, always within the harmonic approximation. The results are not strongly dependent on the method used (see Ref. [17] and references therein). To obtain what is indicated as the "calculated" frequency of $^{18}$O samples, we have multiplied the experimental Raman frequency of the $YBa_2Cu_3^{16}O_{6+x}$ of the same mode for a proportional factor relative to the

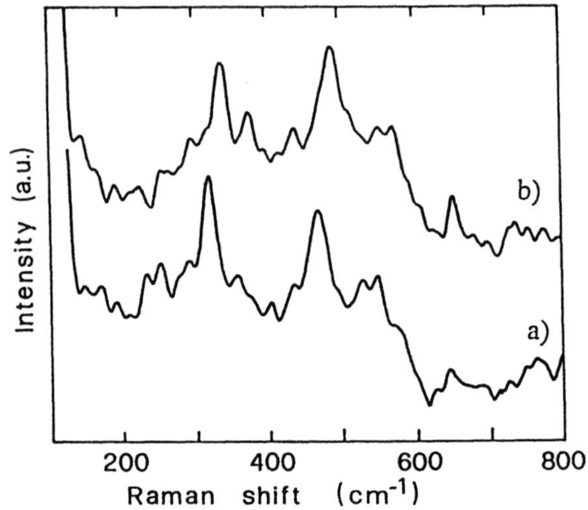


**Fig. 5.2.** RT Raman spectra of ceramic $YBa_2Cu_3^{18}O_{6.8}$ (a) and $YBa_2Cu_3^{16}O_{6.8}$ (b).

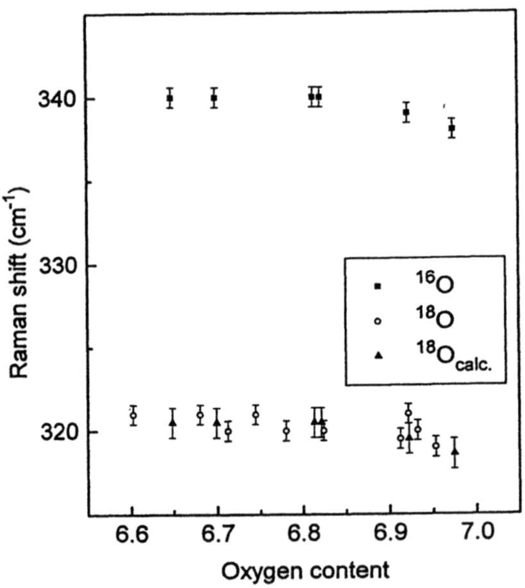


**Fig. 5.3.** Frequency of the Raman $A_g$ mode associated to the out of phase motion of O(2)/O(3) oxygen atoms along the c axis; full squares: experimental value of $YBa_2Cu_3^{16}O_{6+x}$ samples; open dots: experimental values of $YBa_2Cu_3^{18}O_{6+x}$ samples; and full triangle: computed value of 96% of $^{18}O$ substituted samples from the $YBa_2Cu_3^{16}O_{6+x}$ experimental value isotopically shifted using the harmonic approximation. The lines are guide to the eyes.

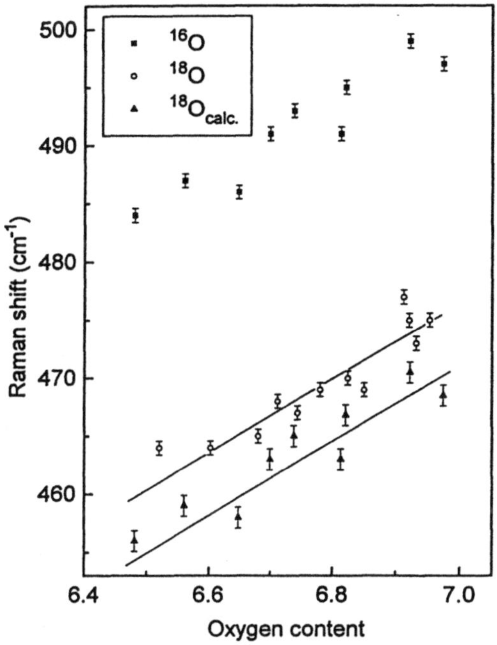


**Fig. 5.4.** Frequency of the Raman $A_g$ mode associated to the displacement of the O(4) oxygen atoms along the c axis; full squares: experimental value of $YBa_2Cu_3^{16}O_{6+x}$ samples; open dots: experimental values of $YBa_2Cu_3^{18}O_{6+x}$ samples; and full triangle: computed value of 96% of $^{18}O$ substituted samples from the $YBa_2Cu_3^{16}O_{6+x}$ experimental value isotopically shifted using the harmonic approximation. The lines are guide to the eyes.

frequency isotope shift obtained from the "shell model" calculations [18], taking into account that our $^{18}O$ samples are not completely isotope substituted. The Raman frequency obtained in this way coincides, in the case of the $A_g$ O(2)/O(3), with the frequency observed experimentally for the $^{18}O$ substituted samples. This confirms that the harmonic approximation is valid in the case of oxygen atoms in the $CuO_2$ planes.

On the contrary the observed Raman shift, induced by isotope substitution, of the $A_g$ O(4) mode for all oxygen concentration (see Fig. 5.4) is much smaller than the one computed in the case of an harmonic oscillator. The accuracy in the determination of the oxygen content (the error bars are smaller than the symbols used in Fig. 5.4) and in the determination of the peak position are high enough to exclude that this effect can be ascribed to experimental errors.

Similar behaviour of the isotope induced shift of the two Raman bands has been observed by J.C. Irwin et al. [17]. They have observed a larger isotope shift of the $A_g$ O(2)/O(3) band than for the O(4) band. To explain such a difference they suggested the hypothesis of a different substitution efficiency for oxygen in O(2)/O(3) and O(4) sites (higher mobility of the oxygen in the $CuO_2$ planes than in the BaO planes). As mentioned before, our sample have been prepared from oxides already isotopically enriched with oxygen $^{16}O$ or $^{18}O$, so it is not possible in these conditions to obtain samples with different site occupancy even if different kinds of diffusion processes with different activation energies in the different oxygen sites exist. Moreover, it has been demonstrated by low temperature annealing [7],[19] that the oxygen isotope substitution and therefore the oxygen diffusion proceeds from basal (chain) plane to BaO plane (apex) and then to $CuO_2$ plane, leading to the conclusion that a higher percentage of isotope substitution in the O(2)/O(3) sites with respect to the O(4) sites is not possible.

The small isotope shift of the $A_g$ phonon energy observed for the apex oxygen is consistent with the existence of an asymmetric anharmonic potential. As has been suggested in several papers, the existence of an anharmonic potential related to the apex oxygen can have a strong impact on the superconducting properties of high $T_c$ materials [20]–[23]. There is evidence from several experiments that suggest the existence of an anharmonic potential for the apex oxygen [24]–[26]. The apex oxygen potential have been shown to be anharmonic also by calculations using LAPW method [28].

Moreover, recently J.H. Nickel et al. have shown by site selective oxygen isotope substitution in $YBa_2Cu_3O_6$ that apex oxygen O(4) vibrations contribute to the pairing mechanism. Such results are emphasizing the validity of models that describe the Cooper pair formation in the conducting $CuO_2$ plane as induced by the coupling of correlated electrons to anharmonic vibrations of the O(4) atom, as for example, the model proposed by Bishop et al. [27] in which a strong charge transfer coupled with the apex oxygen vibration along c within a double well anharmonic potential is the source of superconductivity in the close $CuO_2$ plane of $YBa_2Cu_3O_{6+x}$.

## 6. Untwinned 1-2-3 Crystal: Electronic Background

Several Raman studies of twin-free $YBa_2Cu_3O_7$ crystals have been reported [29]–[36]. These investigations were carried out with visible light excitation and revealed differences in the relative intensities of the $A_g$ phonons depending upon whether the polarization of the incident and scattered light was along the $x$ or $y$ direction. The complete polarization study of Raman active phonons were reported in Ref. [32] and the properties of the background in the Raman spectra of YBCO were investigated in Ref. [34]–[36].

Here we report a Raman spectra of untwinned $YBa_2Cu_3O_7$ crystal obtained with excitation by laser energy of 1.16 eV.

The single-phase YBCO $T_c \sim 90$ K crystals with dimensions $1 \cdot 1 \cdot 0.3$ mm$^3$ were grown and detwinned according to a method described in detail elsewhere [37]. The characterization of the crystal was made by means of inductive measurements, polarized microscope and reflectivity measurements of

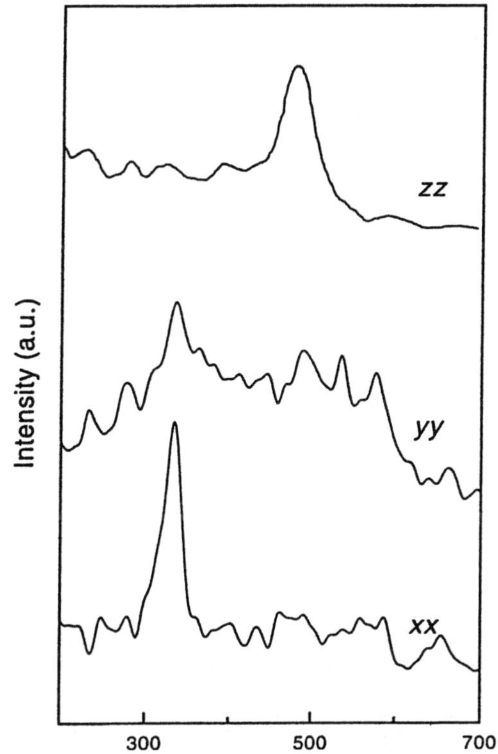


**Fig. 6.1.** Raman spectra of $YBa_2Cu_3O_{7-y}$ untwinned single crystals for different scattering geometries exciting at 1.16 eV.

plasma edge. Room temperature Raman spectra for three scattering geometries are shown on Fig. 6.1. The (yy) spectrum (y corresponds to Cu-O chains direction) is rather complicated, it consists of intensive broad maximum at $\sim 400$ cm$^{-1}$ serving as a background for relatively sharp and lower intensive phonon lines. There are three well pronounced lines in the (yy) spectrum at frequencies 337, 491, and 575 cm$^{-1}$. The (xx) spectrum contains only one phonon line at 337. The (zz) spectrum in Fig. 6.1 contains only one well pronounced line at 498 cm$^{-1}$. Our measurements in crossed (xy) polarization did not reveal any structure in the spectrum, so all the features in the spectra shown in Fig. 6.1 should be regarded as having $A_g$ character.

The visible Raman spectra of the same crystal is shown in Fig. 6.2. The most pronounced difference with respect to the IR excited spectra is the absence of this rather broad maximum in the (yy) spectrum around 400 cm$^{-1}$. A similar broad feature is observed in the ceramic samples with $x > 0.5$ (see. Fig. 4.2–5.2). The relative intensity with respect to the phonon band and the width of this background depend on oxygen content $x$: at $x \sim 1$

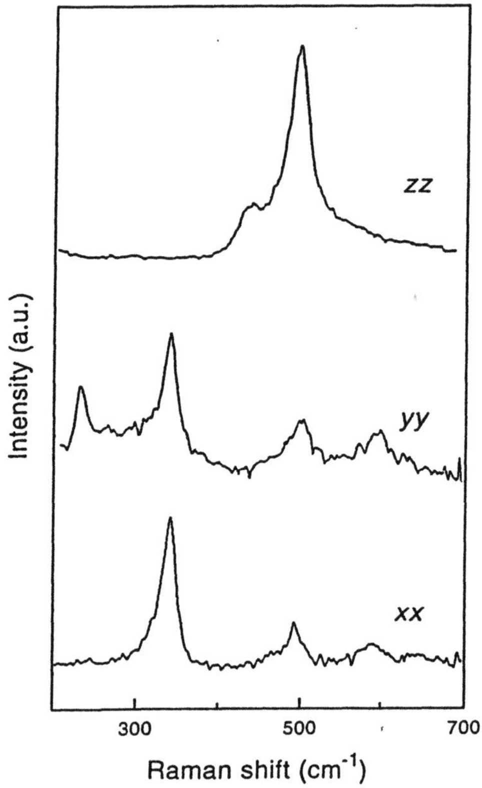


**Fig. 6.2.** Raman spectra of $YBa_2Cu_3O_{7-y}$ untwinned single crystals for different scattering geometries exciting at 2.51 eV.

it is approximately two times larger than at $x \sim 0.5$. In Fig. 6.3 the IR excited Raman spectra of the untwinned single crystal in yy geometry with together those obtained in 1-2-3 ceramic samples with similar oxygen content are reported. It is clearly seen that the broad peak is identical in all the three samples. The nature of this peak is clearly non phononic: in fact the comparison of the Raman spectra of the two ceramic YBCO samples with the same oxygen content but different isotopes shows no detectable shift. If this structure had a phononic origin, its relatively "high" energy would imply the oxygen nature of the mode with the consequent shift related to change in the mass of the isotope.

Recent measurements on $^{18}O$ substituted samples have shown the existence of some features below 800 cm$^{-1}$ that is independent of the isotope substitution [38]. These structures have been interpreted, by I.I. Mazin et al. [39] as related to low energy electronic interband transitions on the base of accurate LDA energy bands calculations. In particular one of these has been assigned to the transition between two bands at the S point in the Bril-

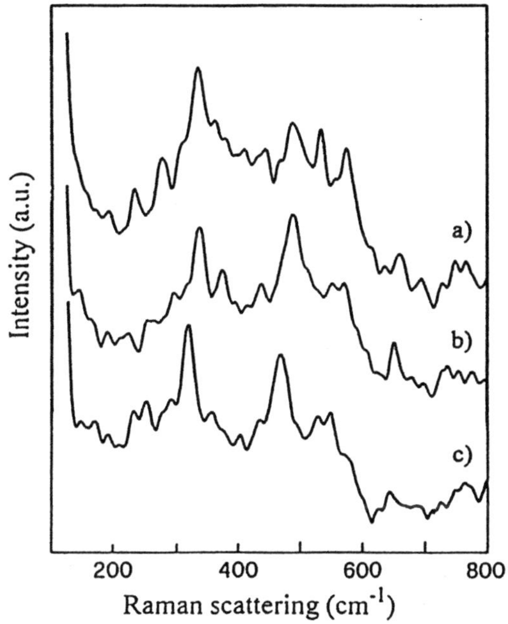


**Fig. 6.3.** Raman spectra of $YBA_2Cu_3O_{7-y}$ (a) untwinned single crystals obtained in yy geometry; (b) and (c) same spectra reported in Fig. 5.2 (b) and (a) respectively

louin zone around the Fermi level $E_F$. The Upper curve, the one crossing the $E_F$ has prevalently $O(4)(p_y)$ in character and the lower one mainly $O(4)(p_x)$ in character. In accordance with LDA calculations [39], such a transition should be observed in resonant conditions also by Raman scattering; in particular the authors claim that such a resonant condition would be matched at energy smaller than 1.9 eV or higher than 0.8 eV. This would explain why this intraband Raman scattering is observed only by IR excited Raman.

In conclusion, the confirmation of the existence of low energy electronic states associated to apex oxygen around the $E_F$, i.e. belonging neither to $CuO_2$ planes nor to chains, and the observation, outside the experimental error, of the anharmonicity of the apex oxygen O(4) in YBCO superconducting system, underlines the peculiar role played by this atom in the formation of the superconducting state. Moreover the similarity of the O(4) in YBCO with the anharmonic highly polarizable $O^{2-}$ anion in the highest $T_c$ ferroelectrics strongly supports those theories that consider a possible common origin for ferroelectricity and superconductivity in oxides.

*Acknowledgement.* This work have been done in collaboration with

- M. Muccini, C. Taliani, I.S.M. - C.N.R., Bologna, Italy
- V.M. Burlakov, V.N. Denisov, A.G. Mal'shukov, Inst. Spectr., Acad. Sci., Troitsk Russia (chapt. III, IV, VI)

– E. Schönner Max- Planck Institut, Stuttgart, Germany (chapt. III)
– P. Radaelli, C.U. Segre, Dep. Phys., IIT, Chicago, Illinois (chapt. IV)
– A. Erb Kristal and Mat., University Karlsruhe, Germany (chapt. VI)
– H.P. Geserich, Inst. Angew. Phys., University Karlsruhe, Germany (chapt. VI)
– K. Conder, E. Kaldis, ETH Hünggerberg, Zurich, Switzerland (chapt. V)
– H. Keller, D. Zech.Inst. Phys, University of Zurich, Switzerland (chapt. V)
– K.A. Müller, IBM Res. Div., Zurich, Switzerland (chapt. V)

I wish also to thank Prof. N.M. Plakida and Dr. M. Zoli for helpful discussions. Financial support by National Research Council of Italy, C.N.R. under the Progetto Finalizzato "Superconductive and Cryogenic Technologies" and EC contract SCI*CT91-0751 is acknowledged.

## References

1. E.T. Heyen, R. Liu, C. Thomsen and M. Cardona in "Electronic Properties of High-$T_c$ Superconductors and Relative Compounds", edited by H. Kuzmany, M. Mehring and J. Fink, (Springer-Verlag, Berlin Heidelberg 1990), p.324.
2. E.T. Heyen, S.N. Rashkeev, I.I. Mazin, O.K. Andersen, R. Liu, M. Cardona and O. Jepsen, Phys. Rev. Lett. 65, 3048 (1990).
3. H. Eskes, M.B.J. Meinders and G.A. Sawatzky, Phys. Rev. Lett. 67, 1035 (1991).
4. R. Zamboni, G. Ruani, A.J. Pal and C. Taliani, Solid State Commun 70, 813 (1989).
5. G. Ruani, R. Zamboni, C. Taliani, V.N. Denisov, V.M. Burlakov and A.G. Mal'shukov, Physica C 185-189, 963 (1991); V.N. Denisov, C. Taliani, A.G. Mal'shukov, V.M. Burlakov, E. Schönherr and G. Ruani, Phys. Rev. B 48, 16714 (1993).
6. C. Thomsen, R. Liu, M. Bauer, A. Wittlin, L. Genzel, M. Cardona, E. Schönherr, W. Bauhofer and W. König, Solid State Commun. 65, 55 (1988).
7. K. Conder, E. Kaldis, M. Maciejewski, E.F. Steigmeier and K.A. Müller, Physica C, in press.
8. R. Nishitani, N. Hoshida, Y. Sasaki, Y. Nishina, H. Yoshida-Katayama, Y. Okabe and T. Takahashi, Japan. Journal of Appl. Phys. 29, L50 (1990).
9. C. Thomsen, M. Cardona, W. Kress, R. Liu, L. Genzel, M. Bauer and E. Schönherr, Solid State Commun. 65, 1139 (1988).
10. M. Bauer, L.B. Ferreira, L. Genzel, M. Cardona, P. Muragaraj and J. Maier, Solid State Commun. 72, 551 (1989).
11. M.K. Crawford, G. Burns and F. Holtzberg, Solid State Commun. 70, 557 (1989).
12. M. Cardona in "Light Scattering in Solids II", ed. M. Cardona and G. Güntherodt (Springer, 1982), p. 1
13. M. Cardona, Physica C 185-189, 65 (1991).
14. P.G. Radaelli, C.U. Segre, D.G. Hinks and J.D. Jorgensen, Phys. Rev. B 45, 4923, 1992.
15. R. Derenzi et al., Conferenza SATT, Riccione, 1993.
16. G. Ruani, C. Taliani, M. Muccini, K. Conder, E. Kaldis, K.A. Müller, H. Keller and D. Zech, submitted to Physica C.
17. J.C. Irwin, J. Chrzanowski, E. Altendorf, J.P. Franck and J. Jung, J. Mater. Res 5, 2780 (1990); E. Altendorf, J. Chrzanowski, J.C. Irwin and J.P. Franck, Phys. Rev. B 43, 2771 (1991).

18. M. Cardona, R. Liu, C. Thomsen, W. Kress, E. Schönherr, M. Bauer, L. Genzel and W. Konig, Solid State Commun 67, 789 (1988); C. Thomsen, Hj. Mattausch, M. Bauer, W. Bauhofer, R. Liu, L. Genzel and M. Cardona, Solid State Commun. 67, 1069 (1988)
19. J.H. Nickel, D.E. Morris, J.W. Ager III, Phys. Rev. Lett. 70, 81 (1993).
20. K.A. Müller, Z. Phys. B 80, 193 (1990).
21. T. Galbaatar, S.L. Drechsler, N.M. Plakida, G.M. Vujicic, Physica C 176, 496 (1991); T. Galbaatar, S.L. Drechsler, N.M. Plakida, G.M. Vujicic, Physica C 185-189, 1529 (1991).
22. D. Mihailovic, C.M. Foster, K.F. Voss and N. Herron, Phys. Rev. B 44, 237 (1991).
23. M. Frick, I. Morgenstern and W. von der Linden, Z. Phys. B 82, 339 (1991).
24. J. Mustre de Leon , S.D. Conradson, A.R. Bishop and I. Batistic, Phys Rev. Lett. 65, 4675 (1990).
25. S.D. Conradson, I.D. Raistrick and A.R. Bishop, Science 248, 1394 (1990).
26. D. Mihailovic and C.M. Foster, Solid State Commun. 74, 753 (1990).
27. A.R. Bishop, R.L. Martin, K.A. Müller and Z. Tesanovic, Z. Phys. B 76, 17 (1989).
28. C. Ambrosch-Draxl, this volume.
29. C. Thomsen, M. Cardona, B. Gegenheimer, R. Liu and A. Simon, Phys. Rev. B. 37, 9860 (1988).
30. F. Slakey, S.L. Cooper, M.V. Klein, J.P. Rice and D.M. Ginsberg, Phys. Rev. B. 39, 2781 (1989).
31. L.V. Gasparov, V.D. Kulakovskii, O.V. Misochko and V.B. Timofeev, Phisica C 157, 341 (1989).
32. V.D. Kulakovskii, O.V. Misochko and V.B. Timofeev, Sov. Phys. Solid State 31 (9), 1599 (1989).
33. K.F. McCarty, J.Z. Liu, R.N. Shelton and H.B. Radovsky, Phys. Rev. B 41, 8792 (1990).
34. F. Slake, M.V. Klein, J.P. Rice and D.M. Ginsberg, Phys. Rev. B 43, 3764 (1991).
35. D. Rein, A. Kotz, S.L. Cooper, M.V. Klein, W.C. Lee and D.M. Ginsberg, Proc. Int. Workshop on Electronic Prop. and Mech. of High-$T_c$ Superconductors (IWEPM), Tsukuba, Japan (1991).
36. D. Reznik, S.L. Cooper, M.V. Klein, W.C. Lee, D.M. Ginsberg and S.W. Cheong, in "Electronic Properties of High-$T_c$ Superconductors", edited by H. Kuzmany, M. Mehring and J. Fink, (Springer-Verlag, Berlin Heidelberg 1993), p. 215.
37. A. Zibold, M. Dürrler, H.P. Geserich, A. Erb and G. Müller-Vogt, Physica C 171, 151 (1990).
38. A.V. Bazhenov and K.B. Razchikov, Physica C 192, 411 (1992).
39. I.I. Mazin, S.N. Rashkeev, A.I. Leichtenstein, and O.K. Andersen, Phys. Rev. B 46, 11232 (1992); I.I. Mazin, O. Jepsen, O.K. Andersen, A.I. Leichtenstein, S.N. Rashkeev and Y.A. Uspemskii, Phys. Rev. B 45, 5103 (1992).

# Photoexcited Electons and Holes in $YBa_2Cu_3O_{7-\delta}$: A Study by Picosecond Raman Scattering

D. Mihailovic and I. Poberaj

Institut Jozef Stefan, University of Ljubljana, Ljubljana, Slovenia

**Summary.** To enable the direct investigation of the electron-phonon scattering for individual phonons, carriers are photoexited by an ultrashort laser pulse and the subsequent scattering of these carriers by phonons is studied by measuring the phonon occupation number with Raman scattering. The density of excitation from the intense picosecond laser pulses is high enough to dope insulating $YBa_2Cu_3O_{6.2}$ into the metallic state, so the photo-doping process can be compared with chemical doping and we find clear evidence of an electronically driven phase insulator-metal transition. The coupling of the lattice to photoexcited carriers is shown to be highly selective in $YBa_2Cu_3O_{7-\delta}$. The interaction with the apical O ions is found to be particularly strong, while the interaction with planar vibrations show strong coupling only in the insulating state. The results are discussed in terms of a two-fluid model of itinerant planar carriers and chain polarons.

## 1. Introduction

Numerous experiments in the seven years since the discovery of cuprate superconductors have shown that the lattice has some interesting properties in these materials. In relation to superconductivity however, there is a wealth of evidence of relatively weak phonon coupling to the superconducting fluctuations excluding a particularly strong electron-phonon interaction and thus shedding doubt on the possibility of a purely phonon superconductivity mechanism. Most notable experiments in this group are Raman scattering experiments on planar O *c*-axis vibrations which show a clear, but small (~1%) phonon renormalization anomaly visible near $T_c$ [1]. Similarly, once other effects like structural changes with doping are taken into account or eliminated, changes in carrier density in the CuO planes appear to have only a relatively small (~1%) effect on c-axis phonon vibration frequencies [2]. In seeming contradiction to these observations, there is clear evidence that the c-axis polarizable perovskite structure plays an important role in superconductivity [3]. Indeed, this point can be illustrated very simply. $La_2CuO_{4+\delta}$ and Tl-2201 ($Tl_2Ba_2Cu_2O_6$) have identical octahedral structure surrounding the CuO planes, but have $T_c$s which differ by nearly a factor of 3 (35 K vs. >90 K respectively). Not only that, both are "single $CuO_2$ plane" structure superconductors, showing that $T_c$ is not related to the number of planes, but to the polarizability of the lattice structure *in between the planes.* This point is nicely echoed by the new Hg containing compounds, where $T_c$ of the one-

Cu-layer material (Hg-1201) is nearly the same as the 4-Cu-layer material (Hg-1234) [4].

Regarding the lattice polarizablity and lattice anomalies, many experiments, including ultrasonic attenuation [5], anelastic relaxation [6], specific heat anomalies [7] and pyro- and piezo-electric measurements [8] are available which all point to a highly polarizable lattice, mostly in the c direction, i.e. perpendicular to the conducting planes with both O and rare-earth ions playing a role, but the apical O vibrations being particularly important, acting as a bridge between the two structures.

Since the discovery of an anharmonic potential for the apical O ion by Conradson et al [9] from EXAFS measuerements, interest in the apical O vibration has been steadily increasing. Anharmonicity of this ion's vibration potential has been confirmed independently from temperature dependence measurements of the Raman vibration frequencies and linewidths [10], although these measurements were not able to confirm a double well potential found in EXAFS. Recently a calculation of the shape of the vibrational potential for the apical O ion (see C. Ambrosch-Draxl, this volume) confirmed the findings of Raman measurements and an anomalous frequency shift due to the isotope effect (see G. Ruani, this volume), has been found providing ample evidence for non-adiabatic effects of the apical O ion. Nevertheless, direct information about e-p coupling and how this changes with doping has so far not been available.

## 2. Photoinduced carriers

A particularly effective way of investigating the coupling of individual carriers to the lattice in the low density limit is by photo-injecting them into an insulating sample with a low power CW laser, and seeing the changes in either a) the infrared vibrational spectrum or b) the mid-infrared response signifying a change in dielectric constant, $\epsilon(\omega)$ in the mid-infrared.

The advantage of the method a) is that the results are fairly simple to understand if the new modes which appear in the spectra can be identified, the way they have been in $La_2CuO_4$ [11], YBCO and Tl2212 [12]. The apical oxygen ion vibration and the Tl-O plane c-axis vibrations are clearly observed as PI local modes, suggesting that a dipole moment is induced in this vibration by the presence of a carrier, this in turn implying the possibility of similar local symmetry breaking also in the metallic, doped materials. Although in the case of YBCO the apical O mode could be coupled to carriers in the CuO planes as well as carriers in the chains (or dumbells in undoped $YBa_2Cu_3O_6$), the unambiguous presence of the c-axis Tl-O plane O vibrations strongly suggests that the carriers are localized in the polarizable layers in between the CuO planes rather than in the CuO planes themselves. This leads immediately to the notion that there may be two fluids simultaneoulsy present [13, 16] in the metallic state: quasi-2D carriers in the Cu planes and more localized carriers in the chains (in YBCO) or more generally in the intermediate structure of the high $T_c$ cuprates.

The photoinduced mid-IR response measured by now in many different cuprates has been shown to be rather well described by a polaron hopping "shake-off" in addition to a temperature-dependent Drude contribution [14, 15] with the effective mass gradually decreasing with increasing $T_c$. Importantly, the strong a-b anisotropy of the mid-infrared $\sigma(\omega)$ for $YBa_2Cu_3O_{7-\delta}$ clearly implies that chain O(1) and Cu(1) $p_y$ orbitals are involved in the transitions [16]. By implication the MIR is associated with a polaron hopping process in the chains. In our present experiments we find evidence that this picture may be correct.

## 3. Photoinduced (super)conductivity and the effects of carriers on Raman spectra

Two types of experiments have hitherto implied that the metallic (and possibly superconducting) state could be photo-induced:

- transient photoconductivity (PC) experiments on insulating YBCO and $La_2CuO_4$ of Heeger and Yu [17] where sufficent carriers are excited to achieve metallic densities and
- CW (photo)-conductivity measurements of Kudinov [18] on YBCO samples which have an O concentration very close to the I-M transition.

While in the latter case it appears to be quite probable that photon-assisted O (re)ordering is responsible for induced superconductivity, in the transient PC experiments this mechanism is unlikely to be as important. Instead PE carriers may be thought to some extent to play the same role as chemical doping. The advantage of photoinduced experiments is that carriers are introduced without chemical or structural changes to the material and all the consequences are electronic in origin.

The present experiments are done on an even shorter, one picosecond timescale and are designed to give more specific information of the electron-phonon interaction of the high-frequency O vibrations in the insulating and PI metallic phases of YBCO. The idea is simply that non-equilibrium carriers will scatter from the lattice and will excite certain phonons more than others. The number of excited phonons is then measured by our Raman experiment. Two things make this possible: the first is the fact that the band-structure of the material is known in sufficient detail from LDA or LAPW calculations [20], optical experiments like ellipsometry [19] and resonant Raman spectroscopy [21] to identify which transitions are taking place in optical excitation and the second, the availability of sufficiently stable ultrashort laser pulses with the appropriate photon energy to be able to probe the non-equilibrium carrier-lattice interactions.

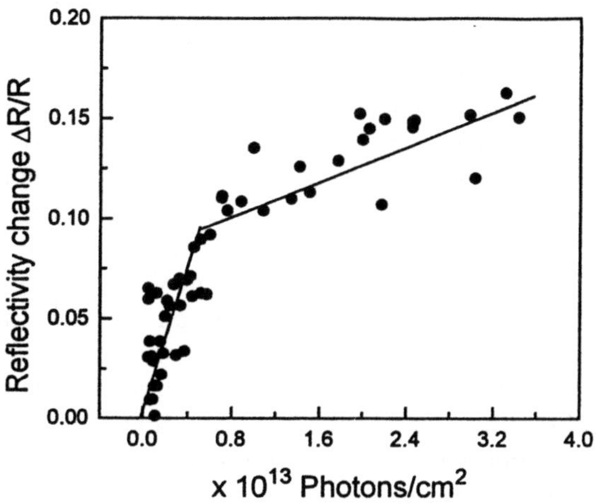


**Fig. 3.1.** Reflectivity change $\Delta R/R$ for unpolarized light at 2.33 eV incident on the crystal *ac* plane, as a function of incident photon flux density. The line is a guide to the eye. A break in $\Delta R/R$ is clearly visible at $\sim 5 \times 10^{12}$ photons/cm$^2$.

In these experiments we also have the opportunity to investigate the insulator-to-metal transition by varying the density of the excited electron-hole plasma by varying the incident laser photon flux. For these experiments we use a 2.33 eV beam of 1.5 ps laser pulse train from a Nd:YAG laser with a repetition rate of 100 MHz. The experimental details have already been described in detail elsewhere [22].

We investigate the electron-phonon scattering for two different phonons: (i) the 340 cm$^{-1}$ *planar* buckling and (ii) the *out-of-plane* apical O vibrations at different incident photon fluxes in both insulating and metallic YBCO starting material. The eigenvectors of these vibrations are shown schematically next to the data in Fig. 3.2. The phonon occupation number is determined from the ratio of Stokes to anti-Stokes Raman scattering intensities through [23]

$$n_{\rm ph} = \left[\frac{I_S}{I_A} - 1\right]^{-1} . \tag{3.1}$$

We now summarize the results:

- A photoinduced I-M transition in $YBa_2Cu_3O_{6.2}$ is clearly observed as an abrupt change in the reflectivity $R$ of the sample (Fig. 3.1). (The change in $R$ upon photoexcitation is similar to the change observed upon O doping.) The transition occurs at a calculated density of $n_c \sim 10^{19}$ carriers per cm$^3$ corresponding to an incident flux of $5 \times 10^{12}$ photons per cm$^2$. In calculating $n_c$, we have assumed a carrier lifetime of more than 1.5 ps (but much less than 10 ns). Although the carrier density $n_c$ is hard to determine accurately and may deviate by as much as a factor of 5, the important observation is that the change in behaviour with increasing carrier density occurs *abruptly*, implying a well-defined electronically driven phase transition.

- $n_{ph}$ as a function of incident photon flux when measured with 1.5 ps pulses is plotted using open squares in Fig. 3.2, for both $YBa_2Cu_3O_{6.2}$ and $YBa_2Cu_3O_{6.9}$ together with a sketch of the ionic displacements of the phonons involved. The estimated maximum photoexcited carrier density is given in the top axes. Three cases are compared. The top and middle figures (Fig. 3.2a and b) are experiments on $YBa_2Cu_3O_{6.2}$ and show that $n_{ph}$ is not constant, but depends strongly on incoming photon flux density. At small PE carrier densities $n_{ph}$ rises very rapidly for both phonons studied, and then changes slope at approximately $5 \times 10^{12}$ photons/cm$^2$. For the apical O vibration (Fig. 3.2a), $n_{470}$ continues increasing with increasing photon flux, while $n_{340}$ for the plane buckling phonon (Fig. 3.2b) drops to the equilibrium value near $n_{ph} \sim 0.3$ (determined by the ambient temperature). The two phonons clearly show very different behaviour and we can infer very different electron-phonon coupling to PE charge carriers for the two cases.
  Fig. 3.2c shows the results of the same experiment in $YBa_2Cu_3O_{6.9}$. $n_{500}$ here keeps increasing near-linearly with increasing photon flux. Indeed, the occupation number for the apical O vibration in chemically doped metallic $YBa_2Cu_3O_{6.9}$ shows a similar increase with increasing laser flux as $YBa_2Cu_3O_{6.2}$ in the photoinduced metallic phase i.e. beyond the I-M transition. Apparently the e-p coupling in the PI metallic phase and the chemically doped metallic phase is rather similar.
- The contribution to $n_{ph}$ due to local heating of the scattering volume is eliminated experimentally by measuring $n_{ph}$ also with a continuous wave (CW) laser as well as with 100 ps pulses at the same wavelength and of the same average power (but obviously lower peak power). These data are also shown as full dots and triangles respectively for all three cases shown in Fig. 3.2, together with a calculated value of $n_{ph}$. The heating is calculated using a very detailed model, including the effects of heat diffusion, accurate boundary conditions as well as taking into account effects of heat build-up due to the finite time between pulses. The agreement with the CW measurements is seen to be excellent, especially since there are no variable parameters in the calculation, all the thermal constants for YBCO being taken from the literature.

## 4. Discussion

There are two ways to understand the transient Raman experiments. The simpler model makes no assumptions about the form of the carrier-phonon interaction and assumes simply that the PE carriers lose energy by scattering from phonons via e-p scattering. Obviously, the carriers will scatter most effectively from phonons which are most strongly coupled to them. Hence the

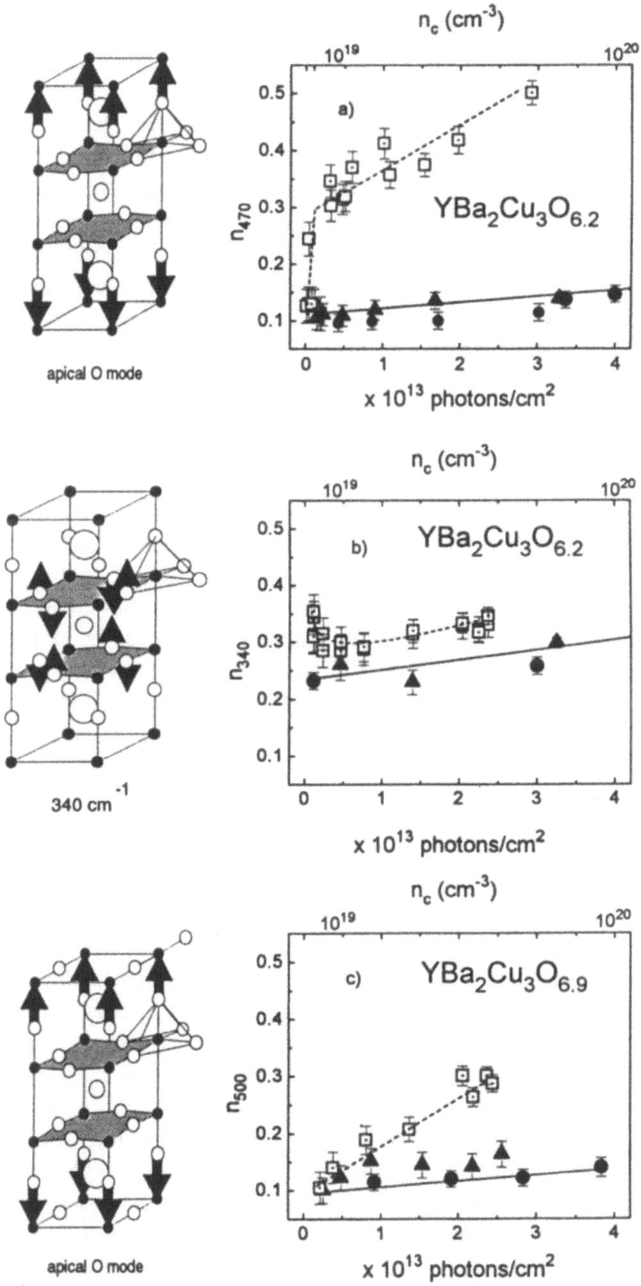


**Fig. 3.2.** Phonon occupation number, $n_{ph}$ as a function of incident photon flux density for (a) the 470 $cm^{-1}$ apical O mode in $YBa_2Cu_3O_{6.2}$, (b) the 340 $cm^{-1}$ phonon also in $YBa_2Cu_3O_{6.2}$ and (c) for the 500 $cm^{-1}$ apical O mode in $YBa_2Cu_3O_{6.9}$. All data in open squares were taken with 1.5 ps pulses. The calculated PE carrier density (assuming uniform density) is shown on the top axis. The solid lines show the calculated sample temperature. Dashed lines are guides to the eye. For comparison, the full triangles are data taken with quasi-CW (100 ps) excitation and the full circles are taken with 514.5 nm CW excitation, all with the same *average* laser power. The ionic displacements for the three cases are shown on the left.

observation of a large phonon occupation number for the apical O phonon implies that it is strongly coupled to the carriers. The interpretation of the data is then straightforward.

An extension of this model has to be made however, if the e-p interaction is strong enough to cause the PE carriers to self-trap and form polarons. Then these polaronic effects need to be considered in addition. In a non-equilibrium situation, polarons make non-adiabatic hops and the extra phonon $n_{ph}$ we observe corresponds to the phonon "shake off" when the polaron moves about. The relatively large $n_{ph}$ observed for the apical O vibration especially implies that polaronic effects need to be taken into account, at least to describe the carriers which are coupled to the apical O vibrations.

The present experiment apparently moves towards resolving the problem of whether there are polarons in $CuO_2$ planes or outside the planes, (e.g. in the Cu-O chains). We know for certain that the apical O is involved in strong coupling to carriers, but this could be as part of either a plane or chain polaron, so we need to analyze the data further. The fact that the planar 340 $cm^{-1}$ vibration shows a strong reduction in coupling in the metallic phase implies that the planar e-p scattering is strongly reduced in the metallic phase, making the formation of planar lattice polarons rather unlikely. On the other hand, evidently the large $n_{ph}$ for the apical O in both materials are clear and direct evidence of PE carriers being strongly coupled to the apical O ion even in the metallic state. We hence infer that the polaron-like behaviour in YBCO comes from the coupling of localized carriers in the Cu-O dumbbells, chains or semi-chains to apical O ion vibrations.

## 5. Conclusion

The photoexcitation technique has been shown to work quite successfully in mimicking chemical doping, and using resonant Raman spectroscopy we are able to obtain some qualitatively new information about carrier-lattice interactions, both in the insulating and metallic regimes. The interpretation that in the metallic material, PE carriers form both itinerant charge carriers in the $CuO_2$ planes and simultaneously semi-localized carriers in the chains is very appealing because it naturally explains the 2-component mid-infrared spectra, the temperature-dependent Drude contribution and the relatively temperature-independent mid-infrared corresponding to the delocalized and localized carriers respectively. The reason why the I-M transition occurs at lower carrier density than expected is still to be understood.

## References

1. M. Cardona in *Physical Properties of High Temperature Superconductors I*, Ed. D.S. Ginsberg (World Scientific, Singapore 1989) and references therein.

2. T. Mertelj et al, Phys. Rev. B **47**, 12104 (1993)
3. D.Mihailovic et al, **45** 8016 (1992)
4. E.V. Antipov et al, (to be published)
5. K. Fossheim and T. Laegrid, IBM J. Res. Dev. **33**, (1989) 365, also Wu Ting et al., Physica C **162-164**, 448 (1989) , H.M. Ledbetter and S.A. Kim, Phys.Rev.B **38**, 11857 (1988) and Y. Mi et al., Physica C**172**, 407 (1991)
6. S. de Brion et al, Europhys. Lett. **12**, 281 (1990), G. Canelli, R. Cantelli and F. Cordero, Phys. Rev. B **38**, 7200 (1988).
7. Steglich et al (to be published)
8. D. Mihailovic and A.J. Heeger, Sol. Stat. Comm. **75**, 319 (1990), D. Mihailovic and I. Poberaj, Physica C **185-189**, 781 (1991).
9. J. Mustre de Leon, S.D. Conradson, I. Batistic and A.R. Bishop, Phys. Rev. B**44**, 2422 (1991), S. Conradson and I.D. Raistirk, Science **243**,1340 (1989) , and J. Mustre de Leon et al., Phys. Rev. Lett. **65**, 1675 (1990).
10. D. Mihailovic, K.F. McCarty and D.S. Ginley, Phys. Rev. B **47**, 8910 (1993), also D. Mihailovic, K.F. McCarty and D.S. Ginley, Ferroelectrics, **130** 107 (1992), D. Mihailovic and C.M. Foster, Sol. Stat. Comm. **74** 753 (1990).
11. Y.H. Kim et al, Phys. Rev. B **36**, 7252 (1987)
12. D. Mihailovic et al, Phys. Rev. B **44** 237 (1991)
13. C.N. Foster et al, Sol. Stat. Comm. **76** 651 (1990)
14. D. Mihailovic, C.M. Foster, K. Voss and A.J. Heeger, Phys. Rev. B **42**, 7989 (1990)
15. Xiang-Xin Bi and P.C. Ecklund, Phys. Rev. Lett. **70**, 2625 (1993)
16. L.D. Rotter et al, Phys. Rev. B **67**, 2714 (1991)
17. G. Yu et al, Phys. Rev. Lett. **67**, 2581 (1991)
18. V.I. Kudinov et al, Physica C **185-189**, 751 (1991), *ibid.* p.1241
19. J. Kircher et al, Phys. Rev. B**44**, 217 (1991)
20. O.K. Andersen et al, Physica C**185-189**, 147 (1991).
21. E.T. Heyen et al., Phys. Rev. Lett. **65**, 3048 (1990), E.T. Heyen et al., Phys. Rev. B **45**, 3037 (1992).
22. I. Poberaj and D. Mihailovic (to be published), also D. Mihailovic and I. Poberaj, Proc. Int. Conf. on Novel Spectroscopies of Superconductors, Santa Fe, (1993)
23. For example: M. Cardona in Light Scattering in Solids II. Ed. M. Cardona, (Springer, 1980)

# Anharmonic Effects Related to the Apical Oxygen Vibration in $YBa_2Cu_3O_7$ by LAPW Calculations

C. Ambrosch-Draxl

Institut für Theoretische Physik, Universität Graz, Universitätsplatz 5, A-8010 Graz

**Summary.** We have calculated the lattice dynamical potential of the apical oxygen in $YBa_2Cu_3O_7$ by frozen phonon calculations using the full potential LAPW method. Even for large displacements there is no evidence for a double well, but we find the potential to be anharmonic with a relevant third order term up to 1000 K. Eigenstates are given and the temperature dependence of the corresponding phonon frequency is estimated.

## 1. Introduction

The microscopic mechanism of high temperature superconductivity is still unclear and there exists no common picture of the relationship between phonons and superconductivity in these materials. Therefore, investigations of the structural and vibrational properties of the copper oxides are very important. Among the common structural features of the layered high temperature superconductors are CuO chains and $CuO_2$ planes. For the interplay between the chains and the planes the bridging (apical) oxygen, forming a Cu-O pyramid with the plane atoms, should play an important role. In particular, dynamic Jahn-Teller effects of distorted Cu-O octahedra were thought to be essential for superconductivity [1]. However, the expected strong electron-phonon interaction of the Cu-O bond may cause anharmonic effects. EXAFS experiments [2, 3] have been interpreted in the sense that there should be a structural instability due to a double well potential for the apical oxygen vibration. From Raman measurements [4, 5, 6] some authors have concluded that there could be a rather strong anharmonicity of the lattice dynamical potential. In order to investigate this problem we have performed frozen phonon calculations by first principle band theory. It has been demonstrated by many papers that band structure calculations give reliable results for structural and vibrational properties [7, 8, 9] of the high temperature superconductors as well as for the charge distribution [10, 11]. Therefore, this investigation is well founded. First results have been published elsewhere [12]. Now we present a more detailed analysis of the potential including eigenstates and an estimation of the temperature dependence of the corresponding phonon frequency.

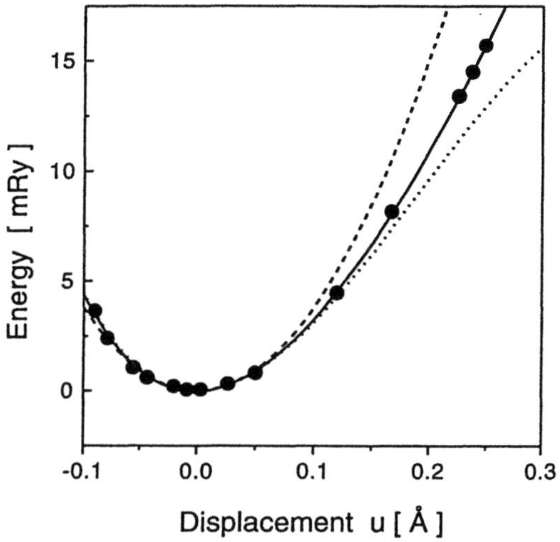


**Fig. 3.1.** Total energy in mRy versus displacement $u$ in the c-coordinate of O(4) in Å. The full circles indicate the calculated data points, the full line a polynomial fit up to the fourth order. The dashed (dotted) line shows the harmonic (harmonic plus cubic) part of the fit.

## 2. Method

The calculations were performed by the full potential LAPW method using an optimized version of the Wien code [13]. It is based on the density functional theory; exchange and correlation effects are treated by the local density approximation. Within this formalism the unit cell is divided into non-overlapping atomic spheres (where the wave functions are linear combinations of radial functions times spherical harmonics) and the so-called interstitial region (where the basis functions are plane waves). Up to 1168 plane wave basis functions were used. The Brillouin zone integrations were carried out with 192 special k-points. We have calculated the change of the total energy by varying the c-coordinate of the apical oxygen O(4). The chain atoms are denoted by Cu(1) and O(1), the plane atoms by Cu(2), O(2), and O(3). As the distance O(4)-Cu(1) is smaller than O(4)-Cu(2) the displacement of O(4) versus the chain was limited by the choice of the atomic sphere radii (1.79 a.u. for Cu(1), 1.9 a.u. for Cu(2), 2.74 a.u. for Y, 2.9 a.u. for Ba, and 1.55 a.u. for all oxygen spheres). The total energy is converged to an accuracy of 0.1 mRy.

## 3. Results

Fig. 3.1 shows the total energy of $YBa_2Cu_3O_7$ as a function of the O(4) position. The displacement $u$ is positive for the movement towards the plane and negative for the movement towards the chain. The full circles indicate the calculated data points. The curve does not show any evidence for a double well potential, but has a single minimum. The calculated points are fitted by a fourth order polynomial

$$\Delta E \; [mRy] = 380u^2 - 688u^3 + 655u^4$$

where $u$ is the displacement of O(4) in Å, and $\Delta$E is the total energy increase in mRy with respect to u. $\Delta$E is given for half a formula unit, according to the movement of one O(4) atom.

The harmonic approximation of the potential is only valid for small displacements (about 0.05 Å). In Fig. 3.1 this harmonic part of the potential is shown as a dashed line. For larger displacements one has to take into account higher terms in the polynomial fit. The negative $u^3$-term is assumed to arise from the different distances of O(4) to the neighboring copper atoms. The increase of the energy is steeper for the motion towards the chain (negative $u$, smaller distance). The polynomial up to the cubic part (dash-dotted line) describes the potential rather well up to displacements of about 0.15 Å (energies of about 1000 K) and is mainly responsible for the observed anharmonic behavior.

The phonon frequencies for the five $A_g$ modes have been determined by two groups [8, 9] diagonalizing the dynamical matrix. Both calculations show very good agreement between calculated and measured values. They find a strong coupling between the vibration of the apex oxygen O(4) and the plane oxygen atoms O(2) and O(3). Since the latter atoms move in opposite direction than O(4), the bare O(4) motion is expected to lead to a frequency which is smaller than the experimentally observed 500 wavenumbers. From the quadratic term of our polynomial fit the vibrational frequency of 419 $cm^{-1}$ is derived. Contributions of O(2) and O(3) would enhance this $u^2$-term increasing the frequency. Thus our calculations are in good agreement with the results of the other groups [8, 9]. However, in the following analysis we only take into account the effects arising from the O(4) motion.

Since the potential has anharmonic contributions one has to solve the anharmonic oscillator equation exactly rather than using the harmonic approximation. Fig. 3.2 shows the energies of the two lowest transitions for the calculated potential (full line) compared to the transitions in the harmonic part of it (dashed curve). The anharmonicity (third order term) affects even the first transition reducing the phonon frequency by 7 wavenumbers. The second transition is reduced by 14 wavenumbers. From these findings we can estimate the temperature dependence of the phononmode. The shift of the Raman frequency of the apical oxygen mode with increasing temperature was measured by Raman spectroscopy for several materials [4, 5] and was found to soften significantly for $YBa_2Cu_3O_7$. Taking into account even higher transitions via a Bose factor we find a softening of the frequency of a few wavenumbers up to 1000 K. With these findings one could explain

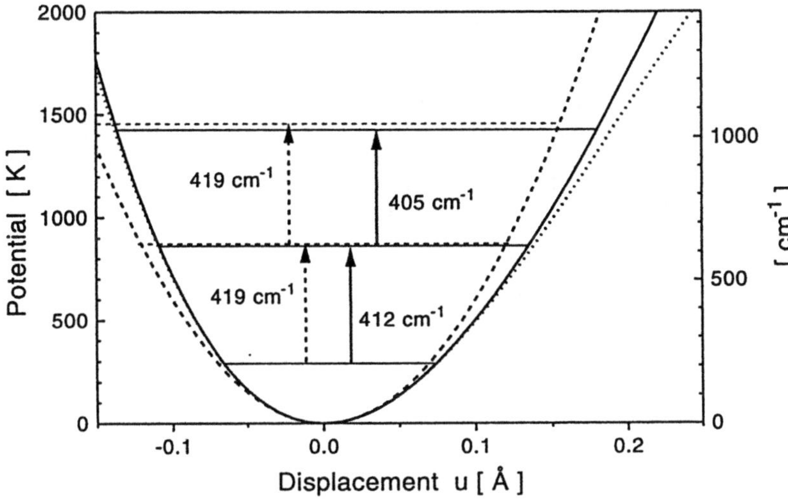


**Fig. 3.2.** Eigenstates for the O(4) potential (full lines) and for the corresponding harmonic part (dashed lines). The arrows indicate the two lowest transitions for both cases.

a softening as measured for the Tl compounds [4], which is much smaller than in $YBa_2Cu_3O_7$, where the measured decrease up to 1000 K is about 90 wavenumbers. So we conclude that there is an additional effect arising from an oxygen loss (in the chains) during the process of heating. However, one has to consider that other contributions, as lattice expansion, are also relevant for this question.

*Acknowledgement.* This work was supported by the Jubiläumsfonds der Österreichischen Nationalbank, Projekt Nr. 4420.

## References

1. K. A. Müller, Z. Phys. B - Condensed Matter **80**, 193 (1990).
2. J. Mustre de Leon, S. D. Conradson, A. R. Bishop, and I. Batistic, Phys. Rev. Lett. **65**, 4675 (1990).
3. J. Mustre de Leon, S. D. Conradson, I. Batistic, A. R. Bishop, I. D. Raistrick, M. C. Aronson, and F. H. Gavion, Phys. Rev. B **45**, 2447 (1992).
4. D. Mihailovic in *Springer Series of Solid-State Sciences* **113**, 452 (1992), edts. H. Kuzmany, M. Mehring, and J. Fink.
5. D. Mihailovic, K. F. McCarty, and D. S. Ginley, Phys. Rev. B **47**, 8910 (1993).
6. G. Ruani, this volume; G. Ruani, C. Taliani, M. Muccini, K. Conder, E. Kaldis, H. Keller, D. Zech, and K. A. Müller, preprint.
7. R. E. Cohen, W. E. Pickett, and H. Krakauer, Phys. Rev. Lett. **62**, 831 (1989).

8. R. E. Cohen, W. E. Pickett, and H. Krakauer, Phys. Rev. Lett. **64**, 2575 (1990).
9. C. O. Rodriguez, A. I. Liechtenstein, I. I. Mazin, O. Jepsen, O. K. Andersen, and M. Methfessel, Phys. Rev. B **42**, 2692 (1990).
10. K. Schwarz, C. Ambrosch-Draxl, and P. Blaha, Phys. Rev. B **42**, 2051 (1990).
11. C. Ambrosch-Draxl, P. Blaha, and K. Schwarz, Phys. Rev. B **44**, 5141 (1991).
12. C. Ambrosch-Draxl and P. Knoll, Physica B (in print).
13. P. Blaha, K. Schwarz, P. Sorantin, and S. B. Trickey, Comp. Phys. Commun. **59**, (1990).

# Thermodynamics of the Emery Model

K. Dichtel, J. Fritzenkötter, and J. Carstensen

Institut für Theoretische Physik, Universität Kiel, D 24118 Kiel, Germany

**Summary.** Some calculations on the Emery model are presented. In the strong coupling regime a previous variational solution of the three band model is extended to include nearest neighbour repulsion and spiral magnetic order. In the weak coupling regime frequency and momentum dependent real and imaginary part of the self energy and the density of states are calculated in a self consistent Green's function expansion. The most important result in the DOS is some onset of splitting of the antibonding band into subbands with increasing correlation. In the neutral case the chemical potential is in a region of reduced DOS. Upon doping it moves into the peak of the upper subband.

## 1. Introduction

The correlated models favoured in the past years to explain the properties of the carriers in the copper oxygen planes of HTSC materials [1] show a variety not only of homogeneous but also of mixed phases of different carrier concentration and magnetic order. The latter fact has been invoked amongst other ideas to provide an understanding of the unusual properties of the HTSC cuprates as caused by spatial regions of a highly concentrated carriers separated by neutral domains [2],[3]. The short range approximation of the Coulomb repulsion in the Hamiltonian of the Hubbard model and of related models without long range Coulomb tails to restore charge neutrality, enforces the tendency of the carriers to separate into different phases. In calculations phase separation usually results from a Maxwell construction in some region of thermodynamic instability. This causes discontinuities in the derivative of the curve chemical potential $\mu$ versus doping $n$ and a part of constant $\mu$ in between which means infinite electronic compressibility $dn/d\mu$. A similar phenomenon namely a critical $\mu(n)$ curve with continuous derivative but with a part of constant value $\mu_c$ over a finite range of carrier densities recently was found in the dimension $D \to \infty$ case [4]. The occuring pinning of the chemical potential $\mu$ in a peak of the DOS at some value $\mu_c$ within a range of increasing number of carriers can be caused by a change in the DOS (transfer of spectral weight instead of a rigid band model) upon doping. Some splitting and moving of the oxygen states in the doped case e.g. was seen in the oversimplified DOS of the strong coupling variational calculation [5]. To address this topic for a more realistic DOS in the 2D three band model, we present in the last chapter some perturbation calculations which show the changes in the DOS with increasing correlation and doping.

## 2. Phase Diagrams Including Separated Phases

We start with some phase diagrams, which show separated phases in the one and three band Hubbard models. The Hubbard model in the standard form is

$$\mathbf{H}_{\text{Hubbard}} = -t \sum_{<n,m>,\sigma} (\mathbf{x}_{n,\sigma}^{+}\mathbf{x}_{m,\sigma} + \text{h.c.}) + \frac{U}{2}\sum_{n,\sigma} \mathbf{x}_{n,\sigma}^{+}\mathbf{x}_{n,\sigma}\mathbf{x}_{n,-\sigma}^{+}\mathbf{x}_{n,-\sigma} \quad (2.1)$$

where the operator $\mathbf{x}_{n,\sigma}^{+}$ creates at site $n$ a hole; $< n,m >$ are next neighbours, $t$ is the hopping and $U$ the on-site correlation constant.

The first to propose phase separation in this model was Visscher [6]. He performed a numerical high temperature cumulant expansion [7] and observed that for large on-site correlation $U/t = 25$ at some doping concentration $n \geq 1.35$ the next neighbour density correlation functions become positive as well for parallel as for antiparallel spin thus indicating some clustering of the carriers. An attempt to improve these computer calculations of cumulants nowadays still could not confirm this observation [8]. The qualitative argumentation however in [6] is a fundamental aspect: It could be energetically favourable for additional carriers in an antiferromagnetic (AF) system to concentrate themselves spatially within some domains to break as few AF bounds as possible. Ferromagnetic (FM) order in the simple picture of splitted Hubbard bands on the contrary is independent of the location of the carriers. In the large correlation limit of the Hubbard model, the t-J model, phase separation is evident from the spin frustrated motion of a hole through AF regions [2],[9]. A phase diagram comparing several results on phase separation in the 2D t-J model is given in [10].

The first quantitative calculations of phase separation in the Hubbard model were performed in [11],[12]. In Fig. 2.1 we report the phase diagram of strong coupling approximations of the Hubbard model from a slave boson dimension $D = 2$ calculation [13]. In [14] a phase diagram from a Gutzwiller variational function calculation for $D \to \infty$ (with scaling $2Dt^2 = 1$) was given. It is remarkable that both different (the latter one rescaled to $D = 2$) results nearly coincide in the extension of the region of AF-PM (paramagnetic) phase separation up to doping $\delta = n - 1 \approx 0.23$ and up to $U/t \approx 20$ and of the regions of AF-FM phase separation for larger values of $U/t$.

For the most relevant orbitals in the Cu–O planes a generalized Hubbard model was proposed in [15]

$$\begin{aligned}\mathbf{H}_{\text{Em}} = {} & \frac{U}{2}\sum_{n,\sigma} \mathbf{x}_{n,\sigma}^{1+}\mathbf{x}_{n,\sigma}^{1}\mathbf{x}_{n,-\sigma}^{1+}\mathbf{x}_{n,-\sigma}^{1} + U_{\text{pd}} \sum_{\substack{\sigma,\sigma' \\ \langle n,(m,j)\rangle}} \mathbf{x}_{n,\sigma}^{1+}\mathbf{x}_{n,\sigma}^{1}\mathbf{x}_{m,\sigma'}^{j+}\mathbf{x}_{m,\sigma'}^{j} \\ & + \epsilon_p \sum_{m,j,\sigma} \mathbf{x}_{m,\sigma}^{j+}\mathbf{x}_{m,\sigma}^{j} + \epsilon_d \sum_{n,\sigma} \mathbf{x}_{n,\sigma}^{1+}\mathbf{x}_{n,\sigma}^{1}\end{aligned}$$

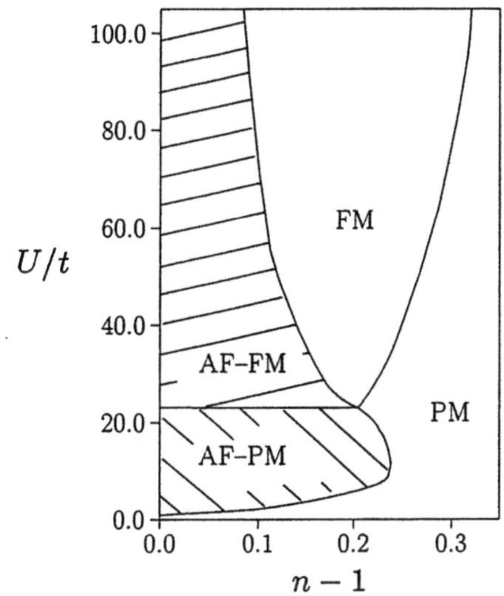


**Fig. 2.1.** Phase diagram of the Hubbard model correlation $U/t$ versus doping $n-1$ in strong coupling approximation from [13]. Hatched areas give AF–FM and AF–PM mixed phases respectively.

$$+\gamma \sum_{\substack{\langle n,(m,j)\rangle \\ \sigma}} \left(t_{n,(m,j)} \mathbf{x}_{n,\sigma}^{1+} \mathbf{x}_{m,\sigma}^{j} + \text{h.c.}\right) \tag{2.2}$$

The operators $\mathbf{x}_{n,\sigma}^{j+}$ create for $j=1$ holes on copper sites $n$ in $\mathrm{Cu(3d}_{x^2-y^2})$ orbitals and for $j=2,3$ holes on adjacent oxygen sites $m$ in $\mathrm{O(2p}_x)$ and $\mathrm{O(2p}_y)$ orbitals respectively. The constant $U$ denotes the on-site Coulomb correlation on copper places, $U_{\mathrm{pd}}$ between copper and oxygen states, $\gamma$ denotes the absolute value of the transfer integral. The summations are restricted to nearest neighbour pairs. The sign of the transfer integral is $t_{n,(m,j)} = \pm 1$ corresponding to the difference in symmetry of the Cu and O orbitals. The parameter $\epsilon_p$ and $\epsilon_d$ are the atomic Cu(3d) and O(2p) level energies respectively. We take $\epsilon_p = 0$ and $\epsilon_d = -\epsilon_0$, the charge transfer energy. The weak coupling thermodynamics of this model was explored by [16],[17],[18]. It is remarkable that contrary to the Hubbard model for this model even in the weak coupling a region of phase separation is found [18].

## 3. A Strong Coupling Ansatz for the Emery Model

We refer to some previously published [19] temperature dependent strong coupling calculations using the thermodynamic variational principle of Bogoljubov [20]

$$\Omega \leq \Omega_0(\mathbf{U}) = \Omega_T + \left\langle \mathbf{U}^+\mathbf{H}\mathbf{U}\right\rangle_T - \left\langle \mathbf{H}_\mathrm{T} + \mu\mathbf{N}\right\rangle_T, \tag{3.1}$$

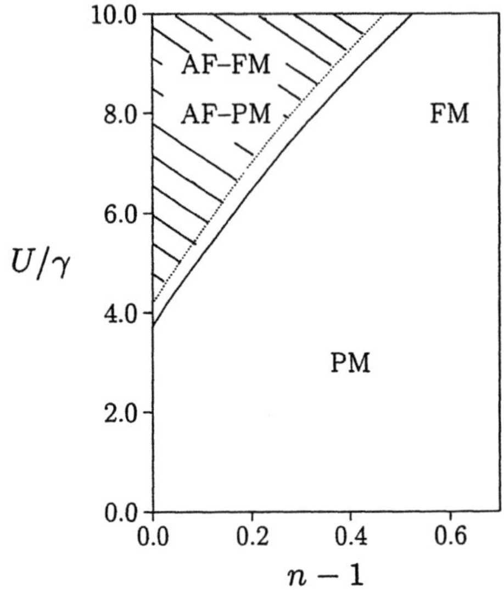


**Fig. 3.1.** Phase diagram of the Emery model: correlation $U/\gamma$ versus doping $n-1$ in strong coupling approximation ($n^*$ from ansatz below). Hatched areas show separated phases (a) $\epsilon_0/U = 0.5$ (b) $\epsilon_0/U = 0.4$ (dotted).

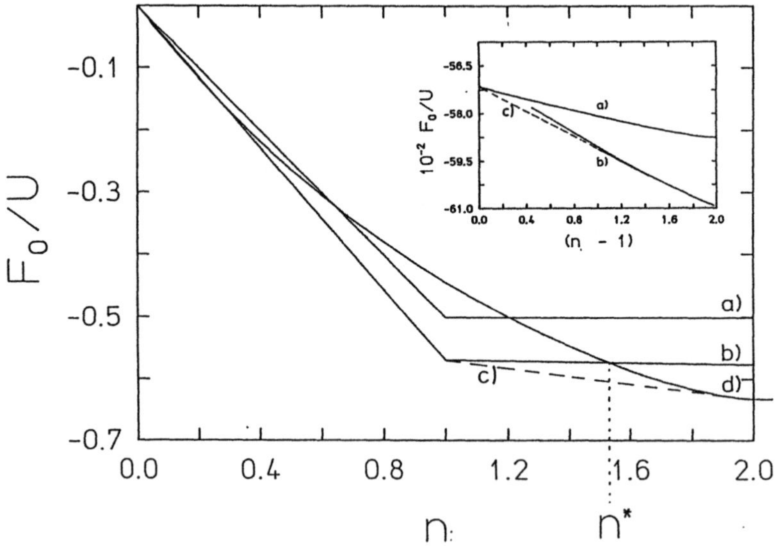


**Fig. 3.2.** Free energy of the Emery model $F_0$ versus hole concentration $n$ for $U/\gamma = 10$, and $\epsilon_0/\gamma = 5$. (a) atomic limit, (b) our ansatz for AF order, (d) Hartree Fock solution, and (c) (dashed) common tangent of Maxwell construction between (b) and (d). Inset: Free energy for $T > T_N$ where no AF phase exists. (a) PM phase, (b) FM phase, and (c) (dashed) Maxwell construction between both.

where the test–Hamiltonian $\mathbf{H}_{\mathrm{T}}$ and the canonical transformation $\mathbf{U}$ can be chosen arbitrarily with some variational parameters. ($\langle ... \rangle_T$ means the thermodynamic average with $\mathbf{H}_{\mathrm{T}}$ ). Usual Hartree Fock (HF) approximations correspond to the choice of $\mathbf{H}_{\mathrm{T}}$ as a sum of single particle operators (for the Hubbard model see e.g. [21],[22]). For a strong coupling approximation of the Emery model we choose

$$\mathbf{H}_{\mathrm{T}} := \sum_{c_1,c_3} \tilde{u}_{c_1 c_3} \frac{1}{2} \sum_{\sigma,n} \mathbf{x}_{n,\sigma}^{c_1 +} \mathbf{x}_{n,\sigma}^{c_1} \mathbf{x}_{n,-\sigma}^{c_3 +} \mathbf{x}_{n,-\sigma}^{c_3} - \tilde{\epsilon}_0 \sum_{n,\sigma} \mathbf{x}_{n,\sigma}^{1+} \mathbf{x}_{n,\sigma}^{1} + \sum_{n,c_1} h_n^{c_1} \, \mathbf{s}_{z,n}^{c_1} , \tag{3.2}$$

where $\tilde{u}_{c_1 c_3}$, $h_n^{c_1}$, $\tilde{\epsilon}_0$ and some chemical potential $\lambda$ are variational parameters. The choice of this simple local Hamiltonian in real space is improved by the introduction of a canonical transformation $\mathbf{U} = exp(-i\varphi \mathbf{S})$ which is optimized to second order of $\gamma/U$ with $\varphi$ as an additional variational parameter. It is only this transformation which provides the coupling between different lattice cells. The ansatz was proposed by Jelitto and Heise [23] for the one band Hubbard model. In view of the recently proved exact mapping of the $D \to \infty$ Hubbard model onto the single site Anderson impurity model in a generalized mean field (see e.g. [24],[25]) it seems that some essential features of highly correlated models are met by this ansatz. For details of the calculation of the expectation values with this canonical transformation, first in real–, then in $\boldsymbol{k}$–space see [19],[26]. Fig. 3.1 gives the phase diagram $U$ vs. $n$ (from this solution and $n^*$ defined below) for comparison with the Hubbard model in Fig. 2.1. Fig. 3.2 gives the resulting curves free energy $F_0$ versus hole concentration $n$ for a) the atomic limit, b) the strong coupling variational approximation for $\gamma/U = 1$ and c) the standard PM HF approximation. For hole concentrations $n \geq n^*$ HF yields a lower (better) free energy than the strong coupling limit. Analysis of the chemical potential $\mu$ vs. doping $n$ curve gives regions of thermodynamic instability to be bridged by a Maxwell construction and thus phase separation for $n \leq n^*$. Besides this "intrinsic" phase separation of our solution another Maxwell construction (dashed in Fig. 3.2) between the two different solutions strong coupling ansatz and HF solution also is possible (as usual in chemical thermodynamics between the potential minima of different phases). A "common tangent" phase separation of this type with a very similar energy curve was proposed early for the Hubbard model [12]. From these as well as from our curves it is evident that the low value of the AF free energy in the neutral case causes the phase separation. In a similar way the low values of the FM free energy for temperatures above the Néel temperature $T_N$ causes PM–FM phase separation as shown in the inset of Fig. 3.2.

A cluster calculation of [27] providing upper and lower bounds for the free energy encloses the values of our solution for small doping values $n \approx 1$ and improves the solution for larger values of $n$.

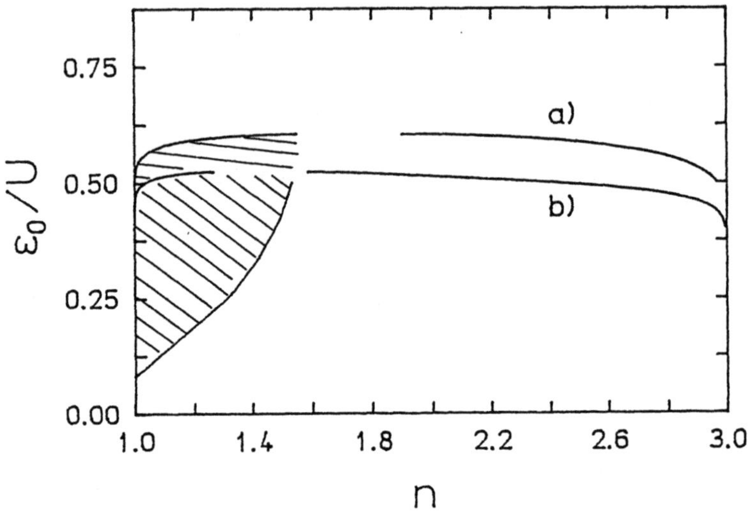


**Fig. 3.3.** Phase diagram of the Emery model charge transfer $\epsilon_0/U$ versus doping $n$. (a) $U_{\rm pd}/U = 0.01$, (b) $U_{\rm pd}/U = 0.02$.

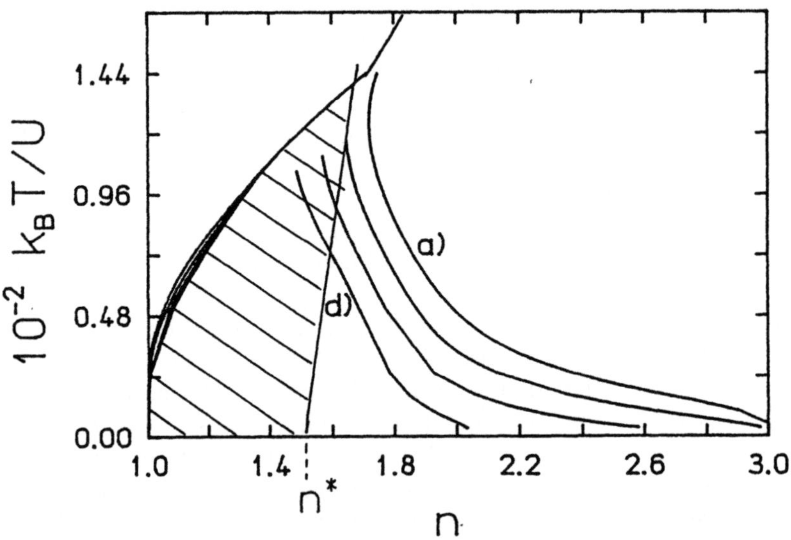


**Fig. 3.4.** Phase diagram AF–FM phases temperature $k_BT/U$ vs. doping $n$ for $U/\gamma = 10$, and $\epsilon_0/\gamma = 5$ and from (a) to (d) for values of $U_{\rm pd}/U = 0$, 0.01, 0.02, and 0.03. In the regions left to the nearly vertical line at $n^*$ which limits the validity of the strong coupling solution phase separation occurs whereas right to this line the standard homogeneous HF regime extends.

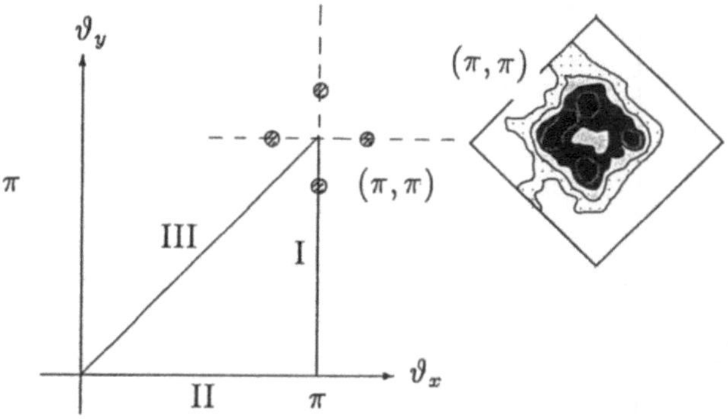


**Fig. 4.1.** Variation of $\boldsymbol{\vartheta}$. Optimal value at $(\pi, 0.8\,\pi)$ and corresponding symmetry points compared with (inset) neutron data from [31] for $n-1 = 0.075$ centered about $(\pi,\pi)$. I: $\boldsymbol{\vartheta} = (\pi,\vartheta)$, II: $\boldsymbol{\vartheta} = (\vartheta, 0)$, and III: $\boldsymbol{\vartheta} = (\vartheta,\vartheta)$.

Fig. 3.3 gives the dependence of the phase separated regions of our solution on the charge transfer energy $\epsilon_0$ . The curve in [28] shows (even for quite different parameter sets) a qualitatively similar behaviour. As an example of our temperature dependent phase diagrams we present in Fig. 3.4 the regions of AF–FM phase separation. Fig. 3.4 shows also a strong reduction of the phase separated regions even for small values of $U_{\mathrm{pd}}$. (Concerning the occurrence of ordered magnetic phases in 2D for $T > 0$ contrary to exact theorems the usual assumption is made of some small additional coupling to the third dimension as it is the case in real materials.)

## 4. Spiral Magnetic Order

In most correlated models the extension of the AF phase is reduced when spiral order is considered too. For spiral phases in the Hubbard model e.g. see [22],[29],[30]. Spiral phases can be described by a magnetization vector

$$\boldsymbol{M}(\boldsymbol{R}) = h \begin{pmatrix} \cos(\boldsymbol{\vartheta R}) \\ \sin(\boldsymbol{\vartheta R}) \end{pmatrix} \tag{4.1}$$

The field $h$ and the pitch $\boldsymbol{\vartheta} = (\vartheta_x, \vartheta_y)$ are now variational parameters (instead of $h$ only in the AF or FM case. A local rotation of the spin–quantization axis is introduced under which the operators transform

$$\mathbf{y}_{n,\sigma}^{\ j+} := \cos(\frac{\boldsymbol{\vartheta R}_n}{2})\mathbf{x}_{n,\sigma}^{\ j+} + \mathrm{sign}(\sigma)\sin(\frac{\boldsymbol{\vartheta R}_n}{2})\mathbf{x}_{n,-\sigma}^{\ j+} \tag{4.2}$$

with sign $(\sigma) = \mathrm{sign}(\uparrow,\downarrow) := \pm 1$. In the ansatz of the test–Hamiltonian the correlation part now is written in terms of these unitary transformed operators. For details of the lengthy evaluation of the variational equations see [26].

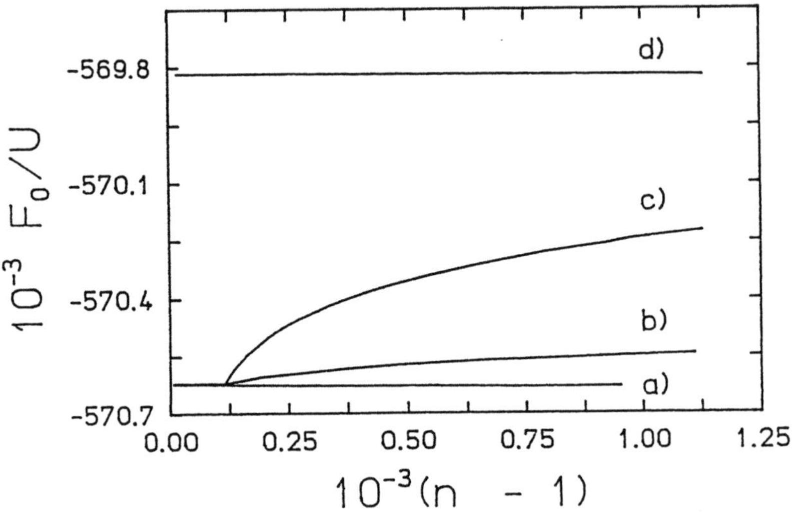


**Fig. 4.2.** Free Energy $F_0$ for different magnetic order vs. $n-1$: (a) AF order, (b) spiral phase with $\boldsymbol{\vartheta} = (\pi, \vartheta)$ (Curve I in Fig. 4.1), (c) spiral phase with $\boldsymbol{\vartheta} = (\vartheta, \vartheta)$ (Curve III in Fig. 4.1), and (d) paramagnetic case.

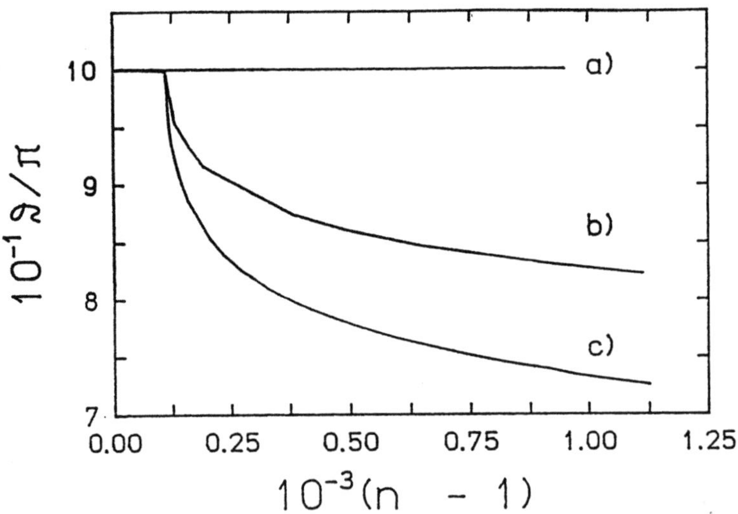


**Fig. 4.3.** Optimal $\vartheta$ in units of $\pi$ vs. $n-1$ for different magnetic ordering: (a) antiferromagnetic order, (b) spiral phase with $\boldsymbol{\vartheta} = (\pi, \vartheta)$ (Curve I in Fig. 4.1), and (c) spiral phase with $\boldsymbol{\vartheta} = (\vartheta, \vartheta)$ (Curve III in Fig. 4.1).

For sake of simplicity only three directions of pitch parameter variation were investigated (see Fig. 4.1). From Figs. 4.2 and 4.3 we conclude: The AF phase is always the most stable, but the energy difference to the parameterization I is small. The optimal spiral phase is $\boldsymbol{\vartheta} = (\pi, \vartheta)$ with $\vartheta \approx 0.8\,\pi$. As shown in Fig. 4.1 this value is in good agreement with recent neutron scattering data of [31] for $La_{2-x}Sr_xCuO_4$.

Thus phase separation within this approximation also takes place if we allow for spiral magnetic order.

# 5. Self Energy and DOS in Weak Coupling

Since Hubbard's first paper on the band splitting [32] the density of states for correlated models across the metal-insulator transition is one of the mostly discussed quantities. Quite recently remarkable progress has been made in the dimension $D \to \infty$ case [33] and for $D = 2$ [34] showing the opening of the insulating gap and the evolution of a temperature dependent Kondo like resonance near to the chemical potential. First indications of these correlation features already were seen in the self consistent second order perturbation theory [35],[36].

In the following we shall extend the latter method to the Emery model. The Hamiltonian (2.2) (now we take $\epsilon_{\rm d} = 0$ and $\epsilon_{\rm p} = \epsilon_0$) is diagonalized for $U = U_{\rm pd} = 0$ to the well known three bands $E_\kappa(\boldsymbol{k})$ for $\kappa = (-, 0, +)$ with the corresponding transformation factors $A_\kappa(\boldsymbol{k})$. The $3 \times 3$ Green's function matrix in these band operators is denoted $\mathcal{G}^0$. In standard Green's function technique up to second order only the graph shown in the inset of Fig. 5.2 contributes. Partially selfconsistency is achieved dressing the lines in the graph i.e. inserting the HF first order $\mathcal{G}^{HF}$ instead of the bare Green's function $\mathcal{G}^0$ as lines. The three Hartree Fock band energies are

$$\widetilde{E}^\sigma_\kappa(\boldsymbol{k}) := \frac{U\widetilde{n}_{\overline{\sigma}} + \epsilon_0}{2} + \kappa\sqrt{\left(\frac{U\widetilde{n}_{\overline{\sigma}} - \epsilon_0}{2}\right)^2 + \lambda(\boldsymbol{k})}\,, \tag{5.1}$$

with

$$\lambda(\boldsymbol{k}) = 4\gamma^2 \left(\sin^2(\frac{k_1 a}{2}) + \sin^2(\frac{k_2 a}{2})\right)\,,$$

$\kappa = (-, 0, +)$, and $\widetilde{n}_{\overline{\sigma}} = \langle \mathrm{x}_{n,-\sigma}^{1+} \mathrm{x}_{n,-\sigma}^{1} \rangle$ as site independent expectation value of the density at the copper sites, determined selfconsistently.

The $3 \times 3$ first order Matsubara Green's function matrix is

$$\mathcal{G}^{HF}_{\mathbf{k}}(i\omega_n) = \left((i\omega_n + \mu)\mathcal{I} - \mathcal{E}(\boldsymbol{k}) - \mathcal{M}^{HF}_{\mathbf{k}}(i\omega_n)\right)^{-1}\,, \tag{5.2}$$

with $\mathcal{I}$ the unity matrix, $\mathcal{E}$ the diagonal matrix of the three bare band energies $E_\kappa(\boldsymbol{k})$ (from (5.1) for $U\widetilde{n}_{\overline{\sigma}} = 0$) and $\mathcal{M}^{HF}_{\mathbf{k}}(i\omega_n)$ the self energy matrix (with the index pair denoted $\mathbf{k} = (\boldsymbol{k}, \sigma_k)$). The components of the self energy matrix are $M^{HF}_{\mathbf{k}\kappa\nu}(i\omega_n) = A_\kappa(\boldsymbol{k})A_\nu(\boldsymbol{k})U\widetilde{n}_{\overline{\sigma}_k}$ instead of the simple self energy $M^{HF}_{\mathbf{k}}(i\omega_n) = Un_{\overline{\sigma}_k}$ of the Hubbard model. The diagonal elements of the self energy matrix show that a hole with spin $\sigma_k$ in the state $(\mathbf{k}, \kappa)$ feels from all $n_{\overline{\sigma}_k}$ holes with the other spin direction in the elementary cell only the repulsion of those which reside on copper sites $\widetilde{n}_{\overline{\sigma}_k}$ times the probability $A^2_\kappa(\boldsymbol{k})$ that the hole itself resides at a copper place. These transformation factors $A_\kappa(\boldsymbol{k})$ cause a $\boldsymbol{k}$-dependence in first order self energy, Green's function and DOS contrary to a merely constant shift of the DOS by $U\widetilde{n}_{\overline{\sigma}}$ in the Hubbard model. With the components of the self energy matrix $M^{HF}_{\mathbf{k}\kappa\nu}(i\omega_n) = A_\kappa(\boldsymbol{k})A_\nu(\boldsymbol{k})\Sigma_{\mathbf{k}}(i\omega_n)$ we obtain after the analytical continuation $i\omega_n \to \omega \pm i0^+ \equiv E - \mu \pm i0^+$ the retarded (advanced) Green function

$$\mathcal{G}_{\mathbf{k}}(E \pm i0^+ - \mu) = \left((E \pm i0)\mathcal{I} - \mathcal{E}(\boldsymbol{k}) - \mathcal{M}_{\mathbf{k}}(E \pm i0^+ - \mu)\right)^{-1}, \tag{5.3}$$

and in standard manner the DOS

$$\rho_\sigma(E) = -\frac{1}{\pi N}\sum_{\boldsymbol{k}} \mathrm{Im}\ \mathrm{Tr}\ \mathcal{G}_{\boldsymbol{k}\sigma}(E + i0^+ - \mu)\,. \tag{5.4}$$

The self energy up to second order:

$$\Sigma^2_{k\sigma}(\omega) = U\widetilde{n}_{\overline{\sigma}} + \left(\frac{U}{N}\right)^2 \sum_{\boldsymbol{p},\boldsymbol{q},\boldsymbol{r}} \sum_{\kappa,\nu,\zeta\in\{-,+\}} \widetilde{A^\sigma_\kappa(\boldsymbol{r})}\widetilde{A^{\overline{\sigma}}_\nu(\boldsymbol{q})}\widetilde{A^{\overline{\sigma}}_\zeta(\boldsymbol{p})}\delta_{\boldsymbol{k}+\boldsymbol{p},\boldsymbol{q}+\boldsymbol{r}} *$$
$$* \frac{f_{\boldsymbol{r},\sigma\kappa}f_{\boldsymbol{q},\overline{\sigma}\nu}(1 - f_{\boldsymbol{p}\overline{\sigma}\zeta}) + (1 - f_{\boldsymbol{r},\sigma\kappa})(1 - f_{\boldsymbol{q},\overline{\sigma}\nu})f_{\boldsymbol{p}\overline{\sigma}\zeta}}{\omega + i0^+ + \mu + \widetilde{E}^{\overline{\sigma}}_\zeta(\boldsymbol{p}) - \widetilde{E}^{\overline{\sigma}}_\nu(\boldsymbol{q}) - \widetilde{E}^\sigma_\kappa(\boldsymbol{r})}, \tag{5.5}$$

is evaluated numerically. With the Fermi Distribution $f_{k\sigma\gamma} := f(\widetilde{E}^\sigma_\gamma(\boldsymbol{k}) - \mu)$ at $T = 0$ and the first order band energies (5.1) $\widetilde{E}^\sigma_\kappa(\boldsymbol{k})$ and the transformation factors

$$\widetilde{A^\sigma_\kappa(\boldsymbol{k})} := \frac{\widetilde{E}^\sigma_\kappa(\boldsymbol{k}) - \epsilon_0}{\widetilde{E}^\sigma_\kappa(\boldsymbol{k}) - \widetilde{E}^\sigma_{-\kappa}(\boldsymbol{k})}. \tag{5.6}$$

The latter ones and the summation over the band index ($\kappa = -, 0, +$) are the only difference to the corresponding formulae of [35] and [36] for the simple single band Hubbard model. For the numerical calculation of (5.5) the same algorithm including some neighbour shells beyond the local approximation in real space as in [36] was used to reduce the numerical effort.

The results for the DOS calculated from (5.4) are summarized in the following figures (increasing hole states from left to right).

In Figs. 5.1 and 5.2 an extremely large correlation $U = 20$ (unit $\gamma$) is used in order to show clearly the resulting features. Compared to the free bands: antibonding(AB), nonbonding (NB) and bonding (B) band now the AB band (predominantly Cu) is depleted and nearly completely splitted into some lower peak (her denoted as I) and some upper peak (denoted as II). Thus in the neutral case of Fig. 5.1 for $n = 1$ the chemical potential $\mu$ (calculated

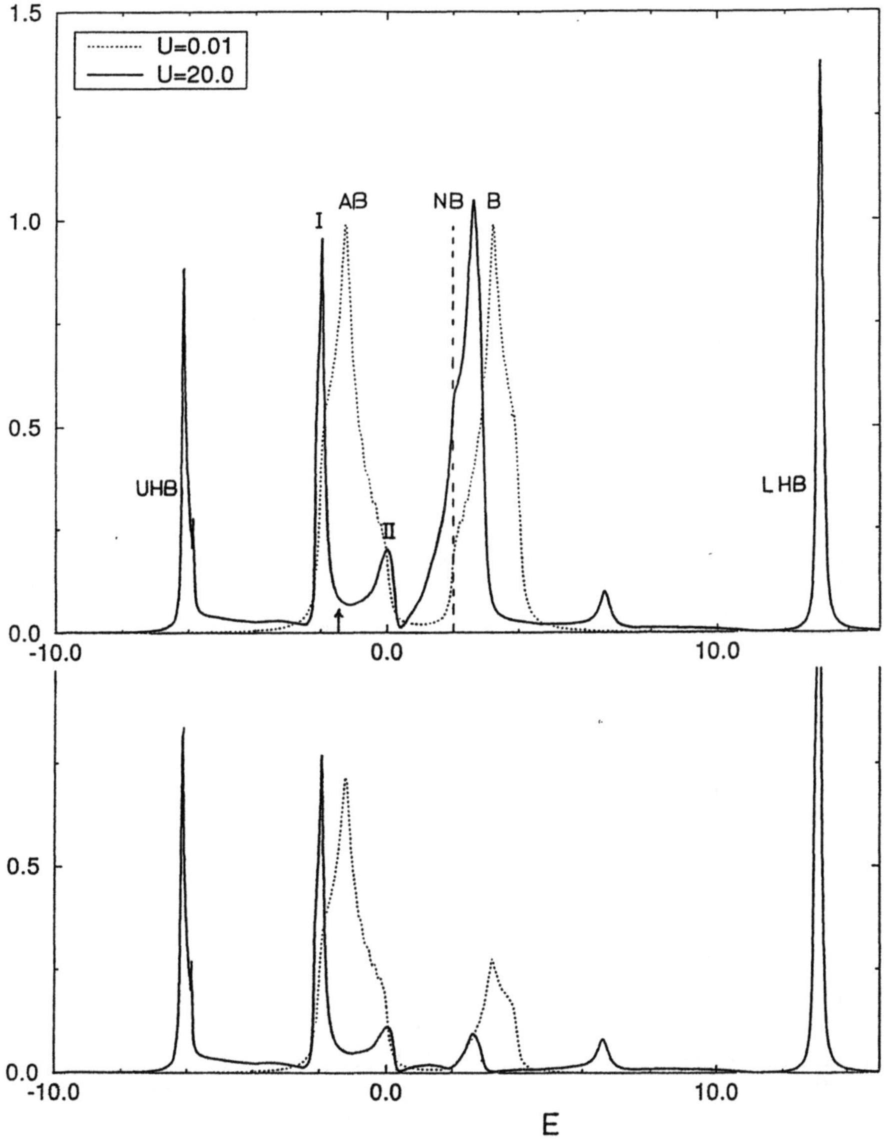


**Fig. 5.1.** DOS (lower part Cu DOS) vs. $E$ for $U = 20$ and $n = 1$, dotted lines: nearly free antibonding (AB) and bonding (B) bands; dashed nonbonding (NB). Arrow: chemical potential; I and II mark peaks of the subbands.

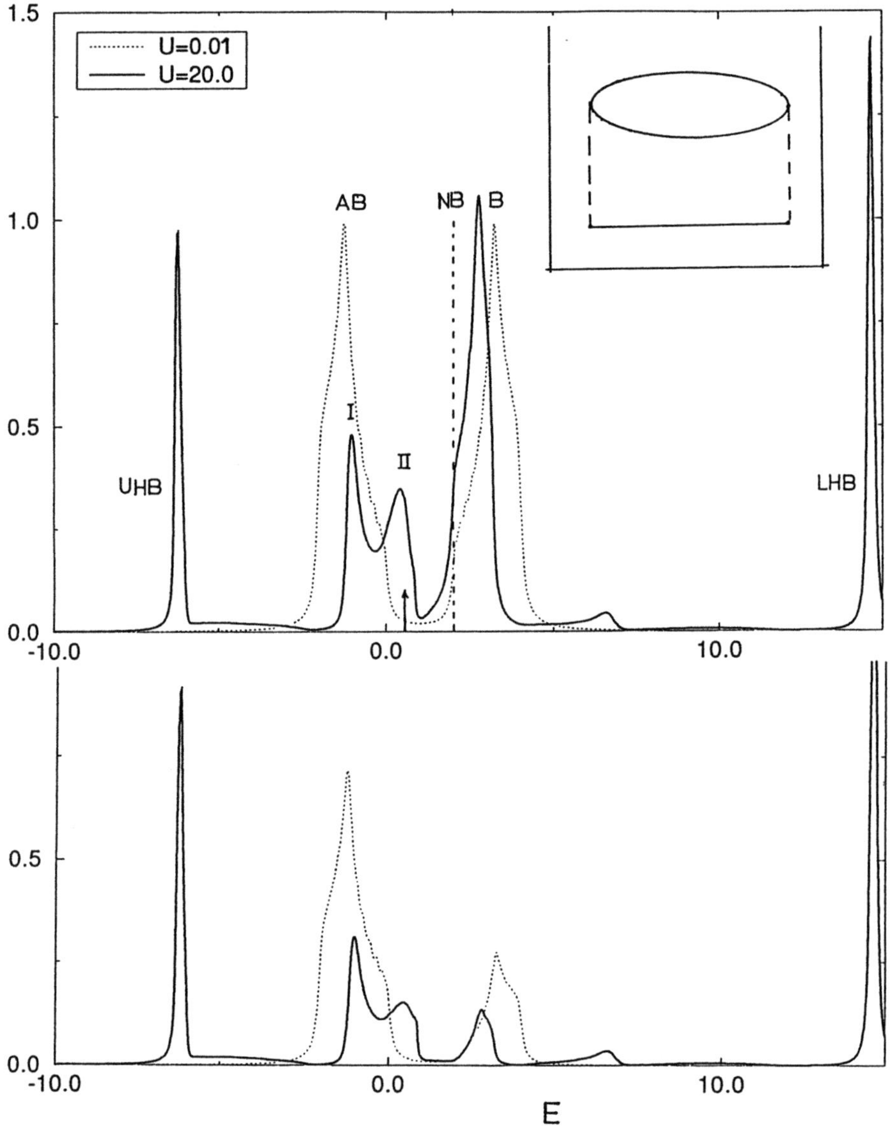


**Fig. 5.2.** DOS (lower part Cu DOS) vs. $E$ as in Fig. 5.1 for doping $n = 1.4$. Inset: Skeleton graph which alone contributes to second order self energy.

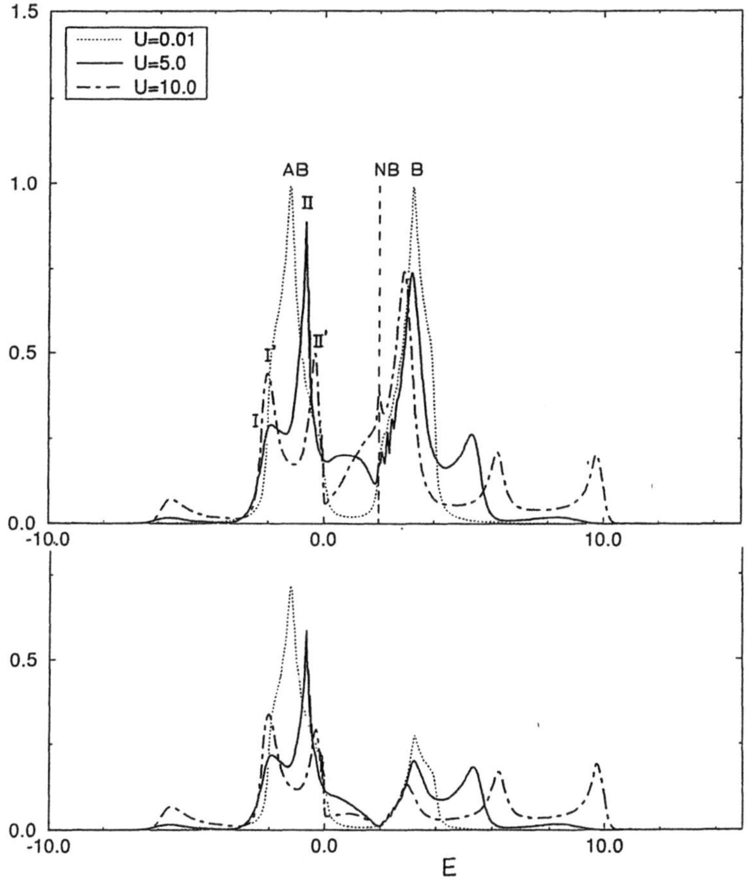


**Fig. 5.3.** DOS (lower part Cu DOS) vs. $E$ for increasing correlation $U$: dotted lines: bands for $U = 0.01$, full lines $U = 5$, dashed dotted lines $U = 10$.

from (5.4)) is situated in a region of drastically reduced DOS instead of the maximum DOS in the van Hove peak of the uncorrelated AB band. In this way the insulating state is at least approximated. On the contrary in the doped case $n = 1.4$ the chemical potential moves up to some value where now instead of the charge transfer gap of the free bands peak II provides sufficient amount of DOS for at least semiconducting behaviour. Furthermore from the AB band spectral weight is pushed outside and Cu satellites are formed which can be interpreted as upper (UHB) and lower (LHB) Hubbard bands of Cu. On the contrary the free band B consisting mostly of oxygen states nearly remains unchanged; only some small shift occurs with increasing $U$.

From the results for this extremely large value of $U$ we see that the strong coupling picture: upper and lower copper Hubbard states and oxygen states in between can be reached principally from the weak coupling side. Fig. 5.3

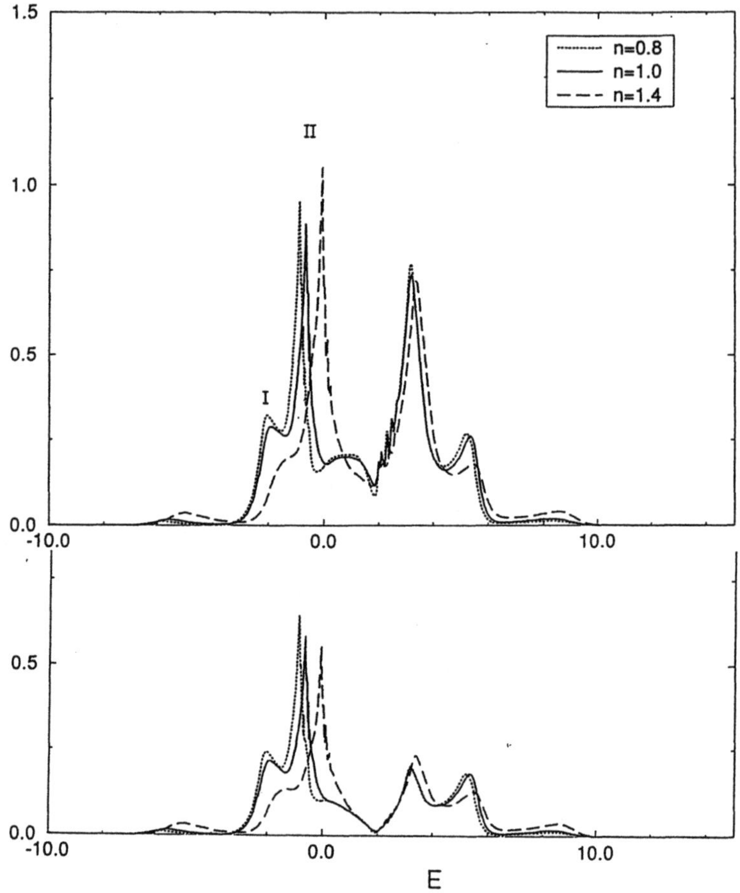


**Fig. 5.4.** DOS (lower part Cu DOS) vs. $E$ for increasing hole concentration $n$: dotted lines: $n = 0.8$, full lines $n = 1$, dashed lines $n = 1.4$.

shows the change in the DOS with increasing $U$ and the tendency towards the insulating state. Fig. 5.4 shows the change for three different values of doping. Peak II shifts, but remains sharp. Thus from this point of view the van Hove scenario (see e.g. [37]) in these lectures attributing the superconducting properties of the cuprates to a singularity in the DOS is supported. The strong doping dependence of peak II suggests some special physics (e.g. Kondo scattering at the copper spins, formation of copper oxygen singlets) as origin of this feature in the DOS. The oxygen states consist of three main parts: The nearly unchanged band B of the free case, the NB deltapeak located always at $\epsilon_0$ (dashed) and the peak II which contains, to nearly equal parts, both oxygen and copper. This is in qualitative agreement with other results from strong coupling approximations e.g. [38].

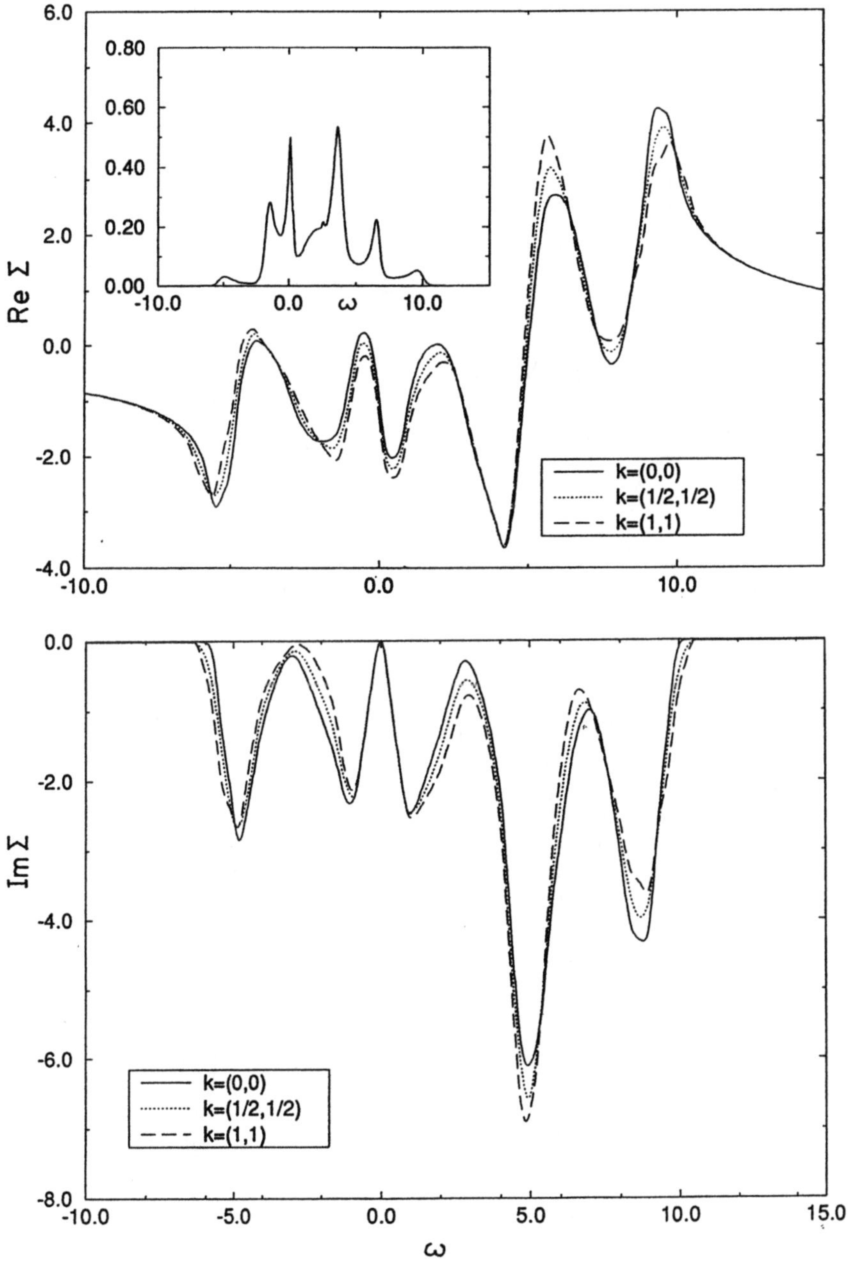


**Fig. 5.5.** Self Energy vs. $\omega$, Top: Real Part Re($\Sigma$), Bottom: Imaginary Part Im($\Sigma$), Inset: Corresponding DOS for $n = 1$, $U = 7.5$ and $\boldsymbol{k} = (0,0), (\pi/2, \pi/2), (\pi, \pi)$.

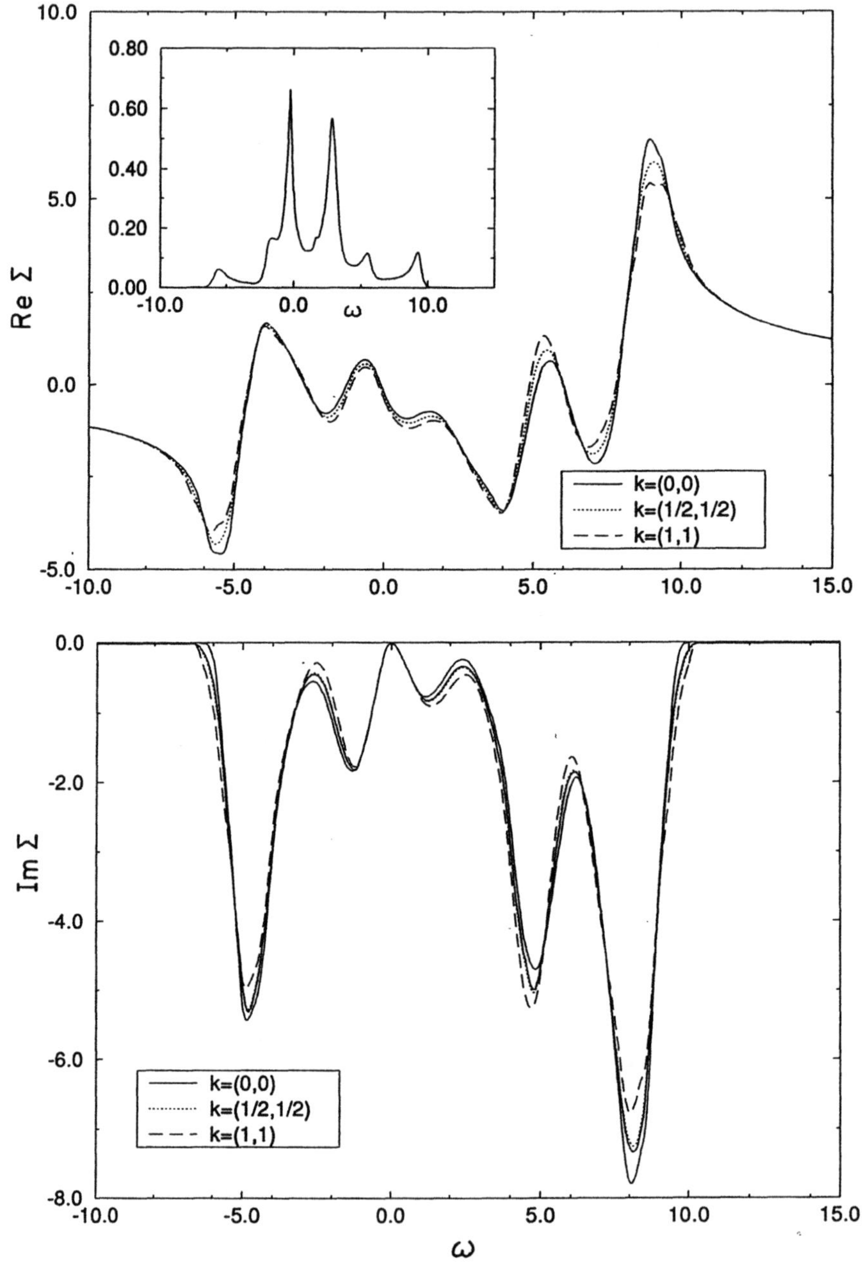


**Fig. 5.6.** Self Energy vs. $\omega$, Top: Real Part $\mathrm{Re}(\Sigma)$, Bottom: Imaginary Part $\mathrm{Im}(\Sigma)$, Inset: Corresponding DOS for $n = 1.4$, $U = 7.5$ and $\boldsymbol{k} = (0,0), (\pi/2, \pi/2), (\pi, \pi)$.

Figs. 5.5 and 5.6 display real and imaginary part respectively of the self energy as a function of $\omega$ for some values of $\boldsymbol{k}$. The imaginary part is always negative (as it should be) and has several minima according to elementary excitations and to the peaks in the DOS. Near to the minimum at $\omega = 0$ or $E = \mu$ respectively the curves for different $\boldsymbol{k}$-values nearly coincide thus indicating that the self energy only depends on $\omega$ as in the $D \to \infty$ limiting case [39]. The $\omega$ dependence of the imaginary part in this region looks — apart from numerical rounding ($\mathrm{Im}(\Sigma)$ is very sensitive to numerical accuracy there) — approximately quasilinear in the neutral and quadratic in the doped case, thus indicating marginal Fermi liquid behaviour in the former and Fermi liquid behaviour in the latter.

We still cannot give the curve $\mu(n)$ here because the values of the chemical potential are not very accurate. Complete thermodynamic selfconsistency with respect to $\mu$ and $\rho_\sigma(E)$ is not yet achieved. Moreover there is a tendency to smear out the features in the self energy upon iterating eqs. (5.4) with (5.5). This may be due to the fact that there are different fixpoints of the iteration on the Fermi liquid and on the insulating side as seen in the $D \to \infty$ limiting case [40]. Nevertheless we believe the first step of iteration to show clearly the tendencies of the change in the DOS when passing with the value of $U$ to the Mott Hubbard side. In the same way the satellites of the lower and upper Hubbard band appear from the free DOS in an analogous calculation in [36] for the simple Hubbard model.

## 6. Conclusions

The standard spectroscopic picture of the correlated density of states, with upper and lower copper band and oxygen states in between, is shown to evolve continuously with increasing correlation from the bonding and antibonding hybridized oxygen-copper bands in the band picture. A strongly doping dependent nearly complete splitting of the antibonding band into some Cu Hubbard band and some correlation peak occurs. Doping shifts the chemical potential from a region of strongly reduced DOS into the correlation peak and increases this peak, which is probably important for the physics of the system.

*Acknowledgement.* We are indebted to the "Rechenzentrum der Universität Kiel" and to the "Regionales Rechenzentrum der TU Hannover" for providing us with ample CPU time for the numerical calculations.

## References

1. J.G. Bednorz and K.A. Müller. *Z.Phys.* **B 64**, 189 (1986).

2. V.J. Emery, S.A. Kivelson, and H.Q. Lin. *Phys. Rev. Lett.* **64**, 475 (1990).
3. V. Hizhnyakov and E. Sigmund. *Physica* **C 156**, 655 (1988).
4. Q. Si and G. Kotliar. *Phys. Rev.* **B 48**, 13881 (1993).
5. K. Dichtel and J. Carstensen In K.A. Müller and G. Benedek, editors, *Workshop on Phase Separation in Cuprate Superconductors, Erice 1992.* World Scientific, (1992).
6. P.B. Visscher. *Phys. Rev.* **B 10**, 943 (1974).
7. P.B. Visscher. *Phys. Rev.* **B 48**, 933 (1974).
8. S. Dombrowski. *Eine Methode zur Berechnung von Hochtemperaturgrössen bei Hubbard und Emerymodell.* Diplomarbeit, Universität Kiel, 1994.
9. V.J. Emery and S.A Kivelson. *Pysica* **C 209**, 597 (1993).
10. W.O. Putikka, M.U. Luchini, and T.M. Rice. *Phys. Rev. Lett.* **68**, 538 (1992).
11. J. Florencio and K.A. Chao. *Phys. Rev.* **B 14**, 3121 (1976).
12. K. Ogawa, T. Kanda, and T. Matsubara. *Progr. Theor. Phys.* **53**, 614 (1975).
13. M. Deeg, H. Fehske, and H. Büttner. *Z. Phys.* **B 88**, 283 (1992).
14. P. Fazekas, B. Menge, and Müller-Hartmann E. *Z. Phys.* **B 78**, 69 (1990).
15. V.J. Emery. *Phys. Rev. Lett.* **58**, 2794 (1987).
16. S.N. Coppersmith. *Phys. Rev.* **B 41**, 8711 (1990).
17. S.N. Coppersmith and P.B. Littlewood. *Phys. Rev.* **B 42**, 3966 (1990).
18. Y. Bang, G. Kotliar, C. Castellani, M. Grilli, and R. Raimondi. *Phys. Rev.* **B 43**, 13724 (1991).
19. J. Carstensen and K. Dichtel. *Phys.Rev.* **B** , 11040 (1992).
20. H. Koppe, in *Werner Heisenberg und die Physik unserer Zeit, F. Bopp (ed.).* Vieweg, Braunschweig, 1961.
21. K. Dichtel, R.J. Jelitto, and H. Koppe. *Z. f. Phys.* **251**, 173 (1972).
22. G. Fischer. *Zur Stabilität magnetischer Ordnungszustände in einem Modell der Übergangsmetalle.* Dissertation, Universität Frankfurt, (1987).
23. M. Heise and R.J. Jelitto. *Z. Phys.* **B 25**, 381 (1976).
24. U. Brandt and C. Mielsch. *Z. Phys.* **B 75**, 365 (1989).
25. M.J. Rozenberg, X.Y. Zhang, and G. Kotliar. *Phys. Rev. Lett.* **69**, 1236 (1992).
26. J. Carstensen. *Das Emerymodell im Fall starker Kopplung.* Dissertation, Universität Kiel, (1992).
27. J. Stolze and T. Wittmann. *Phys. Rev.* **B 48**, (1993).
28. M. Grilli, R. Raimondi, C. Castellani, C. Di Castro, and G. Kotliar. *Phys. Rev. Lett.* **67**, 259 (1991).
29. R. Fresard and P. Wölfle. *J. Phys. Cond. Mat.* **4**, 3625 (1992).
30. E. Arrigoni and G.C. Strinati. *Phys. Rev.* **B 44**, 7455 (1991).
31. S.M. Hayden, T.E. Mason, H.A. Mook, S.-W. Cheong, and Z. Fisk. In K.A. Müller and G. Benedek, editors, *Workshop on Phase Separation in Cuprate Superconductors, Erice (Sicily).* World Scientific, (1992).
32. J. Hubbard. *Proc. Roy. Soc. (London)* **A 276**, 238 (1963).
33. Th. Pruschke, D.L. Cox, and M. Jarrell. *Phys. Rev.* **B 47**, 3553 (1993).
34. N. Bulut, D.J Scalapino, and S.R. White. *Phys. Rev. Lett.* **72**, 705 (1994).
35. G. Bulk and R.J. Jelitto. *Phys. Rev.* **B 41**, 413 (1990).
36. H. Schweitzer and G. Czycholl. *Z. Phys.* **B 83**, 93 (1991).
37. G. Bok. In E. Sigmund and K.A. Müller, editors, *Phase Separation in Çuprate Superconductors, Cottbus 1993.* Springer, (1994).
38. P. Unger and P. Fulde. *Phys. Rev.* **B 48**, 16607 (1993).
39. Müller-Hartmann E. *Z. Phys.* **B 76**, 211 (1989).
40. X.Y. Zhang, M.J. Rozenberg, and G. Kotliar. *Phys. Rev. Lett.* **70**, 1666 (1993).

# The Effect of Coulomb Interactions on Phase Separation

U. Löw[1], V. J. Emery[1], K. Fabricius[2], and S.A. Kivelson[3]

[1] Dept. of Physics, Brookhaven National Laboratory, Upton, NY 11973
[2] Physics Department, University of Wuppertal, 42097 Wuppertal, Germany
[3] Dept. of Physics, UCLA, Los Angeles, CA 90024

**Summary.** The effect of Coulomb interactions on phase separation is studied in a spin-one lattice gas model which is a coarse-grained representation of the charge distribution in high-temperature superconductors. The ground states are determined for the complete range of parameters by using a combination of numerical and analytical techniques. A crossover from phase separated to Wigner crystalline states proceeds via a rich structure of highly symmetric striped and checkerboard phases. There is no devil's staircase behavior because mixtures of stripes with different period phase separate.

## 1. Introduction

Over the past few years, it has been established that phase separation is a pervasive feature of the phase diagram of models of doped holes in an antiferromagnet [1, 2, 3, 4]. Since the values of the model parameters are not easily determined *ab initio*, it is necessary to study the physical properties of a given material in order to establish whether phase separation is of importance in practice. The answer to this question is not obvious at the outset, especially since the doped holes are charged, and long range Coulomb interactions surely would suppress *global* phase separation if the neutralizing background were uniform.

This conference is devoted to the proposition that phase separation is a reality for high-temperature superconductors which are examples of doped insulators, typically obtained by chemically adding charge carriers to highly-correlated antiferromagnetic materials. The most direct evidence is provided by oxygen-doped $La_2CuO_4$, in which the $O^{2-}$-ions are mobile below the convolute temperature and therefore are able to maintain charge neutrality [5, 6, 7].

The systematics of the behavior of $La_{2-x}Sr_xCuO_{4+\delta}$ [5, 6, 7] dispel any doubts that the phase separation is driven by the mobile holes rather than, for example, elastic interactions between oxygen ions. Therefore we may go on to examine the consequences for immobile dopants, such as Sr in $La_{2-x}Sr_xCuO_4$. In this paper, we address this question by studying a classical model of the competition between the long range Coulomb interaction and the short-range forces that favor phase separation. It will be seen that the charge density of the typical ground state is locally inhomogeneous and that charge inhomogeneity should be a general feature of the properties of high temperature superconductors. In the classical model the inhomogeneity is static and ordered. Elsewhere [2] it has been argued that, as the doping increases, the inhomogeneity becomes dynamical and is manifested as low-energy collective modes which are responsible for the anomalous properties of the normal state and also provide the mechanism of high temperature superconductivity.

Our starting point is a spin-one Ising model in which the spin variable $S_j$ at each site can be thought of as a coarse-grained representation of the local density of mobile holes [2]: each site $j$ represents a small region of space for which $S_j = +1$ and $S_j = -1$ correspond to hole-rich and hole-poor phases respectively, whereas $S_j = 0$ indicates that the local density is equal to the average value. The Hamiltonian is given by:

$$H = K \sum_j S_j^2 - L \sum_{<ij>} S_i S_j + \frac{Q}{2} \sum_{i \neq j} \frac{S_i S_j}{r_{ij}} \tag{1.1}$$

where $L, Q, K \geq 0$ and $S_j = \pm 1$. The long-range antiferromagnetic Coulomb force of strength $Q$ competes with a short-range ferromagnetic interaction of strength $L$ [8]. The quadrupolar field $K$ controls the concentration of sites at which $S_j^2 = 1$. Independently of this interpretation the model also is of great interest as a problem in statistical mechanics.

The fully phase separated state has $S_j^2 = 1$ and is ferromagnetically ordered, with $S_j = +1$ in one half of the volume, and $S_j = -1$ in the other, so as to maintain overall charge neutrality. This ferromagnetic state is the ground state of the Hamiltonian for $Q = 0$ and $K = 0$. As we shall see, the long range Coulomb interaction does not destroy all semblance of phase separation by forcing a state with $S_j = 0$ for all $j$. On the contrary, provided there is a short-distance clustering sufficient to justify our coarse-grained model, the ground state is typically inhomogeneous and shows a rather rich phase structure as the strength of the Coulomb interaction is varied. Even though the model has a quite simple Hamiltonian it shows a remarkably complex phase diagram due to this competing short-range and long-range interaction.

## 2. The Phase Diagram

In the following we study the zero temperature properties of the model in the thermodynamic limit. We used a combination of analytical and numerical techniques. A general survey of the ordered phases was made first by using a Monte Carlo simulation to heat up the system, followed by single spin flip dynamics to cool it to zero temperature. This method is particularly useful, because the low energy states obtained thereby are independent of the chosen initial states. In this way it was found that all ground states of finite systems consist of simple periodic structures for all values of the parameters L/K and Q/K. The pure short range interaction favors the completely phase separated ferromagnetic state and the pure Coulomb interaction favors a Néel state. As

$Q$ decreases, the crossover to the ferromagnetic state proceeds via a series of striped phases of ever increasing width. Consequently we were able to *assume* specific periodic ground state configurations for which the energy could be written:

$$E = KE_{\text{self}} - LE_{nn} + \frac{Q}{2}E_{\text{coul}} \tag{2.1}$$

where the self energy $E_{\text{self}}$, the nearest neighbor energy $E_{nn}$, and the Coulomb energy $E_{\text{coul}}$ are constants, characteristic of the assumed state. It was then straightforward to obtain the phase boundaries as straight lines in the $(Q/K, L/K)$ plane by comparing the energies of the different configurations. As explained below, $E_{\text{coul}}$ may be reduced to a finite sum which may be calculated numerically to arbitrary accuracy. The precision of our results rests heavily on the use of these analytical expressions, which allow a very fast numerical determination of the energies of the periodic spin configurations. The other two constants may be easily evaluated analytically for each given configuration.

In describing the phase diagram obtained by this procedure, the states will be characterized by the periods $(2m, 2n)$ along the $(x, y)$ directions. In this notation, the Néel state is $(2, 2)$, whereas a phase consisting of positive and negative vertical stripes each of width $m$, as shown in Fig. 2.1a, corresponds to $(2m, \infty)$ for $m = 1, 2 \ldots$. The checkerboard phase shown in Fig. 2.1b is specified by $(2, 2n)$.

First consider $K = 0$, for which the phase diagram for $K = 0$ is shown in Fig. 2.2. It can be seen that there are no stable phases with average spin equal to zero. As the system has only one free parameter, the phase diagram in L and Q is radially symmetric. We see that, when $L/Q$ is decreased from infinity to 0.122, the system passes through a succession of striped phases $(2m, \infty)$, from $m = \infty$ to $m = 1$. If $L/Q$ is decreased further, the system goes from $(2, \infty)$ to the Néel state $(2, 2)$ via a series of $(2, 2n)$ checkerboard phases, illustrated in Fig. 2.1b. This second staircase is much narrower, being confined to a region $0.109 < L/Q < 0.122$. We show the stripe width m of the striped phases versus $\sqrt{L/Q}$ in Fig. 2.3.

A systematic study did not reveal any intervening phases between the regular stripes and the $(2, 2n)$ checkerboards. This will be discussed in greater detail later.

When $K > 0$ it is necessary to use two parameters, $Q/K$ and $L/K$, to specify the phase diagram. In this case, new phases, consisting of stripes of width $2m$ separated by stripes of zeros, as illustrated in Fig. 2.1c, appear for $Q/K$ and $L/K$ small. When $Q/K$ and $L/K$ are large, the phase diagram is similar to the case of $K = 0$. The complete phase diagram for $K > 0$ is shown in Fig. 2.4.

$$\underbrace{\begin{array}{cccccccc} + & + & \cdots & + & - & - & \cdots & - \\ + & + & \cdots & + & - & - & \cdots & - \\ + & + & \cdots & + & - & - & \cdots & - \\ + & + & \cdots & + & - & - & \cdots & - \end{array}}_{2m} \qquad \text{(a)}$$

$$2n\left\{\begin{array}{cccccccc} + & - & + & - & + & - & + & - \\ + & - & + & - & + & - & + & - \\ \vdots & \vdots & \vdots & \vdots & \vdots & \vdots & \vdots & \vdots \\ + & - & + & - & + & - & + & - \\ - & + & - & + & - & + & - & + \\ - & + & - & + & - & + & - & + \\ \vdots & \vdots & \vdots & \vdots & \vdots & \vdots & \vdots & \vdots \\ - & + & - & + & - & + & - & + \end{array}\right. \qquad \text{(b)}$$

$$\underbrace{\begin{array}{ccccccccc} + & + & \cdots & + & 0 & - & - & \cdots & - \\ + & + & \cdots & + & 0 & - & - & \cdots & - \\ + & + & \cdots & + & 0 & - & - & \cdots & - \\ + & + & \cdots & + & 0 & - & - & \cdots & - \end{array}}_{2m+1} \qquad \text{(c)}$$

**Fig. 2.1.** Types of ground states: (a) Phases consisting of stripes of $m$ up and $m$ down spins. (b) Checkerboard phases of size $(2 \times 2n)$. (c) Stripes with one row of zeros.

## 3. The Structure of the Ground States

Since the relevant states are doubly periodic with periods $M_x$ and $M_y$, it is possible to obtain a deeper understanding of the phase diagram on the infinite lattice by reducing the problem to an effective model within the $M_x \times M_y$ unit cell. The argument will be presented for the striped phases, which are of greatest interest here. They correspond to $M_x = 2m$, $M_y = \infty$ and, in-cell, the model is a one-dimensional spin-half Ising model on a periodic lattice of $2m$ sites. The effective Hamiltonian may be written:

$$H = \sum_{i,j=1}^{2m} \sigma_i \sigma_j V(i-j) \tag{3.1}$$

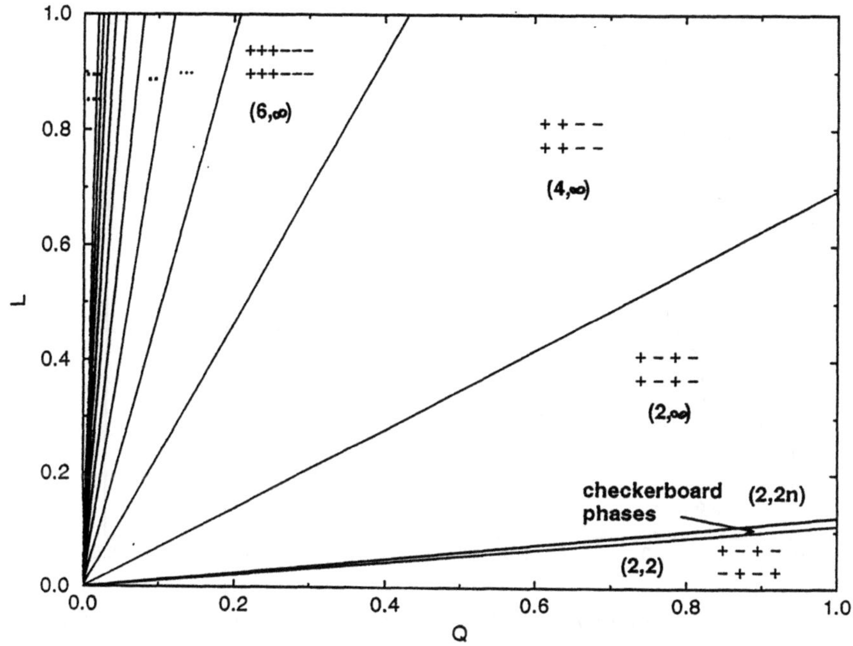


**Fig. 2.2.** Phase diagram for $K = 0$.

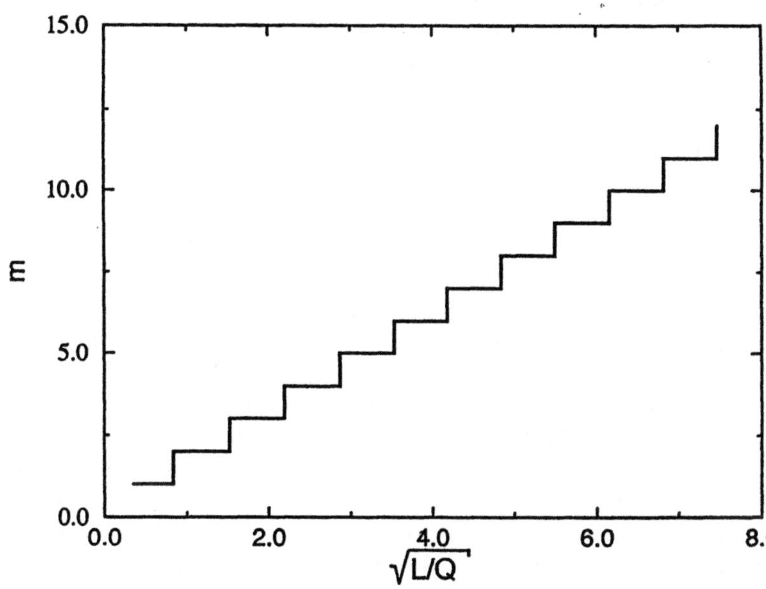


**Fig. 2.3.** The half-width $m$ of a striped phase plotted versus $\sqrt{L/Q}$ for $K = 0$.

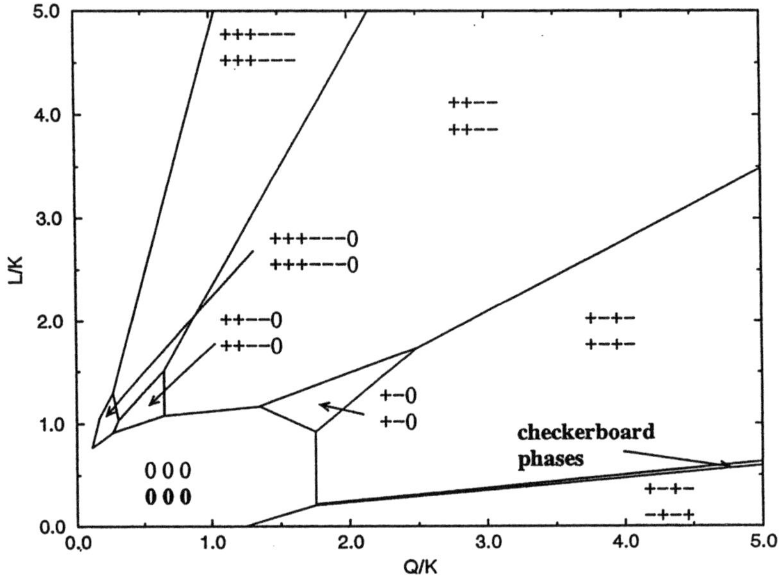


**Fig. 2.4.** Phase diagram for $K < 0$. The structures of the ground states are marked symbolically.

where the spin variables take the values $\sigma_i = \pm 1$. The Coulomb contribution $V_c(x)$ may be obtained by using the Poisson sum formula to evaluate the potential due to the rows of charges along the $y$-direction, and then carrying out the sum over unit cells in the $x$-direction:

$$V_c(x) = -2\ln\left|\sin\left(\frac{\pi x}{2m}\right)\right| + 4\sum_{l_x=-\infty}^{+\infty}\sum_{n=1}^{\infty} K_0\left(2\pi n(2ml_x + x)\right) \tag{3.2}$$

for $x \neq 0$. Here $K_0$ is the modified Bessel function of order 0 and infinite terms have canceled in virtue of the assumption of charge-neutral states. The first term in this equation comes from the uniform component of the charge distribution in a row, and it is the usual potential between columns of charge or line vortices: when $x = 0$, it should be replaced by $2(\ln(m/\pi) + C)$, where $C$ is Euler's constant. The second term in Eq. (3.2) comes from the finite wave vector components of the charge distribution of a stripe, and it gives the short-distance transient contributions to the potential. Using Eq. (3.2) we find for the total energy $E(m)$ of the striped configurations

$$E(m) \sim c_0 Qm + \frac{2L}{m} + c_1, \tag{3.3}$$

where $c_0 = 0.85$ and $c_1 =$ are constants. For the details of the analytic evaluation of $c_0$ we refer to [9]. We find that the Coulomb energies are linear in $m$ for essentially the whole range of interest. The period of the ground state for any $L/Q$ may be obtained by minimizing $E(m)$ with respect to integral values of $m$. The transition from $m$ to $m+1$ occurs when $E(m) = E(m+1)$.

This one-dimensional formulation of the problem enables us not only to understand the simple structure of the phase diagrams, but also to address the question of whether there are additional more complex ground states, possibly giving rise to a devil's staircase [10], *i.e.* regions of the phase diagram in which any two steps are separated by an infinity of extra steps. Bak and Bruinsma [10] showed that, for a one-dimensional Ising model with an infinite range convex interaction, the average magnetization displayed a complete devil's staircase as the applied magnetic field was varied. For the problem defined in Eq. (1.1), charge neutrality forces the average magnetization to be zero. But a devil's staircase might arise in the concentration of spin *flips* as its conjugate variable, the ferromagnetic coupling $L$, is changed. Consequently a systematic study of the Coulomb energies within subspaces $S_{m,l}$ with $2l$ spin flips was undertaken. For such a subspace the near-neighbour energy per site is fixed and equal to $-LE_{nn} = (2l/m - 2)L$ and the Coulomb energies may be obtained from Eqs. (3.1) and (3.2) by using a computer algorithm to generate states in the subspaces $S_{m,l}$. In practice, we considered all states with $m \leq 13$ (several hundred thousand states), and found that, when $l$ is a divisor of $m$, the ground state is given by a periodic pattern of $m/l$ up-spins followed by $m/l$ down-spins (stripes of width $m/l$). In other words, the spin flips are equally spaced. When $l$ was not a divisor of $m$, the ground state in the subspace $S_{m,l}$ separated into two phases: one an extended striped phase of period $2n$, the other an extended striped phase of period $2(n+1)$, where $n < l/m < n+1$. For finite $m$, the energy of such a state is higher than the energy of each of the two coexisting phases, because of a finite boundary energy.

According to this discussion it is clear that the ground states are simply periodic, except possibly at the values of $L/Q$ for which the period jumps (as shown in Fig. 2.3) where two phases may coexist. For the simply periodic phases, $m$ may always be chosen so that the ground state for any $L/Q$ has $l = 1$, *i.e.* within the unit cell it is "ferromagnetic", with total spin zero. From Fig. 2.3 it is also clear that, in considering states with period up to 26, the ground state has been identified for $L/Q$ less than $\sim 50$. However the same structure will persist for larger values of $L/Q$, since the "ferromagnetic" state is favored by $L$. It should be emphasized that these conclusions did not involve a finite-size approximation: equations (3.1) and (3.2) are *exact* representations of the Hamiltonian for $(2m \times \infty)$ periodic states, and the computer algorithm was simply used to carry out a systematic examination of the states.

As we have seen, the possibility of a devil's staircase, in which the states would consist of a mixture of the adjacent phases of period $2n$ and $2(n+1)$ [10] is pre-empted by phase separation, and the transition from one periodicity to another is first order. A physical feeling for the origin of the phase separation may be obtained by the following argument. In a phase of period $2n$, each unit cell may be regarded as a column of dipoles. It is straightforward to show from Eqs. (3.1) and (3.2) that the interaction between dipoles pointing in the same direction is attractive. Moreover we find that, starting from a phase-separated state: 1) it costs energy to transfer a $2n$-dipole into the $2(n+1)$ phase and vice-versa and 2) there is an attractive interaction between two $2n$-dipoles in the $2(n+1)$ phase and vice-versa. Consequently, the phase separated state is stable and, conversely, the mixed state tends to phase separate.

Stability of the "ferromagnetic" unit cells for all $m$ requires that the "compressibility" (or, equivalently, the second derivative of the total energy with respect to the volume) is positive. Since $m$ is an integer, this condition is equivalent to the requirement that the second *difference* of the total energy $2mE(m)$ with respect to $m$ is positive. It is evident that this condition is satisfied by the approximation (3.3). We have verified numerically that the condition is also satisfied for small $m$, where there are small deviations from (3.3).

## 4. Conclusion

In conclusion, we have shown that a classical coarse-grained model of the competition between long range Coulomb interactions and a short-range tendency for phase separation has an interesting and remarkably complex phase diagram, in which the charge density of the typical ground state is locally inhomogeneous. Charge inhomogeneity is the characteristic consequence of frustrated phase separation, and it is apparent in the physical properties of high temperature superconductors, as demonstrated in the experimental talks at this conference. In generalizations of the model studied here, the ground states should be inhomogeneous but they may not be simply periodic. In particular, introducing quantum fluctuations or allowing the charge to be different for $S_i = +1$ and $S_i = -1$ should introduce time-dependence and disorder. Frustrated phase separation may provide an explanation of specific effects such as the cluster spin glass behavior and may also be the fundamental reason for the anomalous normal state of high temperature superconductors as well as superconductivity itself [2].

UL was supported by the Max Kade Foundation. VJE and, in part, UL were supported by the Division of Materials Science, U.S.Department of Energy under Contract No. DE-AC02-76CH00016. SAK was supported in part by the NSF under Grant No. DMR-90-11803.

## References


1. V.J. Emery, S.A. Kivelson, and H.-Q. Lin, (1990) Phys. Rev. Lett. **64**, 475; S.A. Kivelson, V.J. Emery, and H.-Q. Lin (1990) Phys. Rev. **B42**, 6523.
2. V.J. Emery and S.A. Kivelson, Physica C**209**, 597 (1993)
3. V. Hizhnyakov et al, lecture at this workshop
4. C. DiCastro, lecture at this workshop
5. J.-C. Grenier, lecture at this workshop
6. C. Hammel et al, lecture at this workshop
7. D.C. Johnston et al, lecture at this workshop
8. It should be noticed that our definition of the coupling constants is different from that used in ref. [2]
9. U. Löw, V.J. Emery, K. Fabricius and S.A. Kivelson to be published Phys. Rev. A**4**, 1071 (1971)
10. P. Bak and R. Bruinsma, Phys. Rev. Lett. **49**, 249 (1982)

# Options for the Effective Electron-Electron-Interaction from Unitary Transformations

M. Wagner

Institut für Theoretische Physik III, Universität Stuttgart, Pfaffenwaldring 57, D-70550 Stuttgart

**Summary.** If an electronic system is strongly coupled to other subsystems (phonons, excitons, spins), unitary transformations provide an excellent tool to establish and discuss effective electron–electron interactions. Two classes of transformations, which can be given in a closed form, are presented. They amount to the choice of a *moving base* for the electronic 2nd quantization. Physical implications are discussed and an illustrative example relating to high–$T_c$–systems is given.

## 1. Introduction

An increasing number of workers in the field of high-$T_c$-superconductivity tend now to consider phase-separation [1] as a necessary prerequisite to establish metallic behaviour in cuprate materials. Most of the contributions to this volume are concerned with this concept both theoretically and experimentally. But after the mechanism of the advent of metallic behaviour has been established, the second crucial analysis remains to be done, which is the understanding of the superconducting mechanism itself.

The theory of superconductivity which applies to conventional metals is governed by a very simple Hamiltonian which can be found in any textbook on solid-state theory. It is the BCS-Hamiltonian

$$H = H_0 - V \sum_{q} \sum_{\boldsymbol{k},\boldsymbol{k}'} c^{+}_{k+q} c^{+}_{k'-q} c_{k'} c_{k} \tag{1.1}$$

where the interaction constant V is taken to be finite only for a narrow $\boldsymbol{k}, \boldsymbol{k}'$–region close to the Fermi surface. For $\boldsymbol{k}' = -\boldsymbol{k}$ the Hamiltonian describes Cooper-pairing [2] and eventually leads to the standard formula for the critical temperature of superconductive behaviour

$$T_c = 1.14 \Omega_D e^{-1/(V \cdot \rho_F)} \tag{1.2}$$

where $\Omega_D$ is the Debye temperature and $\rho_F$ the one-electron density of states at the Fermi surface. From (1.1) one also derives the relation

$$2\Delta = 3.5 T_c \tag{1.3}$$

between the gap $\Delta$ and $T_c$, and the groundstate energy is given by

$$E_{\mathrm{g}} = -\frac{2\rho_{\mathrm{F}}\Omega_{\mathrm{D}}^{2}}{\mathrm{e}^{2/(V\cdot\rho_{\mathrm{F}})}-1}\,. \tag{1.4}$$

Superconducting materials of the type first found by Bednorz and Müller [3] are rather complicated both regards structure and composition. Therefore, one carefully has to reconsider all approximations and simplifications made in the 1950's to establish the BCS-formulation. Hamiltonian (1.1) has been derived in a seminal paper of Fröhlich [4] by means of a unitary transformation, which in our times presumably is regarded as the most famous of all unitary transformations.

In the present contribution I advocate the use of unitary transformations in the investigation of possible alternative or modified mechanisms of superconductivity in the cuprate materials. These transformations have the beneficial property to transmute the shape of a physical problem into another shape, and if one is clever enough one may end up with a shape which displays the dominant dynamical features of the physical problem in its most natural or suggestive manner. In an illustrative way one could say that unitary transformations are something like mirrors in the mirror halls of entertainment parks, where each of them reflects a different appearance of the observer, and not necessarily the plane reflection of the original displays the optimal characterization.

## 2. Unitary Transformations in Physics

For reference we recall here the basic definitions and properties of unitary transformations [UT]. More detailed background and applications are given in a book by the author [5]. We define the UT of an operator $A$ as

$$T : A \equiv \tilde{A} = U^{+}AU \tag{2.1}$$

where $U$ is a unitary operator, $U^{+} = U^{-1}$, and is most conventially established by means of an exponential form, $U = \exp S$, where any antihermitian operator, $S^{+} = -S$, then guarantees the unitarity of $U$. It is understood that $U$ (or $S$) are characterized by certain parameters, which may be optimized by means of certain physical extremal principles. To give an example of a possible optimization strategy, consider an electron-phonon Hamiltonian $H(c^{+}c, P, Q)$ where $c_i, c_j^{+}, P_r, Q_s$ stand for the electronic and phononic variables respectively. Then also $U$ is a function of these variables, and the transformed Hamiltonian e.g. may be separated in two parts,

$$T : H \equiv \tilde{H} = H^{(1)}(c^{+}c, f(P,Q)) + H^{(2)}(P,Q,g(c^{+}c)) \tag{2.2}$$

such that $H^{(1)}$ predominantly comprises the dynamics of the electrons and $H^{(2)}$ that of the phonons. An effective Hamiltonian of a simple form then may intuitively be chosen, which separates the two subsystems and which is assumed to simulate the dynamics of the true Hamiltonian $\tilde{H}$ by means of adjustable parameters $u_\alpha, v_\beta$,

$$H_{\text{eff}} = H_{\text{e}}^{(1)}(c^+c, u) + H_{\text{p}}^{(2)}(P, Q, v) \,. \tag{2.3}$$

These parameters, as well as the parameters incorporated in the transformation exponent $S$ may be fixed by means of an extremal principle, e.g. via the Bogoliubov inequality,

$$F \leq F_{\text{eff}} + \left\langle \tilde{H} - H_{\text{eff}} \right\rangle_{\text{T}}^{\text{H}_{\text{eff}}} \tag{2.4}$$

where $F_{\text{eff}}$ is the free energy pertaining to $H_{\text{eff}}$ and the bracket denotes the thermal expectation value with respect to $H_{\text{eff}}$.

The major difficulty for the application of unitary transformations is posed by the fact that they often can only be given in a series expansion,

$$T : A \equiv \mathrm{e}^{-S} A \mathrm{e}^{S} = A + \frac{1}{1!}[A, S] + \frac{1}{2!}[[A, S], S] + \ldots \tag{2.5}$$

which cannot be fully summed up. In this case an optimization can only be achieved in a perturbative sense. The most prominent of this type of procedures is the removal of perturbation terms *in lowest order*. Writing the Hamiltonian in the form

$$H = H_0 + W \,, \tag{2.6}$$

where $W$ is supposed to be small of 1st order, one may invoke the Fröhlich condition, i.e. choose $S \equiv S_{\text{F}}$ to satisfy the relation

$$[H_0, S_{\text{F}}] = -W \tag{2.7}$$

which yields

$$T : H = H_0 + \frac{1}{2}[W, S_{\text{F}}] + O(W^3) \tag{2.8}$$

A drawback, but at the same time a flexibility option of this procedure, is the observation that condition (2.7) is *not unique*, which one easily verifies by considering the three contrasting unitary operators $U_F = \exp S_{\text{F}}$, $U_{12} = (\exp S_1)(\exp S_2)$ and $U_{21} = (\exp S_2)(\exp S_1)$ which all yield the same Fröhlich condition $[H_0, S_{\text{F}}] = -W$, if $S_F = S_1 + S_2$, but generate different 2nd order terms,

$$(T_{\text{F}} - T_{21}) : H = (T_{12} - T_{\text{F}}) : H = \frac{1}{2}[H_0, [S_1, S_{\text{F}}]] + O(S^3) \,. \tag{2.9}$$

In this case an additional optimization condition is necessary.

## 3. Fröhlich's Transformation [4]

In most textbooks on superconductivity the theoretical ab initio consideration is based on the Fröhlich-Hamiltonian, which is written in the form $H = H_0 + W$, where

$$
\begin{aligned}
H_0 &= \frac{1}{2}\sum_q (P_q^+ P_q + \Omega_q^2 Q_q^+ Q_q) + \sum_k \omega_k c_k^+ c_k\,, && (3.1)\\
W &= \sum_{kk'} G(\boldsymbol{k}', \boldsymbol{k}-\boldsymbol{k}') Q_{k-k'} c_k^+ c_{k'}\,, && (3.2)\\
G^*(\boldsymbol{k},\boldsymbol{q}) &= G(\boldsymbol{k}+\boldsymbol{q}, -\boldsymbol{q})\,. && (3.3)\\
& && (3.4)
\end{aligned}
$$

It is then subjected to a transformation $U_{\mathrm{F}} = \exp(S_1(Q) + S_2(P))$, where

$$
\begin{aligned}
S_1(Q) &= \sum_{kq} \alpha(\boldsymbol{k},\boldsymbol{q}) Q_q c_{k+q}^+ c_k && (3.5)\\
S_2(P) &= \sum_{kq} \beta(\boldsymbol{k},\boldsymbol{q}) P_q^+ c_{k+q}^+ c_k\,. && (3.6)\\
& && (3.7)
\end{aligned}
$$

The transformation parameters $\alpha(\boldsymbol{k},\boldsymbol{q})$, $\beta(\boldsymbol{k},\boldsymbol{q})$ are found via relation (2.7),

$$
\begin{aligned}
\alpha(\boldsymbol{k},\boldsymbol{q}) &= -\frac{(\omega_{k+q}-\omega_k) G(\boldsymbol{k},\boldsymbol{q})}{(\omega_{k+q}-\omega_k)^2 - \Omega_q^2} && (3.8)\\
\beta(\boldsymbol{k},\boldsymbol{q}) &= -i\frac{G(\boldsymbol{k},\boldsymbol{q})}{(\omega_{k+q}-\omega_k)^2 - \Omega_q^2} && (3.9)\\
& && (3.10)
\end{aligned}
$$

and the transformed Hamiltonian reads

$$
\begin{aligned}
T_{\mathrm{F}} : H = \tilde{H}^{\mathrm{F}} &= H_0 \\
&- \frac{1}{2}\sum_{kqq'} \left[ \frac{(\omega_k - \omega_{k-q'}) G(\boldsymbol{k},\boldsymbol{q}) G^*(\boldsymbol{k},-\boldsymbol{q}')}{(\omega_k - \omega_{k-q'})^2 - \Omega_{q'}^2} \right. \\
&\qquad \left. - \frac{(\omega_{k+q} - \omega_{k+q-q'}) G(\boldsymbol{k}-\boldsymbol{q}',\boldsymbol{q}) G^*(\boldsymbol{k}+\boldsymbol{q},-\boldsymbol{q}')}{(\omega_{k+q} - \omega_{k+q-q'})^2 - \Omega_{q'}^2} \right] \\
&\qquad \cdot Q_q Q_{q'}^+ c_{k+q}^+ c_{k-q'} \\
&- \frac{1}{4} i \sum_{kqq'} \left[ \frac{G(\boldsymbol{k},\boldsymbol{q}) G^*(\boldsymbol{k},-\boldsymbol{q}')}{(\omega_k - \omega_{k-q'})^2 - \Omega_{q'}^2} c_{k+q}^+ c_{k-q'} \right.
\end{aligned}
$$

$$
\begin{aligned}
& - \frac{G(\boldsymbol{k},\boldsymbol{q})G^*(\boldsymbol{k}+\boldsymbol{q}+\boldsymbol{q}',-\boldsymbol{q}')}{(\omega_{\boldsymbol{k}+\boldsymbol{q}+\boldsymbol{q}'}-\omega_{\boldsymbol{k}+\boldsymbol{q}})^2-\Omega_{\boldsymbol{q}'}^2} c^+_{\boldsymbol{k}+\boldsymbol{q}+\boldsymbol{q}'} c_{\boldsymbol{k}} \Bigg] \\
& \cdot (Q_{\boldsymbol{q}} P^+_{\boldsymbol{q}'} + P^+_{\boldsymbol{q}'} Q_{\boldsymbol{q}}) \\
+ \; & \frac{1}{4} \sum_{\boldsymbol{k}\boldsymbol{k}'\boldsymbol{q}} G(\boldsymbol{k},\boldsymbol{q}) G^*(\boldsymbol{k}',-\boldsymbol{q}) \left[ \frac{1}{(\omega_{\boldsymbol{k}+\boldsymbol{q}}-\omega_{\boldsymbol{k}})^2-\Omega_{\boldsymbol{q}}^2} \right. \\
& \left. + \frac{1}{(\omega_{\boldsymbol{k}'-\boldsymbol{q}}-\omega_{\boldsymbol{k}'})^2-\Omega_{\boldsymbol{q}}^2} \right] \\
& \cdot c^+_{\boldsymbol{k}+\boldsymbol{q}} c_{\boldsymbol{k}} c^+_{\boldsymbol{k}'-\boldsymbol{q}} c_{\boldsymbol{k}'} \\
+ \; & O(G^3) \,. \qquad (3.11)
\end{aligned}
$$

The last of the terms given explicitly is an electron-electron interaction term which, however, displays both attractive and repulsive contributions. The latter in the advent of BCS-theory have been considered irrelevant. Also, the one-electron terms of the form $QQ^+c^+c$ and $(QP^+ + P^+Q)c^+c$ have been discarded in the BCS-Hamiltonian, where for high-$T_c$ materials at least the first should be rechecked carefully, since it initiates a thermodynamic coupling of the two subsystems. Except for Q this general drawbacks expr. (3.11) involves resonance denominators, which may cause divergence problems even in the case of small coupling.

It is of interest to have a view onto the non-uniqueness of the Fröhlich-condition, as described in Sect. 2, and to consider a transformation $T_{21}$ generated by $U_{21} = (\exp S_2(P))(\exp S_1(Q))$, where $S_1(Q)$ and $S_2(P)$ again are given by (3.5, 3.6) to satisfy condition (2.7). Then the effective electron–electron interaction reads

$$
\begin{aligned}
H^{21}_{\mathrm{ee}} = & -\frac{1}{2} \sum_{\boldsymbol{k}\boldsymbol{k}'\boldsymbol{q}} G(\boldsymbol{k},\boldsymbol{q}) G(\boldsymbol{k}',-\boldsymbol{q}) \\
& \cdot \left[ \frac{(\omega_{\boldsymbol{k}+\boldsymbol{q}}-\omega_{\boldsymbol{k}})(\omega_{\boldsymbol{k}'-\boldsymbol{q}}-\omega_{\boldsymbol{k}'}) + \Omega_{\boldsymbol{q}}^2}{\left[(\omega_{\boldsymbol{k}+\boldsymbol{q}}-\omega_{\boldsymbol{k}})^2-\Omega_{\boldsymbol{q}}^2\right]\left[(\omega_{\boldsymbol{k}'-\boldsymbol{q}}-\omega_{\boldsymbol{k}'})^2-\Omega_{\boldsymbol{q}}^2\right]} \right] \\
& \cdot c^+_{\boldsymbol{k}+\boldsymbol{q}} c_{\boldsymbol{k}} c^+_{\boldsymbol{k}'-\boldsymbol{q}} c_{\boldsymbol{k}'} \qquad (3.12)
\end{aligned}
$$

which is to be contrasted with the corresponding term in expr. (3.11)

$$
\begin{aligned}
H^{\mathrm{F}}_{\mathrm{ee}} = & \frac{1}{4} \sum_{\boldsymbol{k}\boldsymbol{k}'\boldsymbol{q}} G(\boldsymbol{k},\boldsymbol{q}) G(\boldsymbol{k}',-\boldsymbol{q}) \\
& \cdot \left[ \frac{1}{(\omega_{\boldsymbol{k}+\boldsymbol{q}}-\omega_{\boldsymbol{k}})^2-\Omega_{\boldsymbol{q}}^2} + \frac{1}{(\omega_{\boldsymbol{k}'-\boldsymbol{q}}-\omega_{\boldsymbol{k}'})^2-\Omega_{\boldsymbol{q}}^2} \right] \\
& \cdot c^+_{\boldsymbol{k}+\boldsymbol{q}} c_{\boldsymbol{k}} c^+_{\boldsymbol{k}'-\boldsymbol{q}} c_{\boldsymbol{k}'} \,. \qquad (3.13)
\end{aligned}
$$

For $\boldsymbol{k}' = -\boldsymbol{k}$ we note that (3.12) becomes totally attractive, whereas the original Fröhlich expression does it only for electronic excitations which are smaller than the Debye frequency $\Omega_D$. But whether a truncated Hamiltonian of the form (3.12) is closer to physical reality than the Fröhlich interaction (3.13) can be decided only, if at least the effect of 4th order terms in $W$ are analyzed.

## 4. Moving Base Transformations

Since a perturbative approach, as described above, is not fully satisfactory, the question arises, whether there are no physically meaningful electron-phonon transformations which can be fully performed, i.e. yielding closed forms for the transformed operators. Indeed, a class of such transformations has been found earlier by the author [6], which are related to the choice of the one-electron base functions $\varphi_\kappa(x)$ employed to define 2nd quantization, $\Psi(x) = \sum_\kappa c_\kappa \varphi_\kappa(x)$ (see standard textbooks). Usually the set $\{\varphi_\kappa(x)\}$ is not assumed to depend dynamically on the coordinates of a second subsystem, say the phonon subsystem. We will consider such electronic bases, which either depend on the vibrational displacement coordinates $Q_q$, i.e. $\{\varphi_\kappa(x, Q)\}$ or on the vibrational momenta $P_q$, i.e. $\{\varphi_\kappa(x, P)\}$. Since these basis sets respectively no longer commute with the vibrational momenta or displacements, the dynamics of the phonon subsystem will be involved in such a base choice.

### 4.1 Moving Base in Q

We consider a transformation operator of the form

$$U_a = e^{S_a(Q)}, \; S_a(Q) = \sum_{\kappa\kappa'} \Lambda_{\kappa\kappa'}(Q) c_\kappa^+ c_{\kappa'} \tag{4.1}$$

which is related to a base transformation $\{\varphi_\kappa^{(0)}(x)\} \to \{\varphi_\kappa(x, Q)\}$,

$$(e^{\Lambda(Q)})_{\kappa\kappa'} = \left\langle \varphi_\kappa^{(0)} | \varphi_{\kappa'}(Q) \right\rangle_e \tag{4.2}$$

where the bracket denotes integration over the electronic coordinates. Then it is found [6] that the basic operators of the two subsystems transform as

$$\begin{aligned} T_a : Q_q &\equiv U_a^+ Q_q U_a = Q_q && (4.3)\\ T_a : P_q &= P_q + \sum_{\kappa\kappa'} c_\kappa^+ c_{\kappa'} \left[ e^{-\Lambda}(-i\frac{\partial}{\partial Q_q} e^{\Lambda}) \right]_{\kappa\kappa'} && (4.4)\\ T_a : c_\kappa &= \sum_{\kappa'} (e^{\Lambda})_{\kappa\kappa'} c_{\kappa'} \, . && (4.5) \end{aligned}$$

This allows for a closed transformation of any electron-phonon-Hamiltonian. Specifically, the transformation of Fröhlich's Hamiltonian (see Sect. 3) reads

$$\begin{aligned}T_{\mathrm{a}}: H = H_0 &+ \sum_{\kappa\kappa'}\sum_{\kappa_1\kappa_1'}(\mathrm{e}^{-\Lambda})_{\kappa\kappa_1}G(\kappa_1',\kappa_1-\kappa_1')Q_{\kappa_1-\kappa_1'}(\mathrm{e}^{\Lambda})_{\kappa_1'\kappa'}c_\kappa^+c_{\kappa'}\\
&+ \frac{1}{2}\sum_{\kappa\kappa' q}P_q^+\left[\mathrm{e}^{-\Lambda}\left(-i\frac{\partial}{\partial Q_q}\mathrm{e}^{\Lambda}\right)\right]_{\kappa\kappa'}c_\kappa^+c_{\kappa'}\\
&+ \frac{1}{2}\sum_{\kappa\kappa' q}\left[\mathrm{e}^{-\Lambda}\left(-i\frac{\partial}{\partial Q_q^+}\mathrm{e}^{\Lambda}\right)\right]_{\kappa\kappa'}P_q c_\kappa^+c_{\kappa'}\\
&+ \frac{1}{2}\sum_{\kappa\kappa' q}\left[\left(i\frac{\partial}{\partial Q_q^+}\mathrm{e}^{-\Lambda}\right)\left(-i\frac{\partial}{\partial Q_q}\mathrm{e}^{\Lambda}\right)\right]_{\kappa\kappa'}c_\kappa^+c_{\kappa'}\\
&+ \frac{1}{2}\sum_{\kappa\kappa'}\sum_{\kappa_1\kappa_1'}\sum_{q}\left[\mathrm{e}^{-\Lambda}\left(-i\frac{\partial}{\partial Q_q^+}\mathrm{e}^{\Lambda}\right)\right]_{\kappa\kappa'}\\
&\cdot\left[\mathrm{e}^{-\Lambda}\left(-i\frac{\partial}{\partial Q_q}\mathrm{e}^{\Lambda}\right)\right]_{\kappa_1\kappa_1'}c_\kappa^+c_{\kappa_1}^+c_{\kappa_1'}c_{\kappa'}\end{aligned} \tag{4.6}$$

which is an exact transcription of the original Hamiltonian. The last two terms of (4.6) are of *nonadiabatic* nature, since they arise by way of the noncommutatibility of the vibrational momenta with $S_{\mathrm{a}}(Q)$. The first two terms are *adiabatic* and may be diagonalized, if $\Lambda(Q)$ is chosen properly. The particular base achieving this *adiabatic* diagonalization is denoted as *adiabatic* base. In this case the electron-electron interaction in the strict sense would be of nonadiabatic nature. We further note that if $S_{\mathrm{a}}(Q)$ is chosen linear in $Q_q$ it assumes the form of the constituent $S_1$ in Fröhlich's transformation (vid. (3.5)).

## 4.2 Moving Base in P

A base transformation $\{\varphi_\kappa^{(0)}(x)\} \to \{\varphi_\kappa(x,P)\}$ is established by the unitary operator

$$U_{\mathrm{b}} = \mathrm{e}^{S_{\mathrm{b}}(P)},\ S_b(P) = \sum_{\kappa\kappa'}\Theta_{\kappa\kappa'}(P)c_\kappa^+c_{\kappa'}\,, \tag{4.7}$$

such that

$$(e^{\Theta})_{\kappa\kappa'} = \left\langle\varphi_\kappa^{(0)}|\varphi_{\kappa'}(P)\right\rangle_{\mathrm{e}} ts, \tag{4.8}$$

and has the properties

$$\begin{aligned}T_{\mathrm{b}}: P_q &\equiv U_{\mathrm{b}}^+P_qU_{\mathrm{b}} = P_q &(4.9)\\
T_{\mathrm{b}}: Q_q &= Q_q + \sum_{\kappa\kappa'}c_\kappa^+c_{\kappa'}\left[\mathrm{e}^{-\Theta}\left(i\frac{\partial}{\partial P_q}\mathrm{e}^{\Theta}\right)\right]_{\kappa\kappa'} &(4.10)\\
T_{\mathrm{b}}: c_\kappa &= \sum_{\kappa'}(\mathrm{e}^{\Theta})_{\kappa\kappa'}c_{\kappa'}\,. &(4.11)\end{aligned}$$

It is the *sister*-transformation of $T_a$. Again, by means of these formulae any electron-phonon Hamiltonian may be fully transformed. The transformed Fröhlich Hamiltonian (see Sect. 3.) reads

$$
\begin{aligned}
T_b : H = H_0 &+ \sum_{\kappa\kappa'} \sum_{\kappa_1\kappa_1'} (e^{-\Theta})_{\kappa\kappa_1} G(\kappa_1', \kappa_1 - \kappa_1') Q_{\kappa_1-\kappa_1'} (e^{\Theta})_{\kappa_1'\kappa'} c_\kappa^+ c_{\kappa'} \\
&+ \frac{1}{2} \sum_{\kappa\kappa' q} \Omega_q^2 \left\{ Q_q^+ \left[ e^{-\Theta} \left( i \frac{\partial}{\partial P_q} e^{\Theta} \right) \right]_{\kappa\kappa'} + \left[ e^{-\Theta} \left( i \frac{\partial}{\partial P_q^+} e^{\Theta} \right) \right]_{\kappa\kappa'} Q_q \right\} c_\kappa^+ c_{\kappa'} \\
&+ \frac{1}{2} \sum_{\kappa\kappa' q} \left[ \left( -i \frac{\partial}{\partial P_q^+} e^{-\Theta} \right) \left( i \frac{\partial}{\partial P_q} e^{\Theta} \right) \right]_{\kappa\kappa'} c_\kappa^+ c_{\kappa'} \\
&+ \frac{1}{2} \sum_{\kappa\kappa'} \sum_{\kappa_1\kappa_1'} \sum_{q} \Omega_q^2 \left[ e^{-\Theta} \left( i \frac{\partial}{\partial P_q^+} e^{\Theta} \right) \right]_{\kappa\kappa'} \left[ e^{-\Theta} \left( i \frac{\partial}{\partial P_q} e^{\Theta} \right) \right]_{\kappa_1\kappa_1'} \\
&\qquad \cdot c_\kappa^+ c_{\kappa_1}^+ c_{\kappa_1'} c_{\kappa'} \\
&+ \sum_{\kappa\kappa'} \sum_{\kappa_1\kappa_1'} \sum_{\kappa_2\kappa_2'} (e^{-\Theta})_{\kappa\kappa_2} G(\kappa_2', \kappa_2 - \kappa_2') \left[ e^{-\Theta} \left( i \frac{\partial}{\partial P_{\kappa_2-\kappa_2'}} e^{\Theta} \right) \right]_{\kappa_1\kappa_1'} (e^{\Theta})_{\kappa_2'\kappa'} \\
&\qquad \cdot c_\kappa^+ c_{\kappa_1}^+ c_{\kappa_1'} c_{\kappa'} \, .
\end{aligned}
\tag{4.12}
$$

Now there are 3 terms which are established by the noncommutativity of $S_b$ with $Q_q$. The *moving* base $\{\varphi_\kappa(x, P)\}$ may be designated as an *anti-adiabatic* or *diabatic* one and the electron-electron interaction (last 2 terms) no longer are of nonadiabatic nature, although the picture would be overstrained to speak now of an adiabatic type of electron-electron interaction. We further note that now, if $S_a$ would be chosen linear in $P_q$, it would be of the form of the second constituent $S_2$ in Fröhlich's transformation (see. (3.6)). There still is another aspect to a linear choice of $S_b$ in $P$: in this case the transformation would be of a displacive nature in the phonon subsystem, i.e. it would hint at a dynamical relaxation of the lattice nuclei which would depend on the instantaneous situation in the electronic subsystem. So, with some care one could say that $T_a$ enforces a concept where the electrons follow adiabatically the nuclei, whereas by contrast $T_b$ enforces one, where the nuclei react *adiabatically* onto the electronic subsystem. Finally, we note a remarkable property of transformation $T_b$: since $S_b(P)$ commutes with the kinetic energy operator of the phonon-subsystem, there is *no isotope effect* in the electron-electron interaction established via this transformation.

## 5. A Didactical Cu – O – Cu Model

To return somewhat to the problem of superconductivity in oxygen-cuprate materials, we consider a simple model unit in which an oxygen ion can move between two fixed Cu-ions (coordinate $Q$). It has been known for a long time that the electronic shell of the $O^2$-ion is very soft [7] and that the effective volume of the oxygen strongly depends on its electronic occupation. Hence the local frequency of the O-ion will drastically change if an electron is transferred to a neighbouring Cu–ion or vice versa. This suggests a Hamiltonian term of the form $Q^2 n_{0\sigma}$ where $n_{0\sigma} = c^+_{0\sigma} c_{0\sigma}$ is the occupation operator at the O-site. For simplicity we characterize the Cu – O – Cu–model by a base of 6 states, $|l\sigma>, |0\sigma>, |r\sigma>$, i.e. a single orbital state with spins $\sigma = \uparrow, \downarrow$ per site. From symmetry we deduce that the dynamics remains in the 4 state base $\{|1\sigma> = [|l\sigma> + |r\sigma>]/\sqrt{2}, |0\sigma\}$, whence we consider the Hamiltonian

$$\begin{aligned} H &= \frac{1}{2}(P^2 + \Omega^2 Q^2) + \sum_{j,\sigma} \epsilon_j c^+_{j\sigma} c_{j\sigma} + \alpha \Omega^2 Q^2 \sum_\sigma c^+_{0\sigma} c_{0\sigma} \\ &+ \Delta \sum_\sigma (c^+_{0\sigma} c_{1\sigma} + h.c.) \end{aligned} \tag{5.1}$$

where $j = 0, 1$. More details may be found in ref. [8]. We now apply the moving base transformation operator $U_\mathrm{a} = \exp S_\mathrm{a}(Q)$,

$$S_\mathrm{a}(Q) = \lambda(Q) \sum_\sigma (c^+_{1\sigma} c_{0\sigma} - h.c.) \tag{5.2}$$

where $\lambda(Q)$ is Hermitian. To be definite we choose $\lambda(Q)$ in such a way that in the transformed Hamiltonian the one-particle term is diagonalized. This then amounts to the choice of a pure adiabatic base (vid. Sect. 4.). This is achieved if $\tan 2\lambda = 2\Delta/(\epsilon_0 - \epsilon_1 + \alpha\Omega^2 Q^2)$ and yields

$$\begin{aligned} T_\mathrm{a} : H &= \frac{1}{2}[P^2 + \Omega^2 Q^2] + \sum_{j\sigma} \tilde{\epsilon}_j(Q) n_{j\sigma} \\ &+ i\left(P \frac{\alpha\Delta\Omega^2 Q}{4\Delta^2 + (\epsilon_0 - \epsilon_1 + \alpha\Omega^2 Q^2)^2} + h.c.\right) \sum_\sigma [c^+_{1\sigma} c_{0\sigma} - h.c.] \\ &+ 2\left(\frac{\alpha\Delta\Omega^2 Q}{4\Delta^2 + (\epsilon_0 - \epsilon_1 + \alpha\Omega^2 Q^2)^2}\right)^2 \sum_\sigma (n_{0\sigma} n_{1\sigma} - 2 n_{0\sigma} n_{1\sigma}) \\ &- 2\left(\frac{\alpha\Delta\Omega^2 Q^2}{4\Delta^2 + (\epsilon_0 - \epsilon_1 + \alpha\Omega^2 Q^2)^2}\right)^2 \sum_\sigma (c^+_{0\sigma} c^+_{0-\sigma} c_{1-\sigma} c_{1\sigma} + h.c.) \\ &- 4\left(\frac{\alpha\Delta\Omega^2 Q}{4\Delta^2 + (\epsilon_0 - \epsilon_1 + \alpha\Omega^2 Q^2)^2}\right)^2 \sum_\sigma c^+_{0\sigma} c_{0-\sigma} c^+_{1-\sigma} c_{1\sigma} \end{aligned} \tag{5.3}$$

where

$$\tilde{\epsilon}_{0,1} = \frac{1}{2}(\epsilon_0 + \epsilon_1 + \alpha\Omega^2 Q^2) \pm \frac{1}{2}\left[4\Delta^2 + (\epsilon_0 - \epsilon_1 + \alpha\Omega^2 Q^2)^2\right]^{1/2} \tag{5.4}$$

and $n_{j\sigma} = c^+_{j\sigma} c_{j\sigma}$. The last 4 terms of (5.3) are of non-adiabatic nature. The 3rd one is a nondiagonal one-electron term, whereas the 4th indicates a *thermodynamic* coupling of the two subsystems, since it involves electronic occupation operators. The 5th term is of pairing type and in the conventional concept of superconductivity is the most interesting one. It displays a remarkable qualitative difference as compared to the BCS-Hamiltonian: its prefactor no longer is a constant, but decisively depends on the vibrational coordinate $Q$. Thus, if one carries over these findings from our simple prototype system to corresponding multiparticle systems, performs a mean-field factorization and then proceeds in the BCS-manner, finally ending up with formula (1.2), both effective quantities $V$ (by way of the $Q$-dependent prefactor in the pairing interaction) as well as $\rho_F$ (by way of the mentioned *thermodynamic* coupling term) would become temperature dependent.

For completeness we finally refer to the last term of eq. (5.3), which is of a spin-exchange type. It thus just would favour the basic process of ferromagnet cluster formation which first has been considered by Hyzhniakov and Sigmund [1].

## References

1. V. Hizhnyakov, E. Sigmund: Physica C, **156**, 655 (1988)
2. L. Cooper: Phys. Rev. **104**, 1189 (1956)
3. J.G. Bednorz, K.A. Müller: Z. Phys. B -Condensed Matter **64**, 189 (1986)
4. H. Fröhlich: Proc. Roy. Soc. **A215**, 291 (1952)
5. M. Wagner: *Unitary Transformations in Solid State Physics*, (North-Holland, Amsterdam 1986)
6. M. Wagner: phys. stat. sol. **107**, 617 (1981)
7. R. Migoni, H. Bilz, D. Bäuerle: Phys. Rev. Lett. **37**,1155 (1976)
8. U. Beuttler, *Diplomarbeit, University of Stuttgart*, (1993)

MIX
Papier aus verantwortungsvollen Quellen
Paper from responsible sources
FSC® C105338

If you have any concerns about our products,
you can contact us on
**ProductSafety@springernature.com**

In case Publisher is established outside the EU,
the EU authorized representative is:
**Springer Nature Customer Service Center GmbH**
**Europaplatz 3, 69115 Heidelberg, Germany**

Printed by Libri Plureos GmbH
in Hamburg, Germany